Endless Frontier

ALSO BY G. PASCAL ZACHARY

Showstopper!: The Making of Windows NT and the Next Generation at Microsoft

Endless Frontier

VANNEVAR BUSH

Engineer of the American Century

by

G. PASCAL ZACHARY

THE MIT PRESS

Cambridge, Massachusetts

London, England

First MIT Press edition 1999

Originally published by The Free Press, New York, 1997

© 1997 Gregg Pascal Zachary

This book was set in Adobe Garamond by Wellington Graphics.

Printed and bound in the United States of America.

Library of Congress Cataloging-in-Publication Data

Zachary, G. Pascal.

Endless frontier : Vannevar Bush, engineer of the American Century / G. Pascal Zachary.

p. cm.

Originally published: New York : Free Press, 1997.

Includes bibliographical references and index.

ISBN 0-262-74022-2 (alk. paper)

1. Bush, Vannevar, 1890–1974. 2. Electrical engineers—United States—Biography. 3. Mathematicians—United States—Biography. 4. Military art and science. 5. Science and state—United States. I. Title.

TK140.B87Z33 1999

621.3'092—dc21

[B]

99-14146

CIP

To my Parents,
Rosalyn and Michael,
and to Stanley Goldberg
(1934–96)

Contents

"Call it a war"

If we are really well armed the Reds will not force a world war on us.
—Vannevar Bush

For these men the war had never ended.

On a quiet Monday night in Washington, D.C., Dwight Eisenhower, Carl Spaatz and Chester Nimitz, the chiefs of the world's strongest military power, slipped into the Carlton Hotel for a private party. Joining them were their civilian bosses, Navy Secretary James Forrestal and Army Secretary Robert Patterson.

It was January 20, 1947. The hot fight against the Germans and the Japanese had given way to an insecure world, in which U.S. atomic bombs symbolized the nation's unsurpassed military and industrial strength but did not guarantee peace.

No one knew when a new world war might come. But if it did—no, *when it did*—the old soldiers dining in the Carlton Hotel knew they must have more than God and Capitalism on their side. They must have Technology, too.

Make no mistake: the outcome of war was now decided, as much as anything, by a nation's scientific and engineering wizards. This was the lesson of World War II. The laboratory, as much as the factory, proved to be the great arsenal of democracy. Radar. Missiles. Radio-controlled fuzes. Mass-produced penicillin. The atomic bomb. Never had a nation at war harvested the knowledge and inventiveness of its people on such a grand scale. Never had scientists and engineers so altered the face of battle.

And never had any army or navy relied so heavily on civilians to make the basic tools of war—and form the very strategies and tactics of battle.

Their great contribution was not lost on the scientists and engineers, whose leaders were also in attendance that evening at the Carlton. Success, even a measure of celebrity, had altered their perspective. Once these self-styled eggheads had had to beg the military to consider their advice. Now they demanded an equal say over the strategy, tactics and technologies of war, and they openly worried that the technological naivete of hidebound generals and admirals posed a grave danger to the nation. At times, these new technocrats talked as if they would be satisfied with merely directing their well-funded labs. But at other times they talked as if they would settle for nothing less than control of the military's lifeblood, the new weapons that poured from America's nascent military-industrial complex.

How life had changed. Barely five years before, in the hectic weeks following Pearl Harbor, the military virtually ignored technology and took for granted that the weapons of the last war would determine the victors of the new one. Officers treated scientists and engineers as mere hired hands or, worse, useless dreamers. But after the success of radar, the proximity fuze and—most dramatically—the atomic bombing of Hiroshima, the military gave star treatment to its researchers. Scrambling to keep up, the military assembled its own cadres of technocrats.

The officers, service secretaries and about 30 other government insiders had a special reason for attending a black-tie affair at the Carlton. They were marking the official closure of the Office of Scientific Research and Development, the legendary war agency that had quietly overseen the creation of many of the powerful weapons unleashed during World War II. Beyond respectfully burying the OSRD, these bigwigs planned to celebrate the partnership between the Armed Services and the nation's top civilian researchers. The OSRD had bankrolled thousands of these researchers during the war and then pressured a skeptical military into using their most compelling innovations.

The OSRD had been winding down for two years now, but it still cast a large shadow over military research. And that was largely because of the vitality of the agency's longtime chief, Vannevar Bush, the official master of ceremonies for the evening's festivities.[1]

Friends called him "Van" because, he joked, they could not pronounce properly his full first name (it rhymed with beaver). Many acquaintances simply called him "Dr. Bush." At the age of 57, he personified military research in America and was the most politically powerful inventor in America since Benjamin Franklin. Among the most influential 20th-century Americans, he had played a crucial role in the Allied victory in World War II. During the war, one popular magazine began a profile of Bush with the simple introduction: "Meet the man who may win or lose the war."[2]

A gifted mathematician and electrical engineer, Bush came from a peculiarly American line of can-do engineers and tinkerers, a line beginning with Franklin and including Eli Whitney, Alexander Bell, Edison and the Wright brothers. Born in 1890, during a tidal wave of American ingenuity, Bush tinkered with gadgets as a boy, cofounded a radio-tube company as a young professor and designed the world's most powerful mechanical calculators in the 1930s, laying the groundwork for the advent of the digital computer and the information revolution made possible by this machine. During World War II, he advised President Roosevelt on science and technology and organized the successful effort to build the first atomic bomb, popularly known as the Manhattan Project. Bush joined in the decision to drop the A-bombs even as he warned that the U.S. couldn't sustain for long its atomic "monopoly" and that an arms race was likely. While unapologetic over the A-bomb attacks, he secretly tried to halt the first test of a hydrogen bomb and, after failing to do so, claimed that "history will show" that H-bomb advocates "have a great deal to answer for" in sending humanity into "a grim world."[3]

World War II was Bush's shining moment. Just as Franklin had seized the Revolutionary Era to enter the public sphere with all the energy and accomplishment of the can-doer, so did Bush grab on to the birth event of the American Century. Despite his technocratic garb, Bush's preoccupation was politics. He balanced the demands of contending scientific factions and handled relations with the military, the Congress and the president. By his own admission, he "was engaged in the political aspect of it more than anything else."

While his influence reached its zenith during the war, he remained an influential personality in postwar America. In 1945, he published two landmark essays that expressed a stunning vision of a future in which technology would serve humanity's highest intellectual and political ends. The first essay, "As We May Think," predicted that new technologies would someday deliver an unprecedented ability to receive and manage information, thus improving the quality of life in untold ways. His words contained the germ of what would become the Internet and won him a posthumous reputation as the sage of cyberspace. The second essay, *Science—The Endless Frontier,* skillfully equated scientific and technical progress with national health—and convincingly made the argument that government must finance independent researchers at levels far above those seen before the war.

To the public, Bush was the patron saint of American science, "one of the most important men in America." He was a wise, dollar-a-year man who helped to keep the country on the track, courtesy of his employer, Carnegie Institution of Washington. When *Time* put Bush on its cover in 1944, the magazine dubbed him the "general of physics." Hollywood cast him as a shrewd hero in a movie celebrating the making the atomic bomb. Every college president in the country knew him intimately, or wished he did. Even the Average Joe could find common cause with Bush, whose zeal for tinkering in his personal workshop drew acclaim from *Popular Mechanics* magazine and Edward R. Murrow's television show.[4]

Though coveting his celebrity, Bush was never entirely comfortable with his public image. He sometimes bristled at being called a scientist, concerned that the achievements of engineers were overlooked. He repeatedly sought ways to limit the military's sway over national security and science, yet led the drive to link soldiers, scientists and industry. He celebrated the superiority of democracy over dictatorship, yet carped about the peril of an America weakened by partisan politics, rampant consumerism and shallow entertainment. Even as he extolled the virtues of self-reliance, he worried that the American emphasis on individualism might leave it vulnerable to totalitarian rivals whose regimentation led to greater efficiencies.

To escape from these paradoxes, Bush believed Americans should freely give public-spirited experts ultimate authority over the nation's security. He ranked the engineer as first among equals, a sort of super-citizen who could master virtually every activity essential to the smooth functioning of a modern nation. What distinguished the engineer from other experts was his breadth. Bush saw the engineer as a pragmatic polymath; the engineer, he once wrote, "was not primarily a physicist, or a business man, or an inventor but [someone] who would acquire some of the skills and knowledge of each of these and be capable of successfully developing and applying new devices on the grand scale."[5]

This realization that the engineer was the *engine* of 20th-century capitalism qualified Bush as the godfather of high technology and a leading proponent of industrial vitality through innovation, not intrigue. He cofounded one company and inspired many others that formed the nucleus of the "Route 128" high-tech cluster near Boston. One of his students, Frederick Terman, used Bush's ideas about academic-industry collaboration in order to spawn Silicon Valley in California. Bush's keen appreciation of the value of entrepreneurs, especially in technical industries, made him a lonely advocate for economic dynamism after the war when most economists welcomed the concurrent rise of big business and big government. At midcentury, he was among the few who realized the curative power of new ventures. The best way to limit monopoly economic power, he insisted, was through "the advent of small new industrial units, for if these latter have half a chance they can cut rings around the great stodgy concern."[6]

Such contrarian views made Bush a divisive figure. His personality didn't always help either. His philosopher-king aura smacked of arrogance, even meanness to some. He struck his critics as imperious, intimidating and at times even a bully who harbored "a relentless, perhaps insatiable, drive for power." Still he had redeeming qualities. His wit and charm prompted comparisons with the folksy Will Rogers. His intelligence, vitality and candor impressed many. He enjoyed a good tussle, refused to back down from anyone and, when opposed, could explode in anger. He rubbed people in authority the wrong way, but he was principled about it. He felt he never angered anyone without good reason. Aware that his penchant for battle cost him good will, he still never shied away from a fight, and he took as much ground as his opponents ceded. "My whole philosophy . . . is very simple," he told a few generals during the war. "If I have any doubt as to whether I am supposed to do a job or not, I do it, and if someone socks me, I lay off."[7]

That night at the Carlton, Bush looked as if he had never laid off for very long. Standing two inches under six feet tall, he weighed 150 pounds and looked lean in a dark three-piece suit. With blue eyes twinkling, his leathery face sported a sly grin. A shock of his hair, once black and now graying, refused to lie down. He insistently puffed on a pipe he had carved himself. His wire-rim spectacles sounded the only wrong note, making him look less of a fighter and more of an egghead. But he rarely took off his glasses outside of his home. When he did, he revealed deeply set, dark eyes. His voice was gruff and gravelly. He spoke in the manner of an earlier New Englander, saying "patt'n" for pattern and describing an upper respiratory illness as the "grippe."

From the start, the evening bore Bush's stamp: the festivities were funny and serious, intelligent yet gritty. Months in the planning, the OSRD celebration was meant to evoke the shadowy, arcane and sometimes bizarre world of military research. It was a world in which military men often asked for what was technically impossible, and researchers displayed shocking naivete about battlefield conditions. Sometimes both sides accepted the same script, but it might be cockeyed—like the time they planned to round up thousands of bats, paint their legs with phosphorus and drop them over Japanese cities in order to make night-bombing more accurate.[8]

"The military men were delighted to come to the party," recalled one junior official who had helped to organize the event. "They all wanted to show their appreciation to the scientists for the help they'd received in the war. And they wanted to kid with the scientists about some of their failures, while the scientists wanted to kid the brass about their attitudes."

On arriving at the Carlton, each guest was handed a set of orders marked, not "Secret," but "Unmentionable." Some were also given genuine U.S. patents covering real weapons, while others received descriptions of fanciful contraptions. Later, scientists rose and asked military officers why these weapons were being ignored. One weapon was a plow that could change into a gun at the push of a lever. Another was a device that could instantly cause a ship to change direction, thus avoiding collisions.

Then Bush took over again. He read aloud a playful message from President Truman naming him commander of an operation codenamed "Payoff." That got a laugh from the generals. Then he teased some of his closest wartime associates in attendance: Harvard's president, James Conant, seated beside Forrestal; MIT's president, Karl T. Compton, who sat next to Spaatz; and Julius Furer, Bush's staunchest wartime ally in the Navy.

"The old town isn't what it used to be," Bush finally quipped. "The corridors where once tramped the embattled scientists are empty.

"The time has come," he said, "to call it a war and quit—to mark fittingly the end of a great adventure." He then added: "But not really to quit, to shift the emphasis."

Bush saluted his comrades as "a sturdy group that gathered in the dangerous days." Their camaraderie, "forged in the heat of Washington," should be treasured, he said.

He still wasn't finished, though. Next he handed out trophies, honorary degrees and diplomas. When he was finished, Admiral Nimitz took the floor. Decked out in his Navy uniform, Nimitz launched into a presidential nominating speech. For whom, it was not clear. Conant and Ike squirmed. Both had been named in the press as being made of presidential timber, and they worried that Nimitz might be lampooning them. Instead the old sailor nominated Bush. The host tried to show his embarrassment by crawling under a table.

The entertainment was a hit. Forrestal had planned to leave at 9:00 P.M., but stayed until midnight. His aide John Connor, who had served as OSRD's legal counsel for a time, considered it "a hilarious evening. The whole thing was carried off beautifully." Furer, the Navy's top research officer in the war, considered it a "perfect" party. Newton Richards, OSRD's medical chief, thought the "privilege of sitting down with Eisenhower, of chatting with Nimitz and Spaatz made [for] an unforgettable experience." Harvey Bundy, the War Department's liaison to Bush on atomic and other matters, believed he had "never attended a party where there was a greater sense of fellowship and friendliness." Only Eisenhower, who retired with indigestion, seemed less than elated.

All in all, the evening was an apt symbol of the partnerships between science and the military, technology and national security. These partnerships, forged by a shooting war, were now sustained by hopes and fears, dreams and nightmares.

Yet appearances were deceiving. Below the surface fellowship between soldiers and civilians lay profound differences regarding the most pressing problem of the age: how best to keep the military's technical edge and so secure the nation against its enemies? The question had spawned a contest between America's mightiest thinkers and actors. And at a crucial time, Bush stood in the eye of this storm.

On his death in 1974 *The New York Times* honored Bush with a front-page obituary, calling him "the engineer who marshalled American technology for World War II and ushered in the atomic age." Jerome Wiesner, science adviser to President Kennedy, judged Bush's influence on American science and technology so great that "the 20th century may yet not produce his equal."

A half-century after the peak of his power, Bush is virtually forgotten, recalled most often as a pioneer in computing and a prophet of the Information Age. His relative obscurity might please him. He refused to write a true memoir, complaining that the exercise would invariably produce "some kind of a fake story." He boasted, meanwhile, that any biographer interested in him would soon give up. "I hope nobody'll ever write a biography of me, because I think it probably would be terrible," he said. What bothered him most was how some biographers luridly analyzed their subject, while giving short shrift to the significant events of his life. "I don't have much use for biographers, is my trouble. There are so few really good ones." He added: "The thing I do like is the story of a man's involvement with something important."9

Acts of importance were the measure of Bush's life, and they are the reason his life deserves study today. His was a political life, wrapped in the enigma of science and invention. An apostle of expertise, he transcended the labels of "liberal" or "conservative" and pursued the progressive ideal of public betterment through the private efforts of people of good will and merit. His own life neatly charted the shift of America from small town to big city, from isolationist to globalist, from weak military power to the world's strongest, from a nation dominated by generalists to one managed by specialists.10

In an age of complexity, Bush's habits of mind transcended easy categorization and prefigured the postmodern embrace of contradictions. He was a contrarian, skeptical of easy solutions yet willing to tackle tough problems without a compass. He looked askance at social status based on wealth, but fervently believed that mass opinion should be directed by a "natural aristocracy" of meritorious Americans. He was a pragmatist who thought that knowledge arose from a physical encounter with a stubborn reality. The mathematician Norbert Wiener called him "one of the greatest apparatus men that America has ever seen—he thinks with his hands as well as with his brain." Despite being drenched in a world of particulars, Bush was ultimately a moral thinker whose grand themes were individual self-reliance, democracy with a small *d* and the absolute necessity for thinking men and women to build—with the help of technology—meaningful patterns from the confusing buzz of facts, ideas and emotions that compose the discourse of any era.[11]

Suspicious of big institutions, whether run by public servants, the military or corporations, Bush objected to the pernicious effects of an increasingly bureaucratized society and the potential for mass mediocrity long before such complaints became conventional wisdom. Yet by institutionalizing the creation of new and ever-more-dangerous weapons of mass destruction, Bush helped to rob the individual of a measure of control over his own destiny by giving an impersonal government ultimate power over a people's very survival.

Bush knew his legacy to be contradictory. By marrying the intellectual resources of an ascendant community of technologists to the bureaucratic imperatives of a security-obsessed state, he had helped create a world in which efficiency triumphed over humanity, raw power trumped compassion and reason mocked sanity. In the end, he felt isolated from this new world yet could not repudiate it. His great failure and his enduring triumph was his realization that the course of modern history would be shaped by large hierarchical institutions, making plans and settling scores behind closed doors, working best when insulated from public opinion. That these institutions lost their energy and legitimacy as the 20th century waned would not have surprised Bush. Whether overseeing the creation of the atomic bomb or lobbying to fund "pure" research without utilitarian purpose, he believed the beleaguered individual was still of paramount importance.

"The individual to me is everything," he wrote on the eve of World War II. "I would circumscribe him just as little as possible." In the murderous years that followed, he never lost his faith in the power of one.[12]

Part One

The Education of an Engineer

Chapter 1

"The sea was all around"

(1890–1909)

When I was young I could follow an underground stream, in Provincetown on Cape Cod where I spent much of my youth, with assurance and
precision. It flowed out of a pond a mile back, traversed part of the town,
flowed under our house, and finally emerged below high water mark.
—Vannevar Bush

Richard Perry Bush, a short man wearing a tall silk hat, was on his way home
from a funeral. If he didn't like funerals, it didn't show. He officiated at two,
sometimes three funerals a week in Chelsea, an industrial city near Boston. In
the 1890s Chelsea was a place where old Americans and new immigrants collided. As he walked down Broadway, Chelsea's main avenue, friends and well-
wishers hailed him. It seemed he couldn't go more than a few feet without
seeing someone he knew. He brought a sunny outlook to life, which was probably why he was in such demand for funerals. He seemed always ready to speak
a comforting word. A tolerant and civic-minded minister, he had a sense of
poetry, yet was practical and no pushover. A fellow minister said, "His manli-
ness was his power."[1]

Perry wore a moustache and long sideburns. He had narrow, dark hooded
eyes, a small mouth, a broad, sharp nose and coarse hair strenuously parted near
the middle of his head. He was born in 1855 in Provincetown, a scenic but declining fishing and trading center at the tip of Cape Cod, smack on the Atlantic

11

Ocean. The Pilgrims had first stopped at the tip of the cape in 1620 before forming a permanent settlement in Plymouth. Fishing drew the Pilgrims back to the tip each year, and Provincetown was incorporated in 1727. By the Revolutionary War, however, it had just 205 inhabitants and 36 families.

Though isolated, Provincetown seemed to give its residents a window on the world. From High Pole Hill in town, some thought they could see the whole world. Visiting Provincetown about the time of Perry's birth, Henry Thoreau wrote that the "dry land itself came through and out of the water in its way to the heavens."[2]

As a boy in Provincetown, Perry felt "the sea was all around [him]. It was his playmate. It was his inspiration. Something of the moods of the sea were always with him." The sea had sustained his ancestors for as far back as he knew. His father, also Richard Perry, was a sea captain whose own ancestors (and those of his wife) stretching back six generations were among Massachusetts' earliest settlers, well established a century before the American Revolution. Most of this self-reliant crowd earned their living from the sea: traveling to Africa, South America and other exotic ports of call; whaling; trading; bankrolling the voyages of others. The elder Richard, born in 1828 and also raised in Provincetown, went to sea at an early age. In his prime he commanded both fishing and cargo vessels, "winning for himself a high reputation for uprightness and attention to business."[3]

The vagaries of the seafaring trade meant the Bush family was comfortable but not wealthy. Perry sailed as a cook on a fishing boat at the age of 14, just four years after the Civil War. He did not aspire to a life at sea, however. He was smitten instead with religious feeling, though he turned his back on his parents' strict Methodist creed. Such religious rifts were common in Provincetown. When Methodism first attracted some residents in the 1790s and they began to build a church, rival faiths were jealous. A mob tore down the frame of the building, built a bonfire with the wood and burned an effigy of the Methodist preacher.

Perry was drawn to a cooler, more temporal spirituality, yet one that was still muscular. His religious journey, however, took him away from Provincetown. Bent on becoming a minister, he attended Tufts College, an academically rigorous school in Medford, Massachusetts, founded by the liberal Universalist faith. A boyhood friend named John Vannevar joined Perry at Tufts, making the break with his family easier.

It took courage for Perry to leave his family and strike out on his own. "I left my home while yet a boy to seek my fortune in the world," he later recalled. "Going out from home I lost the tie that might have bound me" to family traditions. It also meant fending more for himself. To help pay his school bills,

Perry supplied wealthier students with coal for the stoves in their rooms. He carried the fuel himself, sometimes climbing three flights of stairs to make a delivery, the coal on his back.[4]

In 1879, Perry graduated from the divinity school and then moved to the nearby town of Everett, where he spent 13 years as a pastor. Building a life with little family help gave him "a lot of sympathy for anybody struggling with any kind of difficulty." In 1892, he went to Chelsea, becoming the pastor of the Church of the Redeemer, a 50-year-old Universalist church. Perry arrived at his new post with his wife, Emma Linwood Paine, the daughter of a prominent Provincetown family and the mother of their three children. Edith, the oldest, was ten. Reba was five. The third and youngest child was a one-year-old boy. Born on March 11, 1890, the boy was named Vannevar, after Perry's lifelong pal.[5]

In Chelsea, Perry quickly emerged as a civic leader. "A man of strong convictions, he had a remarkable power for making friends." His religious convictions helped. Universalism, a Protestant offshoot with affinities to the Deism of Thomas Jefferson and other revolutionary-era figures, held that all men will be saved, no matter what their earthly actions. The faith flourished in 18th-century England, then spread to the colonies. By the late 19th century, Universalists espoused a belief in a single God and rejected the idea of Christ's divinity and the Trinity. Adherents possessed an ecumenical spirit rare for the times and an appetite for social action. Perry himself "was always loyal to his church, but mankind was more important than any church, and when he was called upon to help he never stopped to ask as to a man's creed, or race, or color, but only as to his need and the way in which he might be comforted or helped."[6]

Perry's Universalist creed made him sensitive to the swift and unsettling changes occurring in his city of 40,000. Through the Civil War, Chelsea remained largely rural and was dominated by a few landowners. Its people intensely supported the Yankees in the conflict, sending 1,000 men into battle by the time of Lee's surrender. After the war, Chelsea emerged as a summer resort, catering to wealthy Bostonians and gaining a reputation as perhaps the poshest of Boston's suburbs. But in the last third of the century Chelsea's population quadrupled, and business, attracted by the easy connections to Boston proper, thrived. By 1880, 150 manufacturing firms were located in the city. Within a decade, the number had doubled and business investment had quadrupled.[7]

The boom drastically changed the character of Chelsea. Many of those wishing larger residences had moved to Brookline, Newton and other nearby towns. By the turn of the century, thousands of immigrants had taken their places, prompting one oldtimer to moan: "How was it possible for a city of wealth, with a population of ten to fifteen thousand, to change in so short a time to a business and manufacturing community with a population of forty thousand, including ten thousand Hebrews?"

Some of the old stock remained, of course. Perry himself lived in the middle-class Irish and Yankee part of Chelsea, located across the Boston & Albany and Boston & Maine railroad tracks from Jewish immigrants and impoverished newcomers. He did not retreat into his besieged Anglo-Saxon world but rather saw Chelsea's social upheaval as an opportunity to break down ethnic and religious walls. He mixed with all kinds. With a local priest, he campaigned to "tame" the city's saloonkeepers. He regularly exchanged pulpits with a rabbi. And Perry's interests weren't always so lofty. He was a sharp pool player and knew his way around the city's seamy side. No teetotaler, he once forbade a friend to indulge in alcohol even as he swigged a drink (on the presumption, apparently well-grounded, that the friend couldn't hold his liquor). And he didn't browbeat his parishioners, but sometimes won them over through guile. Once asked by a mother to counsel her wayward son, Perry gained the boy's respect by beating him at a game of billiards.[8]

Chelsea's schools gave Perry the means to satisfy his desire for social betterment. For 26 years, he served on the city's school committee, helping to manage the ballooning enrollment, which rose by 50 percent in the ten years beginning in 1895. Perry also supported progressive education, taking the unusual step in 1900 of teaching English to foreign-language students from the ages of 10 to 14. Within six years, the program had grown from one class of 25 pupils to four classes with a total of 100 students. "The school is a beehive," the committee wrote in its annual report of 1906. "Nowhere else in the city is found greater intensity of interest on the part of the pupils, or more heart or grateful response to the demands of the teachers."

In general, the quality of the city's teachers was a source of pride. Perry and his fellow board members insured high standards by not allowing "political influence" to contaminate hiring practices. "In no city in the state, perhaps, has this evil been so thoroughly eradicated as here," *The Chelsea Gazette* wrote in 1908.

Perry was an active Mason, reputed to have achieved "literary" success on the strength of his Masonic writings, which one admirer claimed were "accepted as authority in this country and many parts of Europe." He also wrote poetry, usually devoted to spiritual themes. The writing was passionate, but didactic and usually lacking in lyricism. Son Vannevar once confessed, "I fear that my father was not much of a poet."[9]

But Perry could make a point. In "Fame," for instance, he suggested the folly of worldly achievement:

And so I thought, it is in life:
> We write in snow on walls of Fame,
> But other snows come drifting fast
> And for a while another's name

Gleams out before the gaze of men,
> All bright and flowing for a day;
> Then it in turn is lost to sight,
> In turn to others it gives way.

Sometimes, Perry's verse lapsed into sentimentality. In "Children's Sunday," he wrote:

Hail once more this happy Sabbath,
> Gladdest day in all the year;
When about the holy altar,
> Children fair in joy appear.
Little ones, we love them dearly,
> Stars they are in earth's dark night;
Angels sent to us from heaven,
> Bearing messages of light.

These verses were designed to succor the downhearted, commemorate friends and colleagues, uplift spirits and fix minds on the promise of a better day. Perry was an optimist. He saw life as a challenge to be met and overcome; a game to be won. In "Four Pictures of Life," Perry brooded about man's predicament, taking a youth through the "happy time" of hope, the "awful barrier" of destiny, the "tears and white-robed Sorrow" of despair. But the final phase he baldly described as "Victory," when "the storm that beat upon our youth is gone." The "fire" of adversity "purifies" the nature of youth, giving birth to a "manly strength."

And man—not youth—against the wrong hath striven
> Despair lies vanquished at the feet of Love,
And Faith proclaims the victory of Heaven.[10]

Perry's ornamented poems seemed almost understated compared to his arcane speeches. He was in demand as a dinner speaker at Masonic gatherings and once gave the keynote address at a ceremony attended by President Teddy Roosevelt. The occasion was the groundbreaking for the Pilgrim Memorial Monument in August 1907. Perry toasted the Pilgrims who "dared and died for principle" and declared, "We hold it as our conviction that when they went forth from England it was in obedience to a heavenly vision and a divine command." He went on to toast his country (for its "amalgamation of all races and peoples leavened by the spirit of the Pilgrim, the Puritan and the Virginian cavalier") and then Roosevelt. Any U.S. president was "exalted above every other potentate of earth," but the sitting president stood alone. "Never since the birth of our Republic [did a president have] so strong a hold upon the confidence and respect of the American people as the present incumbent of the Presidential chair."[11]

Perry frequently lectured on secular subjects for a fee, waxing philosophical about camaraderie and country. He was a liberal, but believed in frankly admitting the differences between people, not simply hiding them. "I want no man to tolerate me and I do not tolerate any man," he once said. "The word tolerate has no place in [Masonry] because when we enter the Lodge room we put aside our differences and creeds and meet upon a common basis. No—I believe in brotherhood but I do not believe in toleration. I believe in equality of man with man, in manly fashion." While sympathetic to progressive values, Perry was suspicious of "do-gooders." Once on a visit to Niagara Falls, he impressed his son by angrily replying to the suggestion of another tourist that water from the falls not be diverted for electric power. Perry countered that doing so spared people from working as miners (The encounter impressed Vannevar, who grew up thinking that do-gooders "often pose a holier than thou attitude which is maddening.").[12]

Of all Perry's fascinations, Freemasonry was probably the oddest. His own father and grandfather were Mason, and he frequently discoursed on the oddities of the sect's rites and history. As a youth he strayed from Masonry, he admitted, but he returned to the fold and held fast to his allegiance: "Early in my career as a Mason, I think, I doubted somewhat the antiquity of the institution. There are some inconsistencies in our ritual; but, as I have looked into the archives, as I have had a little of access to the lore of our Craft, I am more than convinced that we are the lineal descendants of the dusty sons of old Egypt of long, long before the Christian era."

The mysteries of Masonry might seem strange to others, he allowed, but faith always inspired unusual rituals:

> Always man has worshipped; instinctively the knee is bent and the face is turned towards the blue arch. The heart naturally bows in prayer. But we cannot worship in abstractions; we must have forms, and symbols; and men have sought out these from the rudest carving of the idol-maker to the grandeur and magnificence of the

modern Lodge-room, and of cathedrals. So it was that, as we traced the architecture, we traced also the history of building; and out of that history we find what brought forth Masonry, as also, what brought forth the church.[13]

To later ears, Perry's speeches would seem flowery, almost overwrought. But contemporaries found "his language was choice and his thought was always presented with a clearness and force, a simplicity and conviction that ranked him as one of the most delightful and eloquent speakers of his time." The secret to his patter, he said, was careful planning. One should never start a speech, he advised, "unless you clearly have in mind the sentence with which you are going to conclude."

A good speaker, though, still must think on his feet. "When you are making a speech your mind is in three parts," Perry once advised his son. "One is paying attention to your actual wording at the moment. Another is roaming ahead to plan what you will say next. A third is following behind, picking up slips you may have made. Suppress that third part or it will get you into trouble."

Bush learned much about speaking from his father, even copying his delivery. Once Bush even regaled his father, along with a gathering of friends, by imitating his "language, gestures, subject matter, all of which I knew fully well. Dad was the first to tumble as to what was going on. He caught my eye and then subsided so as not to give me away. Then I could see one member after another nudge his neighbor as he caught on. I ended with a peroration which was my dad all over—gestures, resonances, and all, which I could reproduce with some accuracy by that time. It was not a caricature. It was an imitation, and one that expressed my pride in my father."[14]

The Bushes lived in the church parsonage on Clark Street for much of Perry's tenure in Chelsea. The family was frugal. Mother Emma hailed from a successful Provincetown family—her father, Lysander N. Paine, was an important merchant who formed a bank in town—but she had simple tastes. She was not, for instance, much of a cook. Visitors to her kitchen politely described her meals as "a little bit thin." Towering over her short husband, she was quiet and easygoing and "an unusually fine woman," one friend recalled. "Her face had the beauty that comes only from a kind heart and compassion for others." Still, any minister's wife had it hard; she was invariably scrutinized by her husband's congregation. "The parishioners felt it their perfect right, if not their duty, to act as judge and jury for the minister's wife. Her mode of dress, her housekeeping, her actions, speech—in fact almost everything she did or did not do—was subject to their critical scrutiny." But Emma "was so lovable . . . that no one had anything critical or unkind to say of her."

Emma ran her house on a tight budget because Perry's parish income barely covered basic needs. He often presided over weddings for extra cash, winning himself the nickname "Marrying Minister" among fellow Universalist clergy. The weddings "put him in a different financial class from the rest of us," a colleague recalled. Perry usually held weddings in the living room of his home. The Bush children sometimes served as witnesses or even well-wishers. But they disliked this duty and often fled in advance of a wedding party.[15]

Through his many activities, Perry became one of the best-known men in Chelsea. According to local lore, one couple on their way to the altar simply asked a hack driver to take them to the minister. They seemed to have forgotten his name. No matter. The hack went at once for Perry, "possibly from force of habit or perhaps he too felt that only Perry Bush could perform the ceremony properly."

Perry was sought out in sorrow as well as joy. After a great fire devastated Chelsea in 1913—it destroyed much of the city, including Perry's church—a friend from Boston searched for him amid the confusion. He wandered aimlessly until he asked a youth where he could find Dr. Bush.

"Never heard of him," the man said.

"How long have you lived in Chelsea?"

"All my life."

"And you don't know Dr. Bush?"

"Nope. No Dr. Bush in Chelsea, you can bet your boots."

"Well I happen to know better. Perry Bush has—"

"Perry Bush! Why in thunder didn't you say so? Know Perry Bush. Everybody knows Perry Bush. [He's] up there in the schoolhouse."

And there his friend found him, helping the poor people of his city pull themselves together and carry on.[16]

Perry expected his children to be grateful for what they had and not to dwell on what they lacked. He had "a kind of fearlessness in the conflicts of the world," which made him seem stoic at times. When son Vannevar was five, he joined his father at a funeral, only to break down in tears during the service. On the way home, Perry stopped his son's crying by saying, "We've paid our respects to our dear friend, and we'll have happy memories of him. There's nothing more to be done." The younger Bush never forgot the lesson.[17]

The stiff upper lip suited Vannevar. By his own account, he was "not a particularly husky youngster" and was bedeviled by a series of illnesses. Rheumatic fever, which at first seemed to have weakened his heart, left him "cursed" with rheumatism, so that "for years . . . occasionally I had to drag a leg behind me."

He suffered typhoid fever, possibly from drinking fetid well water. He ruptured an appendix. He also caught "the usual childhood" sicknesses.

All in all, Bush was "ill a good deal of the time." He spent one teenage year bedridden. "At the time I know I thought most about the way in which it interrupted my school and my usual pleasures," he later recalled. But there were "many pleasant days when I could read, and do puzzles, and learn to do new things with my hands, and I remember the friends who came to see me and talk to me." During his forced idleness, he "learned to knit, to make tatting, and do all sorts of queer things," including whittle.[18]

Compensating for a sickly childhood, Bush grew self-reliant, confident and pugnacious. He occasionally fought with other youngsters, some as far away as East Boston, and once returned home with "a somewhat damaged nose." There was more to his combativeness than mere bravado. He was fiercely independent, a budding maverick. "In my youth I had been taught that the most independent thing in existence was a hog on ice, and I emulated a hog on ice," he later wrote.[19]

He also suffered snubs, inspired by class and religious affiliation. This only seemed to embolden him, filling him with an outsider's scrappy pride and an instinctive sympathy for the underdog. The town YMCA, for instance, barred Catholics and Jews as well as liberal Protestants such as his family. "As the net result of that, my boyhood friends were the Catholics and the Jews," Bush said. "I was not only not a Boston Brahmin, I acquired a very considerable set of prejudices against them. . . . My prejudices were all in the direction that I thought I belonged with the Catholics and the Jews, some of the fellows that were out of luck otherwise. I didn't have much use for the gang that lived up at the [wealthier] end of town."[20]

He was too independent to curry favor with schoolmates, though he was elected vice-president of his junior-high-school class. He did not take to those who put on airs and enjoyed "mixing it up with anyone" who showed "a touch of [the] stuffed shirt." More a budding despot than a politician, he did not like to be told what to do. He credited his ship-captain forebears for instilling in him "some inclination to run a show once I was in it."

Bush saw life on the sea as a model for terrestrial society. As a boy he explored the coast along Cape Cod alone on a motorboat. He learned the ins and outs of boat-building from grandfather Lysander, who though in his seventies still ran businesses in Provincetown. He dreamed of sailing his own ship across "the high seas." For entertainment he read old whaling logs over and over. Even more than the sheer adventure of whaling, the logs taught him about leadership and group dynamics. "The relations between the captain and the

; voyages that lasted for years, strained human nature to the ut-
~~m~~ote, "and it also produced some queer by-products."[21]

~~~~ sound lessons too. He learned that successful captains were autocratic, that they met all kinds of people and did so on their own terms; that
they could lose everything on a gamble but were richly rewarded for success;
and that they demanded loyalty, even deference, from subordinates, but were
fiercely loyal and protective of those who stood by them.

Bush had captaincy in his blood. In her prime Perry's mother, who lived in
the Bush home, ran a shipping business with her husband. The couple specialized in trade with the West Indies; Perry's father captained the ship, while wife
Mary Willis kept the accounts. Now and again she took to the sea herself and
once sailed across the Atlantic and up the Amazon River. Mary was an intense
force in the Bush home. Even after losing her sight late in life, "she would not
quit" fighting.

Bush showed his own "spark of belligerency." He was quick to take exception to things. Even his own first name, with its Dutch pronunciation (Vuh-
NEE-ver), irritated him. "The strange name" was "a nuisance," always
requiring an explanation or a quick lesson in pronunciation. Bush wished his
father had named him John, after the first name of his friend, and his sisters indeed called him John at times.

His name may have been the only mistake he ever pinned on his father.
Perry's influence on his son was obvious, and Bush celebrated it. "When I
think of teachers who have molded my own patterns of thought, I think at
once of my father," he later wrote. "I acquired much from him, although I
hardly realized it at the time."[22]

Sister Edith also influenced Vannevar. A math whiz, Edith joined the faculty
of Chelsea's high school after graduating with honors from Jackson College
(the sister school of Tufts) in 1903. A member of the school's math department, she taught trigonometry. One year brother Vannevar ended up in her
class. He was no slouch, grasping her lessons with an uncommon alacrity.
Math was his best subject, but Edith tried to keep her brother humble by calling him only "a good student."

Edith's talents fueled Bush's desire to excel in math. The siblings competed against each other in card and parlor games. They were worthy adversaries. Edith was just as "stiff-necked" as her brother and "inclined to be very
fussy about rules and regulations." She was also something of a tomboy,
often fishing, sailing and swimming. She had her own ambitions too. After
eight years of high-school teaching, she became principal of Provincetown
High School. Two years later, in 1920, she joined Jackson College as a math

instructor. The following year she was named an assistant professor. She taught at the school until 1952.[23]

Bush had a talent that separated him from his smart sister: he also could work adeptly with his hands. Using his hands was as important to Bush as using his mind. In school, he ran track and sang in his father's church choir, but he preferred visiting the shop over any other leisure activity not related to the sea. For a boy with his inclinations, Chelsea proved to be a hospitable place. The city's schools offered an elaborate curriculum in sewing, woodworking, basketweaving, and drawing. Altogether, students spent two to three hours a week on these and other "manual arts."

For educators, this was not window-dressing or simply a vocational program for those destined to earn their living as skilled workmen. Teachers were "convinced that by doing things, the child is developing that part of his brain which can only be developed by using his hands," according to one school report. "Sometimes, for lack of interest in book studies, we find there are periods in a child's life when mental progress seems to be at a standstill. Suddenly he finds he has a special talent for work with his hands. He respects himself and commands the respect of others."[24]

Bush surely respected his ability to shape material into useful things. During high school, he could be found handling test tubes and triggering chemical reactions in the basement of the church parsonage, where his family lived. A rare childhood photograph shows him tapping away with a hammer at what seems to be a dry cell hooked to a clock. He wore a white, long-sleeved shirt, a stiff white collar and a vest. His chemicals stood secure in a Quaker Oats box, and miscellaneous treasures were stashed in salt-cod boxes. No one knew for sure what he was up to; certainly his father did not. Perry had no aptitude for handiwork; he "couldn't drive a nail."[25]

Tinkering in his basement, Bush shared an activity with many brainy, middle-class boys around the country. The romance of invention—or at the very least, of making something—was contagious. Well aware of his family's modest means and the absolute requirement for him to turn an education into a good livelihood, Bush realized that the path of the inventor offered him perhaps the only means of achieving conventional success without sacrificing his maverick leanings.

In the 1890s, an outpouring of technical advances was undermining old patterns in American life—and was a fast path to riches. Bush could not have missed this "technological torrent." In the first seven years of Bush's life, the first gas-powered car was perfected; the German Otto Lilienthal made hundreds of successful gliding flights; the first commercial motion picture was screened; and X-rays were used in the treatment of cancer for the first time. The spread of telephony, the phonograph, electricity and radio contributed to the enthusiasm for technology. It was "an epoch of invention and progress unique in the history of the world," wrote one observer. The period "has been a gigantic tidal wave of human ingenuity and resource, so stupendous in its magnitude, so profound in its thought, so fruitful in its wealth, so beneficent in its results, that the mind is strained and embarrassed in its effort to expand to a full appreciation of it."[26]

All this ignited the curiosity of Bush, who surely noticed how technology was exerting a powerful draw on young, middle-class men eager to get ahead. In the early 1900s, newspapers and magazines extolled the feats of young tinkerers, and dime-novelists picked up on the theme. The appeal of the boy-inventor persona lay in the promise of heroism through inspired ideas and shrewd improvisation. Armed with new gadgets, mere boys could outshine their fathers, performing courageous, even lucrative deeds. Their pursuit of invention, meanwhile, demanded a new concept of manhood, one which conceived of education and expertise as the basis for thrilling journeys into the dangerous technological frontier. For some young men, technical exploration was a middle path between the tired refinements of genteel culture and the "animal magnetism" of sport and fitness enthusiasts.

This new concept of manhood neatly mapped Bush's own evolving sense of self. His future depended on the nation's capacity to absorb college-bound youths. Hampered by periodic illness, impatient with pomp and molded by his own class and religion, he needed a means of advancement. As he graduated from Chelsea High in 1909, he was an outsider who resented the elite of society but hungered for its recognition too. It didn't help that his father had "knocked out the family funds" on the college education of his two older sisters. At his father's urging, Bush decided to attend Tufts. It was an easy choice, of course, but then Bush believed it really did not matter which college he chose because "all of my academic training was circumscribed by the necessity of getting some cash."[27]

Chapter 2

"The man I wanted to be"

(1909–18)

It might possibly be that inheritance has something to do with one's characteristics, for all of [my] recent ancestors were sea captains, and they have a way of running things without any doubt. So it may have been partly that, and partly my association with my grandfather, who was a whaling skipper. That left me with some inclination to run a show, once I was in it.

—Vannevar Bush

Wearing a dark, loose-fitting suit, Bush firmly gripped the handles of what resembled a lawnmower. Head bent, he kept his eyes squarely on the machine as his feet pounded on the yellow grass. From a distance, it looked as if he was clearing one of the fields at Tufts College.

On closer inspection, it was clear Bush was not mowing anything. His contraption consisted of two bicycle wheels linked by a wooden box. Bush had built it from scratch. It was a surveying device, designed to chart the terrain. The box was stuffed with a mechanical recording system that traced out a simple map as the wheels turned. This crude device was Bush's first invention. He anticipated a long period of test and refinement.

Bush's improbable gadget, built the previous summer, would not have surprised his Tufts classmates, who named him best math student of the 125-person freshman class. As a sophomore Bush replaced an ailing instructor for part of a term, and he tutored groups of fellow students in physics and math in order to help pay his way through school. "I'd have a class in the evening, and everyone who came in put fifty cents on the barrelhead," Bush recalled. "Some of these would get the fifty cents back at the end of the hour. These were the ones that really didn't have the money to pay; their job was to drum up trade."

None of this was done with dead seriousness, however. Bush leavened his tutoring with humor. He enlisted the campus football star in a word-of-mouth campaign aimed at drawing new students to his sessions. Once he agreed to give a student math lessons in exchange for tips on improving his tennis game. But soon Bush rebelled. "This isn't working out quite fairly," he said. "I really can see some improvement in my tennis, but I am reasonably sure that *no one* can do much with your mathematics."[1]

Bush's quick wit at times bordered on the malicious. One of his roommates, for instance, was so nearsighted that he was practically blind without his glasses. "We used to move them from the place he left them at night in order to see him gallop around the room squinting at everything that shined and cursing us for moving them," Bush remembered. "As he did so he had the expression of an imbecile to an extent that was most entertaining so we repeated the operation fairly frequently. But when he finally located his glasses, cursed us once more, and put them on his face straightened out marvelously and he became in fact quite intelligent in appearance."

Bush's own glasses were rather thick, of course, and he hated to remove his spectacles. "I suspect I look rather vacuous when I take my own glasses off and I acquire such intelligence in my expression as I can command when I get them on again," he noted. "So it makes me a little worried to see myself without them."[2]

As a smart young man bothered by physical and social inadequacies, Bush eagerly wished to establish himself. To recover what illness and family circumstance took from him, he strove hard but was thwarted time and again. In the spring of 1911, he suffered an appendicitis. He needed an operation and missed a semester. Bedridden for weeks during one stretch, he found consolation in his imagination. He mused about the possibility of perpetual motion, even writing an article in which he explored various ways to achieve it. He also collected information about the Panama Canal, then undergoing construction. Opened informally in 1914, the canal was the biggest engineering challenge of the years before World War I. The project showed how ambitious engineers could leave a mark on the world.[3]

When he returned to Tufts, he still suffered from chronic rheumatism. His 5-foot-11-inch frame carried just 140 pounds. Anxious about his health, he relieved his anxiety through endless activity. He rarely stopped for self-examination. (When asked to state his personal thoughts, he would say, "Let's not spend much time on this.") For a season or two, he ran track, racing at distances of one and two miles without distinction. In his own scrapbook he confessed, "As a miler you're good—for nothing!"

Too frail to play football, he managed the team his senior year and scheduled a contest with West Point (in the game a young Dwight Eisenhower badly injured his knee). Bush was popular, serving as vice-president of his sophomore class and president of his junior class. He joined the engineering fraternity, Alpha Tau Omega. He taught himself the piccolo and the harmonica and sang in his father's church choir. He dated a girl, Phoebe Clara Davis, the daughter of a Chelsea merchant. For amusement, they sang and played music, and they attended parties together at Tufts, where he was active in the Evening Party Association. A night's program might include as many as 20 dances, waltzes being the most frequent.[4]

While not quite frivolous, Bush played hard, smoking, drinking, carousing and cussing. Yet at the same time he managed to essentially pay for his own schooling. Roughly half of his $250 in yearly academic expenses (which included $150 in tuition and $80 room and board) was covered by scholarships, with the rest of his bills covered by money from tutoring and a job as an aide in the mathematics department. Indeed, he earned so much money from the math department one term that the college owed him money when the term ended. The bursar wanted to apply the excess to the next term's bill but Bush insisted on receiving the money then and there; he even threatened to sue Tufts if the bursar didn't pay up immediately. In no time, Bush had his money.[5]

Despite his many outside activities, Bush raced though his studies, achieving straight-*A* grades with the exception of the term interrupted by his illness. Tufts allowed industrious students to gain a master's degree in the same four years it usually took to gain a bachelor's degree. Bush knew his father had difficulty paying his tuition bills on time, so he thought following the master's schedule was a bargain. He took extra courses, or sometimes skipped a course altogether, taking the exams only. Once, he read the textbook for a course in advance and asked the professor, he recalled, if "I could make some time available for other things I had in mind by just taking the final examination in the course when it occurred." The professor refused. Instead, he gave Bush the test on the spot. Bush passed, and was granted credit.[6]

"Now this made an enormous impression on me," Bush later wrote. He admired men of action, despised rules and felt that merit meant everything. No organization should hold a man down purely for reasons of protocol or tradition, he thought. The professor who allowed him to break the rules became "my idol, and the man I wanted to be like."[7]

Bush favored men who set their own rules, and he was bright enough to do so himself. Yet he respected tradition and deferred to his professors at the proper times. He was especially close to a young mathematics professor named William Ransom, who was 12 years older than Bush. Ransom was not brilliant but had a passion for teaching. He could stir an interest in math among even the slowest students. He was impressed with Bush's ability but not awestruck. The year before Bush arrived, Ransom had graduated Norbert Wiener, a child prodigy who would later be seen as a rare mathematical genius. Wiener arrived at Tufts as an 11-year-old, and before long he was lecturing the math faculty. "I used to sit in the front row while [Wiener] worked at the board," Ransom said. "It was easier that way."[8]

Bush did not possess Wiener's mental gifts and, as a math student, rated himself no better than the "fourth or fifth echelon." While certainly a severe assessment of his talents, this judgment did not deter Bush from mathematics. The unbending rules and the cold, hard logic of math appealed to Bush's sense of order and balance. Math shaped his approach to human relationships and gave him the means to achieve practical results. From math, he also gained an inner satisfaction—even a spirituality—that he never found in religion. At his graduation ceremony, he showed his affection for his favorite subject by delivering an address on "the poetry of mathematics."[9]

Mathematics evoked something besides poetry: money. Bush could put mathematics to work. He thought his surveying device, the "profile tracer," could be the basis for a business. The device was a primitive calculator.

> It consisted of an instrument box slung between two small bicycle wheels. The surveyor pushed it over a road, or across a field, and it automatically drew the profile as it went. It was sensitive and fairly accurate. If, going down a road, it ran over a manhole cover, it would duly plot the little bump. If one ran it around a field and came back to the starting point, it would show within a few inches the same elevation as that at which it had started.
>
> The box contained a well-damped pendulum. On this was mounted a disc, driven from the rear wheel. Against this disc rested two sharp-edged rollers. One picked off the vertical distance traveled, and moved a pen. The other picked off the horizontal distance and turned a drum carrying the paper. This much constituted an integrator.[10]

The profile tracer, then, automatically calculated elevations and produced a drawing, or crude simulation, of the terrain crossed by the machine. Bush applied for a patent on his tracer, convinced there would be demand for it. He thought his invention could save surveyors time and money, since it did the work of three men. "The usual method" of creating a profile called for "a surveying party," Bush wrote in a thesis on the device. The surveyors "take the elevation of a sufficient number of points on the line, and by chaining and pacing, locate these points as to horizontal distance, thus establishing the profile. The notes thus obtained in the field are worked up later in the office, and the profile is plotted from them. In this manner, about three miles may be covered in a day. It is easily seen that this method is slow, roundabout, and expensive."[11]

On the last day of 1912, the U.S. awarded patent number 1,048,649 to Bush for his profile tracer. A patent was a powerful symbol to him. It symbolized the way the nation rewarded its inventors—by protecting their creations. Patenting had deep roots in the nation. George Washington signed into law the first patent measure in 1790. It empowered a board to issue a patent, "if they shall deem the invention or discovery sufficiently useful and important." Thomas Jefferson was the first chief of the nation's patent board. Though opposed to monopolies, he asserted "an inventor ought to be allowed the right to the benefit of his invention for some certain time. Nobody wishes more than I do that ingenuity should receive liberal encouragement."[12]

Modified many times over the years, the patent law still protected and encouraged inventors. But markets were increasingly dominated by large companies, which had teams of their own inventors and staffs of attorneys who piled up patents, many acquired solely in order to thwart rivals. Adapting to this new scene, inventors often licensed their patents to large companies, which then brought a product to market (and gave the inventor a tiny share of the sales). Litigation was common as competitors tried to stymie one another through patent-infringement claims. For the lone inventor, the new political economy of patents was a disaster. Even if he was not bankrupted by the byzantine litigation spawned by his patent claim, he rarely reaped the financial rewards of his ingenuity. Big businesses either stole his ideas or paid him a nominal fee for his patents and then sat on them.

Bush knew this fate well. After gaining his first patent, he wrote many companies about the profile tracer, promoting its virtues and offering a license to it. He visited at least one firm in person, accompanied by a friend. No one bought it. Bush was hurt, especially since the effort had cost him dearly at a time "when money was a very scarce article."[13]

But the setback also awakened him to his own naivete. It made him realize the shortcomings of tinkering and underscored his need to find a practical career.

"The trouble was that back in 1913," Bush later wrote, "I was densely ignorant. I knew a bit of physics and mathematics. I had graduated in engineering.

But I was not an engineer. An engineer has to know a lot about people, the ways they organize and work together, or against one another, the ways in which business makes a profit or fails to, especially about how new things become conceived, analyzed, developed, manufactured, put into use. So I charged that invention off to experience . . . and I reoriented my thinking. In fact for the first time, I resolved to become a real engineer. I resolved to learn about men as well as about things."14

Academic life suited Bush, but did not define him. Now 23 years old, he wanted a change. He had never held a job outside school and wanted a taste of industry, if only to gain the "funds to get back to college and try for a doctorate." In the latter half of 1913, he joined General Electric, a giant in the burgeoning electric power industry. GE hired him as a "test man" in its chief facility in Schenectady, New York, near the state capital of Albany. Bush lived in a boardinghouse, visited the local beer parlor and mingled easily with his coworkers. One evening, Perry called on him. Father and son took their dinner in the boardinghouse with a dozen mechanics, who arrived at the table directly from work. Within an hour or two "all embarrassment had disappeared," Bush recalled. "Before the evening was over they were all sitting around telling jokes, the old man included, having a wonderful time." Bush was proud: his father was one of the boys.

After three months, GE put Bush in charge of about 20 test men, without raising his starting salary of $14 a week. Meanwhile, workers struck the plant. Under mutual agreement, the plant's testers, including a sympathetic Bush, crossed the picket line. "I readily took on the task of exploring the works for them, to see whether strikebreakers were being smuggled in," he recalled.15

Bush next moved to a GE plant in Pittsfield, Massachusetts, for the chance to work on high-voltage transformers. His job was to test equipment and insure its safety. Before long, a fire broke out in the plant, burning up the main cables. As punishment, Bush was laid off with the other test men for a few weeks. He left GE in October 1914, returning to Tufts. He tried not to appear chastened, but here he was, a star pupil, with no definite prospects. He told a dean at the school that he was out of work. Bush later noted that the dean politely told him "the only job he had [to offer] was one I would not want." But Bush was in no position to pick and choose. Without asking for details, "I told him I would take it."

The job was no better than advertised: teaching math 14 hours a week to female students at Jackson, the sister college of Tufts. Bush was to be paid $300 for the term, slightly better than his GE salary, and he expected to earn his wages. His predecessor, after all, had quit over the antics in the class. "They tried some of the same games on me," Bush later said. "One of the best was to toss a pants button so it would land near me, then they would all watch it roll across the floor, and stare at me stonily." Bush was unimpressed and told them, "My fun comes later." In the face of his veiled threat to issue poor grades, the students applied themselves, and everyone received a passing grade.[16]

The college dean was pleased. "Mr. Bush has proved himself so valuable that his services should be retained at least for the remainder of the year," he wrote in February 1915. He raised Bush's salary to $400 per term. While the teaching wasn't terribly challenging, it left Bush time to chart his future. He soon decided to pursue a doctorate, which he felt would launch him on a career in academia or industry or both.[17]

Even more than another degree, Bush needed a mentor. He found one in Hermon Carey Bumpus, the president of Tufts and a zoologist by training. Bumpus appealed on Bush's behalf to Clark University in Worcester, Massachusetts, for a scholarship. The school, though small, was distinguished, and Bush wanted to study under one of its engineering professors.[18]

"We have a very excellent man, Vannevar Bush, an instructor in mathematics, who shows signs of having qualifications which will lead to excellence in his chosen field," Bumpus wrote G. Stanley Hall, the president of Clark College, in early 1915. "I think that it would be a very good thing for Clark, and also for Bush, if he could take graduate work in Worcester, particular[ly] with Professor Webster, and I am therefore writing to ask if there is any probability that a fellowship would be available for him, and if available, the amount that it would yield. I would not write this if I did not consider Bush a man of exceptional ability."[19]

Exceptional ability or not, Bush spent the summer as an electrical inspector in the Navy Yard in Brooklyn, New York. Soaking up practical experience, he was confident enough to forge a friendship with the commanding officer in charge of the Yard's material. While his fellow inspectors "went outside to have a smoke" at noon, Bush stayed behind, watching and waiting as the officer later lit his pipe in his office. Since "there were no smoking signs about . . . I stayed at my desk and lit my own pipe. After a while he began to join me and chat. I was now a marked man, having some sort of strange political influence."

His short stay in Brooklyn gave him a taste of the "strange sort of experience" that many newcomers found in this melting-pot city. He rode a motorcycle about town and gave hardly a thought to the widening war in Europe, though in the back of his mind he thought "it was then probably inevitable that we were headed for war. But I can't remember that any of us talked about it."[20]

Bush concentrated on mastering his job but his real concern was advancing his prospects. "The work of subinspector here has been interesting, and valuable in its experience," he wrote Bumpus. "For certain viewpoints, financially for instance, I could well continue in this work for awhile." But the respectability of academia tugged at him. In early September 1915, he quit the Navy Yard and took off for Clark University, convinced he had "an opportunity for larger things" and "that a year from now will see me much nearer where I want to be because of the present sacrifice."[21]

Larger things awaited Bush, and sacrifices too.

Clark University gave Bush a hefty fellowship: $1,500 to defray the costs of obtaining his doctorate under the tutelage of a professor named Arthur Gordon Webster. On arriving, Bush learned that Webster wanted him to devote his doctorate to the study of acoustics. This angered Bush, who resisted taking directions from others. "To hell with it," he said, and abruptly "threw up the scholarship and got out of" Clark.

This act of rebellion revealed Bush's low tolerance for academic servitude. He would not cater to the whims of a patron, even if a scholarship and a doctorate hung in the balance. Indeed, no sooner had he left Clark than he proved he could land on his feet despite his stubborn streak.

Bush next tried to gain admission to the Massachusetts Institute of Technology as a doctoral student in electrical engineering. He wasn't guaranteed a scholarship at first, though one came through in late November. But when the professor handling his admission refused to give him credit for his Tufts course in thermodynamics, Bush fought back. The professor said that the man who taught Bush thermodynamics "didn't know any thermodynamics." "That's correct, he didn't," Bush retorted. "But he isn't trying to enter MIT, I am."[22]

Bush's argument carried the day. He was admitted, though that wasn't the end of his troubles. He wanted a promise from MIT that he would be awarded his doctorate within a year, so long as he produced the thesis. Impatient as always, he now had a specific reason for haste. He planned to marry Phoebe Davis, his college sweetheart, as soon as he finished at MIT. He had barely enough money for one year of study, and he wished to avoid dragging his new wife into a life of penury.

Phoebe, a merchant's daughter, was a respectable and plain-looking woman, one year younger than Bush. They attended fraternity dances together while he was at Tufts, and in their family homes they played music together (she played piano, while he played flute). Phoebe loved Bush's wisecracks and antics, of which there was a steady supply. Visiting her often provided humorous relief from his intense studies. One winter day, he drove his secondhand Stanley Steamer automobile up the hill to Phoebe's house only to hit an icy patch midway up the rise. The car's wheels spun, and the engine exhausted its steam, dropping Bush to the bottom of the hill. The only way past the icy patch was for Bush to "look about for possible cops, pull the throttle way down and roar over the patch, shutting off the steam for a moment as I did so."[23]

Bush thought Phoebe was worth clearing all obstacles for. She gave his ego a wide berth and never tried to compete with him for the attention of others. She was shy, somewhat retiring and very much a creature of her New England upbringing. She found the slow pace of Cape Cod life to her liking and possessed little interest in the wider world. She never tried to understand the technical issues that absorbed Bush, nor did she apologize for her lack of interest. She was at home with herself, happy to be who she was, which was perhaps why she seemed to be the only person who could calm the excitable Bush. He felt they were an ideal match, and he feared she would lose interest in him if he put off marriage for too long.[24]

MIT was the nation's best engineering school, but in the fall of 1915 its electrical-engineering department was only an infant. Founded in 1902 largely by a group of the physicists at the university, the department had awarded only four doctorates in its history. Candidates for a Ph.D. were expected to toil long and hard, especially since a program with Harvard University gave a joint degree from both schools. Additional pressure came from a pending move by the electrical engineers. At the time of Bush's arrival, the department was beginning its final academic year in its original Boston building, and plans were set for a move across the Charles River to Cambridge. MIT had purchased 50 acres of land across the river, and the electrical-engineering department would have two buildings on the plot, one for classes and the other to contain a main laboratory and a ten-ton crane. Professors strained to juggle the move, their research and ordinary teaching duties.

Bush did not make things easier for himself by choosing as a dissertation topic the problem of reducing certain differential equations to more manageable algebraic equations. Based on work by Oliver Heavyside, this new type of mathematics promised to aid in solving sticky problems connected with electric-power transmission. Bush's academic supervisor, Arthur Kennelly, a former colleague of Edison's, accepted his student's aggressive schedule. But fearing Kennelly might renege later, Bush insisted his professor sign a contract specifying the work required. Kennelly agreed, though, as "one of those who did not expect overnight magic," he doubted Bush's ability to finish his thesis in a year.

Yet Bush did finish—and ahead of schedule. In April 1916 he submitted a

169-page thesis to his adviser. He had worked so hard to meet his deadline, he said, "I ended up in better [physical] condition than when I started." Kennelly was unimpressed. He demanded that Bush "increase the scope" of his dissertation, though he offered no specific reason why. Bush protested, and Kennelly stood his ground. Because Bush had "wisely based the theoretical treatment" of his thesis "on a branch of mathematics Kennelly had never studied," his supervisor never realized that his thesis contained a "first-class" mathematical error. The mistake was so arcane that neither Bush nor anyone else discovered it for years.[25]

Bush was outraged by Kennelly's intransigence, but in a tight spot. If he appealed Kennelly's decision to department chairman Dugald C. Jackson and lost, he might never obtain his doctorate. If he delayed taking a job, however, he feared he might lose his sweetheart. For Bush the choice was simple: he would not back down. In a sign of things to come, Bush sensed that he could make MIT's bureaucracy serve his needs. Asking Jackson for a new ruling, Bush was quickly proved right. Jackson overruled Kennelly and granted Bush his doctorate.

This was the start of a fruitful relationship between the two electrical engineers. Bush had found an important mentor in Jackson. Never eager to credit others for his success, Bush regarded Jackson as one of the few "able, inspiring" teachers he ever knew.[26]

Jackson was a rebel. Moody and irascible, he was "a volatile fellow who kept people on their heels," recalls one junior colleague. Students and professors alike feared him. He did not harbor traditional university values either. No academic purist, he thought knowledge must be put to work. Universities graduated engineers, he asserted, who weren't geared toward the new demands of science-based industries. The solution was for the university to cooperate more closely with the leading companies in these industries. He believed electrical engineers, because of their roots in physics, were often too theoretical. They should more willingly address industry's technical problems, not do abstract study for its own sake.[27]

A refugee from industry, Jackson knew the research-oriented corporation from the inside. Companies such as AT&T and General Electric were rapidly formalizing the process of invention by funding research at their own labs. Rather than relying on the muse to strike a lone inventor, these companies sponsored teams of researchers who aimed to improve the corporation's position in the marketplace. A systematic approach to invention—the same problem attacked from different angles—replaced the hit-or-miss tactics of the past. Jackson had participated in this change. He had been chief engineer at a railway and electric-power company before starting the country's second electrical-engineering department at the University of Wisconsin (Missouri opened the first).

After Jackson arrived at MIT in 1907, the institute's electrical-engineering department began offering to perform research for private industry. Jackson himself consulted for companies. MIT allowed him to keep an office in Boston in order to handle demand for his services as well as maintain contact with industry. By 1910, he was telling large corporations that his department stood "ready to undertake some of the more distinctly commercial investigations under the patronage or support of the great manufacturing or other commercial companies." Three years later, the department formed a research division, supported by GE, AT&T, Public Service Railway and other companies.

Jackson's promotion of commercially relevant research "profoundly influenced" Bush. "We recognize it as our duty to contribute men to the industries and perhaps should be cautious about recruiting our staff through robbing the industries," he wrote in 1915. "We need not only train men for the industries but must train them for ourselves," he added, noting that universities "should 'feed' rather than 'feed on' the industries as far as men are concerned."

Academic purists criticized this philosophy, but Jackson countered that professors should not have to apologize for earning consulting fees. He estimated that, through outside jobs, they should at least double their university pay.

This was just what Bush wanted to hear. In need of cash and desirous of influence in the world as well as the academy, Bush was a man in Jackson's mold.[28]

With a Ph.D. almost in hand, Bush began looking for a job in the spring of 1916, quickly narrowing his choice to either Tufts, his alma mater, or American Telephone and Telegraph's research division (later called Bell Laboratories). Bush clearly preferred the university post—Tufts offered him an assistant professorship in its electrical-engineering department—but he played hard to get in the hopes of gaining a higher starting salary. He told Tufts president Bumpus that AT&T had made him "an attractive proposition," with a starting salary of $1,500 and raises of $300 to $600 annually. The deal slightly bettered the offer from Tufts. While Bush did not ask Bumpus for more money, he told him that AT&T was holding a position for him.[29]

Bumpus did not raise his offer, but Bush chose to join Tufts anyway. Even with the higher pay, the AT&T job wouldn't provide him "enough income . . . to live on" because he would be forbidden to sell his services to other companies. By staying in academia, he could supplement his income by consulting for industry. His future brightening, he married Phoebe in August 1916.[30]

The next month Bush joined the faculty of Tufts and quickly found a corporate patron. Housed on campus was the American Radio and Research Corp., a manufacturer of radio parts and complete sets for amateur operators. AMRAD, as the company was called for short, had been founded a year before, in 1915, by Harold Power, a recent graduate of Tufts. As a student, Power had been active in the school's Wireless Society, a club for radio fans. Bush was also a member of the club. By his own account, Bush was smitten with radio. At the age of 19, in 1909, he had experienced one of the great thrills of his life when, while experimenting with a home-made radio set and listening to Morse-code signals, he "suddenly heard a human voice break in on the monotonous staccato of dot-dash, with a cheery cry of 'hello.'"

While Bush and Power shared a zeal for radio, they were otherwise quite different. Where Bush sought to hedge his bets, Power was an unabashed adventurer. During school vacations, he served as a radio operator on various ships. One of these, the yacht *Corsair*, was owned by J. P. Morgan, the investment banker famed for re-ordering American industry. Power wisely got to know the tycoon and his family.

In 1915, two years after Morgan's death, his son and namesake agreed to back Power in a radio venture. AMRAD set up shop at Tufts. Next to its building, the company constructed a 300-foot transmission tower with a range of 100 miles. Besides making radios and parts, Power wanted to produce broadcasts from the Wireless Society's station on campus. At the time, no one regularly broadcast music or other entertainment over the air. Radio operators spoke with one another; simply receiving had little appeal. Power reasoned that for radio to catch on more broadly, it would have to offer something to those who only listened. On the evening of March 18, 1916, the Tufts station broadcast three hours of phonograph music. The station (later known by the call letters WGI) followed by only a few months Pittsburgh's KDKA, which is generally credited with being the nation's first radio station.[31]

Shortly after Bush began work at Tufts, Power hired him to run AMRAD's fledgling lab at a salary that exceeded his Tufts pay. He and Phoebe were "cheered . . . up greatly since we needed cash badly," he later said.[32]

Power had made a shrewd choice. Bush exuded confidence; he thought he could exploit the "fine opportunities" awaiting radio pioneers. He threw himself into AMRAD research, exploring the mysteries of vacuum tubes, which desperately needed improvement if radio was to expand. This was the high frontier of electrical engineering, and Bush realized advances would aid all forms of communication.[33]

Bush was thrilled with the possibilities. In the laboratory, "We dream the life of the future," he wrote. Yet once he almost lost his future when he accidentally received an electric shock. "It knocked me over although it didn't knock me out," he said.

A colleague picked Bush up and dragged him outside. He tried to revive Bush with mouth-to-mouth and, Bush later recalled, "got going so fast at it that I couldn't get a word in to tell him that I was all right breathing without his help."

When he wasn't toiling in the lab, Bush gave business advice to AMRAD because he felt Power "was certainly no manager." Power did not always want the advice, but Bush could not help himself. In his mind, the line between research and business had blurred. The good researcher, he believed, could not act in academic isolation; in order to succeed, he must know something about markets, finance and the organization of a business.

Reflecting on his experience at AMRAD, Bush charted this sea change in the relationship between research and industry. "Not many years ago . . ." he wrote in an article on the company for curious Tufts alumni, "commercial research was looked down upon as undignified and mercenary, and not to be mentioned in the same breath with the study of the swing of the planets in their orbits. To the business man, on the other hand, the pure research enthusiast was a dreamer and a solver of academic puzzles; ornamental, perhaps, but useless and expensive.

"Today all this has changed," he added. "We have learned that no science worthy of the name is so pure as to be entirely devoid of possibilities of service to the needs of a complex civilization."[34]

That service might include improving the tools of war. Since August 1914, war had raged in Europe. The U.S. tried to stay neutral, while still supplying the Allies by sea. Slowly, America was drawn into the conflict. In May 1915, a German submarine sank the British passenger liner *Lusitania,* killing 128 Americans. The U.S. strongly protested, and Germany restrained its submarine attacks until January 1917, when it declared its U-boats would attack all ships headed for Britain. The Germans sank several American ships in the ensuing months, and on April 6, the Congress endorsed President Woodrow Wilson's call for war against Germany.

Consumed by private pursuits, Bush had paid scant attention to the bloodiest war in history. Politics bored him, and he largely ignored international affairs. He was admittedly "provincial." At 27, he was too old to be drafted. His wife had just had a baby boy. His career, after several false starts, was moving.[35]

America's direct involvement in the European war forced Bush from his cocoon. He was not prone to intense displays of patriotism, as was his father, but he was no pacifist either. His country was at war, and he wished to help. At bottom, it was a fresh chance to prove his mettle. He surely had not thought much about how best to apply research to the problems of war. But he had no qualms about trying to do so, just as long as he could set the terms of his engagement.

Bush knew he would not labor alone. Since the outbreak of the war in Europe, hundreds of America's leading scientists and engineers had offered to aid the military. The reception wasn't always warm. When the American Chemical Society first offered its services on behalf of preparedness, the secretary of war demurred, saying his department already had a chemist. Often, neither the military nor the politicians understood science well enough to recognize the gift horse. It was easier for the government to latch on to big names. In 1915, the Navy invited Thomas Edison, the nation's best-known and most revered inventor, to chair a board of consultants. (The offer came after *The New York Times* printed a provocative interview with Edison, which bore the arresting headline: "Edison's Plan for Preparedness: The Inventor Tells How We Could Be Made Invincible in War.") The naval board's aim was to stimulate inventions that would prove Edison's assertion that "modern warfare is more a matter of machines than of men." Edison stocked his board with engineers and industrialists, purposely leaving scientists in the cold. He wanted practical men, not talkers.

Distressed by Edison's snub, the scientists in 1916 formed their own group, called the National Research Council. Formally an arm of the prestigious but moribund National Academy of Sciences, the council vowed to produce innovations in weaponry. With the resumption of U-boat attacks in February 1917, an alarmed Navy asked the council to help it find a way to detect submarines. Robert Millikan, a leading physicist and a future Nobel Laureate, was assigned to direct the effort.[36]

Millikan faced an almost impossible situation. The Navy's best submarine detector could not locate a submerged U-boat further than 200 yards away, and that was not enough. The chances of rapid improvements seemed small. Meanwhile, German submarines were reaping a bonanza in torpedoed vessels: 1.3 million tons in the first quarter of 1917 and perhaps 1 million more in April alone.[37]

Scores of researchers were studying the detection problem. Bush was a late arrival to the field. He had a notion for a crude device that ensnared submarines in a magnetic field. A disturbance in the field should produce a signal. As the submarine came closer, the signal should grow louder. That was the theory at least. Bush was eager to test it.[38]

In early May 1917, he traveled to Washington to meet with Millikan. He hoped to obtain the physicist's endorsement and a broad charter. He acted boldly, grasping for the first time all of his disparate aspirations: inventor, entrepreneur, patriot and insurgent. Cynical about large organizations and loath to cede control, he glimpsed a way to harness his invention to an organization of his own making.

The meeting, which occurred on May 8, went well. Bush described his idea at length. Millikan knew of no one working in "exactly" this vein and thought Bush's model was "well worth trying out on a life-size scale." Reassured, Bush "suggested to Millikan that they consider my problem as being handled by me privately." Millikan agreed, and promised to inform Bush of "any developments which might come in my particular line, and to put me in touch with anyone who might be doing the same work." He also offered to help Bush to obtain "facilities for testing."

Bush was elated by Millikan's encouragement. He had succeeded in staying outside the government's official channels, yet had gained a means of contributing to the war effort. This freelance approach suited Bush. He was skeptical about the research council, as he indicated after his meeting with Millikan in a letter written to Bumpus, his mentor at Tufts. "The Council is extremely busy," Bush wrote. "Their authority comes simply by their influence; and it is not known how long it will be before there are radical changes in organization. For these reasons I do not believe they will immediately do construction work on any device. The Navy might test a new device at this time, but they are pretty well devoted to their particular hobbies. The red tape there looks formidable to me."

Before leaving Washington, Bush obtained one more favor: Millikan agreed to write a letter on Bush's behalf to the son of the late J. P. Morgan, a principal in the House of Morgan, "stating that [he] believed no one else was working in my line, and that my device was worth trying."[39]

If Morgan agreed to back Bush, tests on his detection device could begin in a week. As soon as he left Millikan, Bush wired Power, asking him to contact the investment banker and arrange a meeting. Power did so, and Morgan was an easy sell. Bush sweetened his presentation with the suggestion that his detection technique might be of value in mining.

Morgan's money liberated Bush from the chaotic politics surrounding the research council, which was jockeying with Edison's board for the inside track with a Navy whose officers were skeptical of both teams. The experience left a mark on Bush, who later crowed: "Since I was not in uniform and took no government money, I was a maverick."[40]

Bush's cockiness did not endear him to the Navy's officers. They were so skeptical of him that in May 1917 Bumpus felt compelled to write the commandant of the nearby Charlestown Navy Yard, attesting that Bush "is not an irresponsible enthusiast, but is a man of high scientific standing and ability."[41]

A suspicious Navy would bedevil Bush for the rest of the war. In tests his device, improved with the addition of acoustic filters from Bell Laboratories, detected many submarines, and it withstood depth explosions.

"Then came a shock," as Bush later wrote. "The Navy insisted that the gadget was of no use on a wooden ship; it must be put on an iron ship, a destroyer, for example. I was pretty sure it could not be done . . . but I was young and foolish, I did not get a real chance to argue my case, the decision to use a destroyer had been made by some senior officers who knew even less physics than I did, and so I went to work and wasted six months trying to adapt the equipment to an iron ship."[42]

AMRAD built 100 sets anyway. A handful made it onto British subchasers. Bush insisted his primitive detector could locate enemy submarines, but it never found any before the war ended. This infuriated Bush, who directed his anger at the Navy. He snidely concluded that from World War I, "I learned quite a bit about how not to fight a war."[43]

In the war, Bumpus found a broader lesson that Bush surely would have heard—and endorsed. In a dinner speech in early 1918, Bumpus declared that managerial expertise and engineering proficiency lay at the center of war-making. "The conduct of this war is not merely a military procedure," he said. "It is not an affair of arms. It is not a conflict between armies. It is a huge business proposition. It is a great engineering undertaking."[44]

Chapter 3

"Blow for blow"

(1919–32)

I'm a very peaceful fellow unless there's some reason for starting a row.
—Vannevar Bush

Bush left Tufts in 1919 and rejoined Dugald C. Jackson's electrical-engineering department at MIT. The offer from "DC," as Bush called Jackson's, was irresistible. Few men inspired Bush as Jackson did.

It helped that Bush felt he could handle the moody Jackson, who was famous for intimidating subordinates. "If one wished to visit him in his office, it was well to toss one's hat in first and then, if it stayed in, to follow it," Bush recalled. "He worked well only with those who traded him blow for blow, and I was one of these."

The pattern of their relationship was set from the start. Bush would give no ground to Jackson, who asked him to teach an introductory course on electrical engineering called "601." During the first few weeks of the course, Bush sent a note to his three teaching assistants describing future lessons. Jackson got hold of the note and took exception to it. "Old DC jumped on my neck," Bush said. "Because the method I'd proposed for handling a particular thing was not the way it was handled in his book. So he lit on me like a ton of bricks. . . . I told him I was in charge of the course and he could take me out of that position if he wanted to, but as long as I was in charge of the course by god the people that were teaching it were going to follow my instructions. And he nearly had a fit. That ended the row." There were "no more rows" after that.[1]

Bush returned to MIT just as the Institute faced a severe fiscal crisis. The state of Massachusetts was about to cut off a $100,000 annual subsidy to the Institute from the Gordon McKay fund. To make up the shortfall, MIT asked industry for help. Some educators attacked MIT's appeal for corporate aid as impure, but it was "an immediate success," attracting large sums at first. By the early 1920s, Jackson had built MIT's electrical-engineering department into what was perhaps the best in the world. The university's younger professors viewed him as a role model.[2]

The Institute's embrace of corporate values was a sign of the times. In the decade following the end of the war, society's most vital force was found in the marriage of technology and corporate capitalism.[3] Conservatives were in command. Three Republicans in a row served in the White House; Herbert Hoover, the last of the three, was a mining engineer whose foreign exploits had made him a millionaire by the age of 40. Electrification was remaking American cities, ending the drab isolation of rural life and giving rise to vast utility companies that, like the railroads of the last century, held entire regions hostage. The automobile was the rage. Henry Ford's method of manufacturing was lionized and widely copied; even the anticapitalist Soviet Union saw in "Fordism" a way to harness technology on behalf of the people. Artists, too, were affected by the spirit of the times. In Eugene O'Neill's play *Dynamo,* the hero, an atheist, discovers spirituality in the form of electricity. To some, American civilization was coming to resemble a vast machine. Historian Charles A. Beard asked Americans to "accept the inevitability of science and the machine" in the belief that these forces will not destroy "the love of beauty, the sense of mystery and the motive of compassion."[4]

The era well suited a man of Bush's temperament and drive. Popular with students, he seemed informal and frank at a time when many professors struck students as stuffy and ancient. He wasn't afraid to lighten his sober lessons with a bit of levity. One of his favorite routines involved a pipe wrench. The lecture highlighted the value of precise English, a point often overlooked in engineering classes. Bush opened by holding a pipe wrench before hundreds of freshmen, asking them to describe it. One after another, they did so. Bush tore into each description, eviscerating the vague spots and leaving his students in awe of his critical faculties. He ended this "intellectual free-for-all" by writing in precise English a patent application for the wrench.[5]

Advanced students found Bush's clarity and rigor particularly appealing. After only a few years at MIT, he was widely admired by doctoral candidates for his dedication to a deep level of learning. "Bush was always itching to tackle the job of understanding those facets of a subject going just beyond the point where understanding was firm," recalled Harold Hazen, a gifted graduate student who later became chairman of the electrical-engineering department. "It was the process of coping with the unknown in the attempt to achieve real understanding that excited him and, with him, all his students."

"At the same time, he was a superbly good expositor," Hazen added. "I remember particularly one class in which he was struggling to convey to students the essence of a new concept. We didn't comprehend his first approach to the problem, nor his second. In the end he spent that entire hour in the attempt to get his point across, coming at it in turn from six completely separate and independent approaches, developing each with great expository skill while watching the eyes and faces of his students like a hawk. After each attempt, when he saw that he had not really succeeded with more than a few of us, he backed off and started on another approach. He was so skillful in disguising each new approach to appear as a fresh new problem that few of us were aware that he was, in fact, merely taking another crack at the same old problem on which he had not been successful before."[6]

Bush's horizons extended beyond teaching. Even before leaving Tufts, he had felt the pull of industry. Three weeks before the end of World War I in 1918, he had told Bumpus of Tufts: "I would, of course, not feel like devoting all of my time to teaching." Bush's desire for diversity stemmed partly from his restless intellect, but he also wanted another source of funds. With another MIT professor, Bush wrote an introductory textbook, *Principles of Electrical Engineering*. Published in 1922, the book was well-received, helped Bush's reputation and brought him steady, albeit modest royalties.

Taking his cue from Jackson, Bush devoted a considerable portion of his oversized energies to business. Seeking fame, he would settle for money. After all, he now had a family to support. Phoebe, who had taught school in Chelsea, left her job after giving birth to a son named Richard in 1917. Another child was on the way.

Almost daily, Bush worked on AMRAD matters, judging from the frequent memos he wrote to management in 1920 and 1921. The postwar years had been unkind to AMRAD. In 1917, the concern, still housed on Tufts' campus, had won big orders for radios from the Army's Signal Corps. It expanded to meet surging demand, only to see the government cancel its orders when the war ended. The government's action devastated AMRAD. By 1920, the company was mortally wounded. Bush, still carping about Power's shortcomings as a manager, began looking for other opportunities. An AMRAD mechanic named Al Spencer presented him with one.

Spencer told Bush about an invention he'd made on his own time. As Bush later described it, "It was just a thermostatic sheet which he had dished a little bit by tapping it on an anvil with a hammer, and which would click through like the bottom of an oilcan. Of course, being made of thermostatic metal, it would snap through at a certain temperature." Bush realized Spencer had hit on something "utterly new," a "powerful, inexpensive, thermostatic switch for all sorts of purposes."

Bush first went to AMRAD's attorneys and "stated a hypothetical case," resembling Spencer's, "and asked whether, in such circumstances, the company had any equity in the invention." The attorneys said no and told Bush he was "perfectly free" to shop the invention around. Bush did. He immediately turned to Laurence Marshall, a Tufts classmate who was eager to back new ventures. Trusting Bush's technical judgment, Marshall, a gambler who was enthusiastic about the potential of electronics, agreed to build a company around Spencer's switch. He gathered investors. Among them was Richard S. Aldrich, whose sister married John D. Rockefeller, Jr.[7]

In late 1921, the group formed the Spencer Thermostat Company, hiring Bush as a consultant. While still AMRAD's research chief, Bush ran the new company's labs and aided Marshall in lining up customers for the thermostat. The two men were optimistic because they had the first reliable thermostat for a flatiron. The market was large, but none of the leading flatiron makers would license Spencer's invention. Bush suspected a conspiracy, and he threatened to put Spencer Thermostat into the business of making flatirons. Finally, some manufacturers agreed to license the thermostat, use it in their products and pay royalties to Spencer Thermostat. The payments well exceeded $1 million and launched the company on a prosperous course. Bush's stake in Spencer Thermostat grew in value.

About the same time, Marshall and Bush worked another deal. Bush had stuck with AMRAD through its lingering death, but he was ready to leap. In early 1922, AMRAD was broke. Bush's differences with Power were in the open. In February, Bush presented Power with a list of 12 possible research problems, graded according to their commercial prospects. Power returned the memo, dismissing Bush's advice and noting a new top priority: "Protection of our radio business." By spring or early summer, AMRAD was dead. Power could not convince the Morgan interests to bail him out. AMRAD's facility was closed and its records and equipment moved to New York.[8]

Bush was not through with AMRAD. "When the crack-up became inevitable," Bush "told Marshall that there was something well worth salvage here," namely AMRAD's patents relating to a new "rectifier" tube invented by one Charles G. Smith, a physicist from Texas whom Bush had hired in 1919 as AMRAD's principal researcher. Smith's tube had obvious value for radio, though the Morgan interests, which owned the patents, did not understand them. Then, a home radio set ran on two different types of batteries, and nothing else. Smith's tube eliminated the need for one of the batteries and made it possible for ordinary house current to replace the other.

Smith was a prolific inventor. Bush respected him, thought his tubes were "nothing less than remarkable" and believed he and Smith had made important refinements in the devices. Bush knew that much larger research outfits were attacking similar problems and boasted that his tiny team was besting them. "I noted in a report of the American Telephone and Telegraph Company that they have twenty-five hundred men engaged on telephone research and development exclusively," he wrote on March 10, 1921. "Of these, eleven hundred are engineers. To compare with this, we have two engineers. I trust that we can make considerably more than one-five-hundredth of the progress" of AT&T.[9]

With AMRAD now on the rocks, Bush tried to rescue Smith. He introduced the inventor to Marshall, hoping Smith would convince Marshall to retrieve the tube patents. Instead, Marshall grew starry-eyed over Smith's latest idea—a refrigerator with no moving parts. Marshall paid Smith to build a model, which seemed to work. Then Marshall raised $25,000 and formed the American Appliance Company on July 7, 1922. Bush and Smith counted among the company's five directors.

The refrigerator proved elusive. In November, Bush came to the conclusion it would never work.[10] Smith disagreed. He felt he was on the verge of a breakthrough. Bush was unswayed. The two men argued constantly. Neither backed down. Bush stubbornly held his ground, even at the risk of poisoning a partnership once defined by trust and mutual regard. Smith, clinging to his invention, resented Bush's criticism. He grew suspicious. "I have fears that Dr. V. Bush wants to claim certain ideas as his own that are part of the refrigerator," Smith wrote in his diary on November 27. He also assailed Bush for trying "to cover up the original signs of [his] ignorance."

Smith's insistence swayed Marshall, who gave Smith more time to demonstrate his intuitions about refrigeration. Yet Bush's criticism, which increasingly seemed on target, had its effect. "Not a sign of pleasure has ever been manifest by Marshall when I have tried to tell him that the refrigerator situation is very hopeful," Smith wrote on December 11. Marshall's indifference heightened Smith's paranoia. "The laboratory has had the atmosphere of a morgue for at least the past two months when things have just begun to look encouraging to me," he wrote on December 12. "I have even grown suspicious that Marshall & Bush might have had some scheme in mind whereby in the event of early discouragement in refrigerator field, they might revise the American Appliance Co. in such a way that Bush could come in for a larger share of stock and of power."[11]

For another year, American Appliance floundered. Smith's refrigerator was a failure. Bush consoled him, saying it was "nervy" of Marshall to think he could tackle head-on such appliance giants as Westinghouse and General Electric. Finally, Marshall himself agreed to abandon the refrigerator. In 1924, Bush persuaded Marshall to renew his interest in Smith's tubes (Morgan had done nothing with the patents). In August 1924 Bush and Smith tested duplicates of tubes they had last evaluated four years earlier in AMRAD's lab. Improvements came swiftly, then patent claims were made. In December, Marshall approached the House of Morgan with an offer to purchase the patents and patent applications stemming from the research done at AMRAD by Smith and Bush. Morgan agreed to sell for $50,000, only one-fifth in cash. The rest of the money came in the form of stock in Marshall's company soon to be renamed Raytheon, after a French-Greek concoction meaning "a beam of light from the gods."[12]

Marshall timed his move well. In 1924, the number of homes with radios tripled, a reaction in part to the soaring number of stations, which grew to 556 in 1923 from a mere 30 nationwide in 1922. Within two years, NBC and CBS, the first nationwide networks, were in place. The sudden surge in radio broadcasts, however pleasing to the ears, brought chaos to the airways. Beginning in 1927, the federal government sought to manage this explosion by allocating frequencies to stations, which effectively reduced interference and greatly improved reception.[13]

Raytheon's tubes were destined for success. They brought down the price of home radios and made them easier to use. They took away the sense that mastering a radio required a zeal for gadgetry. The ability to plug a radio into a wall socket, rather than rely on unwieldy batteries, domesticated the radio; it was now no more threatening than an electric lamp.

The new type of radio was an advertiser's dream. "The upkeep is almost negligible," crowed an early ad in *The Saturday Evening Post.* "Even the largest radio set will consume only a few cents worth of power per month. The Raytheon tube, guaranteed for at least a year, costs but six dollars. And the complete power unit, which costs no more than a few of the heavy batteries it replaces, will last for years."[14]

Raytheon flourished. Sales topped $1 million in 1926; profits were an astonishing $320,000. Bush did well too. He earned $5,000 a year as a consultant and held 3.6 percent of the company's outstanding stock. As a sign of his prosperity and faith in Raytheon, Bush paid $5,000 for 100 shares, or about 3 percent of the total, in October 1926.

Raytheon's success, however, drew the wrath of the largest makers of electronics parts in the land: Westinghouse, General Electric, Radio Corporation of America and AT&T. The four companies had pooled their tube patents and jointly licensed the rights to radio manufacturers (AT&T joined, it was believed, simply to clear obstacles in the way of better telephone equipment). The licensing contract, however, required manufacturers to buy all their tubes from the pool, leaving small companies such as Raytheon in a pickle. "It looked like curtains for Raytheon," Bush thought. The company's factory was "going full blast one week and all its orders [were] cancelled the next."

Bush felt the attack was a predatory practice and "as clear a violation of the antitrust laws as one could wish." But the law had few teeth, and an upstart would likely die trying to obtain his just rewards in court. The blow left Bush forever skeptical of monopolies and deeply ambivalent about the growing domination of the American economy by big business. Despite his unwavering support for capitalism, he feared that monopolies would kill the romantic appeal of capitalism by crushing the small man with a bright idea. Marshall also was outraged, and he fought back. Raytheon began making all sorts of tubes, infringing patents willy-nilly. The four big patent holders filed a blizzard of lawsuits. The mess took more than a decade to clean up, but Raytheon survived.[15]

In the midst of this, Bush made his mark as a mentor. At MIT, research in electrical engineering was still a small affair. The department did not even break out a budget for research until 1921, when the funds earmarked for research totaled just $19,500. In 1922, Jackson put Bush, one of the department's 20 faculty members, in charge of graduate study and research. Bush immediately stepped up the department's activity. The number of people earning master's degrees went from four in 1921 to 37 a year later and 45 in 1923. Bush kept up the pace. Over the ten years starting in 1922, the department awarded an average of 51 master's degrees annually, or ten times the number awarded over the prior ten years.[16]

The department's surging productivity did not exempt Bush from Jackson's criticism, however. A cranky DC increasingly relied on Bush to run the department, especially in DC's periodic absences, but this hardly meant the two were chummy. "I remember one day . . . he told me he didn't like the way I was running a research lab," Bush said. "I told him if you don't like it you can stick it up your rear end. We parted on those kind words. And I came home and I told my wife, 'I guess we're all through at MIT. Where would you like to go next?' I said, 'I can go out to Cal Tech if I want to. Do you want to go out to California?' Next day, I went in and met Jackson and you'd never know anything had happened. We walked down the corridor together, you'd think we were old buddies. But if you took it lying down you'd get it in the neck."17

Bush led by example. He tossed off so many research ideas, and so fast, that he could not help but nourish the men around him. Though he might accept more credit than he deserved for inspiring other investigators, he helped them to see their dilemmas more clearly. Uncommonly sure of himself, Bush gave others the courage to face an uncertain future. As Harold Hazen recalled:

> Dr. Bush could practically smell at a distance when a research job was held up by snags of one sort or another and visible progress had ground virtually to a halt. With the utmost casualness he would saunter by and start chatting and questioning about the job. In the next quarter- or half-hour he himself would have sprouted, or elicited from us, at least a dozen ideas as to how to get around the impasse. Of course 11 of the 12 would, on later reflection, prove impractical or unproductive, but one out of that session (or a subsequent session if the problem was particularly refractory) would prove worth following up. The really significant effect of such sessions, however, was the almost miraculous lift in morale and enthusiasm generated simply by his presence and discussion, so that the work in the laboratory was back in high gear under wide-open throttle the moment he left.18

Bush could display tenderness toward colleagues, who were usually men (female students and professors were rare at MIT). When a research assistant named Parry Moon was badly burned in a lab accident in October 1924, Bush kept tabs on Moon while he recovered at his parents' home in California and implored him not to rush his return. Buoyed by an optimistic letter from Moon the following summer, Bush replied immediately, congratulating Moon not only for his "very rapid physical progress" but for his "cheerful outlook on the world in general."

Bush respected men who (like him) faced physical challenges, and he encouraged Moon as best he could. "I want you to be very sure indeed that you have had plenty of outdoor life and gotten yourself into good physical shape," Bush wrote him. "It seems to me offhand that a month is likely to be too little, and that you would benefit by putting a little more time in in the outdoor life, getting back into prime physical shape. If you feel that this is so, do not hesitate to take all the time you need."19

Bush soon faced his own disappointment. On April 2, 1926, his father died suddenly of a heart attack at the age of 71.

Perry's last years had been bittersweet. When he gave up his parish in Chelsea in 1922, he was so distraught that he burned all of his sermons. He had wanted to make a fresh start. "All the preaching I do is new work," he said. "I want to do new things and go along with the progress of the times as long as I live."

But Perry was ill. He was forced to slow down. "It was a stunning blow I got when I knew I had reached the end of my accustomed activity," he told a friend. He tried to accept his lot. "I am getting hold of myself and I hope I'll drop quickly while at work and not lie around long like a useless hulk."[20]

He got his wish. The end came one afternoon in Boston's Masonic Temple, where he had cared for the library since his retirement. His funeral, held two days later at the Tufts College Chapel in Medford, was crowded with mourners from every city and town around Boston. Perry's friends were legion. Ministers and high-ranking Masons attended. So did the mayor of Boston.

Reverend Lee S. McCollester, dean of the Tufts Theological School, delivered the eulogy. "Perry Bush never lost faith in men," he said. "His love of knowledge, of literature, history and philosophy never diminished." Neither did his desire "to carry on his work to the end," McCollester declared. Perry wished that "when the end came it would find him in the midst of work" and that his work, giving "service to humanity," would be found so compelling that those honoring him that day would "take it up and carry it on."

Perry had been a member of Chelsea's school committee for 26 years and taken the lead "in every move for civic betterment." Yet the elder Bush, whom McCollester called an "ardent patriot," was "never so happy as when present at a gathering of the local Grand Army post." His eloquence was uncanny. He was "never at a loss for the right word to say. Always some bowed heart was lifted up; some message of cheer driven home; some conviction of faith forced to a lodgement, some ray of hope thrown through the darkness." His tolerance was admired as well. He addressed men "of all faiths and associations and while always loyal to his own convictions, recognized the sincerity of the convictions of others."

Perry's remains lay in an open, gray half-coach casket, illuminated by the afternoon sunlight. The sun shone on his face and form and shone, too, on the "mass of floral pieces, some of them elaborate, but all of them beautiful," which surrounded him. After the service, the body was driven to the Bushes' old home in Provincetown. The next day Perry was buried in the family plot. Bush made the journey, along with his sisters and mother.[21]

With Perry's death, Bush lost his only hero. He had revered his father in a way, a later observer noted, that "can be defined as hero worship." While Bush had followed a vastly different path than his father, they had shared an outlook and a personality.[22]

They had never drifted apart either. In the last years of his life, Perry joined Bush for an evening meal nearly every Sunday at the sparse home of Bush's maternal grandmother. "The ritual was always the same. . . . Everything edible in the apartment, including the contents of the cupboard and icebox, would be placed on the dining table and in due course all the meager repast would be consumed. Inevitably, as they finished, grandmother would survey the empty plates and say, 'Weren't we fortunate, we had just enough.'"[23]

Perry's death did not retard Bush's growth as a researcher. Pragmatic and mindful of the marketplace, he had a talent for divining inventions of commercial value. His goal, he once said, was "to dream in a rather definite way." His affinity for both mathematics and mechanics was rare, and his imagination respected few boundaries.[24]

Now he took on one of the supreme technological problems of his day: perfecting the networks of power that were supplying electricity around the country.

By the 1920s, more and more people saw electricity as a necessity. But electric-power systems, as they expanded, were troubled by their inability to continuously match supply with demand. This left systems vulnerable to blackouts. Insulation posed another vexing problem: power surges, sometimes even caused by lightning, frequently damaged equipment. Seeking solutions, engineers at Westinghouse and GE built miniature power systems. Bush took another tack: he sought to mimic a complex power system with a mechanical calculator.

Bush was always making offhand comments, the sort of advice that usually could be discarded but might contain a diamond. "He was famous for coming in and sitting down at your desk, looking over your shoulder and making suggestions," recalled one graduate student. Bush's move into mechanical calculation began in 1925 when he made just such a remark to a graduate student named Herbert Stewart. Bush advised him to build a device that, simply put, expressed information in terms of physical measures (in this case, the turning of a shaft). The idea for the device arose from the same mind-set that led Bush to build his profile tracer at Tufts.

When Stewart completed the machine, Bush asked his graduate students to build more powerful devices. This was natural, one student noted, since "Dr. Bush was always encouraging his students to tackle successively more complicated or difficult problems." The result was a new machine, called a network analyzer. It simulated three generating stations, 200 miles of line and six load centers—all contained within an actual space of 50 square feet. Another, an elaboration of the first Stewart device called a product intergraph, won Bush the Levy medal from the Franklin Institute of Philadelphia in 1928.

To his students, it seemed that their research was a casual, modest affair, but Bush actually plotted his moves carefully. Before designing a machine, "I first draw up a statement of what I should like to have the new machine accomplish. Then with this before me, I write a set of specifications, making it as detailed as possible in order to bring out the major points of difficulty. These, with several discussions with some of my associates, I have been revising from time to time."[25]

In 1928, Bush started work on a more powerful machine. Harold Hazen, a graduate student, was so excited about the effort that Bush's request for help "took my breath away." The goal of the new machine was to solve certain kinds of differential equations, the basic equations of calculus. For two centuries, mathematicians had struggled with differential equations, solving them almost by accident. Then in 1876, the British physicist and mathematician Lord Kelvin tried to "substitute brass for brain in the great mechanical labor of calculating" and conceived of a "machine for integrating differential equations." His harmonic analyzer could predict tidal data but "never achieved widespread use," partly because "the British scientific community remained skeptical that a practical machine could assist the theoretician." On the other side of the Atlantic, engineers enthusiastically pursued machines to solve "practical, engineering problems." In a first step toward realizing Kelvin's vision, American Hannnibal Ford built a crucial piece of a differential analyzer, an integrator, for the U.S. Navy in 1919. The device aimed battleship guns, taking into account the flight of a shell and such other factors as the effect of the earth's rotation on its path, the density of the air and the speed and direction of the target.

Unaware of the ideas of Kelvin and Ford, Bush sought to build a device related to but far more complex than Kelvin's original idea. In 1931 he completed the first differential analyzer at an eye-popping cost of $25,000. A typical equation that could be solved by the machine was "X squared plus X equals Y." While not precisely accurate, answers from Bush's machine were exact enough to aid in doing real work.[26]

The differential analyzer was a milestone, Bush explained, "the first of the great family of modern analytical machines to appear—the computers, in ordinary parlance. It is an analogue machine. This means that when one has a problem before him, say the problem of how a bridge that has not been built will sway in a gusty wind, he proceeds to make a combination of mechanical or electrical elements which will act in exactly the same manner as the bridge—that is, will obey the same differential equations—and then by noting how this combination acts he will be able to predict the performance of the bridge. The trick, in a really useful device, is so to construct this model that by

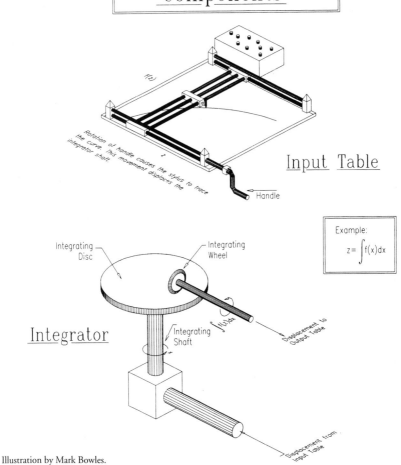

Illustration by Mark Bowles.

shifting some mechanical connections, or better by switching some electrical circuits, one can make it possible to handle a wide variety of differential equations, and hence of practical problems. If one does not know what a differential equation is, perhaps I can make it clear by a very simple example. Suppose an apple drops from a tree. . . . The thing we know about that apple is, to a first approximation, that its acceleration is constant, that is, that the rate at which it gains speed as it falls does not vary. So we just write this fact down in mathematical symbols. That is a differential equation, one very easy to solve, and thus we are enabled to make a plot of the position of the apple at every instant. But suppose we want to include the resistance that air offers to the fall. This just puts another term in our equation but makes it hard to solve formally. We can still very readily solve it on a machine. We simply connect together elements, electrical or mechanical gadgets, that represent the terms of the equation, and watch it perform."[27`]

Proud of his creation, Bush refused to call Kelvin the inventor of the differential analyzer even after learning of the Englishman's earlier efforts. "Inventors are supposed to produce operative results," he insisted, and Kelvin's idea could hardly yield a differential analyzer because the crucial integrator "could not then be built both accurate and capable of carrying a sufficient load." Bush preferred instead to cite Ford as his antecedent, insisting that while the Navy inventor did not build a differential analyzer he "readily could have done so . . . if he had put his mind on it."[28]

Bush was stingy to withhold credit from Kelvin, but he was right to place importance on his ability to actually build a differential analyzer. The machine was, after all, an imposing contraption, requiring its own large room. Electric motors drove an intricate assemblage of gears and shafts. Calculations were carried out by brute force. Metal clanked against metal until a solution arrived. A motor activated a shaft that turned a gear that pushed a rod and so forth. With 18 shafts, the analyzer could take a few days to set up for a problem. With every fresh assignment, it had to be taken apart with screwdrivers and wrenches and reassembled in a new pattern. Once in operation, it required constant care and attention. Graduate students stayed awake nights, stationed near it, in case something went wrong. Even when all went well, the machine needed daily service to insure proper operation.[29]

Despite its brutish qualities, the analyzer was based on simple design. Indeed, the reusable pieces of a Meccano game could be assembled to mimic the actions of the big machine. This fascinated Bush, who loved nothing more than to see things work. It was only then that mathematics—his sheer abstractions—came to life.[30]

The differential analyzer was a potent symbol as well as a means for solving problems. The machine evoked a dream as old as the Greek abacus and as fantastic as the latest fiction from H. G. Wells. While tortuous, the long search for aids to thinking was worth the effort. After all, machines had vastly reduced physical toil. Now why not do the same with mental toil? This possibility struck some as absurd since they viewed consciousness as thoroughly immaterial and mind and body as separate. The calculating power of the differential analyzer, however, called this hoary duality into question. Newspapers invoked enthusiastic images, labeling Bush's device a "thinking machine," "mechanical brain," or "man-made mind." Hazen, no mere popularizer, wondered how the world ever got along without computers. "If all the present-day mechanical aides to computation were somehow suddenly abolished what would happen? Utter confusion," he opined.[31]

Even disinterested observers boggled at the possibilities. The president of the National Academy of Sciences called the differential analyzer "the most complicated and powerful mathematical tool ever devised." And Warren Weaver, director of the Natural Sciences division at the influential Rockfeller Foundation, foresaw more powerful versions of the machine, "the technological problems having been effectively solved."[32]

Bush saw the machine in earthier terms. As early as 1928, he asserted that his analyzer was a physical model of abstract mathematical relationships. By using his analyzer, a man gained "a grasp of the innate meaning of the differential equation," and as a result, "one part at least of formal mathematics will become a live thing." One operator of the analyzer, a mechanic untrained in mathematics, "got to the point where when some professor was using the machine and got stuck . . . he could discuss the problem with the user and very often find out what was wrong." Bush believed that the "fundamentals" of differential equations had gotten "under [the mechanic's] skin."

The differential analyzer got under Bush's skin too. Watching the machine grind away at a solution, turning an abstract notion into something vivid and concrete, was uplifting. "The study of engineering mathematics," he wrote, "becomes soul-satisfying only when one begins to grasp the power that lies in the ability to think straight in the midst of complexity."[33]

Bush was proud of his machine and advertised its virtues. When Philip M. Morse, an MIT physicist, sought to improve calculations on the scattering of slow electrons from atoms, Bush encouraged him to try the analyzer. Morse often visited Bush unannounced, finding him "leaning back in his chair, his feet on his desk, interspersing puffs of smoke from his eternal pipe with bits of dry humor or laconic wisdom, spoken in his Yankee twang."

Morse, who would pioneer the field of operations research during World War II, was both awed and intimidated by the differential analyzer. "It was a fearsome thing of shafts, gears, strings, and wheels rolling on disks, but it worked and it foreshadowed a host of fantastically more capable computers." Morse found it frustrating because it sometimes took several tries to obtain the desired answer: "Sometimes the answer went off the scale and sometimes we found we had connected the wrong shafts."[34]

Within a few years researchers at other universities showed interest in building copies of the differential analyzer. Bush welcomed this interest though he hoped that MIT engineers would make the most crucial refinements to his machine. He never sought patent protection on any of the analyzers nor impeded the efforts of imitators. He was happy to help those who wished to improve on his design, hosting at least one such scientist in the spring of 1933. At the time, Bush was "dreaming in a rather definite way" about improving the differential analyzer himself. He told Weaver, whom he was wooing for more funds, that the "courageous attempt to produce" an improved device naturally should "rest with us" at MIT.[35]

Not only scientists were interested in the differential analyzers; the military was, too. The Army's proving grounds for ordnance in Aberdeen, Maryland, wanted a machine to help it better understand the movements of an explosive shell from the time it left a firing gun until it struck its target. Bush was gratified by the military's interest in his creation, but skeptical that Uncle Sam's engineers, whom he considered second-rate, could do anything more than "produce an exact copy of the existing machine." Nonetheless, his contact with the Army foreshadowed a relationship that would become crucial to the nation.[36]

Bush's fortunes improved despite confusion at MIT. Following the death of President Richard McLaurin in 1919, the Institute spent the next decade in the doldrums. McLaurin's successor was ill and soon died, leaving power in the hands of an ineffective committee. The next president, a former chief of the U.S. Bureau of Standards named Samuel Wesley Stratton, was clumsy and unimaginative. Charitable observers said he was past his prime. During Stratton's reign, one professor noted, "there was no ruler of the Institute with a sure touch, a clear policy, and an unquestioned vigor and understanding." Bush was more blunt. He described MIT's administration as so "perfectly terrible" that it "nearly wrecked the place."[37]

The absence of strong leadership made MIT the scene of warring egos and a refuge for second-raters. "At times," recalled one professor, "it seemed that only the decrepit, the ineffectual, the inept, and a few idealists stayed on." Bush could be counted among the last group. Since Jackson had no head for administration, Bush increasingly ran the department. He even acted as chairman when Jackson was away for the entire academic year of 1929. His salary reflected his broader duties, rising from $3,000 in 1920 to $7,000 in 1929.[38]

Bush was not always politic as he carved out a broader role at MIT. He tended to bully others and could treat harshly those he disliked. He was not subtle about it, either. When Arthur Kennelly, who had been Bush's dissertation adviser, retired, it fell to Bush to find him an office of the type normally given an emeritus professor. Bush balked. He insisted that Kennelly was "a stickler for titles" and only wanted his name on a door. A real office need not be sacrificed to his vanity, Bush concluded. So he put Kennelly's name on a door that opened into a room full of switchboard operators. President Stratton, in a rare display of fortitude, found this unsatisfactory. But Stratton did not stop Bush from continuing his campaign of humiliation. Next, Bush placed Kennelly's name on his own office door, knowing Kennelly would not have the temerity to enter. Bush was right. He never did.[39]

Junior faculty members deferred to Bush. Hazen, who joined the faculty after gaining his doctorate for his work in computing, "absolutely idolized him." Many colleagues and graduate students felt the same way. Bush mesmerized them with his articulateness, his self-confidence and decisiveness and his capacity to toss out at a ferocious pace half-formed suggestions and fully realized memos.

A few men, however, bristled at Bush's directness, transparent ambition and willingness to trade blows in order to achieve his aims. Edward L. Bowles, for instance, had "not a damned bit of use for [Bush's] methods." A native of Missouri, Bowles had studied electrical engineering at Washington University in St. Louis before coming to MIT in 1920. Circumstances made it difficult to pry Bush and Bowles apart; for years, they would travel in the same orbit. With Bush as his adviser, Bowles completed his thesis on vacum-tube design in 1921. Bowles, who thrived on conflict and adversity, was well-suited for a career in radio, which had yet to win the acceptance of academic engineers but was clearly a hot field. "You started with nothing," he said of his early years at MIT, "and had to rely on your qualities as a predator."

Bowles joined MIT's faculty and soon emerged as a force in electrical engineering. He saw himself as a rival to the older Bush. While Bowles was perhaps the only man at the university whose range of interests and organizational talents approached Bush's, he could not match Bush's achievements. "Bush was brilliant, I wasn't," Bowles said. "It made me work all the harder."[40]

Bowles made his mark on MIT. In 1923, he created a new undergraduate program in electrical communications; three years later he took charge of a new MIT research lab, endowed by a wealthy philanthropist named Edward Green, in South Dartmouth, Massachusetts. Under Bowles's direction, the Round Hill Research Station, as it was called, tried to standardize broadcast frequencies and built new antennas and transmission and receiving equipment. Then in 1928, at Green's insistence, Bowles began studying ways to make flying in fog safer.

This project put Bowles in the forefront of airplane navigation research. It also made Round Hill the jewel of electrical engineering at MIT, accounting for anywhere from 40 percent to 71 percent of the department's total research budget for the six years starting in 1929. Green's continued generosity depended on his relations with Bowles. While nominally under Bush's authority,

Bowles could do what he pleased at Round Hill. He asserted his independence. "I set up a miniature MIT," he crowed, with the difference that he encouraged researchers from many disciplines to collaborate. "I didn't worry about the departments," he said. "I made people work as a team. This was all off the beaten path."[41]

Having achieved a measure of success, Bush found that all his striving for knowledge, status and wealth had taken a toll on him. Though he generally succeeded in taming his nervous energies through ceaseless activity, it wasn't always possible to exorcize the demon of anxiety. Hypersensitive to illness because of the extended sicknesses of his youth, he remained susceptible to bouts of "nervous tension" throughout his prime years. These eruptions of anxiety were hard to predict but undeniable all the same.

> One spring [Bush recalled] when I was working very hard at MIT and under a severe nervous tension, my heart started skipping. I got to the point where it was skipping every third or fourth beat and it nearly drove me cuckoo. Fortunately I got hold of a cardiac expert, Bill Reed, an old friend of mine, who looked me over, took cardiographs . . . and said, "very uninteresting heart." I protested and told him that the pump did not tick regularly and that I thought that it was his business to fix pumps. He asked me when I was going to take a vacation, and I told him. He said three days after I relaxed it would start being regular again. Much to my surprise it did just that and, more important, it has not to my knowledge skipped since. The simple fact was that the entire affair was produced by nothing but nervous tension.[42]

With the arrival of Karl T. Compton in 1930, the discord and lack of direction at MIT came to an end. The son of a clergyman, Compton was one of three remarkable brothers, each of whom would preside over a college. "KT," as his many friends called him, came from Princeton University, where he had been chairman of the physics department. He was likable, intelligent and, most important for the job at hand, had great tact. He neither relished the limelight nor demanded credit for his contributions. He had the rare ability to make others feel as if they got their own way even as he got *his.*

Compton moved to cure MIT of its diseased administration. He also strengthened scientific studies at the institute, which suffered from subordination to engineering. He brought in a young scientist to chair the physics department and broadened the charter of the chemistry and mathematics departments. He added more intellectual backbone to the various engineering disciplines, which still retained the "odor of the shop," stressing "mechanical skill" and "mathematics and physics couched in the graphic idiom" over abstract rigor. In the process, Compton gradually brought in hundreds of additional teachers while maintaining student enrollments at roughly 3,000 a year. Finally, he sought to encourage research on campus by slapping tight limits on outside consulting done by professors.[43]

This last act by Compton led to his first meeting with Bush. Upon learning of the consulting limits, Bush had barged into the new president's office, charged that the edict violated a promise made to him "in writing" and threatened to quit. As introductions go, Bush later said, this one was "not ordinarily the sort of thing that would lead to a great friendship, but it did."

At least initially, Bush bent to Compton's will. The new president did not rescind his consulting measure, but justified it so convincingly that, Bush said, "I changed my tune completely" and became a Compton booster. Bush counted the new president as "one of the most likable men I ever knew. Everybody at MIT was so thoroughly loyal to him in every way that he could do anything. He had good sense, he was a kind individual and he was *not* a bloated egotist. . . . No one would hurt him if they possibly could avoid it."[44]

Bush's support for Compton did not go unrewarded. In 1931, Compton raised Bush's salary to $10,000. In March 1932, he made Bush his second-in-command, appointing him to a newly created office of the vice-president. He also named Bush dean of the School of Engineering, one of five MIT divisions, and made Bush a member of the MIT Corporation, the Institute's board of directors. With these promotions, Bush received another boost in salary. He now earned a comfortable $12,000 a year and received $6,000 in expense money that he need not account for.

Compton and Bush made a good team. Seeing in Bush's tenacity and belligerence the ideal foil for his own generous and gentlemanly ways, Compton embraced Bush as his alter ego. Often, this meant that decisions involving money and staff were not made until both men issued their approval. Once a faculty member, seeking funds for a machine for nuclear research, appealed to Compton on a Saturday, finding him dressed in his shirtsleeves in his own physics lab. Compton liked the proposal, but said ominously, "Well, it sounds great to me, but I'm the 'yes' man and Bush is the 'no' man, and you'll have to see Bush too."

The professor made a date with Bush, to whom such appeals were not always welcome. On schedule, the professor arrived at Bush's office only to spend "the whole afternoon there, because [Bush] took all of his phone calls as they came in, and the consequence was [an interruption every] one or two sentences and then another protracted phone call, and one or two sentences, and then another phone call." Bush finally ended the meeting by asking the professor for a memo. He later approved the proposal.[45]

Not everyone at MIT appreciated the Compton-Bush axis. When Compton told Bowles of Bush's appointment as vice-president, Bowles criticized the choice, saying he respected Bush's talents but not his style. It was well known that Bush and Bowles had different styles. "Bush wanted tight organization, clear lines of authority, and control, even while stimulating research," the official historians of MIT's electrical-engineering department have written. "Bowles, on the other hand, preferred operating in a free-wheeling, individualistic way.[46]

In time, Compton decided Bush and Bowles should hash out their differences. As soon as Bowles complained about Bush's selection as vice-president, "what did Karl Compton do but go and tell Bush right across the hall," Bowles recalled. "When I got back to [my] office, there was a telephone call from [Bush] wanting me to come in and tell him why I made the remark to Compton." Bowles went to see Bush. "I opened the discussion by saying that I had been told many times by [D. C.] Jackson that if I got into a fight with Bush, Jackson would fire me. Because Bush was essential to him. So I had been careful. That's what I said to Bush. Now, I said, if you hold your fire I will talk but I don't want to get into a contest." Bush did, and Bowles unleashed a tirade. "It was as if there was a little angel on my shoulder," reminding him of all the "scrupulous and unscrupulous" slights he had endured at Bush's hands. Then Bush called in his secretary and canceled his appointments for the rest of the day. He defended himself but Bowles left feeling "the scrap did some good."[47]

Bowles posed a rare challenge to Bush's authority in 1932. At the age of 42, Bush was a rising power in academia and a fixture in Boston's intellectual firmament. He also was a local boy who had made good, whose success arose from merit and character, not family connections. Thus, his was a morality tale of special interest to Bostonians still in the grip of the Brahmins. *The Boston Globe,* on the occasion of Bush's promotion, captured the mood in a glowing article on Bush that extolled his ethic of hard work, his love of the outdoors and his religious roots. The headline struck all the themes: "Scientific Son of a Minister Named Vice President at Tech. Dr. Vannevar Bush Was Born and Educated in Greater Boston Area—Is Scientist of Note and Yachtsman of Considerable Skill—Tutored Way Through College."

The *Globe* published a family portrait alongside the article. The picture caught Bush and wife Phoebe in a moment of repose, seated in the living room with their children. Bush wore a suit and tie, sat stiffly with his knees wide apart, looked squarely at the camera and stroked Spot, the family's pet water spaniel. Phoebe, her lips pursed and eyes straight ahead, seemed gentle and matronly, though her broad, strong chin contained a hint of stubbornness.

Phoebe's dark, curly hair was combed across her forehead, and she wore a spotted dress, cut low to expose a delicate neck, with the hemline falling over her knees. Richard, age 14, and John, age ten, stood behind their mother and father. The boys wore jackets and ties and looked none too happy about it.

The Bushes had the look of a satisfied, prosperous and proper family. They lived in a modest house at 404 Commons Street in Belmont, an easy drive to MIT in Cambridge. But they spent a good deal of time elsewhere. Phoebe had inherited an ample "cottage" in South Dennis on Cape Cod. It was her preferred spot. The main house consisted of eight rooms (though just one bathroom) and sat on 22 acres of land, some of which bordered the Bass River. There also was a separate guesthouse, a boathouse and a two-car garage. South Dennis was hardly the finest spot on the Cape, but it was 3 1/2 miles from the beach at West Dennis and just six miles to Hyannis, which boasted train and air connections.

For variety, the Bushes could visit Provincetown, where Bush's mother, Emma, vacationed with sister Edith (the two lived together in a house at Jackson College). Or they could visit Phoebe's uncle Walter, who ran a dairy farm in Montpelier, Vermont, with his wife, Rena. Walter was the brother of Phoebe's dad. Bush liked to visit him in the fall, driving up with his family. As Bush's son Richard remembered:

> This in itself was an adventure because the car was subject to mechanical failure and frequent flat tires. Spare tires were stacked on the back and if those were used up, the tubes could be repaired using a clamp and patch with a metal backing which contained a flammable substance. When this was ignited it heated the patch and caused it to stick. One then had to reassemble the tube, tire and rim, fasten the whole to the wheel and inflate it with a hand pump.
>
> The farm was picturesque, approached by a long driveway with a field of cattle corn growing on either side high above the car. There was little mechanization on the farm because Uncle Walter understood animals, including temperamental horses, but he had little insight into machinery. To accomplish some of the work such as chopping corn stalks for ensilage and cutting wood, he had a number of large single-cylinder engines known as "one lungers." When they failed to function, Walter abandoned them and substituted a spare engine. When we visited the farm, there were always a number of these which my father proceeded to repair.[48]

Bush freely shared his views with sons Richard and John, but some of his "dissertations" were pitched over his sons' heads. Other times, the meaning of his advice was clear. John recalled: "My father at times used the phrase, 'justify the space you occupy,' which graphically conveys a message."[49]

The boys were attracted by the ease with which Bush handled tools and gadgets. "He was patient," Richard said. "You couldn't help but get some spin-off from his many interests." Bush taught Richard to use hand tools and to temper steel, for instance. Together, they built a model electric train "from scratch." It was a beauty. It had brass rails and elaborate remote-control switches from which cars could be uncoupled and the locomotive sent into reverse. Lengths of track were fashioned from wooden ties.[50]

"He would pose all kinds of questions that would expose me to things," Richard recalled. In one routine, he asked Richard to identify something from his shop bench.

"A screw," Richard said, correctly.

Bush shot back: "How long?"

"Six inches," Richard said.

Withholding his approval, Bush handed a ruler to Richard, who laid it against the screw. It was three inches long.

Bush also shared his passion for archery with the boys. He enjoyed target shooting and made most of his equipment himself. Once, Richard challenged him to "shoot an arrow over a tall tree." Bush accepted the dare, forgetting he held a pipe in his mouth. When he let his arrow fly, "The bow string caught the pipe, which clattered over his teeth and went sailing over the tree in a shower of sparks."[51]

The accident did no damage to Bush's teeth, and it certainly didn't deter him from smoking his pipe. He smoked incessantly, even in his photographic darkroom. He loved pipes. He loved their clean lines, their precision, their simplicity. He never went anywhere without a pipe. When he wasn't smoking, he often sucked on his pipe, which was almost as much a part of his face as his glasses. Always the engineer, Bush thought much about what makes a good pipe, and he carved many of his own. Sometimes they cracked under the strain of his habit, sending sparks in the air and burning small holes in his clothing, much to Phoebe's displeasure.[52]

Phoebe usually found Bush's pipe-smoking distasteful. She wouldn't allow her husband to bare his pipe at formal gatherings (instead he would smoke a cigarette or cigar). At South Dennis, she made him smoke in his shop.[53] This was one of the few things that ever came between them. Phoebe adored Bush. He made her smile and delighted her. He was affectionate and "terribly protective" of her. He had a sharp tongue and could speak with anger as well as wit. But "he never spoke sharply to Phoebe, not once," said Edna Haskins, a friend. He seemed to have a great reserve of patience for his wife. "She'd be so serious," Haskins said. "Then he'd have the whole household laughing."[54]

Bush's work almost never brought him together with women, and the few women he did know were not professional equals but secretaries or aides or perhaps a much-younger researcher. He had no female friends and did not flirt. "He was almost Calvinist in that respect," a friend said.[55]

Phoebe had taught school after meeting Bush, but never worked after their children were born. She accepted without question her role as her husband's ballast. She stayed in the background of his career, never upstaging him. If she had her own opinions, she kept them to herself in public. If asked for her views, she echoed Bush's own. Among groups of both sexes, she was shy, almost retiring; with women only, she was the same. Neither was she chummy with the wives of other faculty members—with the sole exception of Bush's sister Reba, who was the wife of Ralph Lawrence, an electrical-engineering professor at MIT. The wives of other faculty members addressed her as Mrs. Bush, and Phoebe was as formal with them. Such contact was usually limited to faculty gatherings, which were formal to the point of stiffness. To some, Phoebe's reserve seemed patronizing, as if she were "up on cloud nine or somewhere else," looking down at them, recalled one faculty wife.

Phoebe meant no disrespect, friends thought. Sometimes she seemed prissy, but more often "she was very winning and very charming." She was not outgoing and did things simply in order to show support for Bush. "She was sweet and quite understanding and very much with [Bush] all the way."

Mrs. D. C. Jackson, the wife of Bush's boss, approved of Phoebe's retiring style. Mrs. Jackson made this plain to the wives of electrical-engineering professors by telling "us very clearly that we must attend the monthly meetings of MIT wives—and she was there to know who was missing!" Recalled one faculty wife, "She said my purpose was to support my husband in his career!"[56]

Phoebe did her best to live up to this dictum. But supporting Bush's career was not enough; she also indulged his appetite for travel and serendipity. As hard as Bush worked, he seemed equally intent on relaxing, and he did so by pursuing myriad hobbies. He was a great believer in the tonic effect of hobbies and took his private pursuits seriously. When he fancied something, he did not go halfway. Phoebe enjoyed growing flowers, so Bush built her a greenhouse. Not content, he crafted a pulley system that somehow watered the flowers. "We could be gone for two weeks and the watering would be taken care of at regular intervals," Phoebe said.[57]

On his own Bush explored the sea. He purchased a 40-foot sailboat named *Caribou* in the late 1920s and often cruised the waters of the Atlantic, skippering his ketch along the coast of Maine. Sometimes he stayed out for as long as two weeks, joined by Hazen and other colleagues. Other times he sailed with his sons, sketching out ways to improve his boat's design and pondering the place of technology in an increasingly dangerous world.[58]

Chapter 4

"Versatile, not superficial"

(1932–38)

For better or worse we are destined to live in a world devoted to modern science and engineering. If the road we are on is slippery, we cannot avoid a catastrophe by putting on the brakes, closing our eyes or taking our hands off the wheel. What is the sane attitude of a scientist or layman? Absence of wishful thinking. No emulation of the ostrich.

—Vannevar Bush

Bush preserved his homespun manner even as he acquired a more worldly outlook. For all his accomplishments, he often struck the disarming posture of an unsophisticated Yankee. But beneath his folksy exterior lay the hard-boiled heart of a modern. He could be tough with those who took him lightly.

This was true whether the misjudger was mighty or meek, peer or student. In the summer of 1932, Bush, the newly minted vice-president, attended MIT's camp for incoming freshmen. While taking a break, Bush sat on a log by the lake near the camp. A new student, who had obviously skipped an orientation session, sat down on the same log. As the student later wrote, "Already occupying the other end of the log was a fellow with a shock of hair and a piece of straw in his teeth who, in my infinite wisdom of youth I knew was a New Hampshire farmer." The student struck up a conversation with Bush during which he admitted that he "probably wasn't missing much" by playing hooky. Bush did not object; he kept chewing a piece of straw. Some hours

61

later, however, Bush met the student again at a scheduled activity and needled him about their chance meeting. "My embarrassment at that point was as infinite as my earlier wisdom," the student recalled. "To compound my misery, you reminded me of it a few times during the early school year when we met in the hall."[1]

From students Bush might take disrespect in small doses. He showed less patience with peers, however, displaying flashes of the arrogance and imperiousness that caused some to resent and fear him. When in the early 1930s an administrator at another research institute stated incorrectly that Bush had lent his support to some matter, Bush "corrected him politely." But when the colleague repeated the error, Bush exploded. "The second time, I didn't correct him politely; I corrected him very impolitely, using language that I imagine some of the fellows hadn't heard since they left the steel mills."[2]

Despite the Depression, Bush rode high, escaping the economic tragedy staining the lives of many of his countrymen. The nation's mood hit bottom during the five months from Hoover's defeat in November 1932 to Roosevelt's inauguration in March 1933, but Bush was unscathed. He was financially secure by virtue of his shares in Raytheon and Spencer Thermostat, secure in his status at MIT and celebrated in the technical community as the inventor of the most powerful "thinking machine" ever created. Bush had made it. His success, and his pleasure in it, lent credence to Calvin Coolidge's bald assertion: "Brains are wealth and wealth is the chief end of man."[3]

Bush could not ignore the collapse of the nation's economy and the forces it awakened, however. He had shunned politics, but the Depression challenged his belief in self-reliance and the free market. During Hoover's four years as president, 15 million workers had lost their jobs; 5,000 banks closed their doors, wiping out 9 million savings accounts; wages in major industries fell by 40 percent. The debacle stood as an indictment of laissez-faire capitalism and unleashed an array of critiques of the American economic system.

Among the forces blamed for the Depression was technology, "the god of the 1920s," now "denounced for bringing the curse of plenty."[4] Howard Scott, the chief exponent of this view, argued that productivity gains from automation guaranteed high unemployment and that only a state-run economy could solve the dilemma. Even those unmoved by Scott's plans for a "technocracy" were troubled by the replacement of men with machines. "John Doe isn't quite so cocksure as he used to be that all this science is a good thing," wrote one college dean in *Science* magazine. "This business of getting more bread with less sweat is all right in a way, but when it begins to destroy jobs, to produce more than folks can buy and to make your wife's relatives dependent on you for a living, it is getting a little too thick."[5]

Bush disagreed that America had overdosed on technology, snapping that critics "may as well blame it [the Depression] on medical advance[s]" for increasing population. "Would we go back" to an earlier time with less sophisticated technology? he asked. Of course not, because "the standard of living has gone up, not down, due to technological advance." Besides, humanity cannot separate itself from technology; it is us. "For better or worse," he warned, "we are destined to live in a world devoted to modern science and engineering."

Technology was still Bush's god. Invention required an admirable "combination of ingenuity and nerve and judgement." If anything, industry had made too little use of technology, not too much. He blamed this partly on monopolistic corporations (another fallen god of the Depression era) that restrained innovation merely to protect their hold on existing markets. The larger problem, however, lay in the outlook of the businessman, who was too self-absorbed and arrogant to understand the dynamics of innovation.

> He believes all inventors have long hair and should be shut up in cubicles. He thinks they should have a knowledge of business, and that if they do not they should be instructed in the same way that instruction is given in the art of poker. He thinks there are two kinds of engineers, those who pull throttles and those who push stopwatches. He believes the way to sell any product is to repeat the same slogan millions of times, and paint store fronts in repulsive color combinations. He thinks of [himself] as a super-being apart, who has the only true evaluation of success in the world, to whom homage should be paid.[6]

Bush saw potential advances in many areas: fabrics, fuel, plastics, cars, trucks, radio, dyes, home design. The public's desire for "new things" was "insatiable," the market "unlimited." Still, he admitted the system was gummed up and groped for reasons why. "America never lacks inventors, but usually misunderstands and misuses them," he rationalized. Missing was a commitment to "intensified utilization of scientific methods in advancing the techniques of industry" and an understanding that "standard of living depends on how much or how effective men work and how closely their products are related to needs."

Regarding the basic riddle of the Depression, however, Bush was in the dark. He crudely viewed the crash as an affliction—a "fever by which civilization rids itself of infection." However painful, the Depression brings needed "rest for recovery" and "resets the stage for creative enterprise." There was a danger, he conceded, "that the fever will kill," but "recovery is inevitable if it does not."

Bush confidently (and incorrectly) predicted an economic revival, repeatedly seeing signs in the mid-1930s that "at last [we are] climbing the hill out of the depression." In October 1935, he exuded bullishness, writing his friend Norbert Wiener:

> I am therefore more optimistic than I have been at any time since this show started, although I recognize very definitely that unless such a swing occurs in public sentiment we are bound for much more serious times than we have already gone through. The forces of recovery are powerful things, and they operate to a considerable extent independently of political affairs, in the absence of wars or actual destruction of financial systems by foolish manipulation. Hence I feel sure that we are on the way out and rapidly so, provided the political tinkering can be held down.[7]

Bush was less sanguine about the prospects for American democracy. "I indict the insane inclination of the entire people to try anything once," he fumed. The outpouring of liberal and radical notions unsettled him as did the disrepute into which Hoover had fallen. In these times, "The conservative . . . is labeled a reactionary or worse," he complained. "It is the fashion of the time to be identified with 'isms.' The tendency to demolish the idols of the past has led to the discrediting of an entire generation." At bottom, the citizenry was erratic, "swayed by propaganda" and "emotionally unstable." At times, he saw fleeting signs of sanity; in late 1935, he sensed "a considerable turning away from the New Deal," adding hopefully, "This country being what it is, a swing of that sort once started is likely to go very far."

More often, though, Bush saw populism and the widening participation of citizens in the machinery of government as a recipe for decline. Favoring rule by the well-to-do and highly educated, Bush fretted that "a blind mass rushes on."[8]

Bush's fear of the "blind mass" was widely shared in his circle of senior scientists and engineers. Karl Compton, Bush's boss, espoused almost identical views. Compton was an obvious spokesman for elite research institutes. When in July 1933 Roosevelt called on researchers to help revive the nation, he named Compton as chief of his new Science Advisory Board. A petty jurisdictional dispute hampered the board: The august but ineffective National Academy of Sciences, formed in 1863 with the charge of advising the government, objected to Roosevelt's executive control of the board and placement under the National Research Council, a sister organization. The dispute notwithstanding, Compton eagerly embraced the chance to affect public policy. He sympathized with the plight of the unemployed, agreed that relief measures were needed and was sufficiently independent of the reigning economic orthodoxy to argue that the federal government should sponsor scientific research. For decades leading scientists had opposed government patronage, fearing political control over their research. Instead great foundations, such as those formed by the Carnegie and Rockefeller families, provided the steadiest support for basic research, while the government limited its role to essentially utilitarian pursuits such as studies of the weather and agriculture.

A registered Republican, Compton nonetheless felt the times demanded a break with tradition. In the current crisis, funds for research were depleted, yet economic vitality depended on a steady stream of new ideas and inventions. The science board, Compton believed, gave the research community a historic chance to influence government policy in peacetime and restore luster to the tarnished reputation of science and technology.[9]

Compton's top priority, which he stated at the board's first meeting in August 1933, was to win "emergency unemployment relief" for idle researchers, whose plight was "pathetic." He asked the federal Public Works Administration in September 1933 to spend $16 million over six years on jobs for engineers, scientists and related technical workers. The proposal drew "kind words," but no action.[10]

Undeterred, Compton proposed a more ambitious plan in 1934 in which he called on the government to spend $75 million for research over five years and form a permanent board to advise the government on spending by its civilian technical bureaus. Roosevelt approved an advisory group, but not any money. The president was convinced by aides that funds should go only to the jobless; most researchers weren't on relief rolls. In 1935, Compton asked Roosevelt for $3.5 million in support of research. Again, no action. The Science Advisory Board expired that summer, its potential spent, one historian has noted, because its members acted "largely in a political vacuum. Even Karl Compton was occasionally guilty of ignoring broad social concerns in favor of the concerns of science."[11]

The political defeats suffered by the nation's scientific leadership underscored the deep divisions between advocates of research and New Dealers grappling with pressing social problems of poverty and unemployment. Compton repeatedly attacked critics of science, patiently describing the rewards of research, warning against imposing obstacles to technical change and arguing that the nation would be better off if some federal funds went to academic research rather than "artificial economic control." He even wrote the president a stinging letter, accusing him of unfairly singling out engineers for criticism and lending succor to the "tendency in some quarters to make science the major scapegoat of our social ills, from which social planners will rescue us."[12]

But professors were not generally objects of pity; if they were not envied for their privileged upbringing, they were dismissed as loafers or worse because of their work hours, which seemed exceedingly light. "There was a feeling that life was pretty darn easy [for professors]. You had the whole summer off. What kind of work was that?" asked the wife of one of Compton's professors.[13]

As the Depression lingered, Compton's views rang hollow. He did not see the sources of the crisis any better than Bush. His positions appeared hidebound even to those New Dealers committed to aiding researchers. Secretary of Agriculture Henry Wallace, a successful plant geneticist who rejected the argument that technology was to blame for the economy's ills, nevertheless grew impatient with the conservatism of Compton's crowd. Though he had convinced Roosevelt to form the science board in the first place, Wallace fulminated against researchers who still felt "the good old days will soon be back when an engineer or scientist can be an orthodox stand-patter."[14]

It was harder, however, to dismiss Compton's charge that the government's technical activities were disorganized and largely second-rate. "In the Department of Agriculture there are 18 bureaus, ten of which are presumably scientific, and there is absolutely no coordinating service for these varied and overlapping scientific programs," he wrote an associate. "In the Naval Research laboratories the salaries for civilian scientists are so poor that they have been forced to permit individuals to patent discoveries with the proviso that the Navy has shop rights. This procedure has turned the attention of many of the civilian scientists to researches which promise to lead to profitable patents and it has practically stopped any free cooperation between the Navy and such an organization, for example, as American Telephone & Telegraph. The head of the Naval Research Laboratory, moreover, is a Naval officer on a four-year appointment. The Bureau of Standards is, on the whole, a rather good institution. The Weather bureau probably represents the worst of the governmental scientific activities. Until very recently the qualifications for observers were such that even a high school education was not necessary."[15]

Just as frustrating was the climate of suspicion between government agencies and private researchers. Bush and Compton knew of this condition firsthand. In 1933, the two academics asked the Tennessee Valley Authority, the massive federal power system, to contribute $250,000 toward MIT's study of electrical transmission. The research had potential benefits for TVA. While it had been possible to transmit electricity over distances for about 50 years, technical problems limited transmissions to about 300 miles. MIT believed it might have found a way around these limitations. The TVA was interested in funding a trial system, but the two sides never could come to terms on the rights to any useful inventions arising from the research. The government wanted the right to distribute freely the fruits of the research, while MIT insisted that it retain the right to license the technologies it developed. This position smacked of seeking to use public funds for private gain. David Lilienthal, the TVA's liberal chief, objected, "After all, the money is of the whole people and, in my judgment, the policy should be that any results should be available to the whole people."16

When the contractual negotiations stalled, Bush tried to simplify matters by essentially calling for a good-faith agreement that would insure that the TVA received on favorable terms any equipment flowing from the research. Since it was uncertain that anything useful would come out of the research, Bush felt it was foolish to worry in advance about splitting up the fruits of the effort. "The best way to provide against contingencies," he told the TVA, "is to give broad discretion to a capable organization operating for the public benefit," by which he meant MIT's patent-licensing arm.

Bush was naive. The government was unlikely to grant such latitude to a recipient of public funds at a time when many New Dealers were railing against corporate abuses of the common good. In April 1935 the TVA dropped the matter after two years of sporadic negotiations, leaving Bush and Compton pessimistic about forging technical alliances with the federal government. Indeed, the problem was bigger than MIT. "The TVA was entirely too suspicious to commit the government to contracts based on trust and claims to expertise." The agency's failure to strike a deal with MIT "acted out on a smaller stage disagreements that were echoing through the public places of Depression America."17

Bush had played only a bit part on the Science Advisory Board. He had studied the relation between the patent system and the stimulation of new industries for one of the board's many advisory committees. But the board's failure reinforced his bias against the New Deal. "We are still in trouble in this country and serious trouble," he wrote Norbert Wiener in October 1935. "The next Congress will probably pass a bonus and aid to farmers is running wild. Just how long the national credit can stand this [level of social spending] is anyone's guess. If we should run into serious inflation you would be fortunate in being out of the country."18

Roosevelt's desire for tighter controls on industry irked Bush, who blamed regulations for sabotaging entrepreneurs and creating an atmosphere in which "the making of a large profit has been frowned upon." The New Deal's controls on big business threatened the nation's health by rendering "the creation of truly new industries and products . . . nearly impossible." At the same time, Bush allowed that increased government regulation might have the salutary effect of reducing "ruinous competition."[19]

There was more to Bush's frustration than opposition to the government's economic policy. Bush objected to the president's emphasis on collective action and the attention given to the social roots of class differences. In trying to ameliorate individual suffering, the government must not undermine personal responsibility or it risked creating an authoritarian regime, Bush believed. Echoing Nietzsche, he insisted, "Discipline must be self-applied or it will be externally imposed."

Bush's "own personal religion" stood in sharp contrast to the collectivist ideal. Shunning organized denominations, he raised his faith in individuality into a secular spirituality. A capacity for joy and self-mastery, a sense of tradition and the courage to meet difficulties "with a smile or a joke"—these were the hallmarks, in Bush's view, of a life worth living. "The greatest thing," he told his MIT students, "is to play an effective part in a complete scheme of things. Some insist on knowing where mankind is headed, or [they] won't play. The joy of life is the answer." And echoing religious language, he insisted on the "saving grace of humor."[20]

Bush found romance in adversity and solace in hard work. Sheer activity was his antidote for doubt and the avenue leading to existential satisfactions of all sorts. His faith in activity was unshaken by the Depression, which for many meant enforced idleness. He implored educators to, above all, encourage the student's capacity for self-management. He was against rigid traditions that thwarted a student's ability to get ahead.

Perhaps rationalizing his own decision to race through college, he was enamored with a plan to cram an extra term into the university program, making it possible for students to graduate in three rather than four years. "This would be of particular advantage," he wrote in 1933, "to older students, to the young students who need quickly to become self-supporting, and to the brilliant students who wish as soon as possible to reach the advanced work in which they are really interested." Bush had earlier charged that the "examination system of American engineering schools" was "illogical, antiquated, and damaging to our product." He attacked the reliance on fragmentary courses that students must pass "in a rigid sequence." Students tended "to 'take courses' instead of studying subjects. . . . When the course is passed he forgets it as far as possible. Our system of education degenerates into a sequence of forgetting points."

Bush bemoaned the failure of teachers to show students how to tackle a "comprehensive engineering problem in its entirety, drawing his tools from various sources," ranging across mathematics, physics, chemistry and even economics. Instead, students were usually taught by "a narrow specialist with an interest in the minutiae of a very limited field." As a result, "The student is hounded," Bush insisted. "His hours are crowded and closely scheduled; he has little time for reading or reflection, and he does little such. All but the exceptional students become automatons." The alternative was to adopt a scheme of comprehensive examinations pioneered by schools of liberal arts and medicine. Bush thought that Yale's medical school, for instance, had the right formula: "Here there are subjects, but not courses in the ordinary sense. There is no required attendance, nor are there any examinations in specific subjects. A student has two comprehensive examinations; one at the end of his pre-clinical work, given by the clinical group to determine whether he is ready for advanced medical study; and the other at the end of his clinical work to determine whether he is ready for the degree. No student is required to take either examination until he himself wishes, and he may cut short his period of residence to some extent by a successful early appearance."

For Bush, using freedom wisely was the whole point of life, and so must be the entire thrust of education. He saw, for example, "no reason why the final examination in a subject should not be open to any student who wishes to take it, even if he has not formally taken the subject in course." The teacher's goal, he reminded, was to "extend to the student self-determination in his academic activities" and to aid him in making the "transition from the attitude of the schoolboy to that of the professional man." This maturation was often painful, but distinctions based on merit were inescapable. Education "will eliminate those who cannot stand the freedom of professional life, and [it will] nurture the individuality, initiative and resourcefulness of those who can."[21]

Almost by definition, Bush considered versatility the crucial determinant of practical intelligence. In a telling testimonial to Elihu Thomson, on the 80th birthday of this distinguished inventor and entrepreneur, Bush laid out his vision of the engineer as the most worldly of men and the professional whose values and ideas were most sharpened by the edge of experience. Speaking of Thomson, yet also speaking of himself and those like him, Bush observed: "You have showed us that a man may be truly a professor and at the same time very practical. And one thing more which I wish to emphasize. You have shown us that a scientist or engineer may be, even in this complex modern world, versatile and yet not superficial."

To Bush, this compliment was the highest imaginable. At a time when the pace of scientific and technical advance was accelerating, he worried about the inability of individuals to retain a sense of the whole of knowledge. After all, it was the awareness of how the disparate pieces fit together that was at the bottom of the engineer's method. As technology grew more complex, the pieces became more opaque, yet the need was more pressing than ever to produce technological systems in which the pieces fit snugly together. In toasting Thomson, Bush revealed the core of his anxieties about the dilemma facing technical intellectuals and anticipated the complaint by British scientist C. P. Snow about a widening gap between the "two cultures" of science and the humanities.

As Bush put it:

> In these days, when there is a tendency to specialize so closely, it is well for us to be reminded that the possibilities of being at once broad and deep did not pass with Leonardo da Vinci or even Benjamin Franklin. Men of our profession—we teachers—are bound to be impressed with the tendency of youths of strikingly capable minds to become interested in one small corner of science and uninterested in the rest of the world. We can pass by those who, through mental laziness, prefer to be superficially and casually interested in everything. But it is unfortunate when a brilliant and creative mind insists upon living in a modern monastic cell. We feel the results of this tendency keenly, as we find men of affairs wholly untouched by the culture of modern science, and scientists without the leavening of humanities. One most unfortunate product is the type of engineer who does not realize that in order to apply the fruits of science for the benefit of mankind, he must not only grasp the principles of science, but must also know the needs and aspirations, the possibilities and the frailties, of those whom he would serve.[22]

Here Bush sounded very much like a minister's son. In his mind, the engineer—at his finest—tended to the needs of his flock, "those whom he would serve." In this important respect, Bush displayed the reform spirit that stretched back from the New Deal to the earlier progressive era. Engineers, he believed, must join other professional experts in consciously shaping American society from above. Rankled by the low status of engineering, he urged both citizens and public officials to listen more intently to engineers on a wide array of questions. Along with many others, Bush saw the growing need for experts as posing a fundamental challenge to the prevailing ideas about representative government. Under the influence of journalist and political theorist Walter Lippmann, many people had begun to take a fresh look at the concept of democracy. "In view of the technical complexity of almost all great public questions," Lippmann asked, "it is really possible any longer for the mass of voters to form significant public opinions?"

Lippmann thought not. The issues of the day were too complex; the masses were too swayed by the "big spectacles" of life: murder, love, sport. "The management of affairs tends, therefore, once again to rest in a governing class, a class which is not hereditary, which is without titles, but is none the less obeyed and followed," Lippmann asserted.[23]

Bush agreed, insisting that professionals deserved a greater say in the affairs of their domain "by right of superior specialized knowledge." This knowledge should translate into social and political power, which would be leavened by the profession's commitment to an ethic of service. An engineer, Bush said without irony, "ministers to the people." He is "a strong bulwark against disaster."[24]

In June 1937, Bush described his ideal engineer in a speech laying out his philosophy of expertise. "To be an engineer in these days is to bear a proud title," he declared. "Insistent upon his prerogatives, kowtowing to no man, respected because he speaks the truth the country needs to know, the independent engineer stands as an important member of the professional class." He also shared many traits with the rough-and-ready pioneer and explorer of the 19th century. "The same qualities of courage, resourcefulness and independence which opened the nation are as necessary today as ever," he said. As the raw wilderness of the U.S. retreated under the press of modernity, the engineer would emerge as a new kind of frontier hero, an explorer who, in Bush's words, built "trails in the technological advance."[25]

While he was shrewd to tap into the nation's anxiety about its vanishing geographical frontier (and what instead might serve as an expansionary outlet for American energies), Bush was not the first to identify technological innovation as a substitute. In a 1922 book, *American Individualism,* Herbert Hoover, then secretary of commerce in the cabinet of President Warren G. Harding, compared science to a "great continent" and insisted that "it is only the pioneer who will penetrate the frontier in the quest for new worlds to conquer." Hoover's metaphors caught on with other writers, growing more elaborate over time. For instance, the author of a 1929 book, *Industrial Explorers,* breathlessly described corporate researchers as "trail blazers on the path of progress; and their pioneer work builds the foundation of the road which connects the outposts of industry with the main highways of commerce."[26]

All was not rosy, however. Bush conceded that the individual inventor's power had ebbed. "The growing complexity of life tends to make men cogs," he said in his June 1937 speech. In his youth, the lone inventor—eccentric, ingenious, part tinkerer and part entrepreneur—set the pace for American technology. Eight of ten U.S. patents were granted to individuals at the turn of the century. Edison was the archetype for these innovators. By the eve of the Depression, however, the lone inventor was fading into myth. In his place stood the colorless industrial laboratory, the new locus of invention. Corporations earned nearly half of the nation's patents for the first time in 1930, and their share of patents continued to rise through the decade. Industrial "invention factories" relied heavily on university graduates who pursued research in teams. Engineering refinements were the rule; breakthroughs were rare.[27]

AMRAD, where Bush had worked after the war, was just such an industrial lab. In the decade following World War I the number of these labs grew tenfold, to more than 1,000. To help solidify their reputations, these labs denigrated the methods used by Edison and other inspired inventors. That struck Bush as unwise. He worried that the General Electrics and AT&Ts gave a conservative cast to research. Though supportive of industrial labs, he admired the independent inventor's instinct for novelty and willingness to take risks. While innovation was clearly becoming corporatized, Bush still believed that the "lone researcher often does produce out of thin air a striking new device or combination which is useful and which might be lost were it not for his keenness."[28]

Bush was himself just such an irrepressible inventor. While an astute manager of research teams, he often pursued his grandest intuitions alone. Rapid retrieval of personalized data, stereophotography, typography, internal combustion engines and perpetual motion were just a few of his obsessions. For him, inventing was a calling, a way of life. He flirted with inventions the way some men pursued women, insistently and heedless of the outcome. Inventing was a game. He was good at it and it relaxed and amused him. "O, his head is full of those things," Phoebe once said of his ideas for gadgets. "He has a short cut for everything."[29]

No matter how tight his schedule, Bush never stayed away for long from his own experiments. He once said: "I always keep a piece of personal research going on . . . [so] when I get weary of talking to visitors and shuffling pieces of paper I can slip down and work in the laboratory."[30]

While finding sheer pleasure in tinkering, Bush also craved the recognition brought by a stirring invention. It was no longer money he was after, since he usually granted to MIT any of the proceeds from his patents. Instead, he hoped to open new fields of inquiry. Whereas in the 1920s, he concentrated on the nascent field of electronics, in the 1930s his interests were more exotic. He advised fellow researchers that "sun power warrants serious study" and designed a solar-powered irrigation pump, which was a commercial flop but which he nonetheless thought would demonstrate the potential for devices that tapped the energy of the sun. He also devoted years to an unorthodox but potentially efficient gas-powered engine. No companies took a flier on that, either. Bush often groused about the fate of his idle patents, but he kept tossing off ideas anyway. He dreamed of new gadgets and processes with a vitality some men reserved for romantic affairs. Describing his compulsion to an MIT colleague, he once wrote: "I apparently cannot help inventing. Of the things that I think of three out of four go into the waste basket as soon as I think of them. Of the remainder perhaps one in a half dozen is worth attention."[31]

At his most inspired, Bush imagined novel ways to automate ordinary tasks, though usually he had no interest in working out the details. Typical was a suggestion he made to Frank Jewett, chief of Bell Labs, regarding the potential for automatic dialing equipment. In a letter, complete with sketches of the possible device, he wrote, "It is desirable to provide a piece of telephone apparatus for residence use whereby numbers frequently called may be dialed simply by turning to a name and pressing a single button. The device must be inexpensive, small and reliable."[32]

Bush's most serious affair was with computing machines. The 1931 version of the differential analyzer had been a smashing success. Researchers at Cornell and the universities of Texas and Pennsylvania were building copies; so were colleagues in Norway, Russia and Ireland. For his contribution, Bush continued to garner accolades from peers: the National Academy of Sciences elected him a member in 1934 and the American Institute of Electrical Engineers gave him its Lamme Medal the following year. Meanwhile, Bush's team at MIT, led by Samuel Caldwell, continued to make minor improvements in the machine. By March 1934, Bush was ready to launch construction of a far more powerful analyzer, which would deliver answers with greater precision and contain a "function unit" that translated mathematical functions into electrical signals. The new model also promised to overcome the most glaring inadequacy of the 1931 version, which required mechanical resetting for every fresh problem. This took hours or longer. The new machine (which, on completion, would weigh a staggering 100 tons and consist of 2,000 electronic tubes, 200 miles of wire and 150 motors) would prepare for a fresh problem electrically with the aid of punched tape, bringing a solution much faster simply by slashing the setup time and working on many problems at once. This would be costly to accomplish. MIT built the original analyzer for $25,000. Bush figured the improved model would cost from $65,000 to $75,000, an enormous sum at a time when the research budgets for entire academic departments were often a fraction of that. MIT could not afford to pay for a new analyzer, so work on this massive machine would begin only when "the financial situation permits us to proceed."[33]

For help, Bush turned to Warren Weaver of the Rockefeller Foundation. Weaver, an applied mathematician, had taken an interest in the first differential analyzer after a Norwegian scientist asked the foundation to help pay for a copy of the machine. Weaver first visited Bush in November 1932 and came away "very much impressed by the power and accuracy of the machine." Bush struck up a correspondence and began trying to convince Weaver to underwrite the next version of the machine. After some handwringing, Weaver and the foundation granted Bush $10,000 in 1935. The following year, Bush pressed Weaver for more funds, insisting he saw no barriers to progress: "Our experimental work has this year proceeded to the point where I can say with confidence that all of the unknowns have been removed from the situation to an extent sufficient to see our way clearly. . . . In the new machine we seek to provide a striking step in advance, which will bring powerful aid to all types of research utilizing ordinary differential equations, and we wish this new aid to be so powerful that it will substantially advance the rate of progress in fields of endeavor. This I feel sure we have in sight." With the foundation's money, Bush felt "perfectly sure" he could build a better machine.[34]

The foundation agreed to an $85,000 grant in April 1936, but over the next few years it gradually became apparent that Bush had badly overstated his case. The new analyzer's mechanical elements fell nicely into place, but designing the automatic controls was "extraordinarily complicated." Progress slowed. The machine would not be completed on time.[35]

Calculators were not the only "thinking machines" conceived by Bush. He mused about ways to automate the activity of thinking itself. People trying "to think straight in the midst of complexity," as he often did, needed more than a device that merely crunched numbers. They needed help in disciplining their random ideas. More important, they needed a way to manage the rapidly growing amount of documents and data that threatened to overwhelm the specialist and render whole fields incomprehensible to even the intelligent layman. Bush wondered if some sort of thinking machine might forestall what looked to be an inevitable information glut.

Others were asking this question. In November 1932, Bush received a visit from Watson Davis, an enterprising science journalist who directed Science Service, a small news organization. Inspired by two European lawyers who had proposed a universal "nomenclature for human knowledge," Davis toured the country, appealing to corporations for money to develop a microfilm reader, which would store documents as miniature pictures that could be read when projected. He asked Bush for aid in both solving technical problems and raising funds.

Bush was intrigued by microfilm, but offered Davis little encouragement. Pushing microfilm was an uphill battle. Almost a century old, microphotography had lain dormant until the 1920s when the growth of the movie industry sparked its rebirth. In the 1930s, the technique captivated librarians and archivists, who saw in it a way of making it vastly easier to copy and store books and journals. The goal was monumental: nothing less than unshackling readers from the printed page. But to achieve this, projection technology must vastly improve, or weary-eyed scholars would shun microfilm. Even with inadequate finances, Davis pressed ahead with his campaign and within a few years many libraries were using microfilm to provide copies of their materials or those of other collections.[36]

Bush made no promises to Davis, and they left no record of any further discussion on microfilm readers. But Davis had planted a seed in Bush's mind that would take root in the years ahead. While Bush would gain much credit for promoting microfilm as the medium for storing personalized information—even after his death, Bush remained celebrated among librarians as the "godfather" of the interdisciplinary field of "information sciences"—Davis's effort to achieve a similar end would be largely forgotten.[37]

Microfilm technology appealed to Bush, a photography buff. He naturally took note of the widening interest in microfilm as an information-storage medium. Spurred by his awareness of the ferment around microfilm, Bush published an article in *Technology Review* in which, from the vantage point of the future, he looked back with wonder on the crude technologies of the 1930s. Among other things he envisioned paper books being replaced by microfilm readers. He outlined a device, housed in a desk drawer, that would store and reproduce on a screen thousands of books.

"Many of us well remember the amazing credulity which greeted the first presentation of the unabridged dictionary on a square foot of film," he wrote, without acknowledging Davis. "The idea that one might have the contents of a thousand volumes located in a couple of cubic feet in a desk, so that by depressing a few keys one could have a given page instantly projected before him, was regarded as the wildest sort of fancy."[38]

In his article, Bush skirted the key question regarding any automatic library. This was the source of the software, or the set of instructions that would organize and retrieve the stored information. Ideally, the reader would want to obtain desired information by doing nothing more than typing a few keys.

Retrieval was a mammoth problem that Bush failed to appreciate. He displayed far more insights into the hardware that might make possible a personalized, automatic library. He approached the Federal Bureau of Investigation in the summer of 1936 with a proposal to build a futuristic machine for rapidly locating fingerprints. In April 1937, J. Edgar Hoover personally turned Bush down, even though Bush had promised to deliver a machine that could review 1,000 fingerprints per minute, or two and a half times more than the bureau's current method.[39]

Undeterred, Bush pressed forward with his rapid selector idea. "There appears to be no reason why [the device] can not be built," he insisted. "I believe that by far the best way will be to try it out in a restricted area, such as one of our science departments here. If all goes well, I intend to give some real attention to the subject this summer."[40]

As word spread of Bush's rapid selector, it was clear he had hit a nerve. For large organizations, the handling of documents—whether they contained bank balances or great literature—was becoming a major task. Bureaucrats were especially keen on automation. In a world without electronic computers, few grasped the possibilities of paperless records. A Russian named Emanuel Goldberg had in 1931 applied electronics to rapid searches in a similar way, but Bush was unaware of this work, which had not resulted in a commercial product in any case.[41]

At the end of the 1930s, Bush oversaw the building of four separate rapid selectors, none of which came even close to realizing his hopes. Though marred by myriad technical problems, linked to the limited materials then at hand, Bush's very attempt to build rapid selectors made him stand out. He was the first American to tackle the task of building a personal information processor. His crude results prefigured the personal computer and the Internet, still decades away.[42]

In August 1937, Kodak and the National Cash Register Company contributed $25,000 toward research on a rapid selector. The goal was to build in two years a prototype device capable of rapidly choosing the desired item from vast numbers of microfilmed business records. The device was fast because it relied on improved photoelectric cells and a stroboscopic lamp created by Harold Edgerton, another MIT professor. The lamp was a crucial technology. "By creating a very fast bright flash of light," a historian has explained, "the stroboscopic lamp made it possible to copy a selected microfilm image 'on the fly,' without stopping the film (and the search) to make a copy."

Bush assigned four graduate students to build more rapid selectors under his guidance. The group was enthusiastic. Recalled one member, "We were trying to do what no one else had done." Bush, meanwhile, convinced NCR's president to support other MIT research and to consider the long-term benefits of research on the rapid selector. Taking an unusual attitude, the corporation wasn't pressing for immediate results and even allowed MIT to retain patent rights, asking only for the right to use any inventions springing from the university's work.[43]

Industry was not alone in showing interest in Bush's rapid selection machines. In 1935, the Navy's new Communications Security Group, OP-20-G, called Bush down to Washington to ask his advice on ways to automate the cracking of Japanese codes. At the time, the Navy was just starting to mechanize its cryptanalytic activities, and it installed punch-card machinery by International Business Machines to process code traffic. Bush looked over the Navy's plans and judged that IBM's existing tabulating equipment "was not good enough." Rather than rely on off-the-shelf machines, "I told them that if they were going to mechanize at all, they needed machinery that was special, made especially for their purposes."[44]

Soon afterward, the Navy asked Bush to modify his rapid selector to aid in cracking codes used to encrypt radio and telegraph messages from Japanese diplomats and military officers. Eager to oblige, Bush drew up a plan for a codebreaking "Comparator" that would count the coincidences of letters in different messages. The promise of a $10,000 consulting fee got Bush's attention. But the main reason to accept the Navy's offer was the chance to delve more deeply into the intoxicating subject of codebreaking. Breaking enemy codes was a laborious task that required the statistical analysis of thousands of messages, but Bush thought he had the "canny sort of intuition" needed to build machines that would handle the worst parts of the job.[45]

Reaching an agreement with the Navy proved difficult, however. To start with, Bush had doubts about the Navy's technical sophistication. While serious about deciphering Japanese coded messages, the service had employed just one professional cryptographer as recently as 1930. Then there were financial issues. Bush wanted a simple contract that paid only his out-of-pocket costs, leaving MIT to cover overhead costs. A Navy officer insisted on a complicated contract that Bush thought "tied me down in one way or another. So I said the hell with it."[46]

Nearly a year passed before the Navy's crypto unit offered a deal that Bush liked. The final agreement called for payment in advance and placed no obligation on Bush to build any machines, only to give advice. Despite the loose terms and the delay, the chief officer of OP-20-G's research desk was eager to receive Bush's advice and even considered his $10,000 fee a bargain. By early 1936, Bush was burdened by administrative duties at MIT and work on a new differential analyzer. Still, he drafted a plan for Navy cryptography that envisioned a central role for electronic machinery. The plan called for the creation of a family of optical-electronic devices, called Rapid Analytical Machines, that would be a hundred times faster than existing calculators. The Comparator would be the first in the family.

Officers at OP-20-G were keen on Bush's plan, but the powerful Navy Bureau of Engineering attacked the arrangement, calling Bush's plans unrealistic and his demands unreasonable. After another delay, OP-20-G overcame the bureau's objections, signing a new deal with Bush in January 1937. After outlining his design for the Comparator, Bush ran into trouble finding qualified people at MIT to build it. The need for secrecy made it even harder to staff the project. Other professors were unaware of the nature of the work; Bush told only his friend and boss, MIT president Compton, about it. Even the graduate students ultimately assigned to build the Comparator didn't know what it was for.[47]

With Bush stretched thin, the Comparator got short shrift. In the end, one junior engineer built it, and he found that the machine's parts did not work well together. Repairs were made, and when the Navy finally received the Comparator it did not work. It was fixed, then again shelved after proving unreliable. This was not wholly the fault of Bush's design: to reduce costs and increase reliability, his design relied on relatively few components, but the electronic and memory technologies of the day hamstrung the machine. Indeed, the head of OP-20-G praised Bush's effort in December 1937, concluding, "As matters have turned out it appears we have struck a remarkably good bargain." About the same time, Bush asked to leave the project, informing the director of Naval Communications that his job was finished. For the project to "be of greatest benefit" to the Navy, OP-20-G should take it over, he said.[48]

Bush withdrew, but couldn't break free of Navy cryptanalysis. During 1938, his handpicked engineer kept refining the Comparator. He reported monthly to Bush and won praise from the Navy for a "splendid job clearing up all the various 'bugs' in the new apparatus." In the view of one intelligence officer, the Navy now had "what we consider a fine, reliable machine. We have checked it against all of its specifications and it does everything wanted at five times the speed required." It would take "a long time, a year at least," the officer predicted, for the Navy "to learn how to adapt the machine to our processes, or our processes to the machine." But he was happy enough with the progress that he offered Bush a naval commission. Bush declined the offer.[49]

Besides drawing him into the super-secret world of cryptography, Bush's work on the rapid selector and its siblings put him in the front ranks of a cadre of educators studying mechanical aids to thinking and learning. Radio and television, as well as microfilm and calculators, might make learning easier and more rapid. Seeking to understand the effects of these new technologies on education, the Carnegie Corporation underwrote the formation of the Committee on Scientific Aids to Learning. Bush, invited to join the committee by Harvard president James Conant, wrote its prospectus, which was approved at its first meeting on November 19, 1937. Among the three areas to be examined, Bush was intimately tied to two: "instruments for the storage and selection of data for individuals and libraries"—his rapid selector—and "calculating or analyzing devices." The committee made grants for field work and published such reports as "Auditory Aids in the Class Room," "Central Sound Systems for Schools" and "Equipment and Supplies for Microphotography." These studies examined things "already being pushed by some manufacturer or another," Conant later recalled. "We didn't invent them." But studying the relevance of new media to education was novel. Conant credited the committee's foresight to Bush, who felt "technology was going to have an impact on everything."[50]

But the committee had another purpose: it provided an intimate setting for Bush to discuss international affairs and cement his bond with two of the nation's leading "science statesmen." One was Conant, a distinguished chemist who during the Great War worked without apology on poison gas and was "a square-shooting, level-headed liberal," three years older than Bush. The son of a well-to-do businessman in the Boston suburb of Dorchester, Conant came to know Bush during merger discussions held between Harvard and MIT. The talks were fruitless but lasted long enough for the two men to become unlikely friends. Conant was worldy and urbane; Bush was rough and sly. But each was concerned about Germany's rising militarism.[51]

The other notable member of the committee was Frank Jewett, chief of AT&T's Bell Laboratories, the leading industrial lab in the country. Like Bush and Conant, Jewett had Puritan roots, but he had been born in Pasadena, California, in 1879. A graduate of Throop Institute (the forerunner of the California Institute of Technology), Jewett studied physics at the University of Chicago, where he gained his doctorate. In 1902, he joined MIT as an instructor, but two years later moved to AT&T as a transmission engineer. He rose steadily; by 1912, he was the telephone company's foremost expert on long-distance traffic. In 1925 he took over Bell Labs.

Befitting a senior executive of the country's biggest monopoly, Jewett was a vocal conservative, suspicious of New Dealers. As a member of the Science Advisory Board, along with Compton, he felt the sting of Roosevelt's rebuffs. Jewett first met Bush during the war at the Navy's submarine-detection lab, where he was an adviser. He later drew Bush into the activities of the National

Academy of Sciences. He shared Bush's skepticism about the government's ability to tap scientific talent should the U.S. enter another war.[52]

As the 1930s wore on, Bush and his circle paid increasing attention to the world's resurgent militarism, the flip side of the global economic crisis. Save for a stint as naval reserve officer during the 1920s, Bush had had little contact with the military since the Great War. Yet he recognized the tendency of technology to make war more terrible and the failure of the U.S. to keep pace with advances in weaponry. "Technical men in industry were developing some of the most bizarre gadgetry in the world, but not for war," Bush later wrote of the period. "In this country it was not merely that the people turned aside from the paraphernalia for war. Civilians felt that this was a subject for attention only by military men; and military men decidedly thought so, too. Military laboratories were dominated by officers who made it utterly clear that scientists or engineers employed in these laboratories were of a lower caste of society. When contracts were issued, the conditions and objectives were rigidly controlled by officers whose understanding of science was rudimentary, to say the least. To them, an engineer was primarily a salesman, and he was treated accordingly."[53]

The estrangement between officers and scientists mirrored the larger society's attitude toward military matters. By the early 1920s, many people came to regret America's involvement in the Great War, believing that "merchants of death," bent on war profiteering, had duped the nation. The public was convinced, observed one writer, "that wars were little games arranged by big industrialists in order to get orders for weapons."

Roosevelt fell in line, opposing all but modest increases in defense spending. And even when he sounded more aggressive, there were suspicions that he simply wanted to distract attention from the floundering economy. In January 1938, after witnessing the Spanish Civil War and the Japanese invasion of China, Roosevelt asked Americans to authorize the buildup of the U.S. armed forces, which were then "the merest skeleton of effective military power," wrote journalist Walter Millis. Many of the Army's planes were unfit for combat, and a 1934 survey by then chief of staff MacArthur found his forces with only 12 post–World War I tanks in service. The Navy ships defended only one ocean, the Pacific, and not even that very well.[54]

Both the services showed scant interest in developing new weapons, spending on average just $4 million a year in this area from 1924 to 1933. In 1936, the Army debated whether to increase its spending on research but decided that rather than wait for the results it would simply spend more on the "excellent equipment that has already been developed." As late as 1939 the Army spent $5 million, or just 1.2 percent of its budget, on research and development.[55]

Bush began to speak out in favor of military preparedness in November 1935, a month after Italy invaded Ethiopia. In a lecture at Tufts, he defended the scientist who works on weaponry in peacetime, arguing that the development of secure defenses against, say, air attack might render war unthinkable. "Perhaps the worker on antiaircraft is more effectively a worker for peace than his brother who condemns him?" Bush asked. "I do not make the assertion, I pose the question. Yet when I pose it in conversation I meet the answer many times that all development of war engines should cease. Certainly it should never have begun; but this is wishful thinking; for the application of science to warfare will not cease in the world as it is now divided and governed."

In calling for the nation to prudently improve its defenses, Bush tried to discredit fashionable pacifism. While desiring peace, he decried "the substitution of wishful thinking" for "cold logical reasoning. Worse yet," he insisted, "many otherwise logical individuals, overcome by the sheer weight of the problem, refuse to attempt to think it out at all, but cry 'peace, peace,' when there is no peace, and pitifully hope by their cries to still the storm of human conflict and ambition."[56]

It was not long before Bush's ideas seemed borne out. In 1936, German troops seized the Rhineland in violation of the Treaty of Versailles. The following year, a civil war broke out in Spain. Before long, the country became a laboratory for Russia, Germany and Italy to test their latest weapons. Though Bush did not record his feelings about the widening conflict in Europe, his actions spoke loudly.

With Jewett's backing, Bush had become chairman a year earlier of a sleepy division of the National Research Council, the Division of Engineering and Industrial Research. In the 1920s, the division had been a hotbed of activity, preaching to corporations the benefits of funding their own research. The campaign contributed to a fivefold increase from 1920 to 1931 in the number of U.S. industrial labs. In 1933, the division was reorganized and three years later began serving as a clearinghouse for data on government and university researchers whose work might interest corporations. But the division had just three active committees. Bush wondered whether it "should be discontinued or reduced to a mere paper existence."

With little confidence in the U.S. military's ability to bring new techniques to the battlefield, Bush decided that if he took "a somewhat radical step" the engineering division could help researchers aid the government in time of war. In December 1937, he proposed to revamp the division to do just this. Once the proper links were forged between science, engineering, industry and the military, the division "should quite frankly . . . do practically nothing in time of peace except keep the organization alive," he wrote.

The plan struck a chord. The president of the National Academy, Frank R. Lillie, was impressed. He was "quite concerned" about the ability of the academy and its sister research council to respond well in a "time of stress and emergency." This sentiment was precisely what motivated Bush to consider alternatives to the old ways of mobilizing civilian technologists.[57]

In January 1938, Bush wooed a group of leading corporate researchers at a meeting in New York City, offering his revamped engineering division as a forum for the benefit of corporate laboratories. In attendance were nearly 50 executives from some of the nation's leading companies: Procter & Gamble, Champion Paper, Colgate-Palmolive, Swift, Lilly Research, Burroughs Wellcome and Dodge.

"Gentlemen, I want to emphasize that this is your party," Bush told the industrialists. "We . . . offer you our services, if you want them and if they can be of value to you. In order to get any movement started there are two things always necessary, however. The first is an acute need, and the second is someone to start the ball rolling. We are perfectly willing to act in that second capacity, to get this thing moving, if it is a good thing. But I emphasize that it is your affair."

The research managers, intrigued, showed their approval. The representative from Colgate-Palmolive, for instance, agreed that research managers faced growing problems and needed new methods but he worried about too much friendliness among the participants. "It is not the idea to exchange confidential information," he said. "That isn't the idea of this group at all. It is to make research more efficient."

This was music to Bush's ears. Efficiency in research was his creed. Before adjourning, he said the meeting had gone "very nicely, indeed in just the way I hoped it would." Then he tackled a crucial topic that so far had been ignored. This was the possibility of "a liaison between" government and industry researchers arising "in time of stress." He soon made it clear that the "stress" he worried about was another world war. He hoped that the U.S. would not be drawn into another conflict but if that happened an organization that united government, academic and industry researchers could prove of great value to the nation's military establishment. "I feel quite strongly," he said, "that if this organization [of corporate research directors] gets into healthy operative condition, it would also be an important factor, an important link in the chain between government and the industries of the country in time of stress for the purpose of the national defense."[58]

Seeing signs of war, Bush felt a sense of urgency. But the revamping of the National Research Council's engineering division—hamstrung by the organization's traditions and history—moved too slowly to satisfy him. By the time the division was prepared to mobilize military research, Bush had switched to a new, more nimble organization that would bend to his will. His experience with the engineering division was critical in developing his ideas about mobilization. It also distinguished Bush as the research administrator best positioned to organize researchers on behalf of national defense. Of the leading members of the country's technological elite, he was the most willing to break with tradition in order to accomplish the job.

The looming international crisis ended Bush's days as a world-class inventor. In 1938, the nation's isolationist stance toward events overseas came under increasing pressure. The Japanese announced their intention to unite Japan, China and Manchuria into an Asian empire. Europe, meanwhile, edged closer to war. Franco's forces, abetted by the Nazis, neared victory in Spain. Germany annexed Austria, Hitler's homeland. (Later in the year, war was averted only by a settlement at Munich between the British and the Germans over the dismemberment of Czechoslovakia.) Bush could not tell if the U.S. would be drawn into the war, but he considered moving to the nation's capital in case it was. "Washington is a central point," he thought, "and I might be useful there in time of war."[59]

He did not relish the prospect of living so far from his cherished New England. Washington struck him as alien ground, and even visiting the city was an irritation. On a visit in May 1937, he found the trip from Cambridge to Washington tedious, despite managing to keep "right on moving" over "good" roads on a "beautiful" day. "I do not think I would want to drive down more than once in ten years," he told Jewett.[60]

Soon afterward, an attractive position—for which Bush was well qualified—came open in the nation's capital. The longtime president of the Carnegie Institution of Washington, John Merriam, decided to retire. His job was a plum. Carnegie was a top-drawer patron of science in America, as influential in its fashion as the National Academy of Sciences. Founded by wealthy industrialist Andrew Carnegie in 1902, the institution had an endowment of $33 million and spent $1.5 million annually on research, largely performed at its eight major labs. These included one of the world's premier observatories, in Pasadena, California. Carnegie's president influenced the direction of research in America and was a significant player in Washington scientific and cultural circles. He served on the board of the Smithsonian Institution and informally advised the government on technical matters. He also mixed with Carnegie's esteemed board of trustees. Larded with rich and influential members, the trustees included Herbert Hoover, Army General John J. Pershing, legendary aviator Charles Lindbergh, and Frederic Delano, Roosevelt's uncle and his link to the science community. Bush mentor Frank Jewett also was a trustee.[61]

From the beginning of Carnegie's search, Bush had the inside track. Merriam himself was in no position to select his successor. A solid scholar in his prime, Merriam had aged badly. Bush held him in contempt, privately describing him as "the old fake," an intellectual "poser" whose scholarship was suitable only for "dumb clucks." Others shared this dim view of Merriam. One Carnegie insider called him "a real paranoid. I don't know whether he was hearing voices but his behavior was odd. . . . He thought people were watching him."[62]

With Merriam sidelined, the trustees of the institution held sway. Jewett, a member of the four-person search committee, championed Bush's cause. W. Cameron Forbes, Carnegie's chairman, also liked Bush. A member of a wealthy Boston family and a former governor of the Philippines, Forbes knew Bush from MIT and had once invited him to his family estate.

Jewett and Forbes urged Carnegie to hire him. It wasn't a hard sell. "Bush was as different from Merriam as could be," recalled one Carnegie insider. Whereas Merriam was opaque, aloof and even incoherent at times, Bush "was always a very responsive adviser. You might not like his advice, but he'd tell you, quickly and with great clarity, what he thought."[63]

Carnegie only seriously considered Bush as Merriam's replacement. Jewett especially looked forward to having Bush nearby during what he expected to be a hectic period of military preparation. Karl Compton gamely tried to keep Bush at MIT, offering him the presidency of the Institute (Compton would remain as chairman of MIT's board). But working under Compton did not appeal to Bush, who loathed the thought of someone looking over his shoulder, even if that someone were Compton, the nicest man he knew.

On May 27, Frederic C. Walcott, the Carnegie trustee chairing the search for Merriam's replacement, told Bush that he was the committee's first choice and offered him the job. Bush accepted contingent upon the approval of the full board. This came on June 2. Four days later, Delano sent Bush a crisp letter outlining employment terms. The post would begin on January 1, 1939, and pay Bush the princely sum of $25,000 a year.[64]

Bush quickly started to bone up on Carnegie. He studied the institution's publications and surveyed its operations. "He thinks he has located the soft spots, or at least some of them," a friend noted. "To remedy this situation, he is willing to make *any* necessary move." Bush had yet to realize the full extent of Carnegie's financial woes and the torpor that had befallen some of its programs. Expectations ran high, especially after a spate of newspaper reports heralded his talents. But Bush tried to keep things in perspective. He told *The Boston Globe* that he could only work in short stretches, saying: "Sustained effort of more than an hour or two a day is impossible because the concentration is so intense and the mental processes so involved that you'd go completely crazy if you kept at it much longer." The *Globe* concluded that Bush "has been called a miracle man. Happily he is not."[65]

To Bush, reforming Carnegie was important, but more important still was that he learn his way around the nation's capital. He was especially eager to meet the Army and Navy officers in charge of research. His first real contact with the military came in the field of aviation. Even before moving to Washington, he was appointed to the National Advisory Committee for Aeronautics on August 23. The committee would give him his first taste of the bewildering politics surrounding military technology.[66]

NACA was formed in 1915 "to supervise and direct the scientific study of the problems of flight, with a view to their practical solutions." Once it identified a problem, research usually was performed by a university or government lab. In 1920, NACA opened its own lab in a modest corner of the Army's new air base in Langley, Virginia. Langley's staff was small—100 in 1925—but of high quality. Engineers from all over the country came to the lab, drawn by research director George Lewis and the chance to work on one of the world's best wind tunnels. About five feet in diameter, the tunnel could simulate speeds of 120 miles per hour. The Langley lab also tested various foreign and domestic aircraft. After Charles Lindbergh's solo flight from New York to Paris in 1927, enthusiasm for aviation had erupted, spurring numerous improvements in airplane design. NACA's studies on the drag caused by fixed landing gear led to the quick rise of retractable gear. Looking ahead to faster planes, the committee finished a "full-speed" wind tunnel in 1936.[67]

Since its inception, NACA had worked on both civilian and military applications. In the mid-1930s, research began to diverge. Commercial airlines sought safety improvements and operating efficiencies, while the military wanted faster planes with greater maneuverability. The divergence strained NACA's resources. More alarming, however, was Germany's progress in aviation. Lindbergh, a NACA member since 1931, visited Europe often and had seen the Luftwaffe, the Nazi air force, up close. A national hero, his opinions on German air superiority were profoundly influential. In late September 1938, he shook up the government by telling a U.S. diplomat in London, "I have the certainty that Germany's air power strength is greater than that of all other European countries combined and that her margin of leadership is constantly being increased." Lindbergh added, "If she wishes to do so, Germany now has the means of destroying London, Paris, and Prague."[68]

Bush wanted to know how the U.S. could catch Germany in aviation. Lindbergh was not optimistic. "Germany is far ahead of us," he wrote to Joseph Ames, NACA's chairman, on November 4. German military planes were of better quality and flew faster than the planes of any other nation. "The present quality of German military planes indicates what we may look forward to in the future, and necessitates our devoting much more effort to our own aviation development if we are to keep pace," he observed. But equaling the Germans in research would take a big effort. Germany had aviation labs in at least five cities; one lab alone employed 2,000 people. NACA had only Langley, a few hundred researchers and another lab on the drawing boards.[69]

MIT ruefully dispatched Bush, marking his departure with a mock trial in a hotel ballroom near the Institute. Compton played the judge, and the coauthor of Bush's engineering textbook played the prosecutor. The defense team consisted of four professors of electrical engineering. "Copping a plea," Bush apologized for abandoning MIT. Compton convicted him of "desertion" anyway. Bush begged for mercy, but Compton stood fast.

Following the trial, Bush's colleagues lampooned him. They made fun of Bush's conviction that he could concoct some bit of machinery to improve any process. On a small stage, a man dressed to look like Bush sat in a country field with only a cow for company. "It was plain he felt he could, by the judicious addition of a little machinery, improve that cow," an observer wrote. "After a curtain drop, climax of the scene came when Dr. Bush reached triumphantly under the cow and produced a half-pint bottle of milk, nicely capped and sealed. We glanced over at the real Dr. Bush . . . and observed that he looked quite pleased and proud."[70]

Part Two

Preparing for War

Chapter 5

"The minor miracles"

(1939–40)

It is being realized with a thud that the world is probably going to be ruled
by those who know how, in the fullest sense, to apply science.
—Vannevar Bush

Bush moved to Washington, D.C., fresh from a pleasure trip to Mexico's Yucatan peninsula with his wife, Phoebe. His arrival did not go unnoticed. Just
a few days later, on January 13, 1939, the American Engineering Council
held a dinner in his honor at the plush Mayflower Hotel. Frederic Delano,
the president's uncle and a Carnegie trustee, gave Bush a generous introduction, citing him as evidence that engineers had exchanged their dreary image
for something more up-to-date. When he had graduated school 30 years before, Delano said, "We had the notion pretty generally that the engineering
profession was a profession of men who were to do things and say as little as
possible. That was one explanation of why most of us neither could speak nor
write intelligently."

Four days later, Bush spoke for nearly a day to a congressional committee
studying the concentration of economic power and the impediments this
posed to small business and the lone inventor. Sitting in the shadow of the Depression, the so-called monopoly committee worried whether the economic
system still had room for the little guy with a big idea. Clearly, the nature of innovation was changing: dwarfed by research giants such as AT&T, the lone in-

ventor seemed pushed to the margins of the industrial scene. By the 1930s, corporations were gaining the majority of patents; 20 years earlier, individuals had received nearly three-quarters of patents issued. "A one-man invention isn't very possible these days," Charles Kettering, chief of research at General Motors, told the committee, citing the complexity of most inventions. But Patent Commissioner Conway Coe blamed corporate power for putting the individual at a disadvantage. "My conviction is that the poor inventor, and through him the public, suffer injustice precisely for the reason and to the extent that the monopoly . . . bestowed on him is not fully safeguarded . . . [from] the onslaught of mighty corporations."

Bush calmly countered Coe's gloomy view, extolling the virtues of the country's great corporate laboratories while celebrating the individual inventor. Bush reconciled these two forces, often seen as contradictory, deflecting concern about corporate abuse with the sentimental language of the frontier. He reiterated an earlier claim that a citizen could penetrate the technological frontier just as readily as Daniel Boone had explored the geographical frontier. "That pioneering spirit, that willingness to take a chance has been very important to our industrial advance" and remains so, he insisted. Describing the patent system as "decidedly democratic," Bush saw no reason why the lone inventor could not flourish even as well-heeled corporate laboratories accounted for more and more patents. While these invention factories demanded teamwork and subordinated the identities of individual researchers to the corporate mission, outside operators still had room, provided they possessed "courage and resourcefulness." "I think the day of the individual inventor is not past," he told the committee, "for as fine as these cooperative groups may be and as necessary as they are to our general progress in this country, they do not cover the entire field."[1]

Bush's appealing image of the heroic, lone inventor made a good impression on the committee, which asked him for a followup memo. The press, meanwhile, was charmed by Bush's unfamiliar mixture of braininess and folksiness. *The New York Times,* while incorrectly labeling Bush a "mathematical physicist," aptly described him as "a tall, genial, bespectacled scientist, with a slightly stooped and somewhat stringy figure." He "leads reporters quite out of their depth. They do not seem clear about his abstruse doings, but they never fail to record that he can wiggle both his ears."[2]

Life in Washington agreed with Bush. "He loved it," a friend said. "He loved the glamor of life. The pulse of it. He loved the whole feeling of it." The city's grandeur made Boston seem drab. In the bright light of day, Washington "gleamed white and green in the sun as if Rome had sent its leftover marble columns, arches, plinths, architraves and friezes to be set down there among the trees," one observer noted at the time. "Greco-Roman temples of government rose behind vast ceremonial stairways of a scale and grandeur once intended for emperors and empire."[3]

Bush lived at the Wardman Park Hotel, a short car ride from the Carnegie Institution of Washington (CIW). He had no time for house hunting and expected to reside there for the better part of a year. He found the Wardman agreeable, though "a few minor matters" troubled him. "In my bath room the light is poorly placed, and it would be much preferable to have a light [on] either side of the mirror for shaving," he wrote the manager. "When it is accomplished, there ought to be an outlet for an electric razor installed at the same time. The refrigerator does [work], after a fashion. It is an old model and the cook complains that it does not get cold enough and it is hard to get ice cubes out. If it happens that you can replace this with a later model, I will certainly appreciate it."

"One thing I fear you can do nothing about," Bush added. "It is almost impossible to use the shower bath as the temperature varies widely and erratically. This is somewhat of a trial to me, as I depend upon a shower each morning. I fear, however, that most of the automatic devices would be rather ineffective under the conditions which obtain, as the primary fault is very evidently the major piping system."[4]

The Carnegie Institution was a burden, too. It was in poor financial shape and faced a cash shortfall. Bush wished to ease the crisis with an infusion from Carnegie Corporation, the mother ship of Carnegie institutions around the country. Fourteen years before, Carnegie had "indicated that it would at the proper time consider sympathetically" increasing CIW's endowment by as much as $5 million. However, when Bush approached Carnegie he was told "the Corporation reserves are none too adequate." Carnegie offered "an emergency grant of $750,000," payable in equal portions over five years; a lump-sum increase in CIW's endowment would remain a possibility. Bush was not happy with this scenario, "For it would leave us in an undetermined condition as we face the future."[5] He complained to Carnegie's Frederick Osborn: "With this present [fiscal] emergency, I can hardly think about anything else."[6]

Besides sticky finances, Bush wrestled with nettlesome personalities. He found Cameron Forbes, chair of CIW's board, overbearing. Soon after arriving at Carnegie, Bush received a lesson in Forbes's style during a board meeting. While Bush tried to settle a matter of policy, Forbes took up nearly the entire meeting criticizing the English in Bush's written report on the question. Bush was about to explode in anger when the board member seated beside him whispered, "Keep your shirt on, Van; he does the same thing to me." Challenging Forbes in private afterward, Bush won "a minor battle" in which he made it "clear who was running the show."[7]

Then there was John Merriam, the former president. He was still hanging on, receiving a salary as president emeritus rather than a pension—and looking over Bush's shoulder. Merriam visited Bush often to offer advice, which irritated Bush. He planned to end Merriam's visits by clearing up "this misunderstanding in [Merriam's] mind soon." Worse, Bush thought Merriam was misusing a Carnegie grant of $10,000. The board had hoped the money would be distributed to certain of Merriam's favorite researchers who were to be phased out under Bush. Now Merriam "has taken the surprising position that this was a grant for his own personal research," Bush complained, "so that the problem of these individuals was thrown back" on him.[8]

Merriam and Forbes rated as mere trifles, however, compared to the trouble posed by a researcher named Harry Laughlin, who had studied racial differences in intelligence for nearly 20 years at Carnegie's Eugenics Record Office in Cold Spring Harbor, New York. The office, funded by a private endowment, was the chief scientific institute in the field of eugenics and Laughlin, its director, never tired of espousing the view that certain racial and ethnic groups were biologically inferior to white, Anglo-Saxon stock. An ally of conservative congressmen, Laughlin conducted studies in the 1920s that purported to show that recent immigrants from southern and eastern Europe had poor genes and jeopardized the blood of the nation. Laughlin also was, in the words of historian Daniel Kevles, the "most passionately outspoken advocate of sterilization in America." He favored the compulsory sterilization of people with low IQs and mental illness, sometimes encouraging states to prevent allegedly defective citizens from reproducing.

Laughlin's views were accepted in many places in the U.S. By 1935, five states sanctioned compulsory sterilization and bills for the same purpose were debated by another seven state legislatures. Laughlin provided the intellectual justification for these measures, which won him the gratitude of politicians at home and abroad. In 1936, he accepted an honorary doctorate from the University of Heidelberg. The German government had ordered compulsory sterilization of not only institutionalized people but also those with hereditary diseases, including blindness and physical deformities. Pleased by the recognition, Laughlin cited the award as "evidence of a common understanding of German and American scientists of the nature of eugenics."[9]

Since 1935, staff at the institution had been warning Merriam that many of Laughlin's views were based on "racial or nationalistic sentiments rather than on scientifically ascertained fact." The advisory committee to the Eugenics

Record Office had urged that its activities be "divorced from all forms of propaganda" and such "social reforms" as sterilization and restriction of immigration. Indeed, Laughlin's research was grossly flawed; one tabulation in Carnegie's 1938 yearbook, for instance, compared the IQs of prisoners of "Italian descent" with those "of American blood."[10]

Merriam failed to curtail Laughlin's activities, despite being "much concerned" about the eugenicist "for some years." Bush intended to settle the matter once and for all. Four days after officially starting at Carnegie on January 1, 1939 he told Laughlin to expect a review of his research. In May, Bush advised Laughlin to retire, arguing that his positions on public policy were inconsistent with the role of a scientific researcher. A month later, he officially asked Laughlin to take an early retirement by January 1, 1940, citing his earlier objections to the eugenicist's research and raising a new question about reports of his ill health.[11]

Laughlin countered with a satisfactory report from his doctor and insisted on remaining at Carnegie. He agreed in late June to retire when Bush offered him a lifetime annuity. In October, Carnegie's trustees approved the pension deal. Laughlin soon regretted his decision and in December, with retirement looming, enlisted Senator Robert Reynolds of North Carolina to try to persuade Carnegie's trustees to force his reinstatement. Bush was not about to back down. He considered Laughlin's politics an embarrassment; even his old MIT pal, Norbert Wiener, had called Laughlin's activities to his attention, describing a report by him as "a pretty poisonous piece of fascistic racialistic tripe."[12]

Firing back, Bush insisted that Laughlin's retirement was voluntary and that in any case he was a scientific fraud. Bush saw no need to inquire further into the matter. "I have no doubt," he wrote to a Carnegie trustee on December 11, "that such an investigation [of Laughlin] would show him to be physically incapable of directing an office, and investigation of his scientific standing would be equally conclusive."

Bush's forceful response carried the day. Laughlin's ties with Carnegie were severed for good at the end of 1939. The Eugenics Record Office was renamed the Genetics Record Office, and its budget severely cut.[13]

Despite the awkward situations with Merriam and Laughlin, Bush was determined to put his stamp on Carnegie. To Weaver, a close friend, Bush was "irreconcilably convinced that the only way in which any organization or group can work together satisfactorily is through a centralization of authority and control."[14]

Though only loosely familiar with much of CIW's research and an expert in no single area, Bush had strong opinions about the institution's general direction. He believed he could gain a working grasp of any scientific subject with a few days of study. "He had the self-confidence that he could confront and comprehend any problem," recalled one Carnegie colleague. "He really believed he had as much of a chance as anyone else of coming to grips with it. And that gave him the confidence to wade into questions that other people might avoid."[15]

Bush's penchant for making quick studies—then issuing directives—struck some CIW staff as high-handed. He thought the institution should support only the core, hard sciences, dispensing with "marginal" subjects such as the history of science and archaeology. These core areas, moreover, should "all work together and fructify each other. There is nothing like team research," he said.

Bush's attitudes spelled doom for Carnegie's impressive archaeology program, which for decades had supported important research on Mayan culture in Mexico, Honduras and Guatemala. A series of pioneers in the field had worked at the institution, but they couldn't persuade Bush to support Mayan studies. To historian I. Bernard Cohen, Bush's quick execution of the department "was little short of catastrophic" to the field of archaeology in the U.S., which at the time counted on Carnegie "for most of its financial support."

The same fate awaited the pathbreaking historian of science, George Sarton. As editor of the journal *Isis,* Sarton published articles on the history of science with the aid of CIW funds. *Isis* had been widely seen as one of CIW's real ornaments, but Bush's emphasis on core sciences seemed to leave little room for the journal. Upon Bush's arrival, an anxious Sarton, "very much worried" about whether CIW would withdraw its support for *Isis,* shared his concern with friends and a CIW trustee. When word got back to Bush, he concluded that Sarton "was starting an improper campaign" whose aim was to "bring pressure to bear upon himself." Bush retaliated by halving CIW's support for *Isis* and, for good measure, told Sarton in person "that if he ever did anything like that again, [Bush] would break his neck."[16]

Even if Sarton had kept his own counsel, he had little chance of finding common ground with Bush, who looked down his nose at the methods of historians, writers and the emerging fields of psychology and sociology. "I have a great reservation," he once said, "about these studies where somebody goes out and interviews a bunch of people and reads a lot of stuff and writes a book and puts it on a shelf and nobody ever reads it." By that definition, of course, much of history, the social sciences and the humanities were a failure.[17]

Bush's imperious actions were made possible by Carnegie's structure. "Plainly he could act as he had done primarily because he was not responsible to a larger public in the sense that a college president is responsible to alumni and a powerful faculty," noted I. Bernard Cohen. However crude, Bush's decisive streamlining had a beneficial effect. By the following March, CIW's fiscal situation had markedly improved. Meanwhile, Carnegie Corporation increased its annual outlays to CIW by earmarking the funds Bush had requested.[18]

As Carnegie bent to Bush's will, Europe slid into war. In March, Germany seized the part of Czechoslovakia not already in its grip and Franco's Nazi-backed forces crushed the Loyalists in Spain. In May, Italy and Germany signed a "pact of steel." On August 23, the Soviet Union and Germany signed a "nonaggression" pact, clearing the way for the Nazis to invade Poland on September 1. Great Britain and France countered by declaring war on Germany and Italy.

Despite the widening war, Bush was struck by the widespread view that the U.S. could stay out of it. "I have as yet to find any individual who is not wholeheartedly convinced that we ought to keep out of the war," he wrote James H. McGraw, Jr., on October 9. "There is a difference of opinion as to how this can best be accomplished, of course, just as there is in Congress. There is also in some quarters the distinct feeling that, if this nation states too vigorously that it is bound to keep out of the war no matter what happens, this itself may be one cause tending to bring us in. . . . The sentiment in every group [with] which I come into contact, however, is unanimous that we ought to use our best calm intelligence to prevent this country from becoming involved."[19]

As early as March, however, Bush observed that the weakness of Europe's democracies had allowed fascism to flourish. He thought the U.S. might have to alter its political system, probably by bringing the government, the military and the private sector closer together in order to match the efficiencies of the Nazis. Bush saw only benefits to this type of cooperation. "The totalitarian state can cut rings around the democracy, and ineffectiveness is the price of freedom. The present question is whether the price can successfully be paid," he wrote Fred Keppel, adding: "I wonder whether, if democracy is going to be successful, it has not got to include much more military organization of its units than at present." Anticipating the expansion of what would later be called the "military-industrial-academic complex," Bush envisioned a partnership of three sectors—the military, industry and the universities.[20]

With so audacious a plan, Bush at first took small steps, assaying the strengths and weakness of government researchers, trying to understand the military's peculiar needs and delineating the unique capacities of industry and universities to satisfy the state's imperatives. Wearing many hats (he was still chairman of the NRC's Division of Engineering and Industrial Research, for instance), Bush called upon military men at every level and talked openly and easily with academics and industrialists about collaborations with the Army and Navy. He was dismayed by the absence of a single authority to direct these various actors. The military, meanwhile, was too disorganized even to evaluate its own needs.

This mess could be laid at the president's door. Shrewd yet indecisive, Roosevelt was comfortable with loose lines of responsibility, often assigning aides overlapping duties. He dealt casually with his military chiefs, neither maintaining records of his meetings with the brass, nor requiring the Army and the Navy, his restive services, to make joint plans. A British observer quipped, "The whole organization belongs to the days of George Washington."[21]

In the spring of 1939, Bush concentrated on a narrow but fundamental issue, one whose solution would require much coordination among the military, science and industry. This was defense against air attack. Air power, Bush believed, defined military strength, but its advocates (notably the late Billy Mitchell) had ignored the potential for defensive action against bombers. Bush held "a private conviction that antiaircraft is not receiving the attention it should have." He intended to "stir the Army up on antiaircraft research." He had two paths in mind: exploitation of radio waves, or radar, to detect planes, and the development of better methods to shoot aircraft from the sky. Researchers thought air defense could be vastly improved; that was enough proof for Bush. Aiming to persuade the Army of the import of air defense, he pledged "to get things moving . . . every time I find an opportunity."[22]

The stakes were large. On April 10, 1939, Bush wrote to former president Hoover:

> The whole world situation would be much altered if there were an effective defense against bombing by aircraft. There are promising devices, not now being developed to my knowledge, which warrant intense effort. This would be true even if the promise of success were small, and I believe it is certainly not negligible.
>
> In this country there is a great air program, but it consists principally in developing and building aircraft. I have exerted what influence I could recently toward the emphasis of the long-range research aspect of this matter, but with no success. In the whole program antiaircraft seems to be almost completely overlooked. The Army and Navy have some development going on, but not nearly enough in my opinion to examine into the various possibilities for progress along these lines. The work is distributed through several branches, and anti-aircraft is no one individual's special concern. There is no centralizing agency on research, such as the NACA supplies for aircraft research. When military men are queried they usually reply that

"the answer to a plane is another plane." Even if this were true today, it need not be in a few years. The real reason for lack of intense activity lies in the fact that anti-aircraft matters are deeply buried in bureaus or corps which are primarily interested in something else.

Radar promised to revolutionize warfare by providing a way to track the enemy's moves and to achieve greater accuracy in striking targets. Yet the Army seemed indifferent to radar's potential. The Navy had sponsored radar research for nearly two decades and had built useful detection devices. But its researchers were starved for funds and lacked champions among the admiralty. Congress, meanwhile, took no special interest in radar, and Bush doubted legislators could be persuaded to change. "Having watched the way Congress has handled somewhat similar matters recently," he wrote Hoover, "I am pessimistic as to what can be accomplished in that direction."

Leery of a public campaign, Bush planned to privately press the case for radar, convinced that he could broker a partnership between the Army and the Navy. The two services often pursued contradictory paths to the same technical end; neither shared information nor coordinated spending. For the nation to rapidly improve its air defenses—probably the most pressing military need—the Army and the Navy had to work more closely together. They also had to alter their practices. This would not be easy. The Army resisted change. The Navy, meanwhile, was famously hidebound. Said one former naval secretary, "Old customs and old practices hang on longer in the Navy than anywhere else."

Bush would soon learn the truth of these stereotypes. He had just proposed a joint Army-Navy laboratory on antiaircraft research and convinced an Army officer to promote the idea. It was a long shot, but it was Bush's first attempt to reform the military. He was "not sanguine that much will happen," he told Hoover, because the Army will "probably resent or disregard a civilian scientist's" views. Still, Bush intended to press on, telling Hoover: "If there is any hope of preventing bombers from crossing national boundaries, I would like to contribute to that result, and I believe that in the long run it can be accomplished."[23]

Bush's ambitions extended beyond air defense. He seemed willing to tackle almost any technical problem of interest to the military. He posed as a fixer, a broker; he tried to see the world through a soldier's eyes. In this regard, he was an extraordinary engineer, the first of a new breed of military technocrats. It would be an uphill battle to establish his credentials, since military men generally viewed researchers as uninterested in military problems or immediate results. To combat this prejudice, Bush favored practice over theory. If a general asked him a question, he came back with an answer—and fast. He usually didn't have all the answers himself but relied on his many contacts in universities and industry. When stumped, he first called on professors at MIT, but he pumped experts elsewhere too. He absorbed academic opinions, leavening them with his own sense of what was possible or desirable. Finally he shared his answer with the military, speaking plainly.

Bush fielded many requests over lunch. On March 22, he dined with a general, who had "an interesting problem. The Army has a large stock of powder which is deteriorating and which must be disposed of before it becomes dangerous. Until recently the DuPont Company has bought this and re-worked it in connection with its plastics program. Now, however, they offer him only a small amount and he is actually burning powder to get rid of it the cheapest way. He would like to get someone to study possible outlets for this material." Bush explained the situation to an MIT professor, noting that the general intimated that "he has funds to pay a retainer to anyone who undertakes the work."[24]

Money, pride, patriotism: Bush knew the way to a researcher's heart.

Bush was a new kind of public servant: he organized expertise on behalf of government. At first, NACA provided the best stage for this. Within a few weeks of his move to Washington, the aeronautics committee began making considerable demands on him. Joseph S. Ames, NACA's 75-year-old chairman, was afflicted by various illnesses and "decidedly feeble." Bush, NACA's vice-chairman for just a few months, had no choice but "to be somewhat active in its affairs," he wrote on January 27.

This was an understatement. NACA faced its most severe test ever. The previous August, a committee studying NACA's relations to "national defense in time of war" had recommended that NACA open a second research center to ease the "congested bottleneck" at Langley Field. In December, NACA decided to seek $11 million to build a second lab in Sunnyvale, California. Roosevelt trimmed the request by 10 percent and sent it to Congress on February 3, 1939.

The next month a House appropriations subcommittee, generally sympathetic to NACA's needs, nixed the Sunnyvale lab. The armed services were livid. Henry H. Arnold, chief of the Army Air Corps, and Arthur B. Cook, chief of the Navy's Bureau of Aeronautics, wrote Ames on March 23 in support of the new lab, arguing that a California location made sense since it placed the lab near the country's premier aircraft companies. The Army and Navy planned to spend in the next year roughly $225 million on aircraft; the officers hoped the new planes would incorporate the latest and best techniques.

Ames wrote letters from his sickbed, but Bush had to appear before Congress. On April 5, he visited the Senate Appropriations Subcommittee, asking for five minutes to make NACA's case. It was the first time Bush had ever asked Congress for anything, and his inexperience showed. He was scared, lost his temper and generally acted like a "rank amateur." The senators interrupted him repeatedly. Finally, Bush smashed the table with his fists, insisting on an audience. The senators listened, but were not persuaded. They rejected NACA's request.[25]

It took until August to win enough support in Congress for the new lab. The plan was approved only after the reference to Sunnyvale was deleted from the measure (to satisfy politicians from the East). John Victory, NACA's executive director, did most of the politicking, cinching the compromise by agreeing to formally reconsider its selection of Sunnyvale as the new lab's home. The closed-door wrangling was an object lesson for Bush, who credited Victory with teaching him how to work the levers of Congress, a body buffeted by regional rivalries.[26]

NACA announced its choice of Sunnyale for a second lab on September 22. By then, however, the outbreak of war had convinced the committee to push for a third lab, which would concentrate on the design of airplane engines. NACA had ceded this field to industry, which focused on fuel efficiency and long hauls. That was fine for civilian aviation. But these engines could not match the combat performance of French, British or German planes, which stressed speed and high-altitude flying. The engine lab, planned for Cleveland, Ohio, was swiftly approved by Congress.[27]

The desire for better engines buttressed Bush's view that NACA would benefit from closer ties with industry. He favored a repeal of the agency's longstanding ban on allowing representatives from private industry to serve on its main advisory committee. Proponents of the policy said it protected NACA's independence: too much industry involvement, it was argued, would turn NACA into a mere "consulting service." The need for mobilization, Bush felt, made such concern outdated. After Ames retired in October 1939, Bush took the chair and brought in George Mead, a retired vice-president of the United Aircraft Corporation, as vice-chairman. Bush also worked to break down the resistance of Army officers and NACA's own staff to working with aerospace executives. For a time, George Lewis, NACA's director of research, opposed Bush's program of military-industrial cooperation. But Bush "did a bit of table pounding in a nice way," he told Mead, "and I think George sees the light. . . . He will undoubtedly come along."[28]

In Bush's mind, NACA was "coming to life," and he credited the agency's "unique form," which gave him a measure of political independence, for its vitality. "It is a full-fledged government agency, with its own budget, laboratories, personnel," he observed early in 1940. "Yet it includes civilians who serve without salary, and their opinions may be controlling when there are divergent points of view within government." NACA "draws together rather effectively the Army, Navy and civil interests." It was a useful model for drawing civilians into military affairs.[29]

NACA furthered Bush's political education. He received much advice, for instance, from Charles Lindbergh. In 1939, the two men met privately for an hour on April 18; lunched together on June 7 and June 9; and spoke for three hours on September 8. Bush, immensely respectful of the flier's accomplishments, soaked up his opinions. The two men agreed NACA's "prestige has fallen" and discussed ways to revive the committee. Bush was so enamored of Lindbergh he offered him either the chairmanship or vice-chairmanship of NACA. The aviator demurred, saying he would not renew his membership when it expired at the end of 1939.

On September 15, the discussion over Lindbergh's future became moot. In a national radio address, the aviator asked Americans not to succumb to sentiment or pity but to stay out of the European war and to "be as impersonal as a surgeon with his knife." The speech "rang a loud bell in the heads of every isolationist" and qualified Lindbergh as a staunch opponent of Roosevelt. Bush never recorded his reaction to the speech, though it surely disappointed him. The speech, as one executive wrote Bush, had cost the aviator "his usefulness." Bush disagreed. Two months later, he wrote Lindbergh: "Certainly the NACA is going to have tough problems and certainly if you are about these parts I am going to bring you in them if I can, at least to the extent of asking you to talk them over with me personally when I need some advice."[30]

Lindbergh left a mark on Bush, who did not easily accept influences. A charismatic speaker, Lindbergh showed such respect for German air power that he usually convinced listeners that the Nazis should be granted a wide berth. But Bush reacted differently to Lindbergh's "scare tactics." "He was impelled to action by the very threat which Lindbergh so forcefully presented," noted Robert Sherwood. Not one to retreat from a fight, Bush felt the country could keep its peace only by showing its strength.[31]

Bush had neither been in combat nor studied military history. Yet he wisely asserted that every innovation in war could be stymied by a counter-innovation. Bush's comprehension of the dialectic of military technology lifted him above the humdrum level of research administrator. He glimpsed around the curve of knowledge, exuding a poise and confidence that tomorrow's invention would erase the advantage of today's dominant weapons. He was not unnerved, therefore, by Germany's lead in military hardware. But neither did he accept American weakness. He was eager for technical improvements, no matter how modest. The year in Washington had lowered his expectations while raising his ambitions. He foresaw a large role for research in the second world war even as he worried how to convince the Army and the Navy to push for common solutions to technical problems. Civilians were essential to mobilization, Bush knew, yet no existing organization, inside or outside the military, could produce the required technical advances.

Unwilling to work as a cog in big outfits, Bush devoted much thought to how they worked. He felt structure should triumph over the vagaries of personality, spot judgments and crises, even though in his own life he usually viewed organizational structure as an impediment to inspiration. As 1940 began, he wrestled with the proper structure to coordinate disparate civil and military research. Writing to a friend on January 5, he apologized for writing "a much longer letter than I usually write," but the reason for his wordiness was clear:

> The most important matter, however, remains to provide the liaison between all of this [research] work and the government development. I have been at work on this for some months without a great deal of success. One or two things that I now have under way may produce something along these lines. Whatever the immediate steps, the ultimate procedure should be to establish something resembling the NACA as a definite means of interconnecting the Army, Navy, and civilian interests on this particular problem. I feel quite sure, if the mechanism can be set up, quite a bit of scientific research can be usefully oriented. It is necessary, however, not only that it be done, but that some attention be paid to it after it is done. The Army can do some of this it is true. I am not sure but that another link is needed, namely, some new groups doing intensive development based on the scientific results.[32]

In expressing doubt about the quality of his ideas, Bush was being coy; he saw himself as the missing "link" in new weapons preparation. For the first time since arriving in Washington, he held his future in mind. The details were fuzzy but he knew he wanted to run a research organization, staffed by civilians and independent of the military. Dedicated to defense matters, the agency should cut across the lines dividing the Army and Navy and should have sufficient status to persuade the services to put proposed innovations to use. This was a remarkable admission. Few civilians short of the president had ever imagined so grand a role in military affairs.

But war had changed and so had the elements of a nation's security. The future of the U.S. now depended on an unprecedented show of civilian interest in the machinery of war. Neither military tradition nor political ignorance should rob the country of the chance to apply every bit of American know-how to war. By early 1940, Bush believed America was "sure to get into" a European war "sooner or later." Why not use the most advanced weapons? The nation really had no choice, Bush thought, since the enemy had already demonstrated that the coming war "would be a highly technical struggle." America was unprepared. "The military system as it existed," Bush later wrote, describing the moment, ". . . would never fully produce the new instrumentalities which we would certainly need, and which were possible because of the state of science as it then stood." The officers of the Army and Navy, Bush insisted, were incapable of handling their own affairs in the new age of technical warfare. Without the aid of civilian researchers, working at the highest level of government, the military would never secure the peace.[33]

Henry "Hap" Arnold was the exception who proved the rule. The chief of the Army Air Corps welcomed the help of scientists and engineers. Arnold was perhaps the first senior military man to do so. His unique military background gave him an appreciation for research. At 54, he had been one of the U.S. Army's first two pilots. Orville and Wilbur Wright personally taught Arnold to pilot one of their planes. They gave Arnold "a sense that nothing is impossible." From the Wrights, Arnold learned to revere machines. When bad weather made flying impossible, he studied the construction of his plane. Flight was an imperfect technology; knowledge helped in unexpected ways. Once in the air, a pilot's fate was linked to his equipment. His know-how might prevent an accident and save his life. In 1912, Arnold was reminded of this when his Wright Flyer spun out of control. Only seconds before a crash, he regained control of his plane. It was four years before Arnold flew again.

But Arnold's love of flying won out, and by the 1930s he was among a cadre of veteran military aviators. In 1934, he led ten bombers on a flight from Washington, D.C., to Juneau. When all ten planes arrived intact, Arnold won wide acclaim (no one had ever flown nonstop to the Alaska Territory). At the same time, the vagaries of technology humbled Arnold. He realized the gap between promise and reality could be wide indeed. Earlier the same year, Roosevelt had abruptly canceled the government's airmail contracts with private carriers, giving the job to the Air Corps. Arnold was asked in February to manage airmail traffic in the West. The assignment exposed the biggest vulnerability of American air power: inexperience. Of the 262 pilots who flew the mail, only 122 had flown for more than two years; only 48 had flown for at least 25 hours in bad weather. Lindbergh predicted a disaster. He was right. By June, 12 pilots had died in 66 crashes. The Air Corps was exposed as a shaky service, though Arnold was spared censure because his group, while flying over taxing mountainous areas, suffered the fewest casualties.[34]

When Roosevelt selected Arnold as air chief in November 1938, the Air Corps consisted of less than 20,000 men and a few hundred planes, most of them outmoded. One general described Arnold's outfit as a "fifth-rate air force." This was about to change. Arnold's promotion came just as Roosevelt called for a vast expansion in the nation's air arm. Almost immediately, Arnold was consumed with "building air fields and schools, developing and ordering planes, worrying about the 'bugs' in new models, [and] pleading with Congress for more money." In 1939, Congress responded with a $300-million appropriation for the Air Corps, which enabled it to increase its personnel to 26,000 and expand its force to 800 planes. Arnold's outfit, however, still badly trailed Britain's Royal Air Force, with 1,900 planes and 100,000 men, and Germany's Luftwaffe, with 4,100 planes and 500,000 men.[35]

Simply catching up with the European powers was a monumental task. To his credit, Arnold did not try to make the job easier by standardizing on older aircraft, which could be made in greater volumes but were outclassed by foreign craft. Instead, Arnold kept an open mind about improvements and looked to researchers for ideas, indulging if not wooing them in a fashion uncharacteristic of the military's top brass. Arnold even tried to convince his superior, General George Marshall, the Army's chief of staff, to take research into account. Inviting Marshall to a luncheon with several leading scientists, Arnold paid close attention to the conversation. Afterward, Marshall, who was not easily impressed by new weapons, bluntly asked Arnold what he was doing with these people. "Using them," Arnold replied. "Using their brains to help us develop gadgets and devices for our planes—gadgets and devices that are far too difficult for the Air Corps engineers to develop themselves."[36]

Bush did not mind Arnold's predatory attitude toward scientists. At least Arnold listened. His relationship with Bush was stiff but substantial. Arnold fully supported Bush's bid to revitalize NACA. He liked Bush's decisiveness and favored him as the replacement for Ames, NACA's ancient chairman. He was persuaded by Bush's arguments that NACA would benefit from closer ties with industry. Nor did he discourage Bush from bypassing NACA on certain questions and arranging advice for the Air Corps from others. He agreed that NACA had grown stodgy with age and that in certain areas others might break new ground more quickly.

Arnold's loyalty to the Air Corps did not blind him to the need to compromise with civilians. This set Arnold apart from other officers. Consider his position on standardizing landing instruments. The Army used one system, while civilian aircraft used another. The different approaches were difficult for pilots. When a National Academy of Sciences committee, chaired by Bush, concluded in November 1939 that airplane landing instruments should be designed around a single standard—one different from the Army's—Arnold replied that he found "no fault" with the proposal and would meet the civilians halfway. "We still, of course, believe that the system which we developed is better suited to meet Air Corps needs," he wrote, "but we realize fully that it is an impossible situation to have different aeronautical agencies each develop their own systems, no two of which any one airplane could use. It is absolutely essential that one common system be in use by the whole aviation industry and we are perfectly willing to give and take and compromise in order to arrive at that universal system."

Bush's relations with Arnold gave him confidence that military men would accommodate scientists. But would scientists accept military rules? It was crucial for Bush to win over the National Academy of Sciences, the embodiment of the scientific establishment, whose leading members were suspicious of the government. Splits in the Academy had plagued past attempts to expand the federal government's support of reseach, as the failure of the Science Advisory Board well illustrated. At the very least, Bush had to convince the Academy to stay out of his way. In this he had a head start. Frank Jewett had been elected to the presidency of the Academy in 1939, the first industrial researcher ever to hold the office. Jewett was influential in bringing Bush to Washington as Carnegie's chief, and Bush returned the favor by supporting Jewett's bid for the Academy post.

Jewett endorsed Bush's ambitions for mobilizing science even at the risk that the Academy might lose clout in the process. He was aware of the Academy's limitations too. The most obvious was financial. Military research required great sums of money from the government, but as a practical matter the Academy could not accept such funds because it was bound to cover its own administrative, or overhead, costs. A similar problem hamstrung the National Research Council. The NRC's Division of Engineering and Industrial Research, under Bush's leadership, had been reorganized on the presumption that it would be the Academy's agent in the sponsorship of defense research. The division, in what Bush considered its "most important matter," already was studying the best way to obtain "military aircraft expeditiously in large quantities." But Bush, who had spent three and a half years revamping the engineering division, resigned as chairman at the end of 1939. He no longer saw the division as a means to mobilize civilian research.[37]

Financial constraints aside, Jewett sought to avoid dragging the Academy into the kinds of jursidictional and political disputes certain to bedevil any emergency organization. Turf battles with the military and the executive branch would destroy the Academy by turning it into "just another agency of government." The Academy, he stated, was "in the position of a doctor waiting for clients; it could not adopt the attitude of an aggressive salesman and initiate attacks on what it regarded to be important military problems." But such messy sales pitches were needed. The military was uninformed about which technologies would matter in future wars. Arnold and a vanguard of open-minded officers wanted advice and were troubled by the haphazard means by which they heard from outsiders. The Academy, however, offered few answers. Justifiably resentful over the government's past indifference, members were in no hurry to respond to pleas for help in early 1940. Most preoccupied themselves with university affairs; recalling World War I, they conceived of military mobilization as a leisurely and half-hearted affair. No one expected that the government would soon bankroll the most diverse and gigantic research endeavor in history.[38]

The Academy's sluggish response to a request from the Army in February 1940 revealed these attitudes. At Bush's urging, the Air Corps sought advice on defensive measures from an Academy committee chaired by a professor from the California Institute of Technology, probably the nation's top university for aviation research. A month later, the Army expressed displeasure with the committee. So did Bush. After attending a few meetings of the committee, he wrote Jewett that the group was "a long ways off" from helping. In late March, Bush traveled with Phoebe by train to visit NACA's new lab south of San Francisco and aircraft manufacturers in the Los Angeles area. He also wanted to smooth any hurt feelings at Caltech, which felt it deserved more federal support for its aviation work. "At Pasadena, I got Millikan [Caltech's president] to go with me to see numerous aircraft manufacturers," Bush wrote on April 11 from the Huntington Hotel. "I did so on purpose, so that it would be evident there was no break between NACA & CIT."[39]

On his return to Washington a few days later, Bush was "aghast to find that apparently nothing whatever has been done" to light a fire under the Academy committee. "This seems to me especially unfortunate," he wrote Jewett, "as the Army is now considering whether to call on the Academy for more general advice in connection with national defense matters, and I feel personally that there is a distinct opportunity for service in this connection. I cannot make out what has held the thing up." Bush's frustration with the Academy may have been aimed at convincing Jewett that his outfit should actually take a back seat in war research. The aviation committee's poor performance on a matter so dear to Bush and Arnold indeed augured poorly for any Academy role in military research.[40]

Jewett was only one of several leading science administrators whose backing Bush needed. James B. Conant, the president of Harvard and an outspoken internationalist, was growing ever more appreciative of Bush's ability to handle tough situations. Rather than wrestle with the military behind the scenes, Conant preferred the more visible role of extolling the virtues of an Anglo-American alliance to a wary public.[41] Karl Compton, MIT's president, was hugely influential and a logical choice to head any federal research agency. But Compton's relations with Roosevelt were strained, and his knowledge of the military was limited. Compton, meanwhile, admired Bush and would follow his lead. Also, Bush's ascendance figured to give MIT an inside line on gaining government funds.

Bush also kept Alfred Loomis apprised of his activities. The link to Loomis, a retired investment banker who had a physics laboratory at his home in Tuxedo Park, New York, was especially important. Loomis was a generous patron of leading physicists such as Ernest O. Lawrence; he was also a cousin of Henry Stimson, a member in two Republican administrations and soon to be named by Roosevelt as secretary of war. Stimson, by profession a Wall Street lawyer, was enthusiastic about technology and mystified by the military's hidebound attitudes. Loomis had Stimson's ear; he was in a position to put in a good word for Bush, whom he held in enormously high regard. "Of the men whose death in the summer of 1940 would have been the greatest calamity for America," Loomis believed, "the President is first, and Dr. Bush would be second or third."[42]

With the support of these men, Bush could create a new military research organization and at the same time deflect charges that he was usurping the power of the official scientific leadership. His authority ultimately would rest on his close relations with the military. The current crisis demanded more than a research administrator, and Bush had demonstrated a greater adeptness than his peers at sympathetically weighing the military's technical desires against its needs. Only a new organization, free of the legacy of mistrust that had plagued past collaborations between reseachers and the military, could satisfy the needs of the moment. Only Bush had a neck stiff enough to run it.

Of course, Bush did not need anyone's permission to agitate for a new research organization. By May 1940, he was prepared to seek Roosevelt's support for his plan. He asked John Victory, who had taught him the political ropes at NACA, to draft legislative language calling for the creation of a committee to "coordinate, supervise and conduct scientific research on the problems underlying the development, production, and use of mechanisms and devices of warfare, except scientific research on problems of flight." He named his brainchild the National Defense Research Committee.[43]

Reaching for an explanation of Bush's sudden rise, later observers cited his participation in the Committee on Scientific Aids to Learning, whose members included Jewett and Conant. James Phinney Baxter, the official historian of civilian research during World War II, said the committee was pivotal for Bush. Hunter Dupree, another historian, called the committee "an invisible college" from which sprang "a new partnership" between government and science.[44]

Dupree's assessment gives the committee too much credit. More coffee klatch than college, the committee met infrequently, kept spotty records and seemed increasingly irrelevant as war approached and private industry's interest in the machinery of learning grew. As many as five months could pass between meetings. Minutes of these sessions contain nary a reference to either world affairs or the military mobilization of scientists. "I don't think we ever accomplished much," Bush sourly insisted.[45]

Talks on these subjects did occur informally among the committee members. Conant, Bush, and New York lawyer Bethuel Webster, another member, attested to that. Webster even called the committee "a useful but tiny pilot plant" for the coming war work. But the committee members never moved beyond small talk; they never plotted any strategies. At best, they provided Bush with a patina of consensus with which to mask his personal agenda.

This was never clearer than during the final meeting of the Committee on Scientific Aids, held over a "small informal luncheon" at the Century Association in New York on May 24, 1940. By then, Bush had decided to contact the president on his own. His lunchmates seemed unaware of this. Perhaps Bush was using secrecy to shield himself against the embarrassment of failure. More likely, he simply felt no need to involve the others, choosing instead to play his hand alone. "The meeting that May day was significant," recalled one in attendance. "Aids to learning had been forgotten. The brief noonday discussion was a challenge to action."

But little more. During the discussion Bush tossed out only "a vague proposal" on "how American scientists could assist in the accelerated defense effort," Conant recalled in his memoir. If Bush's ideas sounded vague, it was probably because he was reluctant to set his plans in stone: as he later explained, he was pursuing his goal from "several angles" and was ready to revamp his plans if need be. "We were all convinced that a move was essential and we were scouting around for ways in which such a move could be put into effect," Bush wrote 15 years later. But the effort "was all very informal and very vague," he added, and there were "various false starts." However, the prize

Bush sought was plain. He wanted to sponsor promising research that the military, for whatever reason, would not investigate itself.[46]

None of Bush's colleagues was in a better position than he was to make a strong case to Roosevelt. What drew these men together, Bush noted, was "one thing we deeply shared—worry." But anxiety didn't necessarily breed clear thinking. "We were all convinced activity should start at once," Bush later recalled, but of the people wanting action only he saw the bottom line. "The big problem was how to get control of a significant amount of money," he realized. Only Bush was ready to seek government funds and lead a wholly new organization for defense research. Only Bush was prepared to present himself as the nation's wisest man on matters of military technology.[47]

Finally, and perhaps most important, Bush alone had the gumption to seek the president's endorsement to run his own show. "I was located in Washington, I knew government and I knew the ropes," he later recalled. "And I could see that the United States was asleep on the technical end."[48]

Franklin D. Roosevelt was the linchpin in Bush's campaign to win a decisive role for himself in forming American defense policies. "I knew that you couldn't get anything done . . . unless you organized under the wing of the President."[49]

In early May 1940 Bush asked Frederic Delano, Roosevelt's uncle, to arrange an appointment with the White House. Bush's choice of emissary was a shrewd one. Delano, then 77, was a man of unusually wide interests, an inspired amateur with an outsized confidence in the capacity of experts to shape society to desired ends. A successful railroad executive, he turned to civic causes in the 1920s, quickly earning a reputation as a pioneer in city planning. He was credited with almost singlehandedly halting Washington's haphazard sprawl in favor of what the *Washington Star* described as "an orderly planned expansion that set the pattern for future growth." An early advocate of urban parks and gardens, Delano kept diverse company. He mixed with scientists, architects and businessmen, while finding time to investigate the opium trade in Persia and the Mayan ruins in the Yucatan Peninsula. Known as the "first citizen" of the nation's capital, he was respected and admired to a degree that could hardly be explained by his blood tie to the president.

In the 1930s, Delano emerged as a key champion of New Deal technocrats, chairing the president's National Resources Planning Board, an independent agency whose staff of 50 published what one observer later described as "pioneering reports" on water pollution, mineral resources, public works and state planning. Under Delano, the committee broadly interpreted "resources," gave sometimes "unorthodox" advice to other government agencies and helped form state and municipal planning agencies around the country. Delano's wide-ranging agenda, which included studies of purchasing power and demographic trends, left him open to attack from government critics who equated planning with socialism. One Washington newspaper went so far as to condemn the planning board as advancing "the extravagant shell-pink and dreamy visions of the Delanos."[50]

Delano did not let Bush down. On May 25, he asked for a meeting between Bush and the president, writing to aide Steve Early that the topic to be discussed was "a matter of really great importance" and that after reading a one-page memo describing Bush's plan, "I wish you would let the Pres. see it and later see Dr. Bush."

The Bush memo made the case for a new agency devoted to military research. The National Defense Research Committee, as he called it, "could perform a very valuable function indeed in stimulating, extending and correlating fundamental research which is basic to modern warfare." It also "should supplement, and not replace, the activities of the military services themselves," while relying on the National Academy of Sciences "for broad scientific advice and guidance." Thus construed, the agency might avoid sparking bureaucratic jealousies and "be welcomed, and hence supported by" the War and Navy departments.

Delano's pitch was well timed; "Pa" Watson, a senior Roosevelt aide, almost immediately spoke to the president about Bush's interests and agreed to meet with the Carnegie president "as soon as it can be conveniently arranged."[51]

The Nazis stood poised to overwhelm Europe. On May 10, Hitler's armies invaded the Netherlands, Luxembourg and Belgium. The same day, Winston Churchill replaced Chamberlain as Britain's prime minister. Five days later, Churchill sent the first of many telegrams to Roosevelt, calling for America's entry into the war. Roosevelt, seeking to mollify vocal isolationists even as he cozied up to the British, demurred. On May 25, as the Nazis rolled toward Paris, the president asked Congress for more defense money and an Office of Emergency Management, which would improve military production. The next day, British and French troops began massing at Dunkirk, a coastal town in France, for evacuation to England.

Germany's blitzkrieg at once alarmed and relieved American internationalists. On the one hand, the collapse of Britain was now a frightful possibility. At the same time, the routing of France was a bloody illustration that the Nazis' advance could only be met by force. If America would not enter the war, it must at least do its all to aid the British. That meant producing arms in record amounts. But Bush worried that in the pell-mell rush to work harder his countrymen might forget that they must also work smarter. Addressing the National Aviation Forum on May 29, he declared that the nation's security depended on both the quality and quantity of its research. "For war or for peace," he said, "we must leave no stones unturned in research."

Still in the thrall of air power, he added:

The events of the past few days and weeks have yielded every indication that air power may be the controlling factor in modern war. Strong armies and strong navies are still of primary importance in warfare, but it seems obvious that the nation that can control the air will be dominant.

The fate of a nation, therefore, may depend on its possession of airplanes of superior performance. Superior performance can only be achieved by aircraft designers who are absolutely up to date on their technical information. It is the "know how" that counts, and "know how" in any field, and particularly in aeronautics, is acquired only by constant research.

That there should be some direct relationship between military success and the *quality* of research in aeronautics is fairly obvious. Accurate information and a correct interpretation of research results are necessary. But *quantity* of research is also highly important. A large capacity to do research work is as essential as a large capacity to produce aircraft. It would be foolish to create a bottle-neck for quality production because of lack of ability to obtain research results quickly. Under wartime pressures much may depend upon the ability of airplane designers to get correct answers to new problems quickly.

Adding it all together, it is safe to say that the course of history is largely being influenced by scientific investigations in aeronautical research laboratories. The outcome of the aerial battles of tomorrow are being decided today by the men who are working in wind tunnels, towing tanks, and engine laboratories.[52]

Bush's public fixation with air power was understandable, given his position as NACA chairman. But it was reasonable to assume that vigorous aeronautical research could enhance other military fields. In public, Bush would hardly risk drawing this conclusion himself. Ever careful of embarrassing either the armed services or the administration, he kept to himself his complaints about research organization. Besides, it was by no means certain that the British would hold out long enough for American know-how and productivity to help. On the same day Bush gave his aviation address, Conant delivered a far more memorable speech on radio. "Tomorrow looms before us like a menacing question mark," he said. "A total victory for German arms is now well within the range of possibility."

Delano's request landed Bush a meeting in early June with Harry Hopkins, the president's closest aide. A former social worker, Hopkins was one of the "long-haired idealists or do-gooders" that Bush often derided. An odd pair, Hopkins and Bush clicked. Bush thought they "hit it off well." Hopkins "was immediately impressed with Bush's proposal and Bush himself," Robert Sherwood, a Hopkins intimate, later wrote. "There were certain points of resemblance between the two men," he added. Like Hopkins, "Bush was also thin, quick, sharp and untrammeled in his thinking. He knew what he was talking about and he stated it with brevity and, like Hopkins, a good sprinkling of salt." While Bush counted his sudden alliance with Hopkins "among the minor miracles" he had witnessed since his move to Washington, he recognized that after "various false starts and tentative plans" this was his "break."

Bush had brought Hopkins a short memo describing his plan for a coordinating committee, responsible to the president, that would contract with universities and industrial labs to perform research at the behest of the Army and the Navy. Hopkins immediately realized that Bush, in the words of a later observer, "was tailor-made for exploiting the new atmosphere and for mobilizing and employing the technical comunity for war." There was no time to waste. Hopkins arranged for Bush to see the president.[53]

"The situation at the present time is decidedly chaotic," Bush wrote on June 5. Germany had just invaded France, highlighting anew the need for improved American defenses (Paris would fall nine days later and all of France by June 22). Amid the growing crisis, Bush found the lack of coordination between researchers and the military almost paralyzing. "I have some hope that within a few weeks," he wrote, "there may be steps taken which will bring order out of the present chaos on this particular matter of enlisting the service of scientists in national defense matters. On the other hand, I would not be at all surprised if moves were made which simply rendered the situation worse." Wary of working with government, Bush wondered if he was up for the larger task of mobilizing the research community. Most of his "present time," he wrote on June 6, "is devoted to preventing [NDRC] from getting set up in an ineffective fashion."[54]

Roosevelt was eager to hear how the country could rapidly improve its defenses. Preparing for an unprecedented third presidential campaign, Roosevelt had publicly pledged to shield American boys from the widening European war. But the president sympathized with Britain's plight and saw a showdown with Hitler as probably inevitable. Afraid of alienating isolationists in Congress and the broader public, Roosevelt wished to boost the nation's military power and aid the Allies while formally maintaining neutrality. He was mainly concerned with production of existing weapons, not the invention of new ones. This was understandable. He knew little about technology or science and had no staff member or standing committee to advise him on these subjects. Delano did more than anyone to keep Roosevelt in touch with engineers and scientists, but he did not do much.

Still, the president was a gifted improviser, a talent compatible with the scientist's method. He was also committed to breaking the mold in a crisis. Roosevelt, one historian noted, "wanted an inventive government rather than an orderly government . . . not a team of reliable work horses, but a miscellany of high-spirited and sensitive thoroughbreds," even if this meant he must spend time "handholding" his "prima donnas."[55]

On June 12, at about 4:30 P.M., Bush—a high-spirited thoroughbred and no doubt a prima donna—met Roosevelt for the first time. Joined by Hopkins, he greeted the president, then pulled out a single sheet of paper that contained a crisp description of his plan for mobilizing military technology. Under the title, "National Defense Research Committee," the document listed six items:

1. [The NDRC would be] attached to National Defense Commission.

2. Composed of chairman, members from War, Navy, Commerce, National Academy of Sciences, plus several distinguished scientists or engineers, all to serve without remuneration.

3. Function, to correlate and support scientific research on mechanisms and devices of warfare (except in field covered by N.A.C.A.). Concerned with research rather than industrial development or manufacture.

4. Supplied with funds for office staff, and for financing research in laboratories of educational and scientific institutions or industry.

5. To aid and supplement, and not to replace, activies of War and Navy departments.

6. An Army and Navy officer detailed to work with chairman.

Bush steeled himself to answer tough questions from the president, but Roosevelt had already made up his mind. After a few casual comments, he wrote "O.K.—FDR" on the single sheet of paper. Bush was elated. He had his coveted endorsement. And it had come in less than 15 minutes.[56]

Roosevelt's brevity belied the bond forged between these two men—a bond that in the years ahead would shape the nation in ways neither could anticipate. In minutes, the president had promised Bush a direct line to the White House, virtual immunity from congressional oversight and his own line of funds. The seeds had been sown for their extraordinary relationship.

Four years before, a quick understanding between the president and so harsh a New Deal critic as Bush would have been unthinkable. Now, grappling with a widening war, Roosevelt embraced many enemies of the New Deal, drawing increasing numbers of them into his government, especially figures from industry. Bush had once described "the whole New Deal philosophy" as an "urge for a great bureaucratic organization controlling everything in sight." An autocratic academic who looked down his nose at populist politics, Bush was nonetheless temperamentally well suited to be a Roosevelt insider. In a time of crisis, the president required men such as Bush who could improvise and take bold actions. As Daniel J. Kevles has written, Bush was "prepared to question orthodoxies, defy tradition, master offbeat skills and learn how to deal with people not normally found in his own professional circles." He also was far less concerned than many conservatives about blurring the lines between the public and private spheres. His exhortations to fellow engineers and scientists reflected his belief that "a competent state in a technological age would be required to draw regularly on independent, private experts."[57]

The price of enlisting Bush was right, too. Like many men drawn into the war effort by Roosevelt, Bush would earn no government salary, thus retaining a sense that his opinions and advice could not be bought. He planned to keep his well-paid post at the Carnegie Institution, never mind that it figured to be among the recipients of NDRC grants. Grants under Bush's control also would flow to MIT, his former employer. Nearly every funding decision by Bush might benefit one of his professional friends. Yet Roosevelt showed no concern over what a later age might deem a conflict of interest. To the president, Bush's relationships were invaluable; he trusted that whatever conflicts they posed would be handled with propriety and intelligence.[58]

In Washington, having money and the president's ear were crucial. Yet Bush dared not get carried away with himself. Roosevelt often gave a new aide great authority only to snatch it back fairly soon, and he "habitually set his appointees upon collision courses with one another." He handled matters of great import casually, was bored by details and wished to make only momentous decisions. But at the same time, he was "reluctant to delegate power to those who could relieve him of tedious details. And so issues remained unsettled until they became more troublesome, more expensive, and finally had to be dealt with, usually hurriedly." This crisis mentality especially hampered the armed services, which typically resisted new practices, even practices backed by Roosevelt. Trying "to change anything in the Navy," the president noted in 1940, "is like punching a featherbed." Delighting in giving his aides overlapping assignments, Roosevelt feared making too firm a "commitment to any one adviser or faction." He once told a cabinet member that "a little rivalry is stimulating, you know." Bush knew enough about the president to realize that his charge to mobilize civilian researchers would mean little if he did not show quick results.[59]

At a press conference on June 14, Roosevelt announced Bush's appointment as chairman of the new NDRC. The president did not name the NDRC's seven other members, saying he wished to confer with Bush on the selections. A reporter asked if Bush would keep his job at NACA. "I suppose so, yes," was Roosevelt's desultory reply. Talk quickly shifted to the European war and an appeal from the premier of beleaguered France.[60]

Bush didn't mind the short shrift. He preferred anonymity to the spotlight and believed the best way to accomplish things in Washington was to allow others to receive credit, even for his own achievements. Indeed, one of his first acts as NDRC chairman was to remind his senior aides that they would accomplish the most by drawing little attention to their work.

At a private dinner celebrating his appointment, Bush regaled his science pals with his plans for the future. He held court until past midnight at his favorite watering hole, the Cosmos Club, a Lafayette Square hangout for the science elite. When the group finally broke up, an aeronautics researcher who advised the air forces took Bush aside for a private talk. "Van, you talk too goddamn much," the researcher said. "In your new position you can't talk, you listen. You don't expose your thoughts because people will start quoting, Bush said this or Bush said that."

Bush grew quiet and in his gruff voice told the engineer that he was absolutely right.[61]

Bush quickly assembled his team, winning pledges over the telephone from Compton, Jewett and Conant to serve as his principal aides. Conant eagerly accepted after asking only two questions: was the NDRC real and was Bush its chief? The Harvard president saw "a certain symmetry" in these selections. Compton, a physicist, would carry along MIT, while Conant, a chemist, would enlist Harvard's aid. Together, these two massive institutions covered the intellectual waterfront. Jewett, meanwhile, personified AT&T's enormous industrial might. As chief of Bell Labs, he held probably the most prestigious corporate research post in America, and he had strong ties to the engineering community. To Jewett "fell the lot of calling on many engineers in industry and in the universities."

Richard Tolman, a Caltech physicist who had just moved to Washington, also agreed to join the NDRC, adding a much-needed link to the premier technical school in the West. The government's own representatives were patent commissioner Coe; Rear Admiral Harold G. Bowen, the Navy's chief research officer; and Army brigadier general George V. Strong. Recalling the initial composition of NDRC's leadership, Bush noted that the selections were his alone and that the White House had no real influence in the matter: "I simply got the consent of the men who became members, suggested the list to the President, and the appointments were made. There was no question whatever in anyone's mind as to how the initial group should be constituted for it was made up of the very group that had been working together to get something started."[62]

The NDRC faced two immediate problems: "Find out what research the Army and Navy wished it to undertake and then to place the contracts for research in the best possible hands."[63] Tackling the first item required gaining the confidence of the Army and Navy. Roosevelt helped by explaining to both service secretaries that the aim of NDRC was to supplement the military's own research efforts, not to replace them. Carrying the same message, Bush visited Army chief of staff George Marshall and Admiral Harold Stark, Marshall's Navy counterpart. The Army responded first. On June 17, Bush traveled to Wright Field with Arnold, who wanted NDRC to take over some aviation studies. Three days later, the Army gave Bush a list of its research projects; the Navy did the same within days.

Bush also concentrated on attracting researchers. Conant suggested that NDRC would have to build laboratories and staff them with federal employees. "Not at all," Bush replied. "We will write contracts with universities, research institutes and industrial laboratories." Indeed, the contract would be the key instrument of NDRC's bureaucratic arsenal. It was flexible enough to account for the reality that research does not always work out as planned. "The contractor agreed to conduct studies and experimental investigations in connection with a given problem and make a final report . . . by a specified date."[64]

To Conant, the approach marked a watershed in the government's relations with the scientific community. "I shall never forget my surprise at hearing about this revolutionary scheme," he later wrote. "Scientists were to be mobilized for the defense effort in their own laboratories. A man who we of the committee thought could do a job was going to be asked to be the chief investigator; he would assemble a staff in his own laboratory if possible; he would make progress reports to our committee through a small organization of part-time advisers and full-time staff."[65]

The NDRC's methods profoundly altered the relations between government and those who acted on its behalf. Later called "federalism by contract," the system allowed researchers to remain in their familiar surroundings and freed them from government bureacracy (and, some charged, a degree of accountability too). The contract system made it vastly easier for Bush to assemble a national network of the best researchers. The contract itself, more-over, lent a measure of dignity to the academics working on military projects by creating the impression that the government was a client being served by a consultant at the latter's pleasure. Though the image was partly illusory, Bush considered it essential. "Contracts are made only between independent bod-ies or individuals," he once explained. "Contracts carry responsibility but not subservience."[66]

"Federalism by contract" was not entirely new, of course. For decades, the government had contracted for all sorts of research. NACA, the aviation com-mittee, contracted with some of the very universities Bush now hoped to rely on. What set Bush's new agency apart from past efforts was its potential scale. The government now had a means of drawing on the nation's scientific elite, without challenging the institutional loyalties of the researchers themselves. Even in the course of a single war, this structure might affect the very character of scientific inquiry—and the resulting technical knowledge. Favored fields might expand, while those ignored by the government could wither.[67]

"Federalism by contract" also contained the seeds of a shift in American governance. Decisions were increasingly being made by experts whose author-ity lay beyond the reach of the electorate and was not limited by the rules and habits of civil servants or the military's chain of command. This growing re-liance on essentially freelance "brains" carried obvious benefits but also risks for professionals and the public. With the passing of an emergency, govern-ment might withdraw its support, leaving researchers in the lurch. Science funding had never before been subject to boom and bust cycles, but rather subsisted on a relatively thin but steady diet of foundation grants and gifts to universities. Industrial spending on research was more volatile but it mainly went toward product development. Industry could not be counted on to en-large the "basic" knowledge upon which, many believed, "applied" research rested. Having enjoyed public funding, would scientists ever be content with the insecure standard of the past?

Finally, the NDRC's methods were likely to politicize the research commu-nity in a new way. Bush and his fellow committee members now possessed huge power over which researchers obtained government funds and which did not, which technical mysteries were unraveled and which were ignored. In-evitably, questions would arise over how such a small group of men, whose in-terests were interrelated if not coincident, had obtained such wide authority and why they, rather than others, should decide the manner in which public funds were spent. Bush himself later confessed that the creation of NDRC was "an end run, a grab by which a small company of scientists and engineers, act-ing outside established channels, got hold of the authority and money for the program of developing new weapons." Such a power grab was necessary in order to launch "a broad program . . . on an adequate scale," Bush insisted, but he conceded that it stoked resentment against him.[68]

Bush was too busy to give these political issues more than a cursory glance. The NDRC's reach quickly exceeded even Bush's ambitions for it. In early June, Bush had surmised his committee would spend $5 million over the next twelve months. A month later, he asked the Bureau of the Budget for $10 million and received two-thirds of his request. The press of business robbed Bush of his usual summer vacation. Accustomed to an academic schedule, he ordinarily shuttled between his farm in Jaffrey, New Hampshire, and the South Dennis house. This summer, Washington held Bush captive, leaving Phoebe alone on Cape Cod. Several times, Bush nixed weekend visits, apologizing to his wife.

In late July, Bush faltered under the weight of launching NDRC and Washington's summer heat. He was rushed to a hospital on the advice of physicians at Johns Hopkins, who thought he had been stricken by a bacterial infection. It turned out to be dysentery, a painful stomach disorder accompanied by intense diarrhea. Bush told a friend: "It is . . . merely a bit of a nuisance."

After a brief hospital stay, Bush wrote to Hopkins, Roosevelt's aide, to let him know "I am again back on the job." He was not about to allow dysentery to rob him of the chance to reach the pinnacle of power.[69]

Chapter 6

"Don't let the bastards get you down"

(1940–41)

If we had been on our toes in war technology ten years ago, we would probably not have had this damn war.

—Vannevar Bush

Bush at times protested too much the inevitability of war. Even though he had never fought in combat, war wasn't an abstract matter to him. He tried to remember—and remind other men in laboratories—that winning in war meant killing. A lot of killing.

To drive home this dreadful point, Bush devised a mental exercise for the scientists and engineers who worked for his National Defense Research Committee. Speaking with his Yankee twang and peering through thick glasses, Bush shattered the expectations of these newcomers who saw in his slight frame and professorial appearance no connection with the smoke and din of battle. As one writer observed, Bush was no "absent-minded, dreamy-looking professor. His face is gaunt, his blue eyes are keen and an obstinate lock of coarse straight hair shoots forward above a comparatively low forehead. His movements are quick and angular, his speech terse and salty. One could almost picture him with a birch rod in hand presiding over the pupils in an old red schoolhouse. He might serve as a model for Ichabod Crane."[1]

Startled by his appearance, NDRC novices would be bowled over when Bush presented the following scenario to them:

118

"You are about to the land at dead of night in a rubber raft on a German-held coast. Your mission is to destroy a vital enemy wireless installation that is defended by armed guards, dogs and searchlights. You can have with you any one weapon you can imagine. Describe that weapon."[2]

The question was often a stumper, but then Bush didn't expect it to inspire new lethal instruments. Concerned that his volunteers were too wedded to theory, he sought to quickly introduce them to the messy task at hand. An academic with an appetite for action, Bush dwelled in the peculiar juncture between science and business, the practical and the possible. "I want to be on the firing line and not in a cloister somewhere," he insisted.[3]

Certainly after a life in the relative safety of the laboratory and the administrator's office, Bush found his brush with the military exciting. But the experience also disappointed him. Toiling behind closed doors, the military researcher was cut off from his usual sources of intellectual sustenance, but Bush confessed a growing appreciation for the more primitive rewards available to him now. In Carnegie's annual report, he wrote that "anonymity and isolation" had replaced "the public appreciation and the open scientific fellowship to which" many researchers were accustomed. Along with this came the unappealing substitution of "a narrower national aim" for "an altruistic ideal embracing the whole of mankind." This loss of purity, however, was more than compensated for by what Bush described as "that primal joy that comes from intense group effort in defense of his home." While no researcher was literally engaged in mortal combat, Bush insisted that he could rightfully feel "as though he stood at the mouth of a cave with a few strong men of the clan armed with stone axes against a hostile world."[4]

Bush was hardly working to arm soldiers with stone axes, but he was correct in seeing the world as increasingly hostile. By December 1940, Germany was blitzing England by air and preparing to use its submarines to seal off the island by sea. A Nazi victory seemed inevitable unless the U.S. aided the British and perhaps entered the war itself.

The crisis was so dire that Bush had neither the time nor the desire to move into government quarters. By default, the Carnegie Institution's ornate building at 1530 P Street became the home of NDRC. Eight blocks north of the White House, the building stood at the corner of 16th and P streets. Originally constructed in 1909 and substantially enlarged in 1938, the building was masonry faced with limestone and was best entered from its grand steps on 16th Street. The steps rose steeply from the tree-lined street and were flanked by two Italian marble vases. At the top of the steps was a portico, supported by four sets of Ionic columns. Across the top of the building front was a carved stone balustrade.

From the portico, massive bronze doors led into a spacious rotunda. The floor was marble with inlaid designs, and four pairs of large, Corinthian columns dominated the room. As NDRC's staff grew, the rotunda became home to dozens of makeshift offices built out of temporary partitions and platforms. The crush of workers spoiled the effect of the rotunda, which came to resemble "a rabbit warren. It was chaos."[5]

A driver usually collected Bush each morning from the Wardman Park Hotel and drove him to CIW. He officially presided over the agency from the balcony above the rotunda, where before launching the NDRC he worked in a large, dark-oak-paneled room with a bust of Andrew Carnegie behind his desk. Now Bush preferred instead to do business from Carnegie's more expansive boardroom, allowing his legal staff to inhabit his former office. The boardroom was imposing. The walls were mahogany-paneled and displayed paintings of the institution's past presidents. Large windows allowed generous light. A marble fireplace and a portrait of Andrew Carnegie occupied the east end of the room.

It was not just the boardroom's size and grandeur that appealed to Bush. The room's layout underscored his status. The only entrance was through an outer office, manned by his personal secretary, Samuel Calloway, who had the latitude to send in visitors if his boss was alone and the matter seemed urgent.

Bush leaned heavily on Calloway, who read—and sometimes even answered—his boss's mail. "There was no officiousness about Sam," one colleague recalled. "He was warm and friendly, and he pretty much knew what Bush was doing. You'd ask Sam, 'How am I doing today? Can I get in?'"[6]

Once cleared, a visitor walked through a door behind Calloway's desk, entering the boardroom at the far end from where Bush sat. It could take a dozen strides for a visitor to reach him, and many people felt a rising anxiety as they approached the NDRC's chief. Only when a guest was practically on top of Bush would he notice the small sign in Latin on Bush's desk. If he were especially sharp, he might even understand the Latin phrase that Bush took for a ribald motto. Translated into English vernacular, it meant, "Don't let the bastards get you down."

Puffing on a pipe (which rarely left his mouth), Bush usually could be found dressed in a dark suit and squirming in his chair or with feet on desk. He often saw people unannounced, using the few seconds granted him by the long distance from the door to his desk to assay the state of his visitor's mind. The senior members of his inner circle called him "Van," but younger men or those who knew him less well called him "Dr. Bush." If he was glad to see someone, it showed. "He had a grin on his face the whole time," said one junior aide, re-calling Bush's appearance when he walked into his office for the first time. "I liked him instantly. He was charming when he wanted to charm you."

At 50 years of age, Bush still cut an attractive figure. Formally attired and conscious of rank, he nevertheless made people around him seem special. "Every place I went, as a Bush man, I was welcome at the top," recalled the aide. Once a week, usually on a Wednesday afternoon, Bush served cookies, tea and coffee in his office. The gathering might last an hour and, later in the war, included the screening of films showing NDRC's weapons in battle. The "teas" were consider a morale booster and a useful way for staff members to share information informally.[7]

Younger staff revered him. One aide marveled at his ability to write and think so crisply, calling him "a prodigious producer of memos." His handwrit-ing was atrocious but unmistakable, and he usually signed dictated memos "V. Bush," so as not to draw attention to what he called his "screwball first name." The same clarity evident in Bush's writing was apparent in his speaking. He punctuated his meetings with precise declarations, which lent an aura of cer-tainty to his off-the-cuff decisions.

"He'd outline the situation up to the date, using his precisely accurate recol-lections. Then he'd say, What's new? What are our alternatives? What are the pros and cons of each? We'd kick those around. Then he'd say, Well it looks like we should do so and so. But if we do that what are the pros and cons?" From this winnowing process, a decision would arise with Bush winding up the discussion. "He'd say we'll do that, we'll move. He wouldn't dilly dally, but he'd have a full recognition of the facts and alternatives."[8]

If Bush had a serious flaw, his aides believed, it was his stubbornness. When he seemed off-base, they tried to suggest different approches to a problem without contradicting him. Bush could grow impatient with such challenges, delivering chilling comments, though never raising his voice. Instead his words rumbled out sounding like wheels on gravel. "It was a deadly voice," one listener said.

Bush shifted ground, but on his own terms. "He could change his point of view, if you didn't challenge him personally, if you didn't confront him," one aide said. This was tricky and in the end Bush might dig in his heels anyway. To some, this smacked of arrogance. "He could get pretty well dug in and give someone who wasn't going along with him a pretty hard time," said another aide. But this rigidity struck others as a reflection "not so much out of his stub-bornness but personal conviction."[9]

Bush's convictions were strongly held and they shaped the character of his wartime agency. A believer in delegating authority and responsibility, Bush never took an intimate role in the broad array of research sponsored by the NDRC. He saw himself as more akin to a coach, or a chairman of the board. "My function these days seems to be primarily to attempt to enable others to think and work, and this precludes doing very much consecutive thinking myself," he wrote.[10]

His "first principle of management," he often insisted, was "to get good men about. Get them in the key positions." And the younger the men, the better—provided they were talented and took the initiative. "The smart way to get rapid advance in an unconventional way is to give a group of sound youngsters their heads," he later wrote. "They must distinguish the really practical idea from the thousands of screwball proposals that always abound, but if they are a sound group they will."[11]

Bush's management style was especially valuable on the eve of the U.S. entry into World War II. As the nation's mobilizers soon realized, the war would be decided as much by superior administration of people and resources as by combat strategies and effectiveness. The demand for good administration put a premium on planning. This meant that technocrats—experts, paper pushers, leaders of substance, indeed men very much in the Bush mold—would play a huge role in the mobilization. "The Second World War was the only war in our history which we actually saw coming and undertook to get ready for," Bruce Catton wrote shortly after the war's end. "Eighteen months ahead of time the nation began to mobilize for war. The machinery creaked and groaned a good deal, there was much rushing around in all directions at once, and the job of preparation could have been done much better; nevertheless, the fact does remain that this was the one time in American history when something fairly effective was done about getting ready for an impending fight."[12]

By any measure, Bush had succeeded magnificently in assembling an effective mobilization team. Besides NDRC's distinguished senior leaders—Conant, Jewett and Compton—Bush had drawn a raft of top-flight younger men into the frenzy of assembling what they at the time dimly hoped would become the greatest industrial research organization in history. Among the brightest stars were physicists Lee DuBridge and Merle A. Tuve; Carroll L. Wilson, Bush's chief of staff; and Irvin Stewart, formerly a member of the Federal Communications Commission and the director of the Committee on Scientific Aids to Learning. Stewart now was the NDRC's executive secretary and oversaw its burgeoning administrative and contract-compliance activities.

All of these men had a direct line to Bush. Day to day, Wilson was probably the most crucial. He was Bush's "alter ego," one observer said; a tireless, intense worker who did anything his boss needed. Bush described Wilson as "my right-hand man on whom I depend in connection with practically everything I handle." Besides juggling "any number of things," Wilson had "a balance and a perspective" that Bush had not "seen exemplified in almost any other chap."

An MIT graduate, Wilson had served as Compton's personal assistant in the 1930s and then handled the licensing of the Institute's patent portfolio, which brought him into close association with patent enthusiast Bush. Preparations for war brought the two men closer still. Wilson was "a loner, a straight-arrow and completely loyal" to Bush, one colleague said. Serious yet smooth, Wilson had the ability to "quite often find ways around the personality logjams" that at times made cooperation difficult between Bush and non-technical people. Said a friend, "He had a great sense of dealing effectively with people at a very high level."[13]

Wilson epitomized the organization man. But Bush also relied heavily on prima donnas, especially for the running of his divisions, which actually ran NDRC's projects. He often complained about having, as one observer noted, "to ride herd on as temperamental a bunch of men as any in the world." Probably the key man most difficult to please was Tuve, an iconoclastic physicist who was in charge of designing proximity fuzes, radio-controlled detonators that promised to make bombs more deadly by exploding them at a point where the target would sustain maximum damage.[14]

Tuve was gung-ho about war mobilization. Raised in a small South Dakota town and a graduate of the University of Minnesota, Tuve was a fixture at the Carnegie Institution's Department of Terrestrial Magnetism. A longtime friend of E. O. Lawrence, Tuve set aside his own scientific research in early 1940 to rally support among researchers behind the scenes for Bush's planned NDRC. Unlike many midwesterners, he rejected isolationism, taking the threat of world domination by the Nazis quite seriously. "There were a couple of young postdoctoral fellows in the lab [who] weren't ready to quit," Tuve later remembered. "But I said, 'Let's not do any more research. . . . If those Nazis are going to inherit it, what's the use. I think we've got to find out how we can contribute to stopping this conflagration.'"[15]

Tuve's enthusiasm and creativity made him a huge asset, and the radio proximity fuze had enormous potential: by the end of World War II it was ranked as one of the three or four most important innovations of the war. Tuve insisted on working only directly for Bush, and to satisfy him Bush created an NDRC division in August 1940 solely to house Tuve's effort. Bush slyly named the new division "T." This sop to Tuve's ego didn't quiet his mercurial personality. Though making solid progress on the fuze, Tuve constantly clashed with others, especially with the Navy's Bureau of Ordnance, which badly wanted improved shells. "Tuve had very, very few diplomatic instincts," said one associate. "He made a lot of people mad and was under fire from various directions. That was embarrassing for Bush."[16]

Desperate to quell criticism of Tuve, Bush assigned a junior researcher, David Langmuir, to check out complaints against Tuve. Langmuir wasn't starry-eyed about Tuve, but after a few months with him Langmuir was smitten. "He's an absolute master, totally in charge of the situation and absolutely brilliant. He has so many good ideas his thing just can't fail," Langmuir told Wilson. Langmuir later repeated the gist of this to Bush himself. Looking "deeply relieved to get an independent opinion" of Tuve's fitness, Bush agreed that "of course he is a wonderful leader. [But] damn it, he almost drives me nuts."[17]

Bush might complain privately about his key men, but publicly he backed them. Whereas some war agencies were hampered by turnover, he crowed that he hadn't lost anyone. "One of the glories of Van was that he'd back his own staff to the limit, right or wrong," recalled Caryl Haskins, an aide and later Bush's successor as president of Carnegie. "You could be sure that if you made the sort of dumb mistakes that I made, you'd get defended. That's why the young people who came in felt sheltered by Van. They all did. He'd take care of his own. Part of the deal was that he'd defend whatever they'd done in public and then really give them a bawling out privately."[18]

Whatever troubles Bush had with his own people, they paled in comparison to his tussles with the Army and Navy. Military men had a history of rejecting civilian help, and Bush in particular had a keen memory for the slights he'd received during World War I and, more recently, while directing for the Navy the development of a high-speed retrieval machine (the "rapid selector") useful in cracking codes. Though he had forged a useful alliance with Air Chief Arnold, Bush remained intensely skeptical about the ability of the armed services to adopt innovative weapons and tactics. Now he had to set aside his skepticism about military men and win their attention and respect.

This was certainly Bush's highest priority. In his June 15, 1940, letter creating NDRC, Roosevelt had insisted that Bush reach a "close collaboration with the armed services." If Bush failed to achieve this, his committee would be deemed a failure. Without full cooperation from the Army and the Navy, his researchers would not know what weapons to design and how quickly, nor would they understand the unmet military needs, knowledge of which might spark in one of their brains a novel solution. As one writer later noted, the NDRC "was not simply to fill orders for the military; rather it was to be a source of weapons creativity, unencumbered by what technically untutored military men conceived to be useful and possible."[19]

Bush was prepared to meet this challenge directly. "From the very first days of NDRC, Bush put the highest concentration on relations with the military," historian Richard G. Hewlett has observed. While he advanced from a position of strength, having Roosevelt's imprimatur, Bush knew the paradoxical nature of his vaunted independence. "Bush and the NDR were clearly independent of the armed services, both in terms of authority and budget. There was no way in which the Army or Navy could order Bush or the NDRC staff to do anything. Yet NDRC depended for its existence on cooperation with the military.[20]

This would be a constant sore point for Bush because the need for cooperation from the armed forces exposed a fundamental flaw in the makeup of the NDRC. As its director, Bush could only present his views to the military; he could not insist that the generals and admirals follow his advice. While he and his staff were hardly infallible on technical matters, the military's freedom to ignore the OSRD's advice was problematic. As *Fortune* magazine sourly noted at the time, "Not even OSRD, which has the power to initiate research, can do more than present it to the military. . . . The ultimate decision for action or nonaction in these essentially technological matters [rests] in the hands of men who [are] without the capacity or spirit to make the correct technical decisions."[21]

Bush's passion for military affairs was rather new, of course. He had shown scant interest in military problems at MIT. But the European war had transformed him into a military buff. "He had strong ideas about how the military should be organized," one aide said, adding that he became "a great stickler for clearing everything with the Army and Navy. He didn't want to be put in a position of going over heads." Obeying military protocol wasn't as easy as it sounded. The military was a house divided. Army and Navy officers mistrusted one another and the two services often systematically deprived each other of vital information. For instance, as late as the end of 1941, they refused to share their records of NDRC projects with each other. This made it harder on researchers to grasp the connections between narrowly defined questions. In the first year of the NDRC, Bush often said that "if he made any important contribution to the war effort at all, it would be to get the Army and Navy to tell each other what they were doing."[22]

Bush wasn't interested in just listening to the military. Convinced the top brass lacked the vision to fight the coming war, he came up with his own ideas, displaying a good grasp of basic trends in weaponry. His experience, however brief, as chairman of the NACA exposed him to the latest ideas in aviation and defense against air attacks. His computing background imbued him with the possibilities for improving codebreaking and bomb accuracy. Finally, a crash course in radar lent him an appreciation for the vast potential this technology had to alter the whole course of the war.

Bush was less concerned about gaining a fair hearing from the Army because of his respect for Secretary of War Henry Stimson, whom he described as "the man with the greatest vision in the War Department." But gaining the Navy's cooperation was another matter. Within months of the NDRC's formation, naval officers had mounted fierce criticisms against Bush and his committee. Leading the charge was Harold G. Bowen, director of the Naval Research Laboratory. At first, Bowen quietly nursed his resentment against Bush for insinuating that the Navy, which had its own research lab and a long tradition of technically savvy officers, wasn't up to the task of exploiting scientific advances. Calling Bush and his ilk "Johnny-come-latelys," he suggested that the civilian scientists and engineers streaming into Washington were little more than glory seekers.[23]

By early 1941, Admiral Bowen was openly attacking Bush. Grudgingly praising NDRC for its potential contribution to the fighting capacity of the U.S., Bowen warned Navy Secretary Frank Knox, "Every day it becomes more apparent that the [NDRC] will eventually supplant instead of supplement the research activities of the Army and Navy." The next month, he continued his attack, arguing that the Navy would be better served if it directly contracted with civilian researchers. Jealously eyeing NDRC's first-year budget of $6.4 million, Bowen insisted that "all of this new development work could have been previously undertaken by the Naval Research Laboratory if funds had been available." Moreover, the research sponsored by NDRC, if placed under the direct control of the Navy instead, "would be much further along and some instances completed." Measuring his attack on an institutional rival, Bowen declared that at the "end of the present emergency" the NDRC should cease to function and the Navy and Army should "have charge of their own research." In the meantime, if civilian scientists wanted to help the Navy, Bowen insisted, they must follow directions laid down by naval officers. The only reason to accept independent aid from civilians, he told Knox, "would be on account of pressures exerted by certain well known scientists."[24]

Bush met Bowen's bullying tactics directly. He knew that in order to gain equal standing with military officers, he must teach Bowen a lesson. His own stock rising, Bush set out to discipline Bowen. To start with, he decided to play off one Navy bureau against another. The Navy's bureau system, in place for a century, made this relatively easy to do, since the bureau chiefs reigned supreme over their fiefdoms. Bush shrewdly convinced the Bureau of Ships, Bowen's archrivals within the Navy, to call for NDRC's aid in antisubmarine research. Next, he arranged for the Navy secretary to select an MIT professor,

Jerome Hunsaker, to review the Navy's research organization and its approach to submarine detection. Though a former naval officer, Hunsaker was treasurer of the National Academy of Sciences and had close ties with the civilian science elite. His advice, accepted by Knox, amounted to a stunning defeat for Bowen and provided dramatic evidence of Bush's clout.

As a result of Hunsaker's advice, Bowen's plan to give bureau status to a proposed Naval Research Center was rejected, a new officer was chosen to serve as liaison to civilian scientists and the Naval Research Lab and Bowen himself were placed under the Bureau of Ships. Finally, Knox placed an "unsatisfactory fitness report" in Bowen's personnel file, charging him with being "rather belligerent and temperamental in his contacts outside of the Navy."

Bowen's humiliating defeat clearly reflected "the support scientists had at the highest levels of government," Harvey Sapolsky has written. "Although military officers might continue to resent their participation as equals in the planning of weapons research, they could not openly resist such participation." As a bonus, Bush earned two fresh allies within the Navy: the Navy's new science coordinator, Julius Furer, who was loath to make the same mistake as his predecessor, and Hunsaker, who was asked to advise Furer.[25]

Despite his banishment, Bowen's attack was not forgotten by those worried about the loss of naval independence. After the war, the Navy—and Bowen himself—would attempt to even the score by essentially developing a clone of the NDRC within the Navy. But for the present, Bush had stymied a dangerous rival. He had his marching orders from the president and his own line of funds. Time was short. The NDRC would prove its value, he believed, and then every thinking person would wonder how the nation could secure its borders without something like it.

So the military brass took their lumps, watched and waited. In Roosevelt, Bush had a powerful patron. His relationship with the president was all the more attractive because, as Bush only rarely admitted, Roosevelt had no chance of really comprehending Bush's advice and essentially had to take his views "more or less on faith." Basically, the president bet on Bush's judgment. For an organizer of expertise, this was an enviable position to hold. "Of course, I had kept him apprised of what was going on frequently," Bush once said of the president. But these briefings were by necessity superficial because Roosevelt "couldn't possibly understand the science involved."[26]

So as not to be forgotten by the fickle Roosevelt, Bush relied on Hopkins to deliver "direct advice" to the president "on the scientific aspects of warfare." In a March 1941 memo to Hopkins, Bush conceded that the Army and Navy had "ultimate responsibility" for producing advanced weapons, and he stopped short of arguing that scientists help to plot military strategy. But he insisted that civilian experts and the military must forge a full partnership. "To the grasp of military matters must be joined a grasp of the trends of modern science and technology," he asserted, adding:

> In the last war the sudden advent of poison gas, and of tanks, and the evolution of aerial and submarine technology, were of great significance. With the scientific and technical advances of the past twenty years, it should be expected that comparable modifications of war methods will occur during the struggle which is now going on. To foresee these things as well as may be, to place adequate emphasis upon the right things in time, will require a vision which must be based upon a background of science and engineering, as well as upon a background of military art.[27]

While scientists and engineers often flubbed their predictions of the future, Bush's larger point about the centrality of expertise in a complex world was compelling. Even military officers could not dismiss Bush's views, which made them all the more nervous about him. Indeed, they already were ceding to him some authority over new weapons, a realm they had long dominated. While they recognized that the current war demanded a wider involvement of civilian society, the brass were still burned by Bush's frequent dismissal of them as "technical dunderheads" and his belief that "civilian scientists were single-handedly remaking the defense posture of the Army and Navy." Yet the generals and admirals were too "afraid" of Bush to openly revolt against him, a War Department official said. "Bush had the ear . . . of the President. So I think [the military] did most of their roaring about Bush among themselves."[28]

With the arrival of Harvey Bundy in the War Department in 1941, Bush's relations with the Army, never as bad as those with the Navy in any case, brightened considerably. Bundy was a lawyer with no background in technology who admittedly "didn't know the first thing about science." But right from the start he knew the best way to handle Bush was to clear a path for him. "No man could have been more helpful or more careful not to be regarded as personally ambitious," Bush later wrote of Bundy, whom he credited for "a good part of the fine relations" he maintained with Stimson and General George C. Marshall, Army chief of staff.[29]

As Bush's liaison to the War Department and the upper ranks of the Army, Bundy was all ears. When he hit a roadblock because of Army secrecy or a lack of respect for his researchers, "Bush would needle me and then I would needle the secretary and then the secretary would hit the Army over the head," Bundy recalled.

"My function was one of helping the military understand the civilians, and helping the civilians understand the military and getting action by an educational process really," he added. "Bush was impatient with the military point of view because Bush's vivid imagination could see around the corner, and the Army in his view were way back there dragging their feet. Many of the military officers recognized the extraordinary ability of Bush as a man. But the military don't like to be needled particularly. And they would have naturally the feeling that these damn scientists weren't very practical men; they were visionaries. . . . And they didn't want to waste time on something that wasn't going to win the war."

Neither did Bush. But since his vision was better, he kept railing against military intransigence. The drumbeat of complaints maddened Bundy. After one harsh session with Bush, Bundy went to James Conant and practically begged him to ask Stimson to replace him with another liaison. "I've reached the end of *my* rope here, Bush is so impatient," Bundy told Conant. "Perhaps you'd better get another man to deal with Bush."

Conant was not about to do that. Legendary for his coolness, Conant often tried to persuade the explosive Bush to accept the limits of the possible, sometimes by simple talk and other times by sharing a bottle of scotch with him. In the case of Bundy, whatever was said led to no outward change. Bush remained "impatient . . . impatient with everybody," but Bundy accepted his role. He even took little notice went Bush went over his head, directly to his boss, Stimson. Perhaps Bundy showed such extraordinary patience because he considered Bush, two years his junior, "one of the most important, able men I ever knew." Yet Bundy wasn't starry-eyed. He found Bush "a vain man, but a very, very remarkable human being . . . [who] was right to be vain."[30]

By mid-1941, the financial constraints on Bush were easing. In May, Roosevelt approved the creation of the Office of Scientific Research and Development, or OSRD for short. Under the former setup, the NDRC had been funded from the president's own emergency funds and was constantly strapped for money. OSRD now would receive direct funding from Congress and thus operate on firmer legal ground. Equally important, Bush now had the authority to build small batches of weapons and equipment created by his researchers. Bush expected that this capability would give him greater leverage in discussions with the armed services: if they refused to build a weapon for whatever reason, Bush could go ahead with production himself, demonstrate the weapon and then dare the services to ignore it.[31]

As part of the change, the NDRC became essentially the chief operating unit of OSRD, with Bush retaining the power to reject proposals from the NDRC and to "proceed without" the "recommendation" of its committee members. Conant assumed the post of NDRC chairman, while Bush became the director of OSRD. As part of his expanded domain, he oversaw a new Committee on Medical Research, whose charter extended beyond the obvious task of finding means to both prevent illness in unfamiliar places and treat the wounds of injured soldiers. The CMR also sought to ease the burden of soldiering by, for example, studying ways to better help men adapt to life in hazardous and uncomfortable armored tanks.[32]

Bush handpicked the chief of CMR, and in making his selection displayed some of the devilish wit that charmed some and repelled others. He telephoned his unsuspecting selection, A. Newton Richards, an administrator at the University of Pennsylvania who had provided the first detailed description of the physiology of the kidney. Bush did not know Richards well, "so was quite formal. I presented the plan for CMR and then I told him that for the chairmanship one man alone had been suggested by every group that I had consulted." Unless Richards saw "some reason to the contrary," Bush said, "I proposed to present that man's name to the President the next morning."

Richards then asked for the name, and Bush quickly answered, "His name is Alfred Newton Richards."

The phone went quiet and finally Richards said, "Jesus Christ!"

"Well, we thought of him," Bush replied. "But we thought he was unavailable."

After another long pause, Richards said, "I am coming down to see you this afternoon."[33]

With the promise of bigger budgets, Bush began to take stock of the achievements of his civilian researchers. The NDRC had been too slow at executing new contracts, creating a logjam that was only partly due to an uneven flow of funds. The lack of agreement on a standard contract, covering such contentious issues as who gained what if any patent rights resulted from the research, caused no end of headaches. Stewart, the administrative chief, finally set a goal of signing contracts within 24 hours. This set the right tone even if it wasn't always achievable. Indeed, as late as early 1942 a new OSRD attorney found scores of unsigned contracts "lying around on the floor" in his predecessor's office. Another irritant was working out in practice Bush's "no loss, no gain" rule. The chief insisted that contractors earn no profit on NDRC business, but only cover their costs. The dictum was deceptively confusing, however, because direct costs were usually only part of the expenses incurred in carrying out a contract. What about the costs of the laboratory or clerical help, for instance? Determining a fair reimbursement rate for overhead proved nettlesome. Bush first set the allowance for overhead at 50 percent of the value of a contract for nonprofit institutions and 100 percent for industrial concerns. These percentages were really "a rule of thumb," historian Larry Owens has written, and as time went on Bush drove the percentage downward in some cases.

When contracts with MIT ballooned, for instance, Bush realized that the 50 percent allowance for overhead would lead to a financial bonanza for his old school. Overhead rates on MIT contracts ended up averaging less than 10 percent.

The stickiest issue wasn't actually the percentage of overhead allowed but what contractors were permitted to include as overhead expenses, or what they could get away with. In time, Stewart assembled an in-house staff to monitor claims for overhead expenses. These claims sometimes strained credulity and raised legitimate questions about whether the universities were seeking to profit from war research. "Some of the universities went through some pretty fancy reasoning in an effort to establish the fact that certain generous al-lowances they were writing into the contracts for themselves did not actually represent a profit," recalled one administrative staffer. Indeed, one of Stewart's assistants swore that Columbia University tried to charge the repair of the school chapel bell to the government.

Other examples seemed more troublesome, suggesting that universities saw their government relationships as the basis for future research empires. After Columbia had requested an advance against a contract for submarine warfare, Stewart's auditors found that the university had amassed a large cash reserve from earlier contracts with the NDRC and was using this reserve to finance contracts with the War and Navy departments, which were more restrictive in advancing funds. Harvard University, meanwhile, claimed as an overhead ex-pense the interest lost on the portion of the school's endowment used to cover the costs of war research until monies owed it by NDRC were received.[34]

The administration of OSRD was inevitably a sideshow. The main action in-volved designing new weapons and equipment. On this, Bush's dictates were clear. He had what one observer termed "an iron-clad rule" regarding any pro-ject proposed to his organization. He always asked his researchers: "Will it help win a war; *this* war?"[35]

For scientists and engineers to shed their image as impractical dreamers, they must deliver results quickly. In a top-secret 36-page report to Roosevelt, chronicling the NDRC's first year, Bush cautioned the president that "re-search, by its very nature, takes time" and that "the task of compressing into months a development which, in normal times, would take years is strenu-ous." Nevertheless, Bush happily reported that, "out of the work of the Committee, there are already coming some results which can be applied to in practice to great advantage. More are on the way."

The "results" covered many areas—new vehicles, proximity fuzes and antisubmarine measures, to name a few—but Bush devoted the most space in his report to the area in which the NDRC's "efforts have been the greatest." This was the application of radar to the problem of "aircraft detection and its resulting developments."[36]

The story of radar was one of the most consequential of World War II, since the Allies ended up with a decided advantage over the Germans in this technology. Bush's role in radar was essentially one of guardian rather than principal actor. He oversaw the transfer from England of critical pieces of advances, stimulated research in the U.S. and pressed military authorities to put radar innovations to use on land and at sea.

The U.S. radar effort received a critical boost in August 1940 when Henry Tizard, chairman of Britain's scientific committee on air defense, dined with Bush at the Cosmos Club. Tizard came to Washington at the head of a seven-person delegation charged with cementing technical ties between Britain and the U.S. He arrived with a black box that contained what one historian has called "the most valuable cargo" ever brought to American shores. It was a cavity magnetron, a radar transmitter that was small enough to fit in the palm of a hand and so powerful that "conventional scientific wisdom still put anything like it years off." An astonishing breakthrough, the magnetron emitted ten-centimeter radiation at an intensity "some thousand times as great as the most advanced American tube."[37]

Invented only eight months earlier by two British physicists, the magnetron was immensely valuable. Radar formed the bulwark of British defenses, providing early warning against German bombing attacks. It also was a boon for the offense. Equipped with radar, an aircraft "might pick out U-boat periscopes rising under cover of darkness." Bombers, meanwhile, "could use the extremely short waves the magnetron produced to illuminate the way through the thick cloud cover obscuring Hitler's forces and factories on the European continent, keeping planes flying on days the Royal Air Force would normally be grounded."[38]

Despite the magnetron's superiority, it was hamstrung by glitches, and British scientists feared they lacked the industrial strength and technical wherewithal to fully exploit radar on their own. Hence, their offer to give the U.S. essentially all their know-how about radar devices.

Bush naturally seized on Britain's generosity. He grasped the potential of radar and relished the opportunity to redress his country's big disadvantage in the field. At the time, "The U.S. Army and Navy radar sets all operated at the relatively long wavelengths of one or two meters." Rather than refine the military approach, the NDRC had decided to concentrate on short "microwaves," in the centimeter range. Under the direction of Bush booster Alfred Loomis, the NDRC's Microwave Committee saw its path blocked by "the lack of a vacuum tube capable of generating sufficiently energetic radiation at such short wavelengths."

Now with the arrival of Britain's magnetron that obstacle had vanished. After a series of meetings between British and American experts, the ever-resourceful Loomis quickly arranged for five big manufacturers—AT&T, RCA, Sperry, Westinghouse and General Electric—to supply working magnetrons and other equipment on short notice. Bush, despite having vowed to keep researchers at their home universities, decided to form a central radar laboratory.[39]

The new Radiation Lab, or Rad Lab for short, might possibly be the biggest research plum of the entire war. Whoever snagged the Rad Lab stood to receive an infusion of technical expertise that could power their institution and the surrounding economy for years by supporting many high-skilled jobs and innovations of value to the civilian economy. It wasn't until mid-October, less than a week before Bush and the NDRC formally approved the creation of the lab and its first-year budget of $455,000, that sites were discussed. Loomis urged that the lab be run under the auspices of the Carnegie Institution and based in Washington. But Bush demurred, fearing "disciplinary and administrative problems." Loomis countered by pushing for another site in the nation's capital, perhaps Bolling Field. Bush "protested, and we had a hell of an argument that took half the night and a bottle of scotch." Before the drinking had ended, Bush had convinced Loomis that MIT made the most sense.[40]

It wasn't just an old school tie that drew Bush toward MIT and made Loomis accept the decision. Loomis knew firsthand MIT's capabilities. In early 1939, MIT's president, Compton, had arranged a meeting between Loomis and Edward Bowles, the gifted MIT radio specialist whose own independent streak had led him to clash with Bush over the years. Compton hoped that Loomis would fund joint research with Bowles's team. He agreed. In the summer of 1939, a group of MIT physicists studied propagation of radio waves at Loomis's estate in Tuxedo Park, New York. The work by Bowles was promising enough that in January 1940 Bush told him if "he needed further support, to let me know." Bowles did, and in May Bush directed Carnegie to grant him research funds.

By then, Loomis and Bowles had become something of a radar duo. When Bush placed Compton in charge of the NDRC's "detection" research in June 1940, Compton asked Loomis to organize a microwave committee, and Loomis promptly asked Bowles to serve as his committee secretary.[41]

The overlapping relationships between Bush, Compton, Loomis and

Bowles made it practically inevitable that MIT would snatch the Rad Lab. But there remained one more card to fall. On October 16, Bush met in his office with Bowles and his inner circle at NDRC for the purpose of wrapping up the decision. It turned out that Frank Jewett, head of Bell Labs, had his own ideas about who should run what at the time looked like the most important research project of the entire war. Recalling that Bell Labs had run an antisubmarine lab for the government during World War I, Jewett wondered whether it might now run the Rad Lab. Such a complicated effort would require sophisticated management, and Bell Labs was a proven entity on that score.

"Why, MIT is a wonderful institution," Jewett allowed, "but it doesn't know anything about management." Bush sat back quietly, but Compton bristled, taking Jewett's comment as an insult. Jewett tried to recover by apologizing, but Alfred Loomis broke in. "I'm so glad you approve of the idea of a laboratory at MIT." After briefly deliberating, Compton agreed to house the lab at MIT.[42]

Over the next year, the Rad Lab grew rapidly. Directed by Lee A. DuBridge, chair of the physics department of the University of Rochester, the lab's first staff meeting was held on November 11, 1940. Within a month the lab employed about 35 people, most of them university physicists. "They knew little to nothing about the microwave electronics that would be needed to translate the British 10-centimeter magnetron into a working radar system. But the Rad Lab physicists were not alone: no one knew much about microwave electronics."

DuBridge later recalled the early weeks of the lab as "a blitz." He felt he must quickly demonstrate the feasibility of microwave radar. This would be no mean feat since "there was no adequate account of how the magnetron really worked," a historian of the lab has noted. Yet the lab's physicists advanced rapidly. On January 4, two days ahead of their internal schedule, researchers sent a radar beam from a Cambridge rooftop and detected echoes from buildings in Boston across the Charles River. "In less than two months after they walked into the Rad Lab, a group of physicists . . . had put together a prototype radar system. It was crude, but it worked."

By March 1941, the "blitz" had ended, and DuBridge, convinced his lab had staying power, began planning longer-term projects. His staff now numbered 140, two-thirds of whom were scientists and engineers. On March 27, they tested their first airborne microwave system aboard an Army B-18. Flying over water at 2,000 feet, the plane tracked a 10,000-ton ship heading into port. Flush with success, the physicists on board asked the pilot to fly toward the Navy submarine base in New London, Connecticut. In waters near the base, the airborne radar set detected a surfaced submarine from a distance of three miles.

The Rad Lab was in the hunt, poised to deliver a string of phenomenal equipment. One tantalizing hope had been raised in February 1941, when physicists had begun to build gunlaying radar whose aim was to automate the tracking of enemy targets. Once picked up on the radar screen, the target would be tracked automatically and the necessary information fed into an antiaircraft gun. If this wasn't incredible enough, the device would automatically aim the gun too. At the end of May, the tracking system was demonstrated on an airplane and in December a truck-mounted system was shown off. Adopted by the Army, the system, dubbed SCR-584, was used virtually unchanged throughout the war.[43]

The Rad Lab's fast results epitomized Bush's belief in the ability of civilian scientists to apply their talents to military needs, discard their leisurely academic pace and rush pell-mell for the finish line. Now that his outfit was proving its value, it became all the more important that everyone attached to the OSRD adopt the military's attitude toward secrecy.

This was a tall order. Science and secrecy had never mixed well. Not only did scientists within the U.S. freely exchange information, but scientists from different nations did the same. While lauding peacetime cooperation between scientists, Bush understood instinctively that in their new role scientists must eschew such openness. Of course, this did not mean unthinkingly embracing secrecy. But it meant a sea-change in the habits and routines of researchers.

Bush was naturally given to secrecy. Though he drew frequent comparisons to Will Rogers, Bush's folksy openness masked a desire to keep his own counsel. One acquaintance insightfully pegged Bush as "not a man who easily expressed his deepest thoughts in casual conversation. His geniality perhaps covers an inner shyness."[44]

When it came to cracking down on loose lips, Bush was hardly shy. His instinct was to err on the side of caution. Knowledge was doled out on a need-to-know basis, so that men knew their specific areas but not unrelated areas. When talking to outsiders, NDRC members listened but said little. Even amongst themselves they could be wary. On Bush's personal order, they discussed secret projects only "behind closed doors, under conditions where privacy is assured, and only in the presence of duly authorized personnel."[45]

The stress on secrecy paid off, at least in terms of its reputation. "Of all Washington agencies," *Fortune* observed, NDRC "is the most supersecretive." Not even the Army and the Navy were as tight-lipped. *The Wall Street Journal* described Bush's outfit as "more secretive than the military."[46]

While willing to accommodate the military's demand for secrecy, Bush wasn't always comfortable with the specific forms this took. He was angered by the delays in gaining security clearances for civilian researchers and what seemed to be an excessive examination of their private lives. He often aired his grievances to Rear Admiral Julius Furer, the Navy's easygoing coordinator for research and development. A career officer, Furer had graduated from the Naval Academy in 1901 and received a master's degree four years later from MIT. Believing that the contributions of civilian scientists were an unanticipated boon for the defense effort, he had more patience for Bush's outbursts than any other senior Navy man. His patience was all the more striking since he confided in his diary that Bush "is a difficult man to deal with at times and does not appear to be highly cooperative."

Bush took full advantage of Furer's tolerance. On one occasion he blew his top over over what he viewed as "very unreasonable and arbitrary" treatment by the Office of Naval Intelligence, which he claimed was insisting that scientists with clearances submit additional personal data such as nicknames or membership in professional organizations. The complaint struck Furer as "rather childish and shows how temperamental the scientists are." But giving them the benefit of the doubt, as he was prone to do, he observed that their resistance was perhaps "more an indication of the American disinclination to give personal information about themselves."⁴⁷

Red tape frustrated Bush, but he shared the military's aims and even preferred to keep a low profile. While recovering from a lung infection in March 1941, he wrote Jewett: "There are so many people looking for the limelight that it is rather easy in this town to keep out of sight if one wishes to."⁴⁸

Being out of sight had its advantages. Bush was sensitive about the degree of his self-dealing as director of OSRD and worried about accusations from Congress or the public that he was fattening institutions or companies to which he had close ties. Bush's concerns were hardly unique. As part of war preparations, scores of private executives had joined the government as "dollar-a-year" men. From June 1940 to April 1941 the government had awarded $3 billion in contracts to companies employing such men. Bush's employer, the Carnegie Institution, was among the 15 largest recipients of the OSRD's largesse. MIT, Karl Compton's employer, was the largest recipient of OSRD funds; among the largest was Harvard University, James Conant's employer.⁴⁹

While the whole point of serving the government was to engage the best researchers at elite institutions, Bush still fretted that his cozy deals might come to haunt OSRD. He repeatedly asked for some sort of blanket immunity against future conflict-of-interest claims. On August 21, 1941, the Roosevelt administration finally wrote Bush that at a minimum "you will not be violating" criminal law by giving OSRD contracts to the Carnegie Institution.[50]

Less than a month later, however, Bush again skirted the law, this time regarding Raytheon, the electronics company he had cofounded in the 1920s. The Navy and OSRD were considering signing a research contract with Raytheon. Bush had resigned from Raytheon's board of directors in 1938 but still owned shares in the company. He told Raytheon's president it would be "entirely improper" for him to participate in the negotiations. But as a matter of principle, Bush saw no conflict. "I do not think that I need to be embarrassed because the government has relationships with a company in which I am a stockholder," he wrote his banker on September 18. Twelve days later, after OSRD had reached an agreement with Raytheon, Bush elaborated on his reasoning. "My possible embarrassment is much relieved by the fact that the contracts of the agency under my direction with Raytheon Company are definitely of a non-profit nature, and this is very explicitly stated in the contract," he wrote his banker. "Undoubtedly this removes question as to propriety in the minds of any reasonable individuals."[51]

It didn't stop reasonable people from whispering and gossiping, however. During the course of the war, Raytheon's sales would grow 60-fold; war contracts obviously caused the growth. Some scientists and engineers, meanwhile, resented MIT's privileged status in war research and suspected that Bush's loyalty to his old school had something to do with it. While the OSRD's leaders usually kept silent on this sensitive point, Jewett raised the matter time and again. Stung by the loss of the Rad Lab to MIT, Jewett passed along to Bush a vicious complaint by a group of unnamed scientists that the Bell Labs chief claimed he had overhead on a train but that may actually have been a veiled message for Bush to dispense the government's money more fairly or risk a public squabble. According to Jewett, the scientists charged Bush and Compton with doing an "immoral thing," by "assigning millions of NDRC money to MIT and relatively little to other institutions." More ominously, Jewett raised the specter that "the tide of criticism was mounting rapidly and that before the show was over a thorough airing of the MIT affair was going to be instituted."[52]

While Bush tolerated a growing number of critics in the fractious research community, he did what he could to endear himself to the military. Since behind closed doors he played rough with the top brass, Bush made a point of heaping credit on the Army and Navy in public. In his first major speech after joining the Roosevelt administration, Bush effusively praised the armed services. At a luncheon in New York City on October 24, 1941, he outlined the setup of the OSRD in general terms and described his agency's relations with the military as "very close." The research officers for the Army and Navy, he said, were "some of the brightest, keenest men in the armed services on the technical front."

For the first time since forming NDRC the year before, he publicly revealed that his organization had spent some $10 million on scientific mobilization and enlisted the services of about 2,000 scientists, including about 75 percent of the nation's top physicists and half of the leading chemists. These researchers were doing splendidly, he said, and impressing the military by showing they can "attack a problem in a practical way" and "work long hours and take it with the best." Putting their academic research on hold, they had done in a year what ordinarily might take three years. And they had moved ideas from the laboratory into use despite "the petty inconveniences and annoyances that are inevitable in the confusion of adapting themselves to military ways." Impressively, "they are willing to go into a strange ball park and learn the local ground rules."[53]

Once, scientists were defensive about their role in turning plowshares into swords. After World War I, a wave of recrimination overcame the makers of chemical weapons, and populist politicans heaped abuse on those "merchants of death" who had profited from the weapons of war. Bush sensed that the Nazi enemy was so terrible that during this world war scientists ran no risk of moral condemnation by aligning with the military. He felt free, as he did in his speech, to practically boast about the intimacy between scientists and soldiers. He celebrated this partnership with no pangs of conscience or regret, writing to a friend not long before the speech:

> The idealism [of the interwar years] has largely gone. We are living in a real and tough world. It is no longer regarded as wicked to devise a means for shooting an airplane, in a world where it suddenly appears that men will actually ride in airplanes, and drop bombs. It is not even wicked to work out more powerful bombs to drop on someone else. It is being realized with a thud that the world is probably going to be ruled by those who know how, in the fullest sense, to apply science, whatever their other attributes may be.[54]

Bush's singular contribution to this "real and tough world" would not be to build more powerful bombs, but to organize the experts who would. He would not immerse himself in the making of any special weapon, but become the father, in a sense, to all weapons that would spring from the labs. He would sustain his far-flung researchers financially and organizationally, protect them from arbitrary demands of others in government and in the end shepherd the fruits of their labors onto the field of battle. For such a task, his hours at work seemed modest compared to those of others engaged in war preparation. He usually arrived at the office after 8:00 A.M. and rarely worked evenings and never on Sundays. He refused to broaden his duties beyond military affairs and even curtailed his formal responsibilities, resigning in July 1941 as chair of the National Advisory Committee for Aeronautics. Except for visiting New York or Boston, he almost never traveled on business outside Washington; it wasn't unusual for him to flatly turn down a trip. When Roosevelt raised the possibility of Bush visiting London in early 1941 to cement technical ties with the British, he quickly proposed that Conant go instead.[55]

In many ways, the daily grind of the OSRD weighed more heavily on Bush's chief aides, Carroll Wilson and Irvin Stewart. This was no accident. Bush believed good leaders should "delegate effectively." He thought his personal staff "should not use him, or weary him, or be a constant nag." He also thought every "chief should have leisure." At home, he should not be disturbed "unless the place burns down or the president resigns." Hoarding his free time, Bush kept at many hobbies, so many that one observer concluded that he had "more hobbies than seems any man's legal right." Some of these hobbies, such as his dabbling in computing, pipe-making and newfangled engines, "would be full-time jobs and lifetime pursuits" for other people.

Happiest when tackling a problem—almost *any* problem—Bush found that the stress of mobilization forced him to add late-night basketweaving to his diverse regimen. To relax, he also milled pieces of steel, designed glass-reinforced-plastic fishing rods, toyed with color photography, kept an aquarium stocked with tropical fish and tinkered with a solar-powered pump. In no way defensive about the breadth of his extracurricular activities, Bush declared, "I think hobbies are necessary to anyone who is compelled to work under strain. "[56]

The balance between Bush's professional and private lives paid hefty dividends. Always high-strung, he deserved credit simply for not cracking up under the strain. As it was, his "temper would sometimes snap because of terrible burdens," he told a journalist.[57]

Phoebe Bush helped to keep her husband on a relatively even keel. She was his haven from the hubbub of the nation's capital. She steadfastly stayed apart from political life, clinging to her old-fashioned New England ways and providing a refuge for her husband.

"She was in no way a Washington hostess," said one friend. "She was a quiet American housewife with only one sure thing to say about her: She adored her husband." Uncomfortable around her husband's male comrades, Phoebe stayed in the background, or disappeared altogether. She disliked formal dinners and skipped many of them, letting Bush go alone. "She was not the kind of person you'd expect to say the right thing to the secretary of state. She was Van Bush's very simple and adoring wife."[58]

Phoebe never felt fully at home in the nation's capital. But among female friends she cut a more definite figure. She lunched regularly with one of Bush's few female staff members and formed a group of OSRD wives who convened monthly for lunch in a private club. "She enjoyed it thoroughly, and it developed into a pretty good morale group," said Edna Haskins, the wife of a Bush aide. If wives needed help with children or their homes, Phoebe would look into arranging for aid.

Even women attracted to Phoebe found her an uncomplicated person who left few impressions in casual settings. One acquaintance found her "nice, ordinary, New England." Her indistinct image was probably the result of her shyness: she was intensely private and proper. Said a longtime friend, "Phoebe pretended to be other than she was. Maybe she was a highbrow pretending to be a lowbrow."[59]

Among strangers Phoebe could be moody—at times dour and aloof—but Bush excused her faults. "He just adored her, he absolutely adored her," a friend recalled. Phoebe had a rare talent: "She could ease some of his unease." Her friends found this extraordinary since Bush had plenty to worry about. It seemed impossible to describe how Phoebe put him at ease—her smile, the touch of her hand, the twang of her voice—but friends swore they watched her doing it.

Other women never stirred Bush's romantic affections. Phoebe's intense adoration and her Yankee style seemed to satisfy Bush's desires for female attention and companionship. "He was in love with his wife, and he just wasn't interested in anyone else," a male friend recalled. Work, meanwhile, rarely brought Bush into close contact with women who could speak comfortably about science and engineering, his great passions. So it was hard to conceive of a rival emerging for Bush's affections.[60]

For her part, Phoebe saw her husband as protector and provider. She revealed the extent to which he dominated her world in an amusing anecdote during the height of war preparation. Lunching with a friend in Washington, Phoebe excitedly explained that Bush had given her two exotic birds and that she now hoped they would mate and have offspring.

"Who is going to build them a nest?" her companion asked.

"Why Van is," Phoebe said, acting as if this was the most natural thing in the world for her inventive husband to do.[61]

While not given to romantic outbursts, Bush tried to look after Phoebe even when piled under with work. Just a day after dining with Tizard, Britain's radar emissary, in August 1941, Bush purchased a large house in a fashionable part of Washington. At $35,000, the price for the home at 4901 Hillbrook Lane was steep, but Bush was able to pay about half of it in cash and he'd reached the conclusion that he was "rather likely to live in this town for fifteen years." Besides, his finances were solid; his $25,000-a-year salary from Carnegie was ample and he earned a few thousand dollars extra from his Raytheon shares and other investments. The purchase of the home was a relief too, ending months of searching and essentially living out of suitcases. He and Phoebe had resided in the Wardman Park Hotel since arriving in the capital. While it was convenient, they viewed "the prospect of living in an apartment indefinitely still with some dismay."[62]

The couple, who rented out their former home in Massachusetts, also owned a sprawling farm near East Jaffrey, New Hampshire. Aside from his wife's family home on Cape Cod, the Jaffrey farm was Bush's preferred vacation spot. Phoebe's father lived at the farm for a stretch, and she often visited on her own. Purchased about three years before, not long before Bush's departure from MIT, the farm spread over 300 acres, mostly woodland; perhaps 60 acres were tilled land and pasture. Bush raised hundreds of turkeys, which were sold at market, and co-owned a sawmill with a neighbor. In recent years, he'd made major improvements to the primitive farmhouse, adding a kitchen, bathrooms, a heating system and a new water supply. In mid-July 1941, the day after he submitted a report to the president on NDRC's first year, he telegrammed Phoebe with instructions on the latest renovations: "Just to be sure striped paper goes on guest bedroom also guest sitting room and flowered paper on small boys bedroom. Stop. Please wire."[63]

While Bush now hardly visited Jaffrey, he kept alive the idea of himself as a gentleman farmer and eagerly snapped up the small farming tasks he could handle from Washington. He didn't discourage neighbors from writing him, and in October permitted one to remove some stones from his land. The only caveat was that if the neighbor wanted "quite a few stones," he should "take them out of the stone pile in the field where there are plenty we would like to get rid of."[64]

The new house in Washington was quiet in the fall of 1941. Bush planned to build an elaborate machine shop in the basement but had not yet begun construction. His boys were gone too. The older boy, Richard, was nearly finished with his studies at the Johns Hopkins medical school; Bush expected him to enter the medical corps of either the Army or Navy upon graduation. His younger, John, had just enrolled in his first term at Haverford College, where he also was training to join the Air Forces. While Bush shared a mutual interest in physiology with Richard, he found less common ground with John, who wasn't as strong a student as Richard. Two years before, Bush had become so frustrated by his lack of communication with John that, rather than offer him advice on his future, he told him to speak with the headmaster of his private high school in Kendall, Massachusetts. Somewhat embarrassed by his predicament, Bush confessed to the headmaster, "I have great difficulty in getting at the kid's real ideas anyway. I suggested to him that he . . . seek your advice and I hope he has done so and that you may have found out what the real story is." At Haverford, however, John seemed to have discovered a sense of purpose. So far his first-term grades were solid, and Bush now expected his son's "curve to trend up."[65]

Through his nearly 18 months in Roosevelt's administration, Bush had worked on mobilizing civilian researchers in the belief that the U.S. inevitably would be drawn into a shooting war with the Germans. So far that hadn't occurred. To satisfy domestic critics, Roosevelt had promised during the presidential campaign of 1940 not to commit U.S. troops to a European war. Now he seemed bound by this vow. Yet with each passing month in 1941, the U.S. moved closer to open hostilities with the Axis powers. Even without a formal declaration of war, the U.S. had essentially chosen sides, directing its powerful economy to assist and support Germany's enemies. The mechanism for this was the controversial Lend-Lease Act, approved by Congress on March 11, 1941. The act, approved by the slimmest of margins, permitted Roosevelt over time to equip Britain, Greece and in time Russia, China and other Allied nations with tens of billions of dollars' worth of defense material. Not only did the act create a huge stimulus for the U.S. economy, but it put the Nazis on notice that they might have to outproduce the U.S.—not just the British—to win the European war.[66]

Bush's focus had always been on Germany as the chief threat to the U.S. This was natural given his familiarity with German science and technology. He paid less attention, however, to the deteriorating relations beween the U.S. and Japan, which had angered the president by its aggression in China. After the Japanese pushed into Southeast Asia in July 1941, Roosevelt froze all Japanese assets in the U.S. Four months later, in late November, the president dispatched a harsh note to Japan calling on it to withdraw its military forces from China and Southeast Asia. Insulted, the Japanese cabinet endorsed plans for an attack on the U.S. naval base on the Hawaiian island of Oahu. On December 7, Japanese planes attacked Pearl Harbor, killing some 2,400 U.S. servicemen, wrecking most of the planes on Oahu and putting eight battleships, three destroyers and three cruisers out of action.

Like nearly all Americans, Bush was shocked by Pearl Harbor, but he was not surprised. Ever since arriving in Washington he had spoken about the nation's vulnerability to air attack. Defense against such attacks was simply not up to snuff, which was why the work of the Rad Lab was so crucial. As recently as June, in an address at Harvard University, he had bemoaned that in war, "for the moment at least, attack has outstripped defense." He insisted that defensive measures would come to check offense, but conceded, "it is going to take time." The power of an attack, he said, "rests very largely on surprise, and surprise in turn rests on ignorance." If the nation had a challenge, it was to rid itself of ignorance.[67]

Now that the U.S. was in a shooting war with both Japan and its ally Germany, Bush's advice rang clear. His long period of preparation had ended and his work took on a new intensity. Within days of Pearl Harbor, he wrote, "The nature of the world in which we live has radically altered, and I have hardly a moment to think."[68]

Part Three

Modern Arms and Free Men

Chapter 7

"The man who may
win or lose the war"

(1942–43)

My whole philosophy on this sort of thing is very simple. If I have any
doubt as to whether I am supposed to do a job or not, I do it, and if someone socks me, I lay off.

—Vannevar Bush

Peering through his spectacles, his left hand gripping his bow, Bush pulled
back on his string and held it taut. The target lay before him, squatting above
the withered grasses of Washington's mall, which swept over a mile of trees,
lawn and pools from the Capitol to the Lincoln Memorial.

Bush released his arrow in an instant, but just as he did he lost sight of the
target. This was his "first-class fault" as an archer, he confessed to friends: his
eyes invariably pulled away from the target when he shot and his left hand,
which held the bow, "wants to wave all over the landscape and I can't seem to
make it quiet." As things stood, his shooting was crowded into weekends.
"That is not enough," he complained. If only he had more time for practice,
he might lick his "decidedly bad" habits.[1]

For some months he had come to the mall each Sunday to shoot, joined by
other members of the Potomac Archer club. A dozen years before, when he
lived in Cambridge, near MIT, Bush had become smitten with archery. For a

147

time he had built his own bows and shot arrows in the Massachusetts woods. But then he had lost interest.

"Now shooting at a target, I judge, can get a bit dull," Bush had recently written a friend. But the prospect of shooting in the field excited him. Under the press of war, he spent too much time indoors; perhaps archery was his escape to the outdoors. "I would really look forward with quite a lot of enthusiasm," he confided to his friend, "to an expedition into the woods in the attempt to shoot an arrow into a deer or bear or something that was hard to shoot."[2]

Since Pearl Harbor, Bush thought more and more that his "something" could be a man. A German or a Japanese.

This was not so farfetched. At the age of 52, Bush himself wasn't headed for North Africa or the Pacific, but perhaps his bow and arrow might find a way into battle. In a time of total war, even this hoary instrument of destruction might contribute to the slaughter. Indeed, a full five months before Pearl Harbor Bush had encouraged an NDRC physicist in Illinois, Paul Klopsteg, to study ways to build more lethal bow-and-arrow weapons. By early 1942, Klopsteg had made substantial progress, particularly in refining the critical bow sight. He also found that the English bow was only 50 percent efficient, which meant that only half the energy used in drawing the bow was transmitted to the arrow upon release of the bowstring. Bush helped with the mathematics, so that Klopsteg's improved bows were 90 percent effecient. It now seemed possible that that these improved bows might be used in clandestine attacks on Nazi troops.[3]

Bush's fascination with archery was hard to dismiss as just the latest craze of an inveterate tinkerer but reflected deeper stirrings. Having never been in combat, Bush romanticized the actual battlefield experience. Even as he helped to create impersonal weapons of destruction, he still pined for the time when personal initiative and individual valor mattered decisively in war. He relished archery because it elegantly harnessed a basic force of nature and yet it rewarded the shooter's physical skill and ingenuity. Radar might help a pilot hunt down a submarine, the proximity fuze might help to destroy its target, but the beneficiaries of these weapons might take no personal pride in them. After all, there was far less romance in reading a screen or following the instructions of targeting a bomb. How much more inspiring to rely on your own head and hands to master the bow and arrow.[4]

In later years, his friends were struck by this weird juxtaposition of images. "A psychoanalyst would be fascinated by the archery business in the midst of World War II, and I think he'd have a point," one said. The U.S. czar of military technology "finds Sunday relief by turning to the bow & arrow; weapon of the American Indian and the battles of the 14th & 15th century."[5]

With the U.S. officially at war, the media turned to trumpeting the nation's prospects and whipping up enthusiasm for its leaders. Bush's relative anonymity vanished. Overnight, he became a minor celebrity. "Meet the man who may win the war," crowed *Colliers,* in the most effusive display of Bush-boosterism.

"The chances are you have never heard his name," the magazine went on. "Even in Washington, his headquarters, very few people know him. But he has done a tremendous job, and made less fuss about it than [Interior Secretary] Harold Ickes would make about brushing his teeth."

The article's author, J. D. Ratcliff, gave plenty of credit to other "war brains," but drove home the image of Bush as king of the scientists and engineers in an age when technological advantage determined national security:

> Get the importance of this: The first World War was a war of men and guns. This war is one of instruments and machines—bomb sights, submarine and plane detectors, superchargers. It is Van Bush's job to see that we get the best. If he does his job well—we can assume that he will, he is that kind of man—our planes will fly higher and faster. Our guns will outshoot any others and our tanks will be tougher.[6]

The key to gaining this technical advantage, Ratcliff and a growing number of others realized, wasn't necessarily assigning the smartest individuals to a problem, but assigning the smartest *team* of people. As the great American strategist of sea power, A. T. Mahan, once observed, "War is not fighting, but business." And increasingly the guts of this war business was marshaling expertise.[7]

This task presented some peculiar challenges to American society, which traditionally left the organizing of expertise to individuals and their voluntary associations, whether corporate or civic. As the nation became more urbanized and its problems more complex, especially during and after the 1930s, this voluntarist approach became less tenable. Yet the solution to the management of large-scale modern societies wasn't clear-cut. While governmental agencies were the logical agents for tackling the job, they usually spawned bureaucracies that were incapable of responding creatively and quickly to changing conditions.

Some cynics considered governmental sclerosis an inevitable feature of 20th-century democracy. And this was indeed true, unless unusual steps were taken to forestall it. By the early 1940s, a compelling group of leaders had emerged with an attempted solution. Later labeled "public entrepreneurs" by scholar Eugene Lewis, these leaders sought to blend the best features of the public and private sectors by fusing "the entrepreneurial role with a penchant for action in the public sphere." At once a technical expert and bureaucratic wizard, the public entrepreneur had "the expertise, the charisma, and the will to move burgeoning governmental agencies to undertake actions that might otherwise be stalled by democratic politics." Significantly, this new kind of leader "creates or profoundly elaborates a public organization so as to alter greatly the existing pattern of allocation of scarce public resources. Such persons arise and succeed in organizational and political milieus which contain contradictory mixes of values received from the past. Public entrepreneurs characteristically exploit such contradictions."[8]

Bush neatly fit the description of a public entrepreneur, right down to his capacity for embracing contradictions. His academic achievements gave him unquestioned credentials as an expert; his experience at Raytheon taught him about the highs and lows of entrepreneurship; and his stint as chair of the National Advisory Committee on Aeronautics and NDRC showed the creative possibilities open to government advisors. Only someone with Bush's unique background—one part expert, one part entrepreneur and one part public servant—could have conceived of a military research outfit whose authority sprang directly from the president. Bush's lifetime of experience had taught him that the public would never push for such an outfit, because it lacked an appreciation for science and technology, and that the ossified Army and Navy would never create one on their own.

Then there was the intangible factor of Bush's personality. He had an instinctive drive for power, a knack for seizing it at the right moment and the unabashed self-confidence to use it as he saw fit. The evolution of the NDRC was a case in point. At first, Bush saw the organization "as just working on the outskirts on things that the Army and Navy themselves were not working on, but which we thought were worthwhile." However, he had grander ambitions for NDRC and in time he realized "we were really taking over on war research completely."[9]

While the nakedness of Bush's ambitions galvanized his enemies, his charm and intelligence attracted zealous loyalists. Bush's foot soldiers often felt endowed with extraordinary powers simply by dint of receiving an assignment from him. Bush's outsized expectations made those around him reach higher and drive harder. When H. Guy Stever, a 26-year-old engineer and MIT graduate, joined Bush's staff in 1942, his morale soared. Touring aircraft makers as part of a study of the industry's capacity, Stever was thrilled because "every place I went, as a Bush man, I was welcome at the top."[10]

In his bid to gather "good men about" him, Bush could be relentless. Only days after Pearl Harbor he learned that one of MIT's terrific specialists in acoustics, Richard Bolt, was being wooed by the Navy to join in antisubmarine research. The Army, meanwhile, already had a claim on Bolt, who was a reserve officer. But Bush, keen to retain Bolt, called him on the phone without warning. He unleashed a barrage aimed at hooking the young scientist, whom he envisioned as a good liaison officer in OSRD's London office. After a few moments of listening to Bush, Bolt was woozy but impressed. As he recalled, Bush asked:

"Bolt? Is that you, Bolt? . . . This is Van Bush."

I said, "Yes, sir." I had never talked with Vannevar Bush before, but I knew he was up there reporting to the President.

"I've been looking into your case," he said. "Now you are a reserve officer in the Army."

I said, "Yes, sir. I became a reserve officer when I was in college."

"I know, I know, I know. Now, Bolt, I've been studying your case, and you know what will happen now that we're at war. You'll have to go into the Army, and with your degree in architecture they'll have you designing *latrines!*"

That was just what he said. Then there was a long silence, and then he said, "Bolt, it is my considered judgment that you would serve the country better by staying where you are." Now I had never heard the expression "considered judgment" before. Boy, was I impressed.

I said, "What do I do about it?"

Then he really shouted: "You don't do a damn thing! I'll do it!" Then he hung up.

About ten days later I got a whole sheaf of papers to sign. He had them all made out. I just signed them and mailed them back, and that was the end of my military career."[11]

Popularity with the troops did not guarantee success for Bush and other public entrepreneurs. They invariably ran into roadblocks arising from outmoded traditions or sheer idiocy. Neither recognition from the media of his growing role in prosecuting the war nor a direct line to Roosevelt spared Bush from such aggravations. Of all his difficulties, probably his greatest was keeping his younger scientists and engineers from getting killed—not in the lab but on the battlefield. For it was strangely ironic that as the U.S. public displayed a mania for war-related technology, the nation's draft boards kept on viewing younger researchers as potential draftees.

In the fall of 1940, Congress adopted a draft that obligated men to serve just one year. In August 1941, the term of draftees was extended and, following Pearl Harbor, the pool of eligible men was broadened to include men from 18 to 38. Over the course of World War II, about 10 million men were drafted and more than 5 million enlisted. But the existence of the draft, and the question of its fairness, raised hackles. The House of Representatives approved an extension of the draft by only a single vote; in its first year it had been viewed as temporary, even as a joke. While the continuation of Selective Service made many uneasy, the government was loath to baldly exempt men on the basis of superior intelligence.[12]

Even before Pearl Harbor, Bush tried but failed to win blanket immunity from conscription for draft-age men with talents in engineering and science. With the nation now at war and the net of Selective Service widening, Bush fought to protect some of his wizards of war from actually fighting in one. The Selective Service went to great lengths to appear to treat all men the same, relying heavily on seemingly neutral standardized tests in order to determine military fitness. But inequities abounded, unfairly excluding some men and hurting military performance in the field. Historian Michael C. C. Adams has found that of those rejected for service, 32 percent were deemed unfit for psychological reasons culled from as few as four questions. For instance, a negative answer to the question, "Do you like girls?" could lead to exclusion on the grounds of "latent" homosexuality.[13]

Just a month after Pearl Harbor, Bush conceived a plan he thought would put an end to losing his people to the draft. Appealing directly to Roosevelt on January 9, 1942, Bush asked for an executive order establishing what he called a "scientific corps" in the Army and Navy. Bush had yet to marshal support for the idea—he confessed that "I mention the subject before it is fully studied"— but he hoped that a sympathetic Roosevelt "may instruct me . . . in regard to the manner in which you wish the proposal presented."

While Bush waited for a response from the president, Budget Director Harold Smith went to work. Already wary of what he thought was Bush's loose regard for the budget process, Smith looked askance at giving OSRD more freedom from ordinary rules. Among Roosevelt's most trusted advisers, Smith was believed to see the president more than anyone but Hopkins. A native of Kansas and budget director since 1939, Smith was called "the most important man in the administration," by Vice-President Wallace. Members of Congress were known to joke that they had the power but "Smith writes the laws." The war—and the government's swelling funds—only magnified his job, which the *Washington Star* described as "now a Herculean task."

Legendary for keeping long hours, Smith found time to flatly nix Bush's plan. In a memo to the president on January 16, he drove a stake into the heart of the proposed "scientific corps," then added ominously, "Bush now has adequate authority to finance, stimulate and otherwise promote scientific research." Accepting the advice, Roosevelt told Smith to break the bad news to Bush.[14]

The defeat barely slowed Bush down. Two weeks later, he came back after Smith, urging him to exempt those civilian scientists and engineers whom

Bush deemed "indispensable" to OSRD. "The need for the arrangements suggested is urgent," he said. An unconvinced Smith demurred, coolly referring Bush to the official policy on deferments, which didn't cover his OSRD researchers.[15]

Next Bush tried to evade Smith, appealing to the military directly. In private meetings with senior officers he conceded that Roosevelt had rejected his plan for a "scientific corps," but he was "still trying to work out some scheme to keep these men from being drafted for work other than their specialties." One problem with Bush's pitch was that he really couldn't cite any heinous examples of promising researchers killed in battle or moldering in a tent somewhere. Meanwhile, the military was giving active researchers, even those in the prime 18–25 age group, occupational deferments. Playing along, Bush formed a unit to petition Selective Service for such deferments. By the end of 1942, this unit had endorsed 3,602 requests for scientific and technical personnel on OSRD projects; all but 16 were approved.[16]

Still, Bush was troubled by the principle and the process. His researchers were released from their military obligation on an individual basis, while he preferred a blanket deferment. He feared that in time the nation would need more fighting men and that many researchers would be at risk. More than a year later, he was still fuming over this perilous prospect. In a letter to a senator, he criticized the Selective Service for not guaranteeing that men of special talents would be used to full advantage. Mounting a brief for the view that technologists deserved a special status in the military, he argued, "It seems to me that the prosecution of modern war requires the treatment of scientific men as a special group to be specially allocated for work in the fighting services and civilian research, and that this cannot be done adequately." As things stood, Bush found the necessity to keep defending deferments for researchers "probably the most disagreeable" task of the entire war. But he was unabashed about doing so, insisting, "There is no question here of shielding a special class against the rigors of war. It is rather the question of the *intelligent* use of a nation's great asset."[17]

Of all Bush's wartime relationships, probably the most curious involved the Office of Strategic Services. Created by Roosevelt in June 1942, the OSS was headed by William J. Donovan, a New York lawyer who took Britain's Special Operations Executive as the model for his agency. Mistrusted for its clandestine activities, the OSS spread "black propaganda," or disinformation, analyzed foreign intelligence and directed guerilla attacks, sabotage and espionage against the enemy. Operating worldwide, the OSS had more than 12,000 staff at its peak. It also relied on the ample services of the OSRD.[18]

Never comfortable aiding the OSS, Bush did not take great pride in the agency either. He was publicly silent about his ties to the OSS during and after World War II. He didn't disclose his connection with the OSS in his 1949 book, *Modern Arms and Free Men.* His 1970 memoir, *Pieces of the Action,* never mentions the OSS. Even James Phinney Baxter's 1946 book, *Scientists Against Time,* an official history of the OSRD authorized by Bush, contained just one sentence about the relationship between the two agencies, and that came in a footnote. Even during the war, Bush kept many of his OSS documents in a private safe rather than his office files.[19]

Soon after Roosevelt created the OSS, Bush agreed to allow NDRC researchers to assist in creating unorthodox weaponry and spying gear. "If they want to make a fountain pen that does things no self-respecting fountain pen would ever do, we will make one," Bush said. But serving the OSS wasn't so simple. By Bush's standards, the OSS "was a highly undisciplined outfit . . . a strange and somewhat poorly organized agency, [operating] under conditions where there was very likely to be criticism and disagreement." No sooner had the spy agency formed, Bush later recalled, than the "OSS wild men were running all over my shop and butting into things that didn't concern them, interfering with my contractors and generally making quite a nuisance of themselves."

Often the OSS men were bypassing Bush altogether, and nothing bothered him more than being kept in the dark. "It is deadly embarrassing to have something loosely handled," he said. In late September, he realized the extent to which "OSS men were contacting my organization without my permission" and cracked down. After protests to the offending agents had no effect, Bush sat down with Donovan himself in October. The meeting was rough. Bush opened by calling the OSS "a damn nuisance" and threatened to cut off all aid to it. Then he added that he'd already ordered his researchers to refuse all further OSS requests and merely complete their current assignments.

Donovan dug in his heels, which Bush found understandable since "he was in no position to set up his own shop." After a long talk, they reached a compromise: Bush agreed to continue OSS research after Donovan promised that all requests from OSS would come from one of his own aides. The two men also agreed that if any OSS agent stepped out of line, Bush could throw him "out on his ear."[20]

As his liaison with the OSS, Bush picked Harris M. Chadwell, a mild-mannered chemistry professor who struck Bush as "one of the last chaps in the world that you'd think of as being involved in various types of skulduggery." A stickler for secrecy, Chadwell named his subcommittee the Sandeman Club "in order to cover up its activities" (the official name was Division 19 of the NDRC). These shenanigans worried Bush, who two days after Chadwell's formal appointment confided to NDRC chair James Conant, "This is a some-what difficult matter which has been hard to handle from the outset."[21]

It became clearer just how difficult a few days later when Chadwell sent Bush his first proposed budget for the "Sandeman Club." Of the $400,000 budget, nearly half the funds went toward already approved projects bearing such names as "Rainbow," "Abalone," or "Locomotive," "Brimstone" and "Cannon." Chadwell apologized for not idenitfying the codenames in his budget, but promised to do so in person if Bush wished.

One project, which failed to carry a budget line, especially caught Bush's eye. It was codenamed "Natural Causes," which Bush assumed referred to research on assassinations. In his reply to Chadwell, Bush objected that "Natural Causes" fell outside his agreement with the OSS and that any such research must be personally approved by him. Besides, he "understood from our conversation that no extensive work was contemplated in this field in any case."[22]

Even with such a caveat, Bush avoided thinking much about the uses to which Divison 19's research might be put. How was he to know, for instance, whether a powerful explosive nicknamed "Aunt Jemima"—it looked like flour and could actually be mixed into biscuits—would end up exploding in the bellies of Japanese soldiers? Or what about the division's incendiary pencil, incendiary briefcase and incendiary notebook? Surely, the silent flashless pistol and submachine gun, classified as "top secret," would now and again inflict harm on some men.[23]

To be sure, Bush's objection to research on "Natural Causes" mainly had to do with his desire not to end up holding the bag for any embarrassing OSS episode and to limit his own agency's involvement—at least formally—to providing detached technical advice. In principle, Bush had no moral objection to killing in war; he was no pacifist. Indeed, when an NDRC researcher resigned for such reasons in August 1943, Bush condemned the decision without hesitation. He refused to accept that the researcher, who happened to be one of his MIT research assistants during the 1920s, was following "some ethical code which is in some way superior to my own."

While such resignations were rare, scientists were growing more restive about their role as the war widened and as technologically based weapons grew more destructive. Bush was hardly cavalier about moral questions; he addressed them, but unsentimentally and sometimes too swiftly. Writing to his departing researcher, he seemed to be rehearsing an argument that he might have to use again in the years ahead.

> I, too, have had much thought on the question of the way in which a scientist faces a tough world in a realistic fashion. I do think that your position . . . is logically un-

tenable. If you could persuade the Japanese scientists to join you in your position, you would have a very strong position indeed. The alternative would seem to be to leave my son and the sons of many other Americans to face an aggressive Japanese effort to dominate the world, with their scientists supporting their war effort and ours refusing to do so, which does not in any way make sense. Another alternative would be simply to allow the Japanese to run over us without any resistance on our part. You may be able to join Gandhi in such a point of view, but I am not.[24]

Bush refrained from personally dirtying his hands, but Donovan's aide did not. Handling the messier side of the research for OSS was a flamboyant agent named Stanley Lovell, a chemist turned spy who worked directly for Donovan. A self-styled philosopher of sabotage, Lovell dispensed pithy words of wisdom such as, "The saboteur is a man of violence and action," "Security is a one-way street," or "Often the most simple weapons are the best." Lovell's taste for the bizarre wasn't shared by Bush, but he solved the OSS problem. "Stan had a way about him, and he kept things in order," Bush later wrote.[25]

As the research chief for OSS, Lovell was a loose cannon. Donovan gave him a wide berth: his own guarded building, a generous budget and only the broadest directions. In return, Lovell delivered bombs that looked like crustaceans, buttons and shoes with secret compartments, and reams of false documents, including phony Swiss passports and counterfeit Japanese yen. The OSRD's first gadget created for the OSS, Lovell claimed, was a pocket incendiary. A celluloid case filled with napalm jelly, the incendiary had an ignition that could be set for any time between 15 minutes and three days. Lovell said this device was popular with resistance groups in Europe.[26]

Another unusual OSRD weapon, also favored by Lovell, was a hand grenade that exploded more or less on impact, rather than a set time after its arming lever was pulled. Nicknamed "Beano," the grenade became active during its flight; it needed to travel about 25 feet in order to explode on impact. Lovell loved the new grenade and arranged for a demonstration at the Army's Aberdeen Proving Grounds. The test, however, proved to be a cautionary tale for those who thought the military cavalierly shunned new weapons. The Army engineer chosen to demonstrate the grenade gave what Lovell recalled as "a most enthusiastic lecture" on the device. Then "to the horror of us all, he said it would be handled like any baseball and tossed it high in the air over his head. Of course the throw automatically armed the grenade." When he caught it, the engineer was killed instantly. His death, Lovell reported, caused the Army to "abandon the grenade as unsafe—a most illogical decision."[27]

One of Lovell's more successful ideas involved the use of hallucinogens—drugs that quickly disoriented a victim. Division 19 searched for novel ways to deliver hallucinogens to unsuspecting targets. Bush learned enough about the drugs (though he never tried them) to conclude that "a cigarette loaded with the stuff would give a man the symptoms of schizophrenia for some seven or eight hours." While this capability might appeal to the OSS, the drug "looked like a very dangerous affair" to Bush, who worried that one of his own men might accidentally succumb to it.[28]

While Bush grew to like Lovell, he had no taste for Lovell's boss. Donovan made big claims for his importance, which were hard to verify because of the shadows enveloping OSS. One day Donovan—lying—told Lovell that he'd never met Bush. Lovell, who found this incredible, immediately arranged for the two men to have dinner in a private room at the Carlton Hotel. Lovell tagged along for his amusement. He was disappointed. Since both were enmeshed in an array of highly secretive projects, the conversation was dull. At one low point, Bush tried unsuccessfully to wheedle information from Donovan about OSS spies in Germany.

After dinner Donovan got Lovell alone, thanked him for arranging the meeting and then made a curious comment: "Did you notice he began every single sentence with 'I'? Quite an egoist, wouldn't you say?"

The next morning, Lovell attended a meeting chaired by Bush at Carnegie's office. Before the meeting began, Bush took Lovell aside and told him, "You know it was rather noticeable, I thought, that Bill Donovan talked so much about himself. I couldn't get a word in edgewise."

Lovell wasn't surprised that each man mirrored the suspicions of the other. In Washington at war, he decided, "the humble may inherit the earth, but the egoists run it right now."[29]

Donovan and Bush had more in common than just similar personalities. While Donovan was legendary for his madcap schemes, Bush was no slouch. He had recently asked the Army and Navy about the wisdom of creating "artificial battle sounds" (such as fake explosions, machine-gun fire and orders in the enemy's language) that might disorient enemy soldiers or at least "get their goat." He also considered the possibility of collecting thousands of cave-dwelling bats, chilling them into a coma and then dropping them on Japanese cities.[30]

Bush routinely analyzed madcap schemes relayed to him by other Roosevelt appointees, who saw him as a sort of scientific Svengali. Once, J. Edgar Hoover, the director of the FBI, had sought his advice on the mysterious subject of "death rays." Hoover apparently was intrigued by the idea of killing people from afar with powerful beams of energy, and he wanted Bush to check it out. Since this wasn't the first inquiry about death rays—nearly three years earlier the Army had inquired about a similar fantasy weapon—Bush made quick work of Hoover's query, replying that "it does not seem to me that this disclosure warrants further consideration."[31]

Contacts with the OSS and other wacky schemers were an amusing sideline to Bush's main task of convincing the Army and Navy to adopt the weapons created by his researchers. Bush tended to personalize the military's resistance, blaming it on stupidity or the inclination to apply the lessons of the last war to the current one. While some military officers were obtuse, the resistance to technologically advanced weaponry stemmed from a more profound malady that afflicted all modern societies. Innovations, whether in the civilian or military sphere, tended to gain acceptance only after a period of "lag," during which whole social institutions adapted to the pace of innovation. In short, everyone resisted change, and especially resisted the wholesale changes in habits and outlook increasingly demanded by revolutionary technologies.

Astute observers were starting to grasp the way in which technological change had emerged as an independent force in history. Waldemar Kaempffert, the science editor of *The New York Times,* expressed most clearly the new appreciation for the role of technology in history and the profound stress that technological change placed on societies and the individual psyche. In a groundbreaking article published in January 1941, Kaempffert surveyed the entire sweep of war and technology through history. His conclusions were startling. Professional soldiers at best improve weapons; they don't originate them. But this in no way made them unique. Industry also preferred small improvements over big breakthroughs, and the most successful companies often shunned new technologies with the potential to supplant their existing products. Rather than a unique feature of military life, fear and anxiety about new technologies was a central aspect of modernity. In this sense, Kaempffert concluded, "Military history merely parallels industrial history. Both in industry and in war men are regimented."[32]

The problem, he went on, was that the habits and techniques of men did not stand in isolation; they were embedded in a constellation of habits or techniques that functioned like a whole body and could not be overhauled. This "system," to use his language, stood or fell on its own. "Everywhere," he wrote, "there is system—system in reconnoitering from the air, firing shells from a battery, building an airplane, preparing and packing a breakfast food."

Left to themselves, the parties to a system would resist change; at peace or war, civilian or military systems inevitably lagged in absorbing new elements. "The explanation of technological lag in war and industry is the same,"

Kaempffert wrote. "Expense accounts for some of it, tradition and inertia for much, and standardization for most." Standardization yields great efficiencies, but it also imprisons. Having invested billions in a railway system, for instance, "It is financially hopeless in a capitalistic society to change gauges, signals, tunnel clearances, rolling stock, and brakes" in order to support a vastly faster, safer train.

Kaempffert's vision of the system as the defining principle for cultures, societies and technologies was visionary. But the very logic of his argument spawned a conundrum. If every system was essentially closed, how did radical change occur? His answer made intuitive sense. Outsiders spurred change; it was these interlopers who triggered the collapse of systems and the rise of new ones. In war, he insisted, "all the revolutionary means of killing on a wholesale scale came from 'outsiders,' that is from technologists who were not professional soldiers."

Though hardly written with Bush's dilemma in mind, Kaempffert's article presented powerful reasons for Roosevelt and his military chief to embrace the weapons created by Bush and his tribe. Yet at the same time, Kaempffert undercut his logic by concluding that in times of crisis the forces of lag heightened. By his reading of history, wars triggered technological upheavals, but during a war the prevailing systems were so stressed that they could barely bend. "Innovations," he declared, "can be introduced in the midst of war only on a small and experimental scale."

Kaempffert's conclusion, however well-reasoned, ran against the grain of Bush's ambitions. In the months and years ahead, Bush would put all his prestige on the line for one aim: to revolutionize the fighting of war within the framework of *this* war. Bush believed that not only was this goal attainable, it would be attained so convincingly that technologists would then be catapulted into the front ranks of the nation's leadership. If Bush was right, he and his fellow experts would be given the keys to the kingdom. Experts would win a permanent say over the organization of military affairs and probably many other aspects of American life, too.

But if Kaempffert was right, if radical change in the midst of a war was impossible, then Bush had already embarked on the grandest failure of his life.

To improve his chances of success, Bush pushed for technologists to play a big role in war strategy. After Roosevelt picked him as technology czar in 1940, Bush had concentrated on enlisting researchers into his cause and satisfying the military's immediate needs. About a year later, he urged Roosevelt to create a streamlined "war cabinet" with a scientist—probably himself—on the supergoverning body. By early 1942, when it was clear Roosevelt wouldn't form a war cabinet and that the armed services weren't adapting quickly to new weapons, Bush shifted ground. On March 16, he wrote the president that the OSRD's relations with the armed services had so far been "on the tactical level," but that "the time has come" for the agency to begin considering military questions on "the strategic level to determine emphasis and be sure that striking opportunities are not being overlooked or inadequately pushed."

The basic assumption behind this indirect statement was radical indeed. Scientists and engineers must actually assist in planning war strategy and operations. Bush was most concerned about being shut out of the Navy's deliberations: neither Secretary Frank Knox nor any of his staff exchanged ideas with Bush, leaving him to wrestle with hostile admirals. The Army really listened to Bush, who felt that Stimson "always treated me like a son." Still, Army Chief of Staff Marshall and his right-hand man Eisenhower seemed oblivious to technology. Bush allowed that both men were too old "to recognize that a very large part of what they'd learned had become obsolete" because of new technology. Indeed, he saw Marshall as a prime example of the technological "lag" described so aptly by Kaempffert. The best way to nullify the "lag" factor, Bush insisted, was to encourage scientists and engineers to modify war strategies so as to take full account of the newest offensive and defensive techniques.[33]

Bush's enthusiasm for strategy was the culmination of his long move from military outsider to insider. Uncomfortable as a figurehead, he wanted to participate on an equal basis with the top brass. Fueling his desire was the course of the war. In 1941, the Allies had lost 875 ships in the Atlantic, and the operational strength of German U-boats had quadrupled. U.S. entry into the war at first had little effect on Nazi submarine attacks. In the first six months of 1942, another 490 ships were sunk off the U.S. eastern seaboard despite the presence of no more than a dozen U-boats.

In this dark time, Bush began pushing to give his researchers a bigger role in forming strategies that took full advantage of new weapons. At the very least, he believed that "sharp technical men were needed to help plan key operations—with antisubmarine warfare the prime example—that depended on a sound knowledge of the latest technology." But over time, he presumed that his technocrats would plot out broader military campaigns too.[34]

Bush's ambitions shocked senior military officers. Even Julius Furer, the Navy's research coordinator and a Bush ally, was incredulous. Furer first gleaned Bush's views on the subject at an OSRD meeting in March 1942. Speaking in his usual blunt style, Bush said that scientists "should sit with the heads of the Army and Navy in planning the overall strategy of the war." Furer was dumbfounded, writing in his diary: "He thinks that the scientists come in too far down the line, and are called in only to produce the instruments that the Army and Navy want. It is not clear to me whether Bush is activated only by ambition to get into a bigger field, or whether he thinks the war effort would be accelerated by this method."[35]

While his ambition was often on raw display, Bush had sniffed out a weakness in the U.S. war effort. No less an authority than Roosevelt shared his view. Handing Bush a major administrative victory in May 1942, the president endorsed the creation of a subcommittee to his new Joint Chiefs of Staff that would be devoted to new weaponry. Bush would chair the three-person advisory body, called the Joint Committee on New Weapons and Equipment, whose frank purpose was the "education" of the military's top brass.[36]

Bush was the first civilian outside the cabinet to ever formally have a line into the nation's military chiefs. The arrangement set some brass on edge. Of the top generals and admirals, none was more suspicious of Bush than Admiral Ernest J. King. The first officer to serve as both chief of naval operations and commander-in-chief of the U.S. fleet, King was thought to have the strongest mind among the Joint Chiefs. He also had the harshest personality. "He is the most even-tempered man in the Navy," his daughter once said. "He is always in a rage." King knew how to have a good time; he made no secret of his womanizing and drinking. Yet these twin pursuits never came at the expense of his dedication to duty nor undermined his conspicuous competence. "He made the Navy his whole existence, and he gave to it every energy except those he reserved for a private life of notable gaudiness," one historian has written.[37]

King took a dim view of Bush's maneuvers. Old enough to have served as a midshipman in the Spanish-American War, King had qualified as a naval aviator in the 1920s, but he was skeptical of cutting-edge gadgetry and never encouraged subordinates who had new ideas. In 1941, while commander of the Navy's Atlantic Fleet, he had dismissed a new radar system on one of his ships, declaring, "We want something for this war, not the next one." King put Bush's new weapons committee in the same, irrelevant category. On May 26, two weeks after Bush held his first committee meeting, King privately blasted him for "trying to mess into things in connection with the higher strategy which were not his business, and on which he could not have any sound opinions."[38]

Bush felt King missed the point, accusing the Navy's chief of having "a terrible blind spot for new things—and about as rugged a case of stubbornness as has been cultivated by a human being." Others saw King's complaints in a more charitable light. He could innovate, one military analyst later wrote, "when he understood how new concepts fit into the traditional naval view."

Besides, Bush had never witnessed combat, and most of his opinions on strategy were of the armchair variety. Even members of Bush's own fraternity questioned his insistence that scientists plan war strategy. Aviation engineer Jerome Hunsaker accused Bush of being impractical and said that "scientists should be satisfied to solve the problems that are put to them." Hunsaker's vision of scientists as mere technicians, hired hands who left the traditional structure of military authority undisturbed, was influential. As Bush's replacement at the head of the National Advisory Committee on Aeronautics, Hunsaker had ties with the Air Force and a host of aircraft manufacturers.[39]

But Hunsaker was still a minor player. A bigger challenge to Bush's vision came from radar expert Edward Bowles, who had helped to launch the Radiation Lab but then was forced out by more influential physicists. In between assignments, Bowles had been hired in April 1942 as the radar adviser to Stimson, the war secretary, who wanted the Army to improve the use of this technology. Bush had recommended that Bowles be given the job, despite misgivings that he might "stir up strife between the military and my own sections." A talented organizer, Bowles had more confidence than Bush that the military could directly manage its technological affairs.[40]

Outwardly allied, the two men were often at odds. Personally, Bush tolerated Bowles but saw him as "a strange chap" and too much of a loose cannon to count as a loyal friend. Bowles, meanwhile, respected Bush's abilities but recoiled from his take-charge style and still resented him for murky reasons tied to an academic rivalry at MIT 20 years before. More important, however, were the philosophical differences between the two men. Bush insisted that the chief strength of civilian advisers was their formal independence from the military; hearkening back to his exaltation of the consulting engineer, Bush's technologists had the freedom to dispense their expertise without regard for its effect on military prerogatives. Bowles took the other tack. He believed that civilians could function effectively inside military organizations as essentially guns for hire. Bush suspected that experts, if given the chance, would leave the military's employ in droves. Bowles, meanwhile, wagered that civilians would come to enthusiastically accept military patronage, if for no other reason than that in time the brass would grow adept at wooing them.

With Stimson's support, Bowles set about refining the Army's primitive radar program. The engineer had a flair for diplomacy that Bush, who had no such talent, admired. Bowles "straightened out many things for [Stimson] simply by talking them over with the various Generals before the matter came to a head in such a way that the problem solved itself," Bush observed, calling this knack "an art of a high order."[41]

In the course of winning the Army's respect, Bowles built a mini-OSRD under the wing of the War Department. By late 1943 he had hired 27 scientists as consultants, including such prominent figures as Julius Stratton, Louis Ridenour and David T. Griggs. Bowles and staff also had begun to advise General Henry Arnold, the Air Force chief. Soon afterward Stimson marveled that civilian scientists were "now thoroughly in vogue with our Army."[42]

The reason for Bowles's popularity with the progressive-minded Stimson and Arnold was clear. If intelligently handled, the adoption of new technologies would result in organizational reforms within the Army and Air Force, not just in better fighting performance. Here the contrast with Bush was striking. The OSRD director condemned the armed services for thickening sclerosis. Bowles saw the military poised for a historic renewal, with new technology as both the carrot and the stick.[43]

None of this implied an open break between Bush and Bowles. As usual, Stimson saw the situation clearly, describing both men as possessing "sound strategic judgment" and ranking them as "far wiser than either naval or air force officers who had become wedded to a limited strategic concept."[44]

Still, the differences remained. Bush certainly wanted the armed services to increase their own functional expertise, but not at the expense of depriving civilians of ultimate power within civil-military relations. Bowles, meanwhile, thought the chief danger was that the services were too weak institutionally and saw the Joint New Weapons Committee (JNW) as a way to strengthen the internal expertise of the Army, Navy and Air Force. Taking a different tack than Bush, Bowles told the Army's chief signal officer in August 1942 that JNW was "a military body" and should provide the means of retrieving from the NDRC control over radar research and bringing it "back to where it belongs: within the services."

While this was precisely what Bush hoped would not happen, Bowles was too wily to make plain the contradiction. The Army should not oppose Bush's initiatives, he advised its signal chief, because Bush's "vision has already done much to help win the war through regimentation, coordination and inspiration of civilian scientists."[45]

The JNW was proving largely to be an academic exercise. Bush spent hours in closed-door sessions dutifully tutoring a Navy admiral and an Army general, but his efforts—and the committee's dense reports—had scant practical effect. Since the Joint Chiefs didn't actually have to take Bush's advice, the JNW's reports usually moldered. By April 1943, the JNW had met 30 times, enough to convince Bush this well-placed body would not realize his hopes. But such frank talks with generals and admirals had imbued Bush with a sense of the possibilities of U.S. military power—in this war and beyond. Unenthusiastic about foreign entanglements before the war, Bush now predicted that the nation was destined to become the world's policeman and that maintaining superior military power must be the government's chief aim. Replying to a letter on postwar prospects from his political idol, former president Herbert Hoover, Bush bluntly wrote:

> It is an unattractive idea; but I believe we have got to accept the policy of interfering with the internal affairs of the conquered for some time to come on the same basis as interference with the affairs of backward peoples on a probationary basis is inevitable. I believe the interference has to be military. I hope it can be exercised by agreement among the group of victorious powers. . . . But my great fear is that such an association will be too loose, or too soft, to perform the necessary surgical operations to maintain the health of the world, once the immediate danger [of the Nazis] to civilization passes. Your statement particularly troubles me in this connection . . . that the preservation and advancement of civilization cannot forever be based on force. If I give a mathematical definition to the word "forever," I can then agree. But, for *the next thousand years,* I expect that the preservation of civilization will be based on force if it is preserved at all.

Bush's suspicion bore an eerie similarity to Nazi notions of a "thousand-year Reich." The parallel suggests how much the hard work and ambition of men such as Bush arose out of the global inclinations of U.S. leadership. These inclinations were often buried deep beneath the surface. Perhaps Bush gave them bald expression because he felt that Hoover, whom he greatly admired, failed to grasp the stakes. He politely took aim against what he saw as Hoover's softness, presenting himself as a realist with vision. "It is just the abhorrence of the use of any sort of force" that "tends to allow the storm to gather and break in full force upon nations that are existing in a state of *wishful dreaming.*"

Bush's mission was clear: his job would not end with the defeat of Germany and Japan. Fearing a postwar drift into fantasies of world peace, he planned to aid in the expansion of U.S. military force. It wasn't that he expected the U.S. to have any particular enemy such as the Soviet Union; rather he imagined a postwar world of endless enemies, where "the chances of regulating war between major powers is very small."[46]

Harnessing science directly to the conduct of war was crucial in a hostile world. To achieve maximum advantage from innovations in weapons and equipment, Bush believed, waste and duplication had to be pared to a minimum. He desperately wanted the Army and Navy to coordinate weapons work in much the same way that he oversaw NDRC's far-flung divisions. Central planning, besides saving money, weeded out fruitless projects, which freed scarce researchers to tackle more promising lines of inquiry. As money for weapons research poured into the Army and Navy, Bush suspected that the services were violating his rule of parsimony and neglecting the salutary effects of intense coordination.[47]

Bush wasn't sure the military even wanted to consolidate its internal research, which had ballooned during the war to cover rocketry, vehicles, guided missiles, radar, explosives and communications. Casting so wide a net, the services risked squandering their resources. "The real difficulty is that no really comprehensive programs will be established unless the Chiefs of Staff, or in the Army the General Staff, grasp the need for such programs and see to it that they are formulated, approved, and expedited," Bush wrote to an Army officer in September 1943. "This involves, first, the willingness to spend sufficient time and thought on new weapons and their probable influence upon the course of the war to grasp in general what programs need to be implemented and, second, the thorough backing up of whatever organizational machinery they choose to use for the purpose."[48]

Bush's hectoring paid dividends. In mid-October, after beating up on Harvey Bundy, his liaison in the War Department, Bush won a promise from Stimson to create a "new developments division," charged with bringing new weapons into battle and tying their use into strategy. On October 25, Army Chief of Staff Marshall signed an order that brought into being the sort of overarching technical agency that Bush imagined. The division would encourage "expeditious" use of new weapons, learn about enemy weapons to hasten the creation of countermeasures and discourage duplication of research.[49]

Seizing the momentum, Bush urged Furer the next day to set the Navy on the same course. At first skeptical of Bush's desire to share the table with naval war planners, Furer changed his view after witnessing the Navy resist the innovations in radar and the proximity fuze. Admiral King had "made a serious mistake," he said, "in not inviting Bush to sit in on the discussions of some of the problems connected with the war, especially the anti-submarine campaign."[50]

It was not too late for King to accept Bush into his inner circle, but much more than Furer's sympathy would be needed to change the admiral's mind.

Chapter 8

"A race between techniques"

(1943–44)

We nearly lost both wars by reason of the submarine. The public didn't
understand that.

—Vannevar Bush

Bush suffered through February 1943. Nazi U-boats threatened to nullify U.S.
aid to Britain. Terrible weather in December 1942 and January 1943 had
caused a pause in the Battle of the Atlantic, but in February the U-boats sank
108 Allied ships. In the first 20 days of March, another 107 ships sank, yet
Bush thought the Navy remained too wedded to its conventional tactic of
guarding ships with convoys and too resistant to bringing new weapons to bear
on the U-boat problem.

"As the depredations of the U-boats mounted early in 1943," he later re-
called, "there was no doubt in my mind that we were headed for catastrophe.
It was clear enough that, if U-boat success continued to climb, England could
be starved out, the U.S. could mount no overseas attack on the Nazi power,
Russia certainly could not resist alone." No less an authority than Winston
Churchill agreed, telling the British Parliament in February, "The defeat of the
U-boat is the prelude to all offensive operations."[1]

Bush faced a quandary in responding to the U-boat menace. He knew his
researchers were poised to deliver an array of new weapons—from homing torpedoes to magnetic airborne detection of submerged sumarines—that collectively would doom the U-boat. Of all these devices, the Radiation Lab's
microwave, or centimeter, radar was the most important. The Rad Lab now

employed 2,000 workers on 50-odd projects, yet only a handful of shipboard and airborne radar systems had found their way into service. This was a terrible record, and it vividly proved Kaempffert's technological "lag" thesis. Just as Bush himself had feared, creating weapons did not guarantee their use.[2]

In his nearly three years as Roosevelt's adviser on military technology, Bush had found many ways to charm, coopt and convert skeptics in the armed services. For once, however, he seemed at a loss as to how to transform the Navy into a sub-chasing, sub-killing outfit. In his view, one man's blindness had saddled the country with a botched antisub effort. His name was Ernest King. Impatient with civilians, King held fast to military traditions—and prejudices. Perhaps his most controversial belief was that escort ships would defeat the U-boat. "Escort is not just one way of handling the submarine menace," King asserted, "it is the only way that gives any promise of success." As late as March 1, 1943, he declared that "antisubmarine warfare for the remainder of 1943, at least, must concern itself primarily with the escort of convoys."[3]

Bush had often tangled with Admiral King ever since Pearl Harbor. He had most recently implored King for greater influence over naval strategy, but his real aim had been to gain greater sway over antisubmarine tactics. Bush rejected King's approach to fighting submarines and had told King's subordinates many times in the prior months that "the answer to the U-boat is to be found only in vigorously conducted offensives utilizing all of the new as well as the old methods of detection and destruction," including land- and sea-based attackers. A shift in approach was essential, Bush insisted, because the U-boat menace would likely grow even "more serious" in the months ahead as the Germans adopted "many new technical methods" of their own.[4]

Despite such strong words, Bush made no inroads with the Navy. King's rigidness appalled Bush, but out of deference to the admiral, whom he claimed to admire, he refused to take his case directly to the president. Indeed, Bush even delivered a detailed report on submarine warfare to Roosevelt in February without once noting his dissatisfaction with the Navy's strategy. He went so far as to say the Navy's use of new weapons "is coming on well," which he knew wasn't the case. Unusually restrained, Bush was stymied by King.[5]

But so were many other Washington insiders. The most notable casualty in the war with King was Henry Stimson. A radar zealot, Stimson had emerged by early 1943 as perhaps the most unlikely power in Washington's pantheon. In his mid-70s, he had last served as secretary of war in the pre–World War I cabinet of President Taft. While lacking the vigor of his prime, Stimson possessed an openness to new ideas that shamed younger men. As his biographer, Elting Morison, has written, Stimson's "greatest contribution as Secretary lay in opening up the resistant military to the flow of scientific ideas and applications." At a series of informal meetings in the summer of 1942 Bush had convinced Stimson that the Navy's lack of enthusiasm for radar was "probably the most critical problem that now threatens the war effort," and he had since tried to convince the Navy to wholly embrace this technology.

So far Stimson could boast of little success with the Navy, but he had been instrumental in getting microwave radar on scores of Air Force planes that patroled the Atlantic from bases as far away as North Africa. This was crucial. U-boats spent most of their time on the surface, submerging only to attack or escape—thus their vulnerability to radar. Since the fall of 1942 U-boats had had receivers capable of detecting cruder long-wave radar, which helped them to evade attacks. Planes equipped with short, or micro, wave radar were far more effective, forcing U-boats into other waters and helping to sharply reduce Allied ship losses in the last 11 days of March 1943 and in the first weeks of April.[6]

Disgusted by the losses to U-boats in March, Stimson had finally taken his concern directly to Roosevelt, winning the president's blessing for the Army to take command of an antisubmarine task force. But Stimson still needed King's approval, and King refused to give it, citing the Navy's authority on matters of the sea. But the admiral promised to add air attackers to his convoy escorts when the necessary planes became available.[7]

With Stimson stymied, Bush once again took up the radar cause. In late March, he lunched with the president, who pointedly asked for his opinion on the war against U-boats. This time, Bush spoke candidly, airing his complaints about the Navy. Afterward, he told one of King's staff officers that the president had asked about antisubmarine matters and that he had felt obliged to answer even though this constituted a breach of channels. Bush justified his action by saying, "The President certainly is at liberty to ask the advice of anyone on any subject."[8]

King felt differently. When he learned that Bush had spoken with the president, he blew his top. Fearing Bush planned to use his access to Roosevelt to alter antisubmarine strategy, he dispatched Admiral Furer, the Navy's top research officer, to contain the situation. Furer, probably Bush's lone ally in the Navy's senior ranks, had a long talk with the OSRD director on April 9. Furer found Bush more distressed than usual about both the U-boat menace and King's shoddy treatment of him. In one hopeful point, Bush said that he had "seriously considered at one time going to the mat" with King but that he had decided against this course since it would mean "all of his usefulness to the Navy would be terminated."

Bush's rough talk alarmed Furer, who felt himself that the Navy's antisubmarine tactics exhibited "a certain lack of imagination." He advised Bush to "sit tight and to be patient." Furer promptly visited King and advised his boss to meet with Bush about "the general trend in scientific developments." King agreed, and Furer left thinking he had perhaps once more kept the two headstrong men from "an open break."[9]

The next ten days would tell the tale. On April 12, apparently having exhausted his patience only three days after meeting with Furer, Bush wrote King in what seemed like a last stab at reconciliation. In a six-page letter, he carefully made a case both for his inclusion in naval planning and for a radical reordering of priorities in the antisubmarine war. Taking off his gloves, he then directly attacked King. Citing his effective working relations with the Army, he groused about the Navy's unwillingness to consult him personally or even seek the advice of his researchers on the planning of antisubmarine activities. "I, quite frankly, think this is a mistake," he wrote, then explained the perils of ignoring scientific advice in the U-boat war:

Antisubmarine warfare is notably a struggle between rapidly advancing techniques. It involves, to a greater extent than any other problem I could name, a combination of military aspects on one hand, and scientific and technical aspects on the other. It is certainly true that no scientist could hope to grasp fully the military phases of the problem. This can be attained only as a result of a life spent in close association with the sea, with naval tradition, and with the responsibilities of command. Yet it is equally true that no naval officer can be expected to grasp fully the implications and trends of modern science and its applications. This requires, equally forcefully, a lifetime spent in science, and in the personal utilization of the scientific method.

Bush expected the U-boat problem to worsen, he informed King, and then ticked off a list of new technologies that would make it so. The Allies could respond with new techniques of their own, but they must move quickly. "The point is that we are in combat with a resourceful and technically competent enemy, and are engaged in a race between techniques." This "very core of the problem," Bush noted, wasn't fully appreciated by the Navy, leaving him "exceedingly troubled."

Bush's appeal—tough, logical, convincing—was calculated to persuade King that his vaunted escort-only strategy was obsolete. Calling for "a reorientation of our strategy," Bush offered no solutions to the U-boat menace, saying specific tactics were best devised by naval officers. But he suggested that for the Navy to properly value civilian advice it would require "no small change in procedure." Were this not to occur, he warned in closing, questions would arise about the very nature of civil-military relations "in a democracy in time of war."

Swamped by immediate challenges, few in power had time to analyze the shifting relationship between civilians and soldiers. Suggesting the enormity of the shift, Bush told King that the civilians and the armed services would have to deal with each other in new ways whether they wished to or not—just to survive. "When wars were simple, and methods changed slowly, it did not matter much what relation existed. Now that wars are complex, and techniques change with great speed, it matters a great deal."[10]

Facing criticism inside and outside the Navy, King moved to mollify Bush. After receiving his letter, King arranged for the OSRD director to meet with Rear Admiral Francis S. Low, King's new chief of antisubmarine efforts. Bush brought along a couple of his top radar researchers and "had an excellent discussion." Suddenly feeling more optimistic about relations with the Navy, Bush found Low to be "a highly competent officer" whose "addition is going to make a real difference in many aspects of this whole affair."[11]

Then on April 19, Bush and King met for more than two hours at the Navy Department, breaking for lunch with King's staff. For a change, King solicited Bush's views. Careful not to prescribe specific actions for the Navy, Bush instead emphasized the importance of goals and objectives, insisting that "planning at the top level in the absence of the scientific mind was an incomplete and hence dangerous procedure." King then offered his own idea. Referring to Admiral Low, his new aide, King asked, "Wouldn't our problem be met if one of your scientists sat with Admiral Low and participated in planning with him?"[12]

Bush said the suggestion was "excellent, but could be improved upon by choosing three men rather than one." He also insisted that this trio of advisers have a direct line to King and Low on strategy and that they receive all naval information relevant to the U-boat war. This latter point was crucial because naval secrecy often prevented Bush and other high-level technocrats from knowing even what NDRC's own researchers were doing on the Navy's behalf. The order given a naval antisubmarine unit, staffed by Columbia University scientists, was typical. Though working under an NDRC contract, the Columbia scientists were "to disclose no information to NDRC except when specifically authorized."

King accepted both of Bush's suggestions, the three-person council and the rule on relevant information. Uncharacteristically agreeable, the admiral perhaps thought he could simply ignore the new council (a later analyst called it a "sop" to Bush's ego). But King's motives were less important to Bush than his actions. The pact would take on a life of its own once lower-ranking officers dealt on an intimate basis with scientists. Satisfied, Bush dispatched a memo to King on April 20, the next day, summarizing their agreement.[13]

In May 1943 the final piece fell into place. King formed a new Navy fleet responsible for the transformed antisubmarine campaign. In a statement announcing the formal birth of the Tenth Fleet, King made a small bow in Bush's direction, noting that the new command would include "a research-based statistical analysis group . . . composed of civilian scientists [and] headed by Dr. Vannevar Bush."

The creation of the Tenth Fleet marked a turning point in King's relations to Bush and his scientists. The Tenth Fleet, which was officially commanded by King but actually run by Low, had no ships of its own, but for the first time consolidated all antisubmarine warfare in the Atlantic under a single authority. When soon afterward the Navy's operations research group merged into the Tenth Fleet, it brought all U-boat research and statistical analysis under one roof. These moves were more than a bureaucratic coup for King; they gave the Navy the organizational means to gain the edge over U-boats. This was clear to Stimson, who in mid-June decided that nothing "but further trouble" would result from his insistence that the Air Force help the Tenth Fleet hunt subs. While bitter about withdrawing from the hunt, Stimson realized that King had defeated him only by rejecting his proposal and then shrewdly moving "to raise his own force as quickly as he can, to duplicate our efforts."[14]

For Bush's researchers, the Navy's conversion to radar proved a boon. The Rad Lab was suddenly realizing the ambitions harbored by its leaders. The lab's superior detection equipment, coupled with the Navy's new aggressiveness, put the U-boats on the defensive. In April, 15 German submarines were sunk, and another 30 in May. Germany withdrew its U-boats from the North Atlantic; its submarine commander cited, not "superior tactics or strategy," but "superiority in the field of science" as the the reason the Allies had "torn our sole offensive weapon in the war against the Anglo-Saxons from our hands." By early summer Allied ships were safely traveling through waters that had been dangerous just weeks before.[15]

It wasn't just Allied equipment that helped to defeat the U-boat; it was the minds of the scientists themselves. Bush had long argued that scientists and soldiers must forge an intimate relationship. This fusing of two different societies had to occur at the top, he believed, but not only at the top. Applying scientific methods to war would transform even the daily routines of the lowliest soldiers. The scientist needed the fullest possible data, because the introduction of new weapons and techniques could have wide effects.

In the antisubmarine war, Bush had a prime example to buttress his argument. In organizing the Tenth Fleet, Admiral Low had taken over an "operations research" unit and housed it with other antisubmarine staff in a room near King's office in Washington. Operations research was a new mathematical technique born of the wartime need to analyze vast amounts of information exhaustively and creatively. It was predicated on the compilation of accurate information about the enemy's actions and the effectiveness of Allied attacks. The chief proponent of operations research was a British physicist, Patrick M. S. Blackett, who believed that too much technical effort had gone into the "*production* of new devices" and "too little into the *proper use* of what we have got." Indeed, it was after learning of Blackett's contribution to the British Navy that in March 1942 a captain in the U.S. Navy asked the NDRC to recruit a similar team of American academics. An MIT physicist, Philip Morse, was chosen to lead this new group, which set out to apply statistical and new computational analysis to the task of improving the efficiency of antisubmarine measures.[16]

Admiral King took a dim view of the new unit, but the naval captain backed it to the hilt. At first, it seemed that Morse might not realize the hopes of the captain, who was essentially risking his career in order to demonstrate the value of operations research. The trouble was that the Navy had compiled no body of data on antisubmarine attacks, so his team had to create it. Once they did, they tackled core questions such as how best to search for and destroy U-boats with aircraft. Since German submarines often surfaced to take air, they were vulnerable to aircraft attacks. Indeed, 40 percent of aircraft attacks came when U-boats were on the surface of the sea. The trick was to devise a rule, or search pattern, that would maximize the chances of finding surfaced U-boats. Morse's team expected to find improved methods of searching by analyzing the Navy's reports of U-boat sightings.[17]

"We immediately ran head on into an obstacle that was to hinder us in all our work," Morse later wrote. "The reports failed to answer most of our questions."

The solution, however, wasn't simply to revise the reports, which would always have shortcomings. The scientists "wanted to get as close as possible to the operation we were studying, not to be given data at second or third hand," Morse noted. "We wanted technical data to be collected by technical men."

Putting Morse's men on search planes had three benefits: the scientists saw details that never made it into reports; they reported back to the Rad Lab on reasons for the poor performance of some of the radar devices; and flight crews learned that at least some people cared about the accuracy of their information. As one pilot told the scientists, "Hell, I didn't think anyone ever read those damn reports."

Even after their absorption into the Tenth Fleet, however, Morse's efforts were still undermined by the mistrust of naval officers. By then the civilians had taken over recordkeeping for the entire antisubmarine war. Relying on an IBM data-processing machine, the group daily analyzed a plethora of data on U-boat sightings, shipping losses and attacks anywhere in the Atlantic that had occurred within the last 24 hours.

One particular job involved measuring the accuracy of the radio-interception net over the Atlantic. Each day each enemy submarine talked to its headquarters in Germany—in a burst of high-speed code. A set of receiving stations along the East Coast recorded these bursts and triangulated the position of the U-boat. Sometimes planes actually found the U-boats at or near the predicted location, so that the scientists could compare the electronic estimates with the actual locations. As it turned out, "The compared differences were unbelievably small," Morse concluded. Indeed, they were so small that he brought the matter to the attention of Admiral Low, who with a straight face said he would investigate. The next day, he swore Morse to secrecy and told him that the Allies had broken the German code and that the locations given to his researchers weren't retrieved by electronic triangulation but were the actual positions reported by U-boat skippers to their commanders in Germany.

Morse was both amused and irritated by the incident; it showed "the Navy's refusal to make us members of the family" even at the cost of compromising his group's work. He could understand withholding the codebreaking achievement from the public, "but to keep the matter secret from a group analyzing U-boat behavior meant stultifying the analyses. Technical matters have so many cross-connections that a falsification of one part stands out."[18]

Setting aside his hurt feelings, Morse calmed the nerves of naval officers, keeping their secrets while telling his staff to share credit with the Navy for any scientific successes. Bush credited the efforts of John Tate, an NDRC division leader and one of Admiral Low's three scientific advisers, for insuring that "a tense disordered situation evolved into . . . one of cordial cooperation." By late 1943, Morse's analysts were studying wider problems of strategy than just those associated with the antisubmarine war.

Morse's experience was typical of the new intimacy between soldiers and scientists. In October 1943 Bush further blurred the lines between civilians and the military by creating a major new division of the OSRD called the Office of Field Services. His aim was to move scientists and engineers closer to actual combat in order to better apply new weapons or perform their analytical tasks. Typical was the assignment of two scientists to assist the Air Force in the bombing of bridges. A dozen radar engineers helped the Air Force adapt and install detection equipment in planes. In 1944 the Navy asked for 16 experts in radar countermeasures to aid in the installation of equipment needed in the Normandy invasion; within five days the experts were en route to Europe. At the end of the war, even General MacArthur, the longest holdout against OSRD field aid, had 60 scientists in his Manila headquarters.

The scientists in the field walked a tough line. Some soldiers resented civilian interference, derisively dubbing these roving researchers "longhairs." Bush insisted that no OSRD scientist in the field could actually become an officer in the armed services, so these "combat scientists" wore a military uniform with a shoulder patch that bore only the inscription "Scientific Consultant." The combat scientist was entitled to the same housing, eating and transportation privileges as military officers but was neither "subject to their handicaps" nor likely "to stay and sweat it out with them," noted the official historians of the field service. Something of a gadfly, the combat scientist had, in theory at least, a clear line to the top of any organization to which he was assigned and a positive obligation to break the chain of command in order to insure the proper use of new weaponry or equipment.

The close encounters between civilians and soldiers caused conflicts: some officers, for instance, regarded any scientists who went over their heads as a menace. But the experience also demonstrated the benefits of cooperation between these very different tribes. Bush had long predicted, of course, that a full partnership between soldiers and scientists would prove both desirable and practical. But for this to actually happen in combat theaters required that many military officers essentially undergo a transformation in their attitudes. Of those officers who now viewed scientists and engineers as battlefield assets, none stood out more than Admiral King. The Navy's chief cleared the way for civilian scientists to obtain field assignments in the Pacific. To a fellow naval officer, King now seemed to be "entirely in sympathy with the idea that a civilian scientist can be used to work on operational matters . . . ordinarily handled only by officers."[19]

The Army had been Bush's ally in the fight to widen the use of radar against U-boats. But when it came to another decisive weapon delivered by Bush's researchers, the Army proved more cautious. The weapon was the proximity fuze, a shell with a radio-controlled detonator that for maximum effect exploded near its target, rather than on impact or after a prearranged time. The fuze itself was a remarkable feat of miniaturization and ruggedness. Screwing into the front of an artillery shell, it emitted a steady radio wave. The nose cone served as a receiving antenna. If a target came within a few wavelengths of the fuze, it altered its signal to set off the detonator. "This neat electrical trick" was done with only four vacuum tubes.[20]

James Phinney Baxter, the official historian of the OSRD, ranked the proximity fuze among the three or four most extraordinary scientific achievements of World War II. Only the U.S. used it in battle. Neither Germany nor Japan produced anything like it in the lab. While the Germans experimented with 30 different proximity fuzes, when the war ended none had seen any service.

For those with firsthand knowledge of shells, bombs and other explosives, it was easy to appreciate the importance of the proximity fuze. At the start of

World War II, aircraft held a decisive advantage over ground defenses, in no small part because antiaircraft artillery were inherently inaccurate. In 1940, it was believed that good antiaircraft brought down one plane for every 2,500 rounds. The problem wasn't poor aim but gauging the correct range. Both optical and radar range finders were inherently inaccurate because of the time it took for the shell to reach its target; this lag could be corrected for, but not completely. Gunners instead relied on time-fuzed shells. In theory, the shell would explode when the target was within the cone of its exploding fragments, but in practice the shell might explode anywhere along a thousand feet of its path. This meant hits were few.[21]

The proximity fuze improved the defense against aircraft, erasing the attacker's advantage in many situations when combined with other new techniques. Bush had counted the creation of the fuze as one of his top priorities after forming the NDRC because he thought it might restore balance between air offense and ground defense. The basic work on the weapon was performed by his Division T, under the direction of Merle Tuve, the Carnegie physicist. As with radar, the British gave impetus to the American research. Along with the magnetron, British scientists visiting the U.S. in the summer of 1940 also brought some basic ideas about the radio proximity fuze. Though stymied on many crucial engineering issues, the British, who had been discussing such a fuze, provided a rough design and a research direction. Then Tuve's people tackled the riddles of the fuze. They shrank it to something barely larger than a fingertip. They made it rugged enough to survive storage, transport and the force of spinning hundreds of times per second in the air. They developed a long-lasting battery that essentially didn't start working until the fuze was airborne.

By April 1941, Tuve's team had built working prototypes that performed properly at heights of 150 to 300 feet over water. Later that year, Bush urged the Navy to let out manufacturing contracts for the proximity fuze, even though the kinks were still being worked out. The Navy held up full production until fuzes from a pilot production line—using ordinary assemblers rather than scientists—worked 52 percent of the time in a five-mile test of standard five-inch shells. In late January, such tests results were achieved by a slim margin and the Navy pledged to spend $80 million on the fuzes. No longer under Carnegie's umbrella, Division T moved to a garage in Silver Spring, Maryland, in the spring of 1942. Taking the name of the Applied Physics Laboratory, the team now worked for Johns Hopkins University through a contract with OSRD.

By September 1942, contractors were making 400 fuzes a day; the figure would reach 70,000 a day by the war's end. Two months later, 4,500 shells with proximity fuzes were shipped to the Pacific with instructions that they be given to ships likely to see early action. On January 5, 1943, the fuze saw its first action, taking out a Japanese Achi 99 dive bomber on its second salvo.[22]

From then on, the Navy viewed the proximity fuze as one of its most secret weapons. In the course of 1943, one-quarter of the rounds shot by the Navy's five-inch artillery guns were proximity fuzed, but those rounds accounted for 51 percent of the enemy planes shot down. Bush later boasted that the fuze improved by a factor of seven the accuracy of the Navy's guns. On the sea, he wrote, "the fuze forestalled many a catastrophe."

From the start of the fuze's use in combat, the Navy feared the enemy might recover a dud and reverse engineer a fuze of its own. At the Navy's urging, the Joint Chiefs limited use of the fuze to the sea, banning use over land. The restriction didn't bother Bush at first, since he considered the threat of Nazi reengineering to be "serious business." Overall, the Navy's support for manufacturing and deployment of the fuze was "magnificent," he believed.[23]

But as the massive Allied invasion of northwestern Europe approached, Bush grew more testy about the ban on fuzes over enemy lands. The Normandy invasion would be the biggest land battle of the European war, and Bush wanted OSRD to make a visible contribution. In February 1944, he assailed the ban on fuzes over land at a meeting of the Joint New Weapons committee, telling a high-ranking officer from each service that the policy was undermining the Allies' effectiveness. A few months later, at another JNW meeting, he made the case more forcefully, saying withholding the fuze "means added casualties, added time, longer time to the end of the war." Now sputtering with rage, he regretted accepting the ban on land fuzes for so long. "I would have released [the fuze] a long time ago upon the basis" that the risk of not doing so was too great. If the fuze had been used for the past 18 months, he reasoned, it would have been "advantageous to us." Besides, the possibility of the enemies copying Allied fuzes should be stacked against the chance that they might build fuzes independently. "The Germans and Japanese are perfectly capable of making a bomb fuze," Bush insisted. "It is just a continuing surprise to me that they have been so slow." For all he knew, the enemy was readying their own fuze even as they debated whether the Allies should widen its use.[24]

Despite his angry appeal, Bush lost the argument again; the Joint Chiefs kept the ban on land use of fuzes intact. The inability to force a policy change frustrated Bush. It all came back to raw power. It wasn't enough to build weapons, they must be used, and the service chiefs remained the final arbiters of new weapons. If only Bush had more power. The situation ate away at his nerves. Writing to his closest associates on May 6, a month before D-Day, he confessed:

The stress at the present time is enormous, and I think that the entire organization is on edge. The suspense caused by the [impending Normandy] invasion is intense to us who know some of the elements. The manpower situation has driven us all to the verge of distraction. We are at the end of a winter when academic men normally look forward to respite, and there is no respite in sight. The excitement of broad innovation of new things has been succeeded by the trying job of moving them into use.[25]

Bush's "trying job" grew more difficult on June 13, 1944, when the Germans bombarded London with a terrifying weapon. Called "revenge weapon number one" by Nazi propagandist Goebbels, the V-1 was a pilotless aircraft aimed at inflicting on the British some of the damage the Germans had suffered under relentless Allied air attacks. Armed with a one-ton warhead, this early ballistic missile was a flying bomb with a range of about 150 miles. It covered this distance in about half an hour.

The arrival of the V-1 shook up Bush, but didn't surprise him. Warned by the British that Germany was preparing to launch guided missiles against London and southern England, Bush's Division T began work on defensive measures a full six months before the first V-1 outbreak. Three months before the attack, a shipment of proximity fuzes arrived on British soil. The fuzes would make artillery fire more effective; their use against rockets was approved on the belief that any duds would fall at sea or over British territory.

Bush took the V-1 seriously. He worried especially that the Nazis might arm the rocket with radiological materials, biological agents, or other poisons. When an aide told Bush the rocketry idea "sounds fantastic to me," he retorted, "No, I'm afraid it's for real."

Believing that "the menace of rocket attacks [is] very great indeed," Bush traveled to London in the spring of 1944 to personally warn Eisenhower. Too few defensive measures were in place, he told the general, because the Americans consider "defense against rockets a British job, and vice versa." Without better preparation, he added, German rockets might rain down on Plymouth and Bristol and turn plans for the Normandy invasion into chaos.

"You scare the hell out of me," Eisenhower finally said. "What should I do?"

Bush suggested bombing the German launch sites, which was done. He also urged a greater reliance on the advanced weapons streaming from his research labs.[26]

While the combination of microwave radar and the proximity fuze promised to take out a fair share of German rockets, Bush was painfully aware that the Allies had no match for the V-1 and the more powerful V-2 rocket, which the Germans first fired in September 1944. Weighing 13 tons, the frightening V-2 could travel at supersonic speeds and at altitudes as low as 50 feet. V-2s killed 2,500 Londoners before their launch positions were knocked out in March 1945. The U.S. had no program to produce anything like either the V-1 or V-2 for the current war. Rocketry wasn't the only technical area in which the U.S. trailed its enemies. The Nazis had superior aircraft engines, for instance, while the Japanese, at least for much of the war, had better torpedoes. But in no other area were the strategic consequences so dire as in rocketry.[27]

Riding the success of radar and the related proximity fuze, U.S. researchers sometimes fell into jingoistic breast-beating, citing these weapons as evidence of the inherent superiority of American civilization. Alfred Loomis, the radar expert and close associate of Bush, went so far as to say that the U.S. lead in radar was "convincing proof of the magic efficiency of American individualism and laissez faire." He saw Bush's hands-off way of overseeing the various NDRC projects—"leaving them with complete freedom" on technical matters—as the epitome of the American way. Success, he observed, came from "free agency and free[dom] from politics."[28]

But Loomis made his remarks a year before the first V-1s hit London. German rockets mocked the whole notion that totalitarian science wasn't capable of stellar achievements: Even as Germany was going down to defeat the country's top rocketeer, Wernher von Braun, was writing specifications for a missile with a range of 2,800 miles, or far enough to strike American soil. As the U.S. had not even the beginnings of its own ballistic missiles, perhaps it was not at all obvious that the American character and the nation's economic system would alone guarantee primacy in every seminal military technology. Because of interservice rivalry and jockeying within the Army by its Air Force wing, the U.S. even found it impossible to bring a copy of a recovered V-1 into service. At the end of the war, ballistic missile research was among the most urgent military needs.[29]

Under the circumstances, Bush fended off complaints, as best he could, that his side had not done enough. Mainly, he cited the achievements of U.S. science and engineering in other areas—including effective work done in *unguided* solid rockets—as proof of the country's vitality. This was hardly a hollow rebuttal to claims of Nazi superiority in technology. Despite its lead in rocketry, the Germans were shockingly behind the U.S. in less-futuristic technologies. The country's newest generation of aircraft, slated to replace models designed in the mid-1930s, faced unexpected snags, forcing the Luftwaffe to fly aging models for most of the war. Germany's mobile forces suffered from even greater handicaps because of the failure to build enough tanks and trucks. By 1944, the U.S. and British forces were fully motorized, while the German army still relied on more than one million horses.[30]

Yet on the subject of guided missiles, Bush had to admit that "we are not in this country moving at maximum speed" and essentially had no chance of negating the Nazi lead. In characteristic fashion, he blamed the U.S. deficit on the military's failure to centralize missile research under a single chief, relying instead on a confusing array of separate and sometimes duplicated programs. But rooting out duplication was hard, he confessed, because each project had its defenders. When "you pick out one project and try to terminate it," he once complained to senior officers, "somebody will always appear and think that's the thing that will win the war."

As early as September 1942 Bush had acknowledged that OSRD's approach to missiles was "bad" and told his people "they have got to bring it in order." But the disorganization persisted, perhaps because Bush failed, in his words, "to find one good man to head it up." Yet Bush never considered Robert Goddard, an American rocket pioneer, as the answer to the military's needs, perhaps because he was a loner, not a leader, who was cautious about his results and concerned with potential infringements of his many patents. Even though many contemporaries dismissed Goddard as a crackpot, Bush's failure to solicit his advice during a crisis was puzzling, if not egregious.

To some critics, Bush's desultory attitude toward rockets revealed an unaccountable blind spot. By contrast, Bush had passionately promoted the development of radar, the proximity fuze and the atomic bomb. Were rockets and missiles so much more pie-in-the-sky? Bush certainly thought so. "I don't understand how a serious scientist or engineer can play around with rockets," he had said before the war. To rocket enthusiasts, it seemed that he held fast to this belief.

Not surprisingly, U.S. missile research was bogged down. In May 1944, Bush found the lack of progress disturbing. But he accepted the situation with unusual calm, despite asserting that it was "entirely possible that fully developed guided missiles . . . might even revolutionize" many aspects of warfare. With leading researchers busy on other weapons, the U.S. would take longer than two to three years to build effective rockets. To Bush, it was a question of priorities and likely payoffs. It would take time and expertise to add range and accuracy to missiles, he explained to Furer in August 1944. It made more sense to sink these resources into the A-bomb, which "has a better chance of being developed during this war" than missiles. At best, investigations into missiles "should be done on a long range basis," he said, "and not on the theory that it is a weapon for this war."[31]

Deterred by the disadvantage of the nation's late start in missiles, Bush realized that the U.S. lag partly stemmed from broader cultural conditions. Keenly aware of the way social, political and economic forces shaped technological outcomes, he observed that the U.S. lacked clear incentives to develop rockets because prime targets were so far from its shores. "The enemy has one attractive and obvious target, namely London, whereas we have no comparable situation," he wrote in December 1943. The German effort also had been stimulated by the post–World War I Versailles Treaty, which did not include rockets or missiles on its list of forbidden weapons.[32]

When the Germans finally unleashed their V-1s on Britain on June 13, 1944, Bush at least could boast that the Allied preparation paid off. Relying on an array of defensive weapons, including radar and the proximity fuze used in combination, the Allies limited the damage inflicted by the V-1. On the first day, only one bomb hit London, killing 10 people. To be sure, the Allies' high-tech defense was aided by two factors—the successful raid on Germany's Peen-emunde rocket plant in August 1943, which delayed the V-1 attacks by months, and the Normandy invasion, begun on June 6, 1944, which helped reduce the V-1 threat by driving back the Luftwaffe from positions from which V-1s could strike England, reducing the number of incoming rockets. Of Germany's 35,000 V-1s, only 9,000 were launched against the English.[33]

Finally, a decision by Germany greatly aided the Allies. The V-1s carried conventional explosives, not poison gas or radioactive material. In Washington, Bush learned of the initial attack just as he left his office for Capitol Hill with the Secretary of War. In the car, Stimson put his hand on Bush's knee and asked, "About this V-1 business, Van, how do you feel now?"

Thinking of how the Germans might have loaded their V-1 warheads with poisons, Bush felt the Allies had escaped disaster. The damage and casualties, while regrettable, could have been much worse. "I feel damn relieved," he said.[34]

Indeed, Bush had wagered that Germany would not load its rockets with poisons. Seven months before the first V-1 attack, in mid-November 1943, Bush had reviewed data from the first photographs of German rocket-launching sites in northern France. He concluded that poisonous rockets weren't "a serious threat" because the poisons would be too hard to handle and deliver.[35]

The very utility of the proximity fuze in defending against the V-1 awakened the Army to its value. The Allied armies had landed on the Normandy beaches on June 6, just a week before the V-1 attacks began. Pushing their way across France against fierce German resistance, the Allies needed every advantage they could muster. Yet the ban on use of the proximity fuze remained in effect. To Bush, the rule seemed less rational now that Germany's collapse looked like months, not years, away. His gut belief was that the Nazis could not clone and then mass produce a proximity fuze in anywhere near the time left the Third Reich.

Seeking to persuade the Joint Chiefs of this, he convened a committee of top engineers and asked them one question: If the Japanese or Germans were handed the best proximity fuze, how long would it take either of them to put it into production? The consensus estimate was two years.[36]

Convinced that the proximity fuze should be immediately deployed, Bush made his case to the Joint Chiefs in October 1944, describing the fuze as central in antiaircraft fire and as an antipersonnel weapon fired from howitzers. The latter application, while untested, looked promising. It had long been believed that shells exploded at just the right height over enemy troops would inflict heavy casualties from flying shrapnel. This effect was usually hard to achieve, but tests of the fuze had suggested that it might improve the effect of howitzer fire against soldiers by a factor of ten.[37]

The Joint Chiefs, erring on the side of caution, turned Bush down. Admiral King, it seemed, still insisted that proximity fuzes be used only at sea. Since the Army desperately wanted the weapon, Bush thought he had a chance to reverse the decision. He enlisted the aid of a sympathetic Army officer, General Joseph T. McNarney, the newly appointed supreme commander of the Mediterranean theater. McNarney promised to persuade the Army Air Forces to endorse the use of the proximity fuze on land, if Bush won the same concession from Admiral King.

Bush was willing to try, but he wasn't optimistic. Even after reaching a détente with King on radar, he still considered his nemesis a tough customer. Meeting in late October, King greeted Bush with a scowl and grimly declared, "I have agreed to meet with you, but this is a military question, and it must be decided on a military basis, to which you can hardly contribute."

Bush never backed down from a fight, and his response was equally brutal. The issue of the proximity fuze "is a combined military and technical question, and on the latter you are a babe in arms and not entitled to an opinion."

After this exchange of pleasantries, the meeting settled down and by the end of it King had acquiesced. Bush then joined King on October 25 at a meeting of the Joint Chiefs, who in principle ordered the release of the proximity fuze for use over land. The Chiefs declined to set a specific release date, but Bush was enthusiastic enough to leave 24 hours later for France. Fearing obstacles in the use of the fuze, he wanted to do what he could to smooth the debut of the new weapon on land.[38]

It was Bush's closest brush with combat during the war. The year before, he had visited London to talk with British officials, but that hardly compared with traveling in a country just liberated from the Nazis. Accompanied by an officer attached to his Joint New Weapons Committee, Bush first went to Versailles, where he called on Eisenhower's chief of staff, General Bedell Smith. Bush had wanted to meet the supreme Allied commander himself, but Eisenhower apparently would not meet him. Even General Smith seemed none too happy about accommodating Bush.

"What the devil are you doing over here?" Smith asked. "Don't we have enough civilians in the theater without your joining?"

Bush shrugged off the jibe, and explained that he was here "to try to prevent the destruction of one of the best weapons of the war."

With the sparring out of the way, Smith acknowledged that the proximity fuze was hot stuff and listened as Bush rattled off a plan to get the millions of fuzes already in France to the right places and into the hands of people who knew how to use them. This was no small task, since the very existence of the fuze was so closely guarded that even ordnance officers in the field did not know about it. When Bush finished his litany of requests, Smith barked, "OK. I will do all that. Now will you get the hell out of here and let me get to work."[39]

Bush toured the Western Front for a week, meeting mainly with ordnance officers who had a direct interest in obtaining more effective shells. He spent one evening in a small French town. With a group of ordnance officers he stayed in a French inn that had been hit by a shell but not quite knocked over. The officers asked Bush about the proximity fuze and seemed eager to get their hands on it. As talk grew more feverish, the lights went out and never came back on. In the dark, Bush and men at least twenty years his junior drank a bottle of brandy.[40]

While he learned much from his talks with ordnance officers, Bush resented that no commanding generals met him during his visit to the Western Front. He took the unwillingness of the top brass to see him as a personal slight; it reinforced his sense that they did not appreciate the way new technologies were altering military tactics and strategies. Technical issues were invariably shunted off to the lower echelons. At one stop, General Jacob Devers ducked into a meeting between Bush and some junior officers, but stayed only long enough to say, "I would join you but I have a fight on."

By mid-December 1944 the time had come for the proximity fuze to join that fight. Germany, in a final try at checking the advance of the Allies, counterattacked through the wooded Ardennes region of Belgium and Luxembourg. The thrust led to the largest land battle fought by the U.S. in the war. The Nazi counteroffensive on December 16 prompted the Joint Chiefs to immediately release the proximity fuze, which had not been scheduled for combat until December 25. That first day, the fuze knocked out German planes, and two days later it was first used in howitzers to great effect.

Observers marveled at the proximity fuze, which helped to halt the German advance. Prisoners of war described Allied artillery fire as the most destructive they had ever seen. The fuze, however, was by no means a decisive factor. Military analysts attributed the Allied victory to the speed with which their forces moved; the American units of 1944 were completely motorized, with the U.S. 1st Army bringing 48,000 vehicles into the battle zone during the critical ten days beginning on December 17. Yet while other factors predominated in the Battle of the Bulge, some of the best military minds realized that the wizardry of electronics was rendering combat far more terrible than they had ever imagined.

Writing to his ordnance chief on December 29, General George Patton described "the new shell with the funny fuze" as devastating and evinced a keen sense of the Army's new dependence on advanced research. "I think that when all armies get this shell we will have to devise some new method of warfare," Patton wrote. "I am glad that you all thought of it first."[41]

With the success of microwave radar and the proximity fuze on the battlefield, Bush had amply demonstrated the benefits of placing civilian researchers in the service of military imperatives. In 1944 Bush's OSRD was spending $3 million a week on 6,000 researchers at more than 300 industrial and university labs. The work went well beyond the narrow definition of weaponry. "They are devising mechanical ears and eyes to hear and see what neither the human ear nor eye can detect. New types of projectiles are being planned and so, too, are bigger and better bombs and guns. And while some of these men plot these instruments of death, others equally intent are bending over microscopes and test tubes seeking blood substitutes, growing penicillin and working to develop new methods for saving life."[42]

Bush's grand wager, placed four years before in the privacy of the Oval Office, had paid off handsomely. Not only had he retained the confidence of President Roosevelt, but he had managed an intricate set of relationships, snaking across the confusing landscape of wartime Washington. The details of his achievement were still hidden for security reasons, but the broad outlines of his work were starting to seep out.

Despite his professed desire to remain in the shadows, Bush's reputation was growing. Earlier in the war, the press had depicted Bush as a kinetic egghead charged with melding thousands of American scientists into "a super-brain." As an academic on the lam, he appeared aloof and vaguely intimidating to the American public. Now the press presented Bush in more pragmatic and appealing terms: he was a military asset and a darn important one. With the war unleashing a burst of technological change, he also was viewed as an all-purpose wise man who might help Americans make sense of the new gadgets flooding into their lives. This was an optimistic image of Bush as a benign organizer of expertise who after the war ended might benefit civilian society in ways only dimly perceived.[43]

Bush's new position in the popular mind was ratified by *Time* magazine in its April 3, 1944, issue, which arrived just as he prepared for his struggle with Admiral King over radar. Calling Bush the "general of physics," the newsweekly plastered a drawing of a cheerful and bespectacled Bush on its cover. He looked avuncular, with his shining eyes peering past what looked like a giant vacuum tube and his hair falling slightly onto his brow. His trademark pipe was nowhere in sight.[44]

In the accompanying article, the OSRD's 54-year-old director was unashamedly lionized. His curiosity was "insatiable," his memory "prodigious" and his personality "self-effacing." There was even the suggestion that had the country elevated Bush and his scientific clique sooner, World War II might have never happened. As Bush told the magazine, "If we had been on our toes in war technology ten years ago, we would probably not have had this damn war."

A supreme rationalist who instinctively acted on his convictions, Bush was a tough-minded realist who only reluctantly trained his expertise on the murderous job at hand, *Time* asserted. Yet despite succeeding in the rough-and-tumble big city, the magazine's subject retained his small-town virtues. "Bush feels most at home in a Cape Cod fishing boat," *Time* noted reassuringly, and reminded the reader that, while Bush was now too busy to enjoy his diverse hobbies, before the war "he relaxed by working in a cellar machine shop building boats [or] driving a tractor on his New Hampshire farm."

The hagiographic profile of Bush seemed designed to soothe an anxious nation seeking to personify (and to render less menacing) the impersonal forces altering war and perhaps, all too soon, peace. Bush balked at being fussed over, however, especially when the acclaim came at the obvious expense of dozens of other men whom he felt also deserved it. Within days of the article's release, he dashed off a stinging rebuke to a *Time* staff member. "I am not enthusiastic by any means about publicity of this sort, for this organization has thrived because of its anonymity and we have not been anxious to attract attention," Bush wrote. "I am also disturbed by the considerable amount of personal mention [in the article], for I am simply one of a large team of scientific men, many of whom can not at this time have the recognition that they deserve."[45]

To be sure, *Time* had added considerable luster to the image of OSRD too, saying the agency's was "regarded almost with awe" in the nation's capital as much because of its smooth operations as for the output of its now 18 divisions. The magazine, undoubtedly for reasons of security, sugar-coated OSRD's stormy relations with the Army and Navy, insisting that the "harmony" between the scientists and the military was "silk-smooth." Since a few weeks before Bush had termed these same relations "lurid," he must have at least found amusing the benign description of his bureaucratic in-fighting.[46]

Time certainly glossed over contradictions in its cover story on Bush, but its profile was impeccably timed, catching the OSRD director at the peak of his power. Bush was on top, and it showed. To those who met him for the first time in the summer of 1944, he seemed Olympian and yet endearingly informal. "The man is positively magnetic: friendly, shrewd, frank and made of steel springs," one interviewer wrote. The curtain of secrecy still obscured the full extent of his influence, to be sure. But it was plain that Bush's bark made bigshots tremble, while his straight talk impressed and amused lesser folks.[47]

Perhaps nothing better illustrated Bush's clout than the public battle over draft deferments. Throughout the war Bush had railed against the specter of young scientists drafted into the armed services. He failed to win blanket immunity from the draft for researchers, but his pestering paid off in 1943 with the creation of a Reserve List authorized to exempt as many as 7,500 researchers from military service. This number hardly seemed excessive in light of the pervasive, though low-profile, unease with the draft. For instance by the fall of 1943, politically powerful farmers had persuaded Congress to give 2 million farmworkers occupational deferments.[48]

Late that year, as the Army started to demand more young men for its planned invasion of western Europe, Congress shifted its manpower policy against occupational deferments. With Roosevelt chiding draft boards for their leniency, Bush grew alarmed. Many crucial OSRD projects were dependent on young men. Of the 1,915 male employees at the Rad Lab, 240 were under 25 years old and all but five had avoided military service through special intervention. If these men were drafted, Bush feared the river of new radar devices might slow to a trickle.[49]

In early 1944, he raised his campaign against the drafting of scientists to a new level. He complained to Harvey Bundy, Stimson's chief aide, and hounded the war secretary himself. When Stimson's pestering of Roosevelt failed to satisfy Bush, he grew almost hysterical. His passions inflamed, he suffered bouts of pounding on his desk and cursing. He refused to accept the narrowmindedness of the draft boards. Though just a small share of OSRD's researchers had deferments, Bush cried that their loss would force him to "curtail essential activities." In a long letter on April 3, he even raised the specter of talented young scientists being slaughtered on the beaches of France by pointedly reminding Stimson that Britain in 1914 had stupidly sent one of its greatest physicists to the front lines. "His name was Mosley. He was soon killed in action," Bush grimly wrote.[50]

Fearing the U.S. had not learned the "Mosley" lesson, Bush implored Stimson to handcuff the Selective Service, which he believed was preparing to wipe out the deferments of about 1,000 of his OSRD men. "On the matter of manpower, and in particular the use of young scientists, I have had a continuous and sometimes disheartening struggle," he wrote. Despite OSRD's many achievements, "The young scientist has remained in an unstable state. There have been implications that he was avoiding his full duty to his country. There have been statements that all sound men below a stated age should be inducted into the army." His researchers were with "damaged morale" and "doubt in their minds"; neither helped further work on new weapons.[51]

In the middle of April Bush suffered through one of his most anxious weeks of the war. Gripped by the specter of relentless draft boards snatching scores of vital researchers under the age of 26, he orchestrated a public-relations blitz aimed at shaming the government into reversing course. "As a good soldier," Bush "made no public outcry but outcry there was aplenty from the nation's scientists," who accused draft boards of sabotaging war research, *Time* observed. The scientists argued that research—especially in the burgeoning field of electronics—was a youth's game. One scientist even said that "no man older than 35" could gain a "fundamental grasp" of the subject.

Bush's persistence paid off. Selective Service, at the behest of the War and Navy departments, chose to maintain OSRD's deferments. From the spring of 1944, virtually no one under 26 was drafted; for the entire war, OSRD lost only 64 of the nearly 10,000 men it wanted deferred. Bush's hectoring and the evident value of OSRD weapons had placed scientists and engineers in a new light. As one journalist noted at the time, "Many of these men . . . are worth their weight in generals."[52]

By early 1944, Bush ranked among the nation's most influential government officials. As his many showdowns with Washington bureaucrats and the top brass illustrated, he had left his mark on many of the most vexing issues of the war. His status as a Washington insider, meanwhile, had been assured by the glowing cover story in *Time* magazine. Drawn to the corridors of power, Bush had already been thinking about his role in postwar government. Yet he had an urge to chuck it all and retreat into private life. "When this war is over I want to be a private citizen, able to say what I think, and with that in mind I may pull out of every governmental connection whatever," he wrote a friend in September. "On the other hand, I may be so completely weary that I will not say anything at all."[53]

Uneasy as a member of a government whose scope had greatly expanded in the crucible of war, Bush grasped tightly to the image of himself as an ordinary American. While having a chauffeur and a top security clearance made it impossible for Bush to pass himself off as a regular Joe, he nonetheless was in many ways subject to the limits and obligations imposed on his fellow citizens. His two sons were headed for active duty in the armed services (the older, Richard, would work as a military physician, and the younger, John, would pilot aircraft). Like everyone else, he needed the government's permission to buy a new set of tires. If the numerous perks that went along with high government office seemed to set Bush apart from the common man, he needed no further evidence of his mortality than to recall that his taxes had been audited at the height of war mobilization.

Certainly what Bush came to call his "tax tangle" partly reflected his animus against income taxes. Indeed, the byzantine conflict arose from his rather bald attempt to play fast and loose with his legal residency in a bid to reduce his tax payments.

The origins of the tale stretched back to before the war. Before accepting the presidency of the Carnegie Institution, Bush had purchased a farm in East Jaffrey, New Hampshire. On leaving his home in Belmont, Massachusetts, for Washington on December 27, 1938, he stopped in East Jaffrey for four days. In that time, he declared himself a New Hampshire resident, registered to vote and paid a poll tax. On New Year's Eve, he left for the nation's capital to begin a new life. He made the trip, he believed, secure in the knowledge that he could now legally file tax returns from New Hampshire, a state with no income tax.

The state of Massachusetts was the first to balk at this arrangement. Just as Bush was settling in as Roosevelt's technology adviser in 1940, state tax officials deemed him a Massachusetts resident for the 1939 tax year and insisted on collecting $434.98 from him. Never one to take an insult lightly, Bush paid the tax under protest and filed a complaint with the aid of an attorney.[54]

While the Massachusetts case simmered, the District of Columbia got into the act. Tax officials claimed Bush as a District resident and insisted he owed income tax. Bush resisted, citing his New Hampshire residence—all four days of it—as evidence for his exemption. The dispute sent him into a rage. To start with, there was the transparent incongruity of D.C. officials, who after all were essentially an arm of Congress, hounding him for a few hundred dollars when he controlled millions of public dollars with virtually no oversight. "It seems to me absurd to be having a contest with tax authorities, while trying to win a war," he wrote his D.C. attorney.

Yet Bush was prepared to fight this contest to the finish, stubbornly clinging to his principles. Unburdening himself to his attorney, he exposed his bedrock conservatism. Convinced that his freedom had been trammeled, he fumed:

> Of course I feel strongly that I have a right to pick the place that I will live, my home there, prepare for the time I will retire, establish my relations with my friends and neighbors and contribute my part of taxes to the place. I don't think the burden of proof should be on me in such a situation. I don't think the taxing authorities have a right . . . to demand and collect a tax and force me into expense to get it back, when I have already paid a parallel tax elsewhere. I believe the District should certainly have the burden of proof before it disenfranchises a citizen. I believe the Supreme Court might well affirm that.[55]

Circumstances, however, conspired to rescue the justices of the Supreme Court from ruling on a tax case involving the "man who may win or lose the war." In a judgment that ended his case, the D.C. Board of Tax Appeals sustained Bush's position. As far as they were concerned, the director of the OSRD really was a resident of New Hampshire.

It took until February 1944, but the Massachusetts tax commissioner reached the same conclusion. By then, however, Bush had been forced to present his side in a public hearing, which prompted a Boston newspaper to lay out the details of Bush's finances, including the size of his Carnegie salary. These disclosures enraged Bush anew. Now his victory had been poisoned. "First they demand taxes. Then they demand them at once, and threaten dire results if I do not come across," he howled to his Boston financial adviser. "Then they hold them and pay me no interest on them. Then, when at some labor and expense, I politely try to show they have been in error, they allow public access to some of my private affairs, which would certainly have not been open if they had not dragged me in. Then they are found to have been in error all the time. Then they do not even say 'sorry.'"[56]

In victory and defeat, the incongruity of Bush's "tax tangle" seemed to escape everyone but him. While waging a war against state tax authorities, he was secretly planning the most destructive weapons ever known. It would be hard to imagine Bush's foreign counterparts—in England, Germany or Japan—ensnared in a similar petty imbroglio with the government. That Bush could be pursued as a tax evader and still carry out his military duties unimpeded was a distinctive feature of the American system: rule by law, for the meek and the mighty. Even in a national crisis, federalism lived on, states insisted on their parochial privileges.

There was still another way to read the situation. Bush's tax tangle exposed his distaste for government. That his distaste was both paradoxical and popular makes it no less peculiar. Here, after all, was an organizer of expertise on behalf of national security. Through his actions, he was vastly expanding the state's capacity for governance, not just for the length of the emergency but possibly for generations to come. The precedents he was setting as a military technologist and a civilian manager would not easily be ignored by future presidents. That Bush could assume such a miserly posture toward the very government he sought to defend and strengthen revealed the extent to which he remained faithful to the ethos of an earlier time, perhaps even another century.

Chapter 9

"This uranium headache!"

(1939–45)

I wish that the physicist who fished uranium in the first place had waited a few years before he sprung this particular thing on an unstable world. However, we have the matter in our laps and we have to do the best we can.

—Vannevar Bush

Throughout the war, Bush shared in a secret world, hidden even from most inhabitants of the subterranean society of arms and men. This secret world was like a box within a box, a hidden room behind the wall of war. It had its own rules or, rather, no rules. It mocked all of Bush's other war work, looming in the background like a ghoulish shadow. With the passing of each day, the shadow grew larger and larger until it engulfed everything in sight.

This was the new world of the atomic bomb.

Bush did not cheer the discovery of fission. In early 1939, he was aghast at the public interest in the splitting of the uranium nucleus and the possibility of a fission bomb. For Americans, the first direct evidence that an atom of the uranium element had been split came at a conference in Washington on January 26. At the conference, cosponsored by the Carnegie Institution, the recent discovery of fission by two German scientists was disclosed. Physicists in attendance quickly saw the possibility that a single fission reaction could lead to an uncontrolled reaction, creating an enormous explosion. Physicists at the Carnegie Institution literally ran from the conference room to confirm the revolutionary evidence described moments earlier. The American Physical Society's spring meeting on April 29 attracted even more attention, prompting *The New York Times* to write that conferees argued "over the probablility of some scientist blowing up a sizeable portion of the earth with a tiny bit of uranium."[1]

Bush's reaction to the hoopla over fission was characteristic. A man of sober judgments, he took special delight in pricking inflated technical claims. Of the sensational claims made for atomic explosions, he was intensely skeptical. One Boston writer suggested that "an unscrupulous dictator, lusting for conquest," might "wipe Boston, Worcester and Providence out of existence" with a single bomb dropped from an airplane. Bush thought this an "extremely remote" possibility. He feared such absurd talk raised the "real danger" of a public panic similar to the one that in October 1938 followed Orson Welles's compelling radio account of the fictional landing in New Jersey of invaders from Mars.

At first, Bush tried to debunk "wild" notions about atom bombs. Aided by Carnegie physicist Merle Tuve, he failed to make headway against the fission-inspired hysteria. Bush consoled himself with the view that fission's "great impracticability" meant it was premature for war planners to take it into account. He felt it wise to continue "softpedaling a bit" the possibility of an atomic bomb.[2]

Others were less cautious. A few immigrant physicists obtained the imprimatur of Albert Einstein and, through an intermediary, convinced Roosevelt in September 1939 to form a committee to coordinate research on a fission explosive. The president asked Lyman Briggs, chief of the U.S. Bureau of Standards, to chair the "uranium" commmittee. A political appointee nearing retirement, Briggs was the government's top civilian research official, but he was slow-moving and unfamiliar with atomic science. By the spring of 1940, the committee had funded just $6,000 worth of research. The government's inaction had upset a growing number of physicists.

Bush knew of the frustration with the uranium committee because Tuve was one of its scientific advisers. But Bush was reluctant to take any bold steps, doubting the military value of fission. "I am puzzled as to what, if anything, ought to be done in this country in connection with it," he wrote a colleague on May 2, 1940. His preference was "to do nothing," but he found this course impractical. "The difficulty of doing nothing is that one is not likely to know what others are doing. The whole thing may, of course, fizzle. Someone may discover a barrier to the chain reaction. It seems, however, undesirable to simply sit and wait for this to occur."[3]

Already laying plans to direct war research through his National Defense Research Committee, Bush did not wish to leave out fission. He felt the nascent field would benefit from NDRC attention. "There is no competent organization to handle all aspects" of fission, which is "floating about loose" and "most decidedly cannot be ignored in times like these," he wrote on May 15. "After conferring with the Army and the Navy . . . I now propose . . . to centralize to some extent the work which is going along in various laboratories along these lines."[4]

Even though he regarded an atomic explosion as "remote from a practical standpoint," Bush still wished "there were no such thing" as fission. When he met Roosevelt in early June 1940 to gain the president's approval for NDRC, the two men apparently never mentioned fission research. But in subsequently working out details of NDRC's operation with White House aides, Bush asked to have responsibility for the uranium committee. No one objected, and he took it over.[5]

The shift in control was significant. Under Bush, the uranium committee was now freed of its exclusive dependence on the military for funds, which made it easier for physicists to advance their claims. It did not, however, ease the anxiety of those scientists who felt that Bush and the government had a nonchalant attitude toward atomic weaponry.

Indeed, it was easy for Bush to lose sight of fission. Convinced the U.S. would inevitably be drawn into the war in Europe, he juggled many disparate and pressing priorities related to mobilization. In late 1940, his highest priorities were radar, submarine detection and the proximity fuze. Each of these areas promised advances that would reach the battlefield in time to affect the current war. Atomic enthusiasts, on the other hand, did not know how to build an atomic bomb, when one might be ready or what it would cost. Bush worried about betting too much on the atomic superweapon when he saw rapid advances in more established fields. The failure of a crash program on fission could even jeopardize Bush's whole NDRC program, by absorbing so many men and so much material as to undermine efforts in more promising areas. With Germany threatening to blockade and bomb Britain into submission, Bush naturally valued most the innovations that altered the outcome of battles *next month.* The Allies stood on the edge of a precipice. If Britain fell, America might never reverse Nazi domination of western Europe. Describing the climate, James Conant, Bush's chief deputy, wrote: "To me, the defense of the free world in such a dangerous state that only efforts which were likely to yield results within a matter of months or, at most, a year or two were worthy of serious consideration."[6]

As the war dragged on and Britain's resolve stiffened, criticism of the uranium committee grew. In March 1941 Bush finally began to face the flaws in Lyman Briggs's handling of fission. Ernest Lawrence, a Nobel Laureate in physics and strident critic of Briggs, forced the issue into the open. Lawrence complained that the committee had drawn too few American physicists into the the atomic chase, inhibited progress with its security requirements and failed to explore additional routes to a bomb. He had urged Conant and MIT president Karl Compton to "light a fire" under Briggs. An alarmed Compton telephoned Bush, then later put the gist of Lawrence's views on paper. Briggs was described as "by nature slow, conservative, methodical." The uranium committee, Compton added, "practically never meets." This was "disquieting," because "our English friends are apparently farther ahead than we are, despite the fact that we have the most in number and the best in quality of the nuclear physicists of the world." Could not the Americans do better in the race to build a bomb?[7]

Then Compton gave Bush what amounted to a tongue-lashing, saying the NDRC should not "passively administer" the uranium committee but had "a responsibility for insuring . . . that the project goes ahead not only safely but with the greatest expedition." The problem lay with Bush: he simply didn't have the time to devote to fission. "If you were not so tied up personally with heavier and larger responsibilities it would be quite appropriate for you, for example, to talk to some of the leaders in this field and get their consensus of ideas as to what can be done and how it should be done and also look into the way the existing project has been carried forward. . . . Obviously you cannot do this; neither can Dr. Briggs because of his very heavy assignments in numerous directions."[8]

Bush's options were limited by "personal and administrative complications," as Compton put it. He could do nothing to embarrass Briggs, he felt, because to do so would implicate Roosevelt, who had selected Briggs for the job. He also felt Briggs was not really to blame. In a rare moment of self-examination Bush pointed to his own complicity, asking "whether under the pressure of other matters I have given [Briggs] as complete support in carrying out of this affair as I should have." He tried to protect Briggs's battered reputation by penning flattering descriptions of him. (To one associate, Bush defended Briggs as a "grand person" who "has done exceedingly well to keep his balance.")[9]

To remedy the troubled state of affairs, Bush took a series of steps aimed at working around Briggs, sorting out the web of technical, economic and political questions surrounding fission—and deciding whether the U.S. should do its all to build an atomic bomb.

Before acting, Bush chose to deal with Lawrence, whose criticism of the atomic effort—and appeal to Conant and Compton—struck Bush as a challenge to his authority. Lawrence "decided that when he did not get directly out of me the reaction he wished he would go around and bring pressure, which he certainly did," Bush observed. He believed the physicist's maneuver was aimed at thwarting him.

Still, Bush was tempted to let Lawrence off with a wrist-slap. "Really good men are scarce," Bush wrote, and Lawrence was very good. A titan in the physics community, he surely deserved special handling. The inventor of the cyclotron, he led a group at the University of California at Berkeley that built big machines to accelerate particles in order to study them at high energies. This work had won him the Nobel Prize. A boyhood friend of Tuve, Lawrence was raised in a small South Dakota town and retained his midwestern manners, saying "sugar" and "oh fudge!' when angry. An extrovert, he worked his way through college selling kitchenware. He was still selling: no American physicist before him had raised so much money for equipment. None of Lawrence's achievements, however, seemed to intimidate Bush. When the two men met on March 19, Bush exploded:

> I told him flatly that I was running the show, that we had established a procedure for handling it, that he could either conform to that as a member of the NDRC and put in his kicks through the internal mechanism, or he could be utterly on the outside and act as an individual in any way that he saw fit.

As was often the case, Bush's bark was worse than his bite. Bush told others that after the tongue-lashing Lawrence "got in line," but actually Bush caved in. He desperately needed help and made that plain to Lawrence, whose criticism after all was on target. Following Compton's advice, Bush asked the physicist to act temporarily as Briggs's personal adviser. Lawrence agreed. "I rather think this is preferable to my sailing into the subject in any other way," Bush wrote Compton.[10]

Aggressive and enthusiastic, Lawrence broke the logjam, directing the uranium committee to fund promising research at Berkeley and the University of Minnesota. More important, Lawrence helped Bush grasp certain technical issues for the first time and left him convinced that the uranium committee would greatly benefit from more advice from outstanding physicists. A review committee offering its opinion periodically "would, for one thing, remove some of the weight of responsibility which you and I both feel in dealing with this somewhat intangible and yet exceedingly serious matter which is under your charge," Bush told Briggs on March 27.[11]

The smooth outcome reduced tensions between Lawrence and Bush, men whose outsized egos drew them into conflicts. "Regardless of any friendly arguments we might have, I want you to know that I have great respect for your judgment and that the job you have done in organizing and directing the NRDC is a superb achievement," Lawrence wrote Bush in July 1941. In the same spirit, Bush replied: "I have not been at all disturbed in my own mind about the recent shindig. . . . There is no personality in the group that is not utterly reasonable when it comes down to brass tacks."[12]

An authoritative review of the fission field made more than good scientific and administrative sense. Bush saw his role as that of relayer of reliable opinions from scientists in the trenches to Roosevelt and his aides. "Since I am no atomic physicist, most of this was over my head," Bush later recalled. While he understood some of the technical issues, he never deceived himself into thinking he could reach his own conclusion about the feasibility of various paths to an atomic bomb. Part of his organizational savvy lay in just this raw self-appraisal. He did not mind serving as a messenger, provided he believed in the intelligence and judgment of those who had written his script.

When faced with contending opinions on a technical question, Bush typically sought agreement through a committee. He believed that men of good will and intelligence would invariably find the basis for a fruitful program through the genteel and academic procedure of writing a report. In April 1941, he asked Frank Jewett, president of the National Academy of Sciences, to form a committee to review what was known about fission and to deliver definite advice on how to proceed. Giving the committee much leeway, Bush expressed the hope that the Academy would neatly frame the alternatives, either calling for "a radical expansion of our efforts" or "merely for careful procedure in the way we are now heading."[13]

In its first report of May 17, 1941, the Academy was pessimistic about the chances of making an atomic bomb and advised that, before addressing anew the question of fission's future, the government spend just $350,000 for a further six months of investigation. At first, Bush seemed satisfied with the report, but within a week he took a dimmer view, probably as a result of Conant's criticism.[14]

Cold and unemotional, Conant never shirked tough questions. His skepticism buoyed Bush, who now wanted engineers to join the study for their experience estimating the costs and schedules of large projects. Bush wrote Jewett that "even if the physicists get all they expect, I believe that there is a very long period of engineering work" ahead. He asked Jewett to have the

Academy committee revisit the question of the creation of an atomic bomb and tell him "how quickly results could be put to practical use."

By now, Bush's patience was nearly exhausted. "This uranium business is a headache!" he fumed.[15]

In early July, the Academy sent a draft of its second report to Bush. It left him cold. In a July 9 letter to Jewett, he snapped: "We asked for a review of the situation from the engineering aspect. . . . This report practically says that our request cannot be answered."

The Academy's vagueness angered Bush, who had a lot on his mind. His relations with the military blew hot and cold, so he never knew when a fresh storm would hit. Financial pressure was mounting too. America was not yet at war, and Congress had reduced the president's emergency funds. Bush complained of a recent cut in the NDRC's budget and a threat "to make our allocations quarterly."[16]

Then there was the possibility that the Nazis might build an atomic bomb first. While he could only guess at their progress, he knew the Nazis wanted the bomb. In bad moments, he was "scared to death" of a Nazi bomb and convinced that the U.S. trailed Germany in this crucial field. His suspicion arose from his great respect for German physicists, whom he felt had "a decided head start" by virtue of their pathbreaking insights into fission. He further credited German scientists with the capacity to maintain secrecy over their work: good evaluations of their progress might be impossible to come by. Finally, even if Germany wasn't ahead of the U.S. in atomic research, Bush expected the Germans to be "close competitors."[17]

The confusion over how to approach fission compounded his worries. "I am getting to the point," Bush wrote Jewett, "where I feel that I am being blocked very decidedly."[18]

Indeed he was. His atomic program was trapped between rival factions. On one side, Jewett and other traditional research managers voiced profound misgivings about moving forward without proof that the basic concepts behind an atomic bomb were sound. On the other side, Lawrence and other atomic boosters, convinced of their ability to build a bomb, carped about mystifying delays in government support and begged for a full-scale program even if it meant sacrificing other fruitful research. Still others took Bush's new emphasis on engineering as a sign that he might cut fission studies entirely.[19]

Bush had hoped the Academy would settle this dispute. Now he had no choice but to end the stalemate himself. Once pessimistic about a bomb, he had decidedly shifted his ground. No dramatic moment marked Bush's passage into the atomic faithful, but during the summer of 1941 he became a believer.

Three key factors helped to change his mind. First, Bush had learned in June that Britain's top scientists were convinced a chain reaction was definitely possible, which meant a bomb could be made. Second, Bush felt that a memo from physicist Enrico Fermi in late July gave him for "the first time" just what he needed "in order to have a reasonable basis for judgment" on whether to proceed with a crash project. The memo contained "engineering data," which Bush labeled "good stuff" and crucial to understanding "the economics" of making an atomic bomb. Third, Bush considered the risk of a German A-bomb too great and feared that "the result in the hands of Hitler might indeed enable him to enslave the world." The only way to forestall this dark scenario was for the U.S. "to get there first."[20]

In June 1941, Bush began to seek political backing for a bold atomic program, though he was still unsure of the precise shape and chance of success. On June 29, he spent an evening with Vice-President Wallace, with whom Roosevelt had said Bush could share concern about the atomic program. "Bush says that it is his greatest headache," Wallace wrote in his diary. "In view of the German activity he must absolutely work on it, but at the same time it is one of the most dangerous jobs which he has ever tackled."[21]

In mid-July, Bush drew Roosevelt into atomic matters by suggesting it was nearly time for the president to review "the whole matter." Though more hopeful than in the past, Bush's language remained cautious; he would not promise the president a superweapon. "For some time it appeared that the possibility of a successful outcome was very remote," he wrote, adding that the government faced the possibility "of expending public money on what might eventually appear to have been a wild search." While this chance still existed, Bush acknowledged, albeit cautiously, an important shift. "There has appeared recently, however, new knowledge that makes it probable that the production of a superexplosive may not be as remote as previously appeared."[22]

Next Bush saw Wallace again, this time to discuss the merits of an all-out program and the problem of financing. The two men traveled to Wallace's poultry farm in Maryland on a Sunday afternoon for "an enjoyable time in the open air." Wallace later recalled that Bush "wanted my judgment as to whether we should go full steam ahead. We both realized the frightening aspects of unlocking the power of the atom and we both knew at that time of some of the struggles which would inevitably ensue, but we were certain that the attempt had to be made with all our energy because of the danger that the Germans would get there first."[23]

Bush was pleased to strengthen his link to Wallace, a hybrid-corn inventor and the lone member of Roosevelt's cabinet with an intimate knowledge of technology. "As the matter develops," Bush wrote Conant, "we will wish to share our thoughts on it with some of those high in national councils, and he is certainly in this category."[24]

On October 9, 1941, Bush saw Roosevelt in the White House with Wallace by his side. Though sometimes put off by Roosevelt's nonchalance, Bush was comfortable with Wallace, whom he felt had "faith" in him. Bush began the meeting by reviewing the final version of Britain's report on the feasibility of an A-bomb, which he had received just days before (the so-called MAUD report concluded that a bomb could be ready for use before the war's end). Then Bush talked about the German program to build a bomb (of which he frankly knew little), the whereabouts of uranium sources and the control of atomic power after the war.[25]

Roosevelt's response, never committed to paper and found only among Bush's recollections of this meeting, was brief and sweeping. The president instructed that work on the bomb was to be "expedited . . . in every way possible." However, Bush "should not proceed with any definite steps on this expanded plan" without the president's approval. Policy debates should be limited to a small group, composed of Bush, Wallace, Roosevelt, Conant, Secretary of War Henry Stimson and Army Chief of Staff George Marshall (this group was called the "top policy" committee). Finally, Bush and the president agreed that "a broader program," which might well involve building atomic bombs, ought to be managed by a new, independent organization, since the OSRD could not possibly handle the task.[26]

There was also the matter of money. Building an A-bomb would cost a lot. "To get this money and yet maintain secrecy was a problem which [Budget Director] Harold Smith and the President had to solve," Wallace recalled. They solved it, not by seeking congressional approval, but by creating the first "black," or secret, weapons budget in the nation's history. First coming from "a special source" under Roosevelt's control, funding for the bomb later was buried in the yearly budget of the Army Corps of Engineers.[27]

The October 9 meeting was a milestone in the government's quest to exploit fission. "Bush now had the authority, not to make a bomb, but to discover if a bomb could be made and at what price," the bomb project's official history states. "When this investigation should point the way to a production program, he would need further Presidential sanction. Until then he had virtually a free hand."[28]

Jewett, appalled at the slipshod estimates underpinning the decision to explore the crash program's feasibility, made a last-ditch effort to alter Bush's course. The exchange reflected Bush's rising power in the government and his widening differences with the older, more conservative Jewett. Despite being the National Academy's president and a member of the NDRC, Jewett was still Bush's subordinate. Jewett never tried to go over Bush's head, but he dissented vigorously. Beseeching Bush to slow down, he argued that "no matter how you sliced" the unproven, untested principle behind an atomic bomb, "it is still bologny [sic]."

Jewett missed the point. In wartime, the prospect of a bomb was too alluring for displays of piety about technical or economic judgments that were tainted by expediency. Bush knew this; he had lobbied the Academy to lower its cost estimate.[29]

Because Jewett had been a staunch ally in the past, Bush leveled with him, citing in November 1941 a rationale for his actions that would stand for decades as a justification for government-funded megaprojects whose foundations were technologically questionable. "Now in times of peace the things we are talking about would be almost absurd," Bush wrote. "We are considering full-scale matters when we have not even a test tube, let alone a pilot plant. Nevertheless, the pressure of affairs may well force us to proceed in a way that would not be sound under peacetime conditions." Gamely trying to seem moderate, Bush added, "I do not think this means we ought to run wild, but I think we ought to take all the steps we can . . . take quickly and sanely."

This was an extraordinary admission, and an ominous reminder that experts and organizers of expertise were subject to political pressures. Bush sensed that an unsympathetic reader of his words might bristle at his dismissal of the scientific obligation to seek the best answers as opposed to the expedient ones. He frankly advised Jewett that after reading his letter, "it will probably be best to destroy it and not take the chance it might become incorporated in the files inadvertently."[30]

Over the next month, Bush moved "quickly and sanely" to further the nation's atomic ends. On November 27, 1941, he alerted Roosevelt that he was "now forming a carefully chosen engineering group to study plans for possible production." A week later, he wrote that "there is a great deal moving" on A-bomb matters, especially with regard to the engineering and productions issues of bombmaking that might bedevil the U. S. even if the scientific hurdles were cleared. Both research and engineering work had to move in tandem, Bush believed, in order to complete a bomb swiftly. "I think we will have no problem of drawing the line between research and engineering development," he wrote, "and it will be my job to coordinate these two."[31]

Then on December 6—on the eve of the Japanese attack on Pearl Harbor— Bush talked with his closest colleague, Conant, as if a full-scale project was unavoidable. While lunching at the Cosmos Club, near the White House, he listened intently as Conant and a physicist from Chicago discussed the merits of various technical paths to an A-bomb. Talk was sidetracked briefly when Conant, sipping on his usual glass of milk "took a large swallow, only to discover it was buttermilk," which he loathed. He began sputtering and cursing, which amused Bush tremendously. By the end of lunch, Bush had another reason to smile. It seemed that progress along the various routes to an A-bomb were far enough along that it made sense to at least begin thinking about handing the project off to a new organization. Before leaving the club, Bush proposed to Conant that Marshall, the Army's chief, be asked to assign a high-ranking officer to look after the bomb research programs as they proliferated around the country.[32]

Ten days later, on December 16, Bush told Vice President Wallace of his desire to transfer the daily management of the atomic effort to the Army. "When full-scale construction is started, presumably when pilot plant results are ready, I felt that the Army should take over," he wrote Conant, summarizing the meeting. "There was no objection expressed to this point of view. I also said I felt at the present time there ought to be one officer, of fine technical qualifications, assigned to become utterly familiar with this whole matter. I stated that this would aid me in expediting pilot plant experimentation, which everyone present urged be expedited, and also that it would allow planning to be made so that the take-over by the Army could be handled smoothly." He anticipated no serious problems when it came to the military management of a project essentially staffed and run by civilians. As he had explained to a British official only days before, "The physicists and the engineers understand completely that they are concerned only with technical matters."[33]

As the pace of the A-bomb project quickened, Bush refined his organization, handled political questions and for a short time led the crucial engineering section. But while Conant increasingly oversaw the effort, Bush still considered himself accountable for it and continued to serve as the vital link between the technical community and President Roosevelt. On March 9, 1942, Bush wrote to the president of his continuing responsibility for the project:

> In accordance with your instructions [of October 9, 1941], I have since expedited this work in every way possible. I now attach a brief summary report of the status of the matter.
>
> Considerations of general policy and of international relations have been limited for the present to a group consisting of Mr. Wallace, Secretary Stimson, General Marshall, Dr. Conant and myself. Mr. Wallace called a conference of this group, to which he invited also Harold D. Smith as the matter of funds was there considered.
>
> The technical aspects are in the hands of a group of notable physicists, chemists

and engineers, as noted in the [accompanying] report. The corresponding British organization is also indicated. The work is underway at full speed.

Recent developments indicate, briefly, that the subject is more important than I believed when I last spoke to you about it. The stuff will apparently be more powerful than we then thought, the amount necessary appears to be less, the possibilities of actual production appear *more certain.* The way to full accomplishment is still exceedingly difficult, and the time schedule on this remains unchanged. We may be engaged in a race toward realization; but, if so, I have no indication of the status of the enemy program, and have taken no definite steps toward finding out.

The subject is rapidly approaching the pilot plant stage. I believe that by next summer, the most promising methods can be selected, and production plants started. At that time I believe the whole matter should be turned over to the War department.

In the report accompanying his letter, Bush displayed an awareness of atomic explosives as a new kind of destructive force, not simply a bigger bomb. He wrote:

Fortunately, there is not great danger to personnel at the present stage of experimentation, beyond that inherent in usual atomistics studies, and similarly taken care of. When full-scale production is undertaken, great care will be necessary. Studies are being made of the steps involved. When and if a full-scale experiment is performed it will have to be in an isolated region. There is also much danger from after effects, for the explosive also renders the region exceedingly dangerous to life for a long period.

"Preservation of secrecy on this matter is unusually difficult," Bush went on. "It is not necessary to furnish the experimenters with a confidential military background, so this aspect does not enter. The problem is to keep confidential the progress being made and the plans adopted." But public interest in the field was high, because before the war "the general subject of atomic power was much discussed by physicists, many papers were published and popular accounts were printed, often calculated to terrify. They still appear." But precautions are being taken, Bush noted:

The subject is subdivided, and full information is not given to every worker. It has been considered desirable, in the case of some scientists who are considered as highly able but perhaps not altogether discreet, to join them to the organization under oath, rather than to leave them as independents to speculate with other unattached physicists as to what is being done. The matter is considered to be under control to a reasonable extent, in view of the circumstances surrounding the subject. Yet, it is certainly true that we are more vulnerable to espionage than is desirable. This is an additional reason why it is believed that the whole subject should be placed under rigid Army control as soon as actual production is embarked on.

At the end of the report, Bush delivered a bottom-line assessment for the president:

> Present opinion indicates that successful use is possible, and that this would be very important and might be determining in the war effort. It is also true that if the enemy arrived at results first it would be an exceedingly serious matter. The best estimates indicate completion in 1944, if every effort is made to expedite.

Two days later, Roosevelt replied to Bush in his typically terse fashion. "I think the whole thing should be pushed not only in regard to development, but also with due regard to time. This is very much of the essence. I have no objection to turning over future progress to the War department on condition that you yourself are certain that the War department has made adequate provision for absolute secrecy."[34]

In June of 1942, Bush, whose optimism about the effort was steadily rising, made good on his promise to give the Army responsibility for constuction of bombs. He also predicted to Roosevelt that an "extraordinary" bomb could be available "early in 1944."[35]

In August 1942, for the first time Bush openly endorsed a crash program that overshadowed all else. He wrote Harvey Bundy, his liaison in the War Department: "Faced as I am with the unanimous opinion of a group of men that I consider to be among the greatest scientists in the world, joined by highly competent engineers, I am prepared to recommend that nothing should stand in the way of putting this whole affair through to conclusion on a reasonable scale but at the maximum speed possible, even if it does cause moderate interference with other war efforts."[36]

Assembling the resources required to develop the bomb was difficult; this problem could be easily overlooked in the tense chase for technical solutions. And the new arrangement with the Army.

Bush had reason to fear for his researchers' independence. General Brehon B. Somervell, whose Services of Supply organization was the principal Army contact with the bomb project, wanted bomb scientists and engineers placed in his outfit. Without telling Bush, Somervell also had appointed a new chief of the project (now officially called by its cover name, the Manhattan Engineering District). Bush had thought he would have a say over Somervell's choice, but on September 17 he received a surprise visit from a Colonel Leslie Groves.

The meeting was awkward, perhaps because Bush sensed that it marked the end of his preeminent role in directing atomic affairs. As Groves recalled, Bush "was quite mystified about just where I fitted into the picture and what right I had to be asking the questions I was asking. I was equally puzzled by his reluctance to answer them."

Stunned by Groves's audacity, Bush promptly wrote an official in the War Department that he doubted Groves "had sufficient tact" for the job. To Conant, he huffed that the appointment "may well make a serious tangle," adding that "it would be very bad indeed if we should get off on the wrong foot" with the Army. To Bundy, he wrote: "I fear we are in the soup."

The rift proved short-lived, however. Within days Bush learned more about Groves, who was just finishing the mammoth five-sided military headquarters known as the Pentagon, and cooled off. "I think I was wrong," he concluded, "Groves is going to be OK."[37]

One reason for Bush's calm was that he had succeeded in placing himself at the head of a three-member Military Policy Committee that would oversee Groves. Bush told Roosevelt that he "would feel more comfortable" if he shared responsibility for the project with "something like a board of directors." Bush envisioned a small board, and Roosevelt said, "Fine, we will set it up." The president made Bush chairman—top dog of the bomb project, at least officially.

This post was not aimed merely at feeding Bush's ego. It gave him a hand in the project even as he retreated to its margins (Conant, his deputy, attended to day-to-day issues). It also allowed him to say that scientists retained a direct voice in atomic policy even as the Army's influence grew. That voice, of course, was his, which stoked the resentment of scientists who wanted to directly influence atomic policy. While Bush denied scientists the chance to participate in policy discussions, he did not, as some have claimed, usurp this right. Roosevelt demanded that his atomic circle remain very small.[38]

Bush's authority was limited, as illustrated by Groves's power. Bush never controlled him, and at times they fought. Probably the worst dustup came when Bush realized that Groves had ordered security agents to spy on him. Bush resented the implication that he might be untrustworthy, even though he knew that Groves routinely spied on others. Groves first tried to avoid the issue, but Bush demanded an answer and Groves reluctantly confessed. A furious Bush then declared, "You take steps to see that it doesn't happen again."[39]

On December 2, 1942, the first sustained chain reaction was produced in Chicago's Stagg Field. Two weeks later, a jubilant Bush sent an optimistic letter to Roosevelt. "There can no longer be any question that atomic energy may be released under controlled conditions and used as power," he concluded, adding that there was now "a very high probability" of creating "a super-explosive of overwhelming military might."

While Bush had ceded power by giving the Army management of the bomb project, it had not been a hasty decision. As far back as December 1941, he had voiced his preference for the Army to assume responsibility for making an atomic bomb when "full-scale construction is started." He favored the Army because Secretary of War Stimson, the departmental staff, and top Army officers listened to him. By and large, the naval officers did not.

The decision to exclude the Navy from the Manhattan Project was Bush's alone. The Navy's exclusion from the project was curious since its officers had shown an interest in fission since 1939. More technically savvy than the Army, the Navy had employed the first two people paid by the government to investigate atomic energy (albeit as a source of power for its ships). One of them was a former Carnegie Institution scientist. The Navy maintained its own modest program of atomic research during the war, but its isolation from the main action hardened the animus felt by some naval officers toward Bush. After the war, one admiral described Bush's choice of the Army as "political chicanery" and claimed that the Navy's exclusion slowed progress toward a bomb.[40]

Bush never doubted the correctness of his decision to hand the Army the entire project. Bush's preference for a single service had its benefits. It reflected his basic view that atomic policy should be made by a small group, and he wished to avoid interservice rivalries. Besides, dealing with the Navy was a trial. Frank Knox, the department's secretary, never called on him for advice. Ernest King, the Navy's chief, frequently fought with Bush. Other officers saw the OSRD as a threat. The Navy's patchwork of fiefdoms, meanwhile, did not strike Bush as well-suited to handle a massive project. He may also have thought of his tensions with the Navy before the war over his designs for code-breaking computers. He even might have recalled that during World War I the Navy had snubbed his submarine detector.[41]

Keeping atomic work within a single service also reduced the chances of a public discussion about atomic weaponry. Bush periodically reminded the military's censors of the need to maintain a press ban on the subject. He even saw "a certain amount of harm" in a few newspaper reports on atomic energy. He feared that the media might publicize the bomb project, alerting the country's enemies. But the press played ball. By late 1944, Bush called the cooperation "excellent," noting that "the press on a voluntary basis has prevented the subject being widely discussed in print."[42]

While Bush's penchant for secrecy reflected Roosevelt's own preference, the effect was to deny the public any chance to shape government policy on atomic weapons, even in a limited way. Bush also restricted the information available to his own researchers in order to limit the possible damage caused by espionage and to shield himself from criticism. This sometimes made Bush wary of enlisting the aid of certain scientists, even brilliant ones. For instance, he did not trust Einstein. Even though he had first alerted Roosevelt about the potential for an atomic bomb, Einstein was a liberal, a German and a Jew. Bush approached him cautiously. "I have a problem for Einstein," he wrote a colleague of Einstein's at Princeton University. "If you think that it is entirely safe to put him to work on it, won't you please take it up with him? I have no question whatever about his loyalty, but simply some question as to his discretion. It happens that I would not wish anyone to know that this particular problem is even being worked upon. Hence what I hope is that he can go to work on it personally without communicating to his associates any more than a statement that he is working on a mathematical problem in connection with defense."

The problem examined the flow of gases through a porous membrane; Bush considered it fundamental to gaseous diffusion, one of four paths to an atomic bomb being studied. Einstein was given the problem and wrote a memo about it. But two weeks after his first request, Bush hastily rescinded his invitation to Einstein. "I am not going to tell him any more than I have already told him for a number of reasons," Bush wrote his Princeton contact. "I am not at all sure . . . he would not discuss it in a way it should not be discussed." Bush regretted the situation, but involving Einstein in a matter relating to the atomic bomb "is utterly impossible in the view of the attitude of the people here in Washington who have studied his whole history. It looks to me, therefore, that we might as well let the matter drop. . . . The difficulty and the uncertainty of pursuing the subject further with Einstein seems to me to outweigh the possible advantages."[43]

Even when Bush relied on someone he knew to be reputable he might tolerate other people's raising doubts about the person's character. This happened with Arthur H. Compton, Karl's younger and more outspoken brother. A distinguished physicist, Arthur had chaired the Academy's three atomic studies and led an important group of physicists at the University of Chicago. Naval security men wanted Arthur banned from defense work for spurious reasons. Bush vigorously defended the physicist, labeling him "thoroughly loyal." But he conceded that, if it wished, the Navy could bar Compton from learning its own secrets, which were not related to the bomb project. "I do agree with you that [Arthur] is somewhat lacking in discretion," Bush wrote, adding that he had recently cautioned him about the need to use "proper care" in securing documents and information.[44]

Naturally, Bush's position as keeper of the atomic keys stoked resentment, especially among some independent-minded physicists who felt they should have more sway over atomic policy. Bush described these malcontents as "dissatisfied with the extent and character of the program now being carried on as far as they are acquainted with it. They are inclined to bring pressure for extension of the program in any way possible. . . . Part of their unrest, of course, comes from the fact that they are excluded from the considerations of broad policy."[45]

Bush refused to be drawn into debates about the goals of atomic research, telling researchers they were to concern themselves only with technique, not policy. In a typical brushoff, he wrote Jewett in November 1941: "I have direct instructions from the President to hold consideration of policy on this subject to a very small, named group." The proper procedure for those wishing to make suggestion on policy, Bush noted, was "to give me these ideas and I will see that they are given full consideration, and the same thing applies to anyone else who may wish to express his feelings in regard to policy." This left little room for moral stands. "A scientist could choose to help or not to help build nuclear weapons," Richard Rhodes wrote in *The Making of the Atomic Bomb.* "That was his only choice."[46]

Bush did offer scientists a political consolation prize: he would be their voice. In time, Bush (and his alter ego on atomic matters, James Conant) would advance positions that reflected the views of liberal atomic scientists. But he balked at allowing scientists to go outside proper channels to voice their opinions. He thought it entirely proper to smother dissent. But he paid a personal price for bottling up the views of the atomic rank and file. A gulf emerged between him and scores of researchers whose views he basically shared. Forced to navigate between the demands of the White House, the military and the scientists, he found he could most afford to alienate the last group.[47]

Two Hungarian physicists, pivotal actors in the making of the atomic bomb, vented their anger toward Bush in telling ways. Refugees from war-torn Europe, Leo Szilard and Eugene Wigner were among the first to envision the perils of the atomic age. Experts in atomic science, they had persuaded Einstein to write his famous letter to Roosevelt in 1939, which had been the germ of the Manhattan Project. Now part of Arthur Compton's Chicago team, Szilard and Wigner were driven by a fear that the hated Nazis would build a bomb first; each man had little patience for the needs of organizations.[48]

In May 1942, Szilard wrote Bush to complain that too little had been accomplished since 1939; given the "muddle" of U.S. efforts, Germany must surely lead in the bomb race. "Nobody can tell now whether we shall be ready before German bombs wipe out American cities," he warned.[49]

Szilard's point was well made. Though Bush could not conceive of a German bomb reaching U.S. shores, he worried about the devastation to British cities from an atomic bomb riding a German rocket. Throughout the war, the state of Germany's bomb project was the great unknown. In the days following Szilard's letter, Arthur Compton pondered ways to reduce the threat. "Obviously something can be done about it if we know what they are doing and where," he said. On June 22, Compton wrote Bush about the "exceedingly serious picture" presented by news of German progress, and recommended bombing raids and other "secret" actions to "locate and disrupt their activities."

One of those actions might even be the kidnaping of top Nazi physicist Werner Heisenberg, considered the key to Germany's hopes for an atomic bomb. In October 1942, two U.S. physicists, Hans Bethe and Victor Weisskopf, proposed abducting Heisenberg during a visit to Switzerland planned for December. A giant in the field of physics, Heisenberg had done pioneering work in quantum mechanics and had devised the "uncertainty principle." Bethe and Weisskopf viewed their scheme as "a chance to cripple the German bomb program with a single stroke." While a madcap notion, it was taken seriously enough by Robert Oppenheimer, the scientific chief of the Manhattan Project, that he discussed the possibility of kidnaping Heisenberg with Bush and Groves.[50]

Bush doubted the value of the Heisenberg scheme. He had no moral qualms but thought the kidnaping would be hard to carry off, and he worried about finding the right abductor. He had more interest in conventional countermeasures, such as the bombing of suspected Nazi atomic sites. After Compton's June 1942 letter, Bush and Conant tried to locate facilities of possible use to a German bomb project. They even asked the British to bomb one, which the British declined to do because they planned their own commando raid on the site. In September, Bush wrote General Strong, "The attempt to determine where the Germans might locate a plant of the type we discussed by studying power sources does not seem to get anywhere." But he and Conant continued to try, compiling a new list later in 1942.[51]

At the end of the year, Bush was still in the dark about a German bomb. Writing Roosevelt on December 16, he concluded, "We still do not know where we stand in the race with the enemy toward a usable result, but it is quite possible that Germany is ahead of us and may well be able to produce bombs sooner than we can." This statement merely echoed Bush's past suppositions; his real fears about the stark risks facing the U.S. lay deeply buried. Bush kept in check the panic he felt at the thought of a German bomb. Once, visiting with Vice-President Wallace, he excitedly noted a press report of German threats to "blow up half the globe."[52]

While never in the grip of terror over Germany's destructive potential, Bush felt isolated. He had few people to lean on. Among scientists, he talked frankly about atomic policy with no one but the stiff James Conant. Roosevelt was more jocular, but Bush saw him only infrequently. As McGeorge Bundy, who knew Bush, later observed, "Bush was well aware of the extraordinary loneliness of his own situation." He kept the president informed of the progress toward the bomb, the costs of the project and the little he knew of the enemy's progress, hedging his comments so that he never flatly "told him that we were sure we could produce" an A-bomb. The anxiety was intense, yet Bush refused to trouble the president simply in order to calm his nerves. He later said of his situation: "No one who has not been placed in a post of heavy responsibility can realize what a lonesome feeling it is when there is no equivalent of a board [of review] present, and one reports to a chief who is rarely accessible."[53]

The best antidote for Bush's anxiety was action. Throughout 1943, he pressed the British and U.S. air forces to target suspected Nazi atomic facilites. He even personally briefed the top U.S. air man, General Henry Arnold, on the matter. On June 24, he lunched with Roosevelt at the White House and "told him of German targets and the arrangements now under way." The following month, on a visit to England, Bush importuned General Ira Eaker, commander of the American Eighth Air Force, to bomb a suspected atomic facility in Norway. In November, Eaker directed a raid, which succeeded in shutting the plant down.[54]

Sitting atop the biggest secret project of the war, Bush was a lightning rod for criticism from below. Bush took most seriously the complaints that he did too little to speed completion of an American bomb and not enough to disrupt Germany's effort. But he had little patience for grousing about his autocratic handling of atomic policy and the potentially far-reaching consequences of his decisions for American democracy.

Eugene Wigner challenged Bush on these very points in a November 1942 memo that Bush kept in a private safe for years after the war. Criticizing the decision to place an engineer such as Bush in charge of atomic policy, Wigner argued for the primacy of the scientist in settling political questions about the bomb. "Our process is not something like the tanning of leather which has been known since the time of ancient Greece and to which science has made only relatively small contributions," he wrote. "In such a case a very serious argument can be put up for leaving the practical man in the leading positions, although many tanneries employ scientists for this post. On the contrary, our process has been invented by, and is known to, only scientists, and every step to put others in charge is artificial and deleterious. There is no doubt that engineers . . . must participate and intensively participate in the work. It seems impossible, however, that they can assume sole responsibility."[55]

Wigner's attack, coming two weeks before the first sustained chain reaction in Chicago, reflected a widening gulf between younger, immigrant physicists and the 50-ish Bush, a salty American native. With so much in doubt, Bush was an easy target for those researchers who thought he had sold out their interests or failed to fully appreciate the terrible consequences of not winning the race for the bomb. Yet from Bush's perspective, Wigner could be seen as advancing the narrow claims of his tribe. Perhaps his atomic priesthood could produce a bomb more quickly, if left to run its own affairs. But this missed the point. Making the bomb, as Bush well knew, was fundamentally a political act. To him, the scientist had no more right to determine atomic policy than did the man on the street.

If anyone possessed a privileged standing in the discussion about the bomb, Bush believed, it could only be Roosevelt. The president had asked Bush to act in his name: that was the actual source of Bush's authority and it seemed to him entirely consistent with the nature of power in a time of crisis. As McGeorge Bundy wrote: "Who but the President himself could make the political and military decisions for the United States on this enormous subject? The intimate and inescapable responsibility of the chief political officer for nuclear policy has been recognized in every country and in every decade throughout the nuclear age. In Roosevelt's immediate acceptance of this responsibility there was no usurpation, nor were he and Bush mistaken in believing the undertaking must be managed from the top down."[56]

Yet the objections suggested by Wigner were not easily dismissed. What was the meaning of democratic political rights, if weapons with the potential to alter the whole course of human existence could be built and deployed by a few men surrounding the president? Resentment toward Bush for his presumed usurpation of the rights of technocrats and the whole body politic would only grow. This was hardly adequate recompense for Bush's travails in the atomic shadows. Rather, it was a curse from which he could never really escape.

Before damning Bush, however, a crucial question begged for an answer. Could he have behaved any differently? Was it conceivable that the decision to prove the feasibility of the atomic bomb could have been placed before the citizenry in a referendum, or before Congress? If the public, whether directly or through its delegates, had influenced the course of the bomb's creation, would U.S. interests have been better served?

It seems unlikely that public participation, or even a widening of the circle of discussion to include more scientists and elected officials, would have changed the result. Listening to more physicists would have resulted in more confusion and indecision. As it was, Bush fought off all sorts of advice that a less disciplined man might have seen as justification for delay. As Richard Hewlett and Oscar Anderson, Jr., have written: "Bush had assumed a tremendous burden in June of 1940: the creation of an entirely new relationship between science and government in the interests of national defense. He had to think of personalities and politics as well as technology. He and Conant had to look at uranium in the light of the entire role that science might play in the emergency. They had to turn a deaf ear to blue-sky talk of nuclear power plants and think of weapons. They had to navigate between the Scylla and Charybdis of excessive pessimism and soaring optimism. They had to set a course by the Pole Star of fact.[57]

Bush brought a fuzzy goal into clear view more quickly than anyone had a right to expect. "Without Bush's intervention, the [bomb] project might well [have] been totally abandoned, or at best, continued at [a] relatively relaxed pace." Instead, Bush undertook a crash program that would brook no failures. While taking longer than many scientists liked before embracing the program, he never looked back after making his decision.[58]

When the prize was something as world-shattering as an atomic bomb, even allies were at odds. The British and the Americans had readily shared knowledge of radar, electronics and countless other techniques, beginning in the summer of 1940. Eager for American aid, Britain had approached the United States with an offer of technical exchange, and the tentative cooperation on atomic power flowed out of America's acceptance of this offer.

Bush was a principal player in this "intricate and divisive" affair. He directly negotiated the terms of exchange with the British, and during the early years of the war his effect on Anglo-American relations was large. Bush took a hard-boiled attitude toward Britain, reluctant to commit to future cooperation on atomic power before the bomb had even been built. He worried that the U.S. might inadvertently boost Britain's chances to exploit civilian applications of atomic power. Aware of the dangers of the technology, Bush was nonetheless keen to protect the U.S. stake in its commercial potential. In his talks with the British, he sought to limit exchange to that which would help produce a bomb before the end of this war. Bush's views, however, were often at odds with those of the president, and the conflict caused much embarrassment, confusion, frustration and ill will. On relations with Britain, Roosevelt followed his own muse, sometimes making deals with Churchill without even informing his advisers.[59]

At first, Anglo-American cooperation reaped benefits. Bush's newfound ardor for bomb research in the summer of 1941 had stemmed in part from the feasibility studies performed by scientists on Britain's MAUD committee. Moreover, reports of British progress egged Bush on, the way a friendly rivalry with a classmate might force both students to study harder than normal. "The British," Bush wrote in April 1941, "are apparently doing fully as much as we are, if not more, and yet it seems as though, if the problem were of really great importance, we ought to be carrying most of the burden in this country." In October, Roosevelt, on Bush's advice, wrote to Churchill, suggesting that the U.S. and Britain coordinate or "even jointly" conduct atomic research. Churchill agreed.[60]

A year later, however, Bush changed his mind, influenced by Conant, who saw "no reason for a joint enterprise" with the British on "development and manufacture." In December 1942, Conant elaborated his ideas, arguing that bomb work was now almost entirely based on American ideas and that the arrangement with the British should be modified to limit exhange to well-defined areas. Later that month, the Military Policy Committee, which Bush chaired, passed the essence of Conant's advice on to Roosevelt, who agreed. In January 1943, Conant told the British that from now on they would receive information only if they could "take advantage of this information in this war."

The new policy threatened to poison Anglo-American relations over atomic research. It signaled a new emphasis on the desire for a postwar American monopoly over the bomb, even at the price of delaying its completion. The British, who considered the change "a bombshell," found the situation "quite intolerable" and moved to convince the U.S. to reverse its decision. Among other things, they cited a promise the prime minister believed Roosevelt had personally made to him in June 1942 that the two countries would work together on the bomb and split the rewards (thus freeing Britain from having to build a bomb plant on its own soil, where it might be vulnerable to German attack). Churchill lobbied Hopkins, sending him a detailed account of the contributions made by his countrymen to progress toward a bomb. Bush helped Hopkins reply to Churchill's complaints.

The talks on interchange continued through the spring of 1943. In May, Churchill visited Washington, discussing various war issues for two weeks. At Hopkins's insistence, Bush stood by in case the subject of interchange arose. On the last day of the talks, Hopkins summoned Bush to his office in the White House, where Lord Cherwell presented Britain's case on interchange.

The two advisers parried, then Cherwell said that Britain wanted full exchange of information so it could more easily make a weapon after the war. This was the admission Bush sought. "The matter having gotten very definitely boiled down to this one point," he wrote following the meeting, "I took the point of view, in which Mr. Hopkins joined me, that the delivery of information to the British for after-the-war military reasons was a subject which needed to be approached quite on its own merits."

Hopkins asked Bush to keep the conversation under his hat and gave the impression he would brief Roosevelt. Bush promised to "sit tight." A month later, he lunched with the president and learned Hopkins had not told him about Bush's face-off with Cherwell. Bush recounted the conversation. The president said it was "astounding" that Cherwell "placed the whole affair on an after-the-war military basis." Bush left the lunch believing he had "no instructions to do anything except proceed as we are." Unknown to him, Roosevelt had made a concession to Churchill on interchange during their May meetings. He had meant to tell Bush about this at their lunch, possibly deciding against it after Bush aired his own doubts about British intentions.

By now, Bush was on a collision course with the prime minister himself. Before the war Churchill had loathed Nazi aggression and the territorial concessions his predecessor had made to Hitler. "You were given the choice between war and dishonor," Churchill told Chamberlain after the latter's return from Munich. "You chose dishonor and you will have war." Churchill became prime minister at a low point of Britain's history. With the Germans lusting for his homeland, Churchill rallied the nation simply by refusing to admit defeat. But he was vain and prickly and often overwhelmed his advisers.

In July 1943, Bush visited England for several weeks to learn firsthand of British progress in war technology, especially in the area of antisubmarine warfare. He expected that talk would inevitably turn to interchange, though Roosevelt had given him no instructions. On his arrival, Bush was whisked off to see Churchill for what he thought would be a perfunctory visit. The prime minister was in no mood for formalities. He immediately complained that each time Roosevelt promised to share equally with the British in the atomic project the deal was undermined by an American adviser.

Churchill, who had Bush in mind, then threw down the gauntlet. "For ten or fifteen minutes I received an exposition of the absurdity of the [interchange] arrangement, in typically Churchillian language, with some vocabulary not used in public speaking," Bush later wrote. "He did not like the plan and implied strongly he did not like me either." Bush could not get a word in; he gave up trying. Churchill interrupted his tirade only "to light his cigar, but apparently he had not bitten the end off" so it would not light. He kept tossing matches over his shoulder with an emphasis that suggested he would like to throw Bush after them.[61]

Finally Bush fought back, arguing that Churchill's request for information on reactor manufacturing was a threat to security and that the British really wanted a leg up on postwar commerce. Churchill thundered that he did not give a damn about any of that. So the meeting ended, unresolved.

A week later, the two men tangled again. This time Bush held his own, but his effort went for naught. Two days earlier, on July 20, Roosevelt had confirmed his agreement with Churchill, who had yet to receive the message. The understanding was formalized the following month, August 1943, at the Anglo-American conference in Quebec. The Quebec Agreement left Bush some room for discretion by calling for ad hoc arrangements for exchange of information about crucial manufacturing and operations of bomb plants. But it restored the practice of complete exchange of scientific information and contained a pledge that neither country could use the bomb against a third party without the consent of the other.

The arrangement between the British and the U.S., beyond establishing the basis for a postwar atomic partnership, sent an undeniable message that Roosevelt would have the last word on matters of atomic policy. The president realized that the A-bomb "was a technological breakthrough so revolutionary that it transcended in importance even the bloody work" of prosecuting World War II. While Roosevelt listened to Bush's advice, he would ignore it at his pleasure. As Bush later complained to Conant, there were clauses in the Quebec Agreement "with which we had nothing to do and which had postwar implications." Both men were convinced that an Anglo-American atomic partnership during the war implied a similar partnership afterward. Because they questioned the wisdom of such a postwar partnership, they worried about Roosevelt's striking atomic deals on his own. Yet this was precisely what the president had done and intended to do more of. As historian Martin Sherwin has noted, the Quebec Agreement "led to a new relationship between Roosevelt and his atomic energy advisers. In the dispute over restricted interchange the President did not consult them."[62]

By the summer of 1943, the future of the bomb lay in the hands of the U.S. Army. Over the next two years, two huge production plants, in Hanford, Washington, and Oak Ridge, Tennessee, came on line. A new laboratory opened in remote Los Alamos, New Mexico. Headed by J. Robert Oppenheimer, the lab was charged with making the first atomic bombs. Researchers and production teams faced myriad problems. They continued to pursue multiple routes to the bomb, a strategy that vastly raised the odds of success. It also raised the costs, which now amounted to about $100 million a month. (By the war's end, the Manhattan Engineering District would spend more than $2 billion and require the labor of 125,000 people.)

By contrast, the Germans had never mounted a major program to build a bomb, nor reached any key milestones. While Groves received indications of this from British intelligence in January 1944, he was skeptical and formed a scientific mission to collect information on atomic research in Germany and Italy. Bush approved of the so-called Alsos mission and selected its scientific director, who began gathering data in the spring of 1944. By then, Bush was confident enough that the Nazis would not beat the U.S. to the bomb that he personally assured four U.S. senators of this during a secret June 10 briefing on the Manhattan Project. Still, Bush was startled to learn from Alsos just how far behind the Germans were: for all their vaunted know-how they had little besides a few experimental reactors to show for their efforts. They had failed to separate U 235, the essential fissile element, and never did achieve a sustained chain reaction. For complex reasons, the Germans "had not even reached first base," as Bush later put it. Intense debate went on over whether German physicists, Heisenberg included, had intended to derail the effort. At any rate, the Nazi program was inefficiently spread across as many as a dozen agencies. The German Army, expecting a short war, had decided not to back an all-out effort to build a bomb. When the war dragged on, it was too late to change course.

On his visit to the Western Front in October 1944, Bush spent an evening with the Alsos crew, who made it clear that the Allies had nothing to worry about. Bush concluded that the Germans had made an "incompetent fiasco" of their initial lead in atomic science. As the Alsos team would officially report in May 1945, the Germans at the end of the war were "about as far as we were in 1940, before we had begun any large-scale work on the bomb at all."

The news changed the tone of Bush's visit to Europe. He had put off answering a query about the German bomb from General Bedell Smith, whom he'd met on arriving in France. Smith, Eisenhower's chief of staff, had outlined a timetable for the reconquest of Europe and wondered whether, in light of the German atomic threat, "haste was essential." Bush told him he "could take a couple of more years, if necessary, and there would be no German atomic bomb."[63]

As it became clear that the U.S. alone would possess atomic weapons, the big question became whether the bomb would be ready for *this* war. Much depended on this point, of course. If the war ended without a bomb, Bush thought, "great questions" would be raised about why so much money had been spent on the project. Funds might be halted, and a bomb never finished.

"We might have waked up someday with Russia having the bomb and us still in the background," Bush said later.[64]

Actually, there was little chance that work on the bomb would be halted. By late 1944 the Manhattan Project seemed unstoppable. If the U.S. had atomic bombs ready for use during the war and refrained from using them, the president would likely be pilloried at a later date for consigning more American men to their deaths. Perhaps for this reason, Roosevelt assumed that atomic weapons, if available, would be used during the war just as would any other large explosive. Though he never promised to use the bomb, he never said he wouldn't, either. "If he had any doubts, he never expressed them to me," Bush said later.[65]

Bush himself never appeared to question the necessity of using the bomb. As early as June 1943, Arnold, the Air Force chief, had shown him a thick book of possible A-bomb targets. Bush was impressed enough to ask Conant to "look over" the target book, "realizing full well that this will be rather a long-shot business." A few days later, Bush talked with Roosevelt briefly about the possible use of the bomb against Japan. Writing an account of the meeting soon after, Bush referred to an attempt to explain to the president his belief that "our emphasis on the program would shift if we had in mind use against Japan as compared with use against Germany." Just what Bush meant was never clear. Perhaps he was trying to suggest to the president that the urgency to complete the bomb was much less if the target was Japan, which no one in the government believed could build a bomb on its own.

By 1944, the bomb's schedule had slipped too far for it to be used against Germany. By September 1944, Bush assumed that Japan would be the target of any atomic bombing.[66]

Bush raised no moral issues about using the bomb, though not because he was cynical about weapons of mass destruction. Along with Conant, his atomic deputy, Bush warned that the nation's atomic monopoly might trigger an arms race with other nations. Bush's concern echoed the views of Szilard, who had argued that the spread of atomic bombs would make world cooperation a necessity. Roosevelt, however, paid scant attention to the long-term implications of atomic weapons. Bush was alarmed by the president's tendency to casually strike deals with the British on weighty issues such as whether to tell the Russians about the bomb. Bush also wanted the president to listen to his own atomic advisers rather than such gadflies as Danish scientist Niels Bohr, who had tried to convince both Roosevelt and Churchill to share the bomb with Russia. Bush's patience with Roosevelt was exhausted during a 90-minute meeting held on September 22, 1944, at the White House in which the president nonchalantly said he favored "complete interchange with the British on this subject after the war in all phases." Bush did not feel he could speak his mind because Lord Cherwell, Churchill's atomic adviser, was present. He later told William Leahy, the president's chief of staff, that he "was much disturbed about the post-war handling of the special project, and that in particular I felt that the President needed advice at every point on a subject as difficult as this and that I did not feel at all content to see him handle the matter with the only technical man present a Britisher. The Admiral did not say much, but I gathered he rather agreed."[67]

To be sure, Bush was offended by Roosevelt's clubby relations with Churchill. He wanted the president to lean more heavily on him. He also thought the president's seat-of-the-pants decisions on atomic matters might actually stimulate an arms race.

"Too close collaboration with the British, without considering simultaneously the entire world situation, might lead to a very undesirable relationship indeed on the subject of Russia," Bush wrote after the meeting. Three days later, he elaborated on the problem. Roosevelt "evidently thought he could join with Churchill in bringing about a U.S.-U.K. post-war agreement on this subject by which it would be held closely and presumably to control the peace of the world." Bush felt "this extreme attitude might well lead to extraordinary efforts on the part of Russia to establish its own position in the field secretly, and might lead to a clash, say 20 years from now." By contrast, Russia might restrain itself "if there were complete scientific interchange."[68]

Bush thought an international agreement to control the bomb might limit its use in war. While this seemed far-fetched, it at least "ought to be explored and carefully analyzed."

He was so disturbed by Roosevelt's unwillingness to broaden discussion of the bomb that he complained to Henry Stimson. Acting in concert with Conant, Bush gave Stimson an appraisal of the nation's atomic predicament on September 30. Trying to tutor Stimson on the perils of atomic weapons, Conant and Bush admitted that predicting the future of such "an expanding art" was "difficult," but they advanced several principles that they believed should guide the Roosevelt administration.[69]

The most important was that America's atomic advantage was temporary. "It would be possible, however, for any nation with good technical and scientific resources to reach our present position in three or four years," Bush and Conant predicted. The U.S., they added, could even lose its lead. "Once the distance between ourselves and those who have not yet developed this art is eliminated the accidents of research could give another country a temporary advantage as great as the one we now enjoy."

The next most important consideration was the risk of an arms race with

Russia. "It is our contention," Bush and Conant wrote, "that it would be extremely dangerous for the United States and Great Britain to attempt to carry on in complete secrecy further developments of the military application of this art. If this were done Russia would undoubtedly proceed in secret along the same lines and so too might certain other countries, including our defeated enemies." They further warned that the atomic bomb was only a prelude to more devastating hydrogen bombs. "We see that very great devastation could be caused immediately after the outbreak of hostilities to civilian and industrial centers by an enemy prepared with a relatively few such bombs," they wrote.

Bush and Conant also urged that an international agency dispense scientific information about atomic energy and have "free access in all countries not only to the scientific laboratories where such work is contained, but to the military establishments as well. We recognize that there will be great resistance to this measure, but believe the hazards to the future of the world are sufficiently great to warrant this attempt." Under such an inspection regime, they concluded, there would be "reason to hope that the weapons would never be employed and indeed that the existence of these weapons might decrease the chance of another major war."

In this brief memo, Bush and Conant packaged together issues that had bothered them and many atomic scientists for some time. In one master stroke, they tried to address the political problems arising from atomic weapons, while taking advantage of the U.S. lead. By revealing all but the most sensitive details about the bomb, the U.S. would impress its own citizens and the people of the world with its candor. And candor cost little, they suggested, since "complete secrecy" about the bomb was impossible to maintain. "All the basic facts are known to physicists," they noted, adding that "some outside the project have undoubtedly guessed a great deal of what is going on."[70]

The Bush-Conant memo was a watershed in official thinking on the bomb, and its arguments would influence policy debates for decades to come. But the memo was also notable for its omissions. Bush and Conant did not recommend any specific diplomatic measures or call on the government to unilaterally cease development of atomic weapons. The proposed international inspectors seemed likely to become nothing more than scorekeepers of each nation's atomic arsenal. Finally, the two men did not insist that the U.S. inform Russia of the bomb's existence; that was to occur only after a demonstration of the bomb.

That Bush and Conant were even raising the issue of a demonstration, rather than a surprise attack on Japan, was significant. But they never asked that the bomb be withheld from battle. At most they considered warning Japan in advance of an attack. They believed the bomb would be used against Japan, on some terms, if the war lasted another year. In planning for this event, their chief concern was that the U.S. release adequate information after the fact.[71]

Bush wanted Roosevelt to see his memo, but Stimson would not pass it on. Bush simmered, choosing not to directly address the president. He told Conant in late 1944 that at least Roosevelt ought to form "a good solid group to study the implication of this affair and advise him as to the possible moves." But even that, he noted dourly, "is too much to hope for just at the present time."

Bush found Roosevelt increasingly remote. In December, he pestered Stimson to form a presidential atomic advisory group to study ways to forestall an atomic arms race. By year's end Stimson still had not given Bush's request to Roosevelt. The war secretary was in no rush. While Bush was thinking about the *next* war, he and the president were bent on finishing this one.[72]

Chapter 10

"The endless frontier"

(1944–45)

Roosevelt called me into his office and said, "What's going to happen to science after the war?" I said, "It's going to fall flat on its face." He said, "What are we going to do about it?" And I told him, "We better do something damn quick."

—Vannevar Bush

With an Allied victory in Europe inevitable in late 1944, Bush looked to the future, believing that the nation needed to permanently support civilian research after the end of the war. He certainly wanted to kill OSRD itself, which as an emergency organization deserved a dignified death. Yet technological and military power were now too intertwined for the nation to return to its prewar neglect of scientific research. He believed that a nation could no longer possess military might without technological prowess. Expanding this prowess, meanwhile, might *prevent* future conflicts: after all, had not Allied weakness invited German aggression? As Bush had recently told *Time,* "If we had been on our toes in war technology ten years ago, we would probably not have had this damn war."[1]

Bush envisioned a postwar technology agency that was essentially a peacetime OSRD with a fresh legislative charter. The new agency would help the nation maintain military security through technological strength by handling what Bush felt were the two biggest challenges facing postwar America.

218

The first was the duplication of military research projects, which had grown like weeds in the hothouse of World War II. Though still chafing at the military's blithe ignorance of technology, Bush found that the military officers were converting so rapidly to the scientific religion that they risked drilling too many dry holes, thus wasting resources.

The second was the need to fund research in basic science and engineering. This would address the complaints from physicists at the Radiation Lab, who often growled, "We're not doing science, we're just being engineers." Basic science involved experimentation without utilitarian purpose, as opposed to the Rad Lab's problem-solving mission. Neither industry nor the military had the patience or foresight to support "pure" scientific research, yet both depended on new scientific knowledge to stimulate economic growth and the creation of new weapons.[2]

Basic research, in this view, was the "seed corn" on which technology depended. Bush himself held this linear model of innovation: scientific knowledge nurtured engineering practice and led, in turn, to useful products and processes. Yet he was aware that the linear model was buckling under the weight of growing evidence that no clear line existed between science and technology, and that one often stimulated the other. In fact, the lines between basic and applied and theoretical and practical research had been blurring for decades. Scientists were beginning to focus at least some of their inquiries on technological artifacts: tools, materials and processes. And far from simply applying knowledge gleaned from scientific "discoveries," engineers often succeeded by testing, refining and proving hypotheses. This process seemed wholly scientific despite its practical ends.[3]

While aware of the gradual convergence of basic and applied knowledge, Bush believed the war had exhausted the nation's scientists. Blue-sky researchers needed money, and by making arcane points about the anatomy of knowledge he might only lessen the chances of their obtaining it. Unless the ranks of scientists were replenished and the funds for basic experimentation were increased, Bush argued, U.S. innovation might cease. The country might fall prey to foreign rivals. Only the federal government had the money to pay the bill for world-class research, he believed. Ordinary citizens must understand that researchers deserved tax dollars even in times of peace.

Having laid out his vision, Bush moved to realize it. He drafted a letter, in Roosevelt's name, that asked him to apply the lessons of OSRD to the task of supporting research after the war. In succeeding decades, this letter took on the status of Holy Writ among the nation's science elite. It spawned a Bush report that for decades shaped relations between government and civilian researchers.[4]

In late October, Bush sent a draft of the letter to Oscar Cox, a government attorney who often helped him. Cox gave the letter to Harry Hopkins, Bush's best contact in the White House, with an attached note. "The proposed letter . . . is satisfactory to Bush and his shop. Don't you think it ought to be sent out and released soon?"[5]

Hopkins demurred, complaining that the rather dense letter was far too long. Keep it under two pages, he ordered. The letter also should not say "the government can and should assist research laboratories." Better to "put it out as an idea." He complained about the excessive praise for OSRD. "Don't claim anything for that committee [sic] which the Army or some private outfits can argue about," he warned. Finally, he told Cox to review any changes in the letter with Bush.[6]

Five days later, Hopkins received a new version of the letter and gave it approvingly to Samuel I. Rosenman, Roosevelt's special counselor. After making some minor changes in the letter in the presence of a Bush aide, Rosenman gave it to the president on November 17. A note attached to the letter indicated that Bush had reviewed it "very carefully" and was "enthusiastic about the idea" of a report on postwar science.[7]

On November 20, barely two weeks after winning a historic fourth term as president, Roosevelt publicly released "his" letter to Bush. The president urged him to prepare a report on how OSRD's experience "should be used in the days of peace ahead for the improvement of the national health, the creation of new enterprises bringing new jobs, and the betterment of the national standard of living." The emphasis on the civilian economy was at odds with Bush's laissez-faire attitude and his wartime role, but in line with the political winds. As director of the OSRD, Bush had cast technology as the handmaiden of national security, but future government spending on technology—in peacetime—would have to be justified partly as a boost to the civilian economy. Roosevelt's letter thus reflected the public's fear of a renewed Depression following the war. "New frontiers of the mind are before us," the letter said, "and if they are pioneered with the same vision, boldness, and drive with which we have waged this war we can create a fuller and more fruitful employment and a fuller and more fruitful life."[8]

The letter asked Bush to address four broad questions. Even 50 years later, these questions encompass the debate over the government's obligations to science and engineering.

The first question dealt with "spinoffs": civilian innovations that arise from military research. The mania for secrecy had bottled up thousands of scientific reports, on everything from radar to atomic energy, and Bush felt that a good many of these reports posed no security risk and would stimulate much commercial research.

The second question concerned support for medical research, or the "war of science against disease." OSRD's medical wing had sponsored scores of medical innovations during the war, from better malaria-fighting drugs to commercially produceable penicillin. This work, often ignored by analysts of the OSRD, nonetheless set a powerful precedent for future government aid in disease-related research. While Bush was personally preoccupied with weapons research and policy during the war, he took great pride in OSRD's role in spawning life-saving innovations. In the case of penicillin, the OSRD "succeeded," in the words of one historian, "beyond its wildest dreams." The agency, showing that its institutional arrangements could work for peace as well as war, helped to select a "penicillin czar" who had unusual powers to force rival pharmaceutical companies to collaborate. When industries stopped jockeying for advantage, the result was a staggering increase in penicillin production during the war and a steep decline in its price. By 1945 the drug arguably "had saved more lives than any medication previously devised."

The fourth question was how government could aid the task of "discovering and developing scientific talent in American youth." The draft had interrupted the education of many men and Bush wondered if the war had robbed the nation of a bumper crop of young scientists at just the time of greatest need.

The third question was the most far-reaching and controversial: "What can the government do now and in the future to aid research activities by public and private organizations?" The question was deceptively simple, but it contained critical assumptions. It embodied the belief that the best scientists and engineers would never serve as full-time government employees, so the government must invariably hire private researchers—at universities and corporations—if it wanted the best work. Given these assumptions, there was little worth debating. Embedded in this question was the OSRD model of state-funded but privately executed research.[9]

To answer Roosevelt's four questions, Bush resorted to a common academic maneuver: he formed a committee to address each one. Each committee was composed of about 18 members, giving a patina of democracy to what was essentially Bush's personal agenda. Indeed, Bush carefully controlled the composition of the committees, relying heavily on leaders of elite universities, corporations and foundations, many of whom had been recipients of OSRD contracts. Missing from the committees were women, minorities, consumers—in short, ordinary people. No one served on a committee who was not a corporate executive, a lawyer, a professor or an administrator of a university, research institute or foundation. Bush brushed aside complaints that the committees were too narrow and ignored advice from outside scientists and citizens whom he deemed not part of the elite.[10]

Nominally independent, these committees bent to Bush's will. Only the medical committee followed its own course. In a show of his enthusiasm for the report, Bush pressed his advice on Isaiah Bowman, the chair of the committee studying the crucial third question on government funding of research. On January 10, 1945, Bush told Bowman, "I have no wish to predetermine in any way the course of deliberations of the committee under your chairmanship." He then proceeded to give Bowman, president of Johns Hopkins University, four pages of advice and confessed that he had even spoken with two members of Bowman's committee on topics he cared not to repeat. In closing, he declared it wasn't his intention "to orient the work of your committee or to predetermine the relative importance which it places on various aspects of the problem." Yet Bush indicated that he planned to keep close tabs on Bowman, reserving the right "to pass along to you such comments as come to me directly which seem to be worthy of definite mention." Bush's intensity reflected the importance he placed on the report, not his opinion of Bowman's competence. The Hopkins president was no loose cannon; his university was OSRD's eighth-largest contractor and operated the lab designing proximity fuzes.[11]

Under prodding from Bush, draft versions of the committee's reports began reaching his office by February 1945. They contained few surprises. Having chosen reasonable men of standing within the elite schools of academia and such great corporations as AT&T, Standard Oil and DuPont, Bush received sober statements that supported his own views. The one exception involved the medical committee, chaired by W. W. Palmer of Columbia University. Palmer's committee endorsed the creation of a separate foundation for medical research. Bush considered this unneccessary.[12]

By March 1945, Bush had begun thinking about the introduction to his report. This opening essay was critical, since he hoped to reach a wide audience and he knew the committee reports would be somewhat dry and even tedious. For the introduction, language and image were crucial. At best, the report could become a kind of creation myth, a founding story about the new world conceived by the union of science and government during the war. It seemed the right time for such a fable. After the setbacks of the Depression—when many critics blamed joblessness on too much technology—the prestige of science was rising. The public had not yet learned of atomic bombs, and scientists were unabashedly lionized. Despite secrecy provisions, news leaked out about the frenzy of wartime innovation. Bush was eager to trade on the public's high regard—even awe—for science and engineering while it lasted. After flirting with the title *Science—The Perpetual Frontier,* Bush settled on *Science—The Endless Frontier.*[13]

The title was a gem. It evoked the American character at its most exuberant and audacious. The frontier lay at the core of the American experience, and Bush traded on that, equating the pioneer, blazing a trail in the wilderness, with the scientist and engineer, exploring the frontier of knowledge. As early as 1937, in a June 22 speech to fellow engineers, Bush had said that the quest for knowledge could replace the vanishing geographical frontier as the new source of American freedom and creativity.[14]

While the images evoked by *Science—The Endless Frontier* were compelling, the substance was more complicated. With the help of his staff, Bush crafted a 34-page introduction to the report that elegantly and, at times, passionately made the case for the importance of research in American life. The centerpiece of his essay was a proposal for the creation of a National Research Foundation. The agency would sponsor studies in the physical sciences and medicine, set priorities for long-term military research and lend coherence to the diverse research efforts paid for by other federal departments. Sweeping in scope, the research foundation would elevate private experts to a governmental status heretofore accepted only in periods of national crisis. In order to limit the possibility of political interference, the foundation would allow expenditures to be dictated by researchers themselves. Even the foundation's director would be selected not by the president, but by an independent board of scientists appointed by the chief executive.

Roosevelt's letter in November 1944 and Bush's subsequent drive to produce a blueprint for postwar research were the culmination of nearly a year of private maneuvering. The aims of this closed-door wrangling were complex and confusing. Bush was partly to blame for this. His postwar plan was ambitious, visionary and very much his own, but he sometimes embraced contradictory ideas and ignored crucial distinctions in order to satisfy the many civilian and military interest groups that had a stake in a strong national research community.[15]

Bush had three main goals for the postwar period. First, he wished to guarantee adequate funding for science in order to insure a reservoir of abstract knowledge that could be drawn down freely by industry and the military in order to stimulate economic growth and enhance the nation's security. Second, he wished to improve the quality of the military's research in order to increase the effectiveness of its weapons and strategies. Third, he wished to apply scientific expertise to the task of improving the quality of governmental decisions. This improvement would come initially in the areas of education, research, the military and national security, but ultimately in other aspects of government too.

The three goals shared an underlying unity: science was too important for anyone but the scientist to manage, and its benefits were too compelling to restrict to technical domains. While not quite espousing a scientocracy, Bush envisioned a technologically advanced America governed by the masters of science and technology. If this scientific elite could not actually fill the seats of power, it could at least advise those who did. As the war had shown, "with scientists reaching beyond mere invention to guide military and even diplomatic policy," in practice "once-clear distinctions" between corporate executives, military officials and research administrators "were fading," historian Michael Sherry has noted. "Their roles increasingly overlapped." Bush illustrated this himself. He was simultaneously a patron of Raytheon, a defense contractor he had cofounded; the president of Carnegie Institution of Washington, a civilian think-tank that received government funds; and the director of the OSRD, an executive agency.16

Bush was torn over how to promote this trend. As an organizer of expertise, his actions ran counter to national traditions of self-reliance, direct democracy and local autonomy. The experience of World War II had radically altered his conception of the nation-state. The war was a massive job of administration. As heavily as the Allies depended on innovations in weapons and tactics, they were indebted to their bureaucracies as well.

The material output of the Allies was extraordinary, but, as Richard Overy has noted, the American rearmament and the Soviet revival were all the more striking when contrasted with "the inability of their enemies to make the most of the resources they had." It took expertise, in other words, for the Allies to outproduce the Axis. Germany and Japan were handicapped by an "inadequate level of conversion to munitions production and utterly haphazard planning procedures." By contrast, Allied administrators proved astonishingly competent at churning out useful equipment (and this held despite the drumbeat of complaints against the U.S. War Production Board, justifiably stigmatized as inefficient). While the Axis began the war with a huge lead in weaponry, by 1943 the material balance had reversed. In that year, the Allies outproduced the Axis in aircraft by a factor of nearly four to one. Spared the direct effects of the war, the U.S. achieved unparalleled levels of productivity, especially in aircraft. During the war, the U.S. produced nearly 300,000 planes, fives times Japan's total and more than twice Germany's. In vehicle production, the gulf was even wider. U.S. output of 2.4 million exceeded Germany's production by a factor of six and Japan's by a factor of 15. By the end of the war, half of the world's manufacturing took place in the U.S. "The Battle of Production," one historian has noted, "was virtually a walk-over."[17]

The "battle of production" greatly impressed Bush. While as responsible as anyone for popularizing the idea that superior technology won the war, he could not deny the decisive contribution of the U.S.'s material edge and the faceless administrators who had marshaled resources to good effect. An unshakable believer in the merits of the free market, he thought the Allied advantage lay in the interplay between government goals and corporate initiative. "We were strong in this war because we were industrially strong, and also because our scientists and technical men and industry could work together well and with the military," Bush wrote in early 1946. "But it was because we had a resourceful and extensive industry that we came out as well and as soon as we did."[18]

That no single agency ever directed the nation's vast economy led to waste but also allowed for flexibility. Not only were scientists drawn into close quarters with soldiers, as Bush noted, but experts from other professions were too. The resulting bureaucracies were imperfect; as one observer noted, economic mobilization was "bitched, botched and buggered from start to finish." But the war machine exerted a measure of control over wages, prices, raw materials and the behavior of citizens. Even in democratic Britain and the U.S., severe restrictions were placed on freedom of information and dissenting speech: until 1943, for instance, the government forbade publication of pictures of Americans killed in battle. As writer Walter Millis noted in 1956, "The centralized modern state had developed into an incomparable instrument for waging war."[19]

Even so, the familiar American animus toward central authority remained, but it now coexisted uneasily with a new appreciation for state power. How to manage this contradiction would bedevil Bush for the rest of his life. Even as he later sought to reassert his old ideal of private efforts on the state's behalf, he bowed to the new reality of pervasive state machinery. Many scientists boasted after the war that radar had won it or the A-bomb had won it, but Bush refrained from identifying a single technological factor, insisting more generally that "upon the current evolution of the instrumentalities of war, the strategy and tactics of warfare must now be conditioned." While proudly advertising the achievements of the OSRD, Bush was no technological determinist. He realized that supremacy was not achieved in a scientific or cultural vacuum. Rather than identify winning technologies, he identified the nation's *political inventiveness* as its unique strength. The nation responded to World War II by creating a national system of innovation, tying together laboratory research, mass production, battlefield tactics and boardroom strategies. The scale and rigor of this military-industrial-academic complex were unprecedented and amounted to a revolution in the American way of life. That this revolution was foreign to the country—and indeed unwanted by many of its prime movers, including Bush—in no way undermined its necessity. As Bush explained, the war demanded "a closer linkage among military men, scientists and industrialists than had ever before been required." Upon this web of pragmatic links, he insisted, rested U.S. supremacy—in both war and peace.[20]

Bush's sense of the future, then, reflected his understanding that winning a modern war depended on the links between science, the military and industry. The great challenge of the postwar would be to strengthen these links in ways consistent with a peacetime economy. Many experts realized this, of course, but what distinguished Bush was his belief that scientists were the linchpin in this link. Unlike career military officers or corporate executives, scientists had the intellectual training and discipline to withstand the pressure of trading immediate profits and combat advantages for more durable benefits.

No longer concerned that weapons research would receive short shrift from industry and the military, Bush faced contradictory concerns during the waning months of the war. On the one hand, he worried about the "danger that the country may get over-convinced" of the power to be gained from pursuing "narrow" technological goals. In the rush for new weapons officials might forget that seminal breakthroughs arose from "a sound background of scientific research throughout the country, not aimed directly at military matters, but forming the backlog which is so essential." On the other hand, Bush feared that after a period of peace the military might succumb to a kind of collective amnesia and return to the torpor of the 1930s. "Certainly the situation" before war "was not at all satisfactory," he wrote a colleague in March 1945. While he granted the military seemed zealous about technology now, "I fear that there will be a tendency to slip back to pre-war conditions."[21]

From a philosophical standpoint, Bush saw the end of World War II as an opportunity to permanently alter the role of the expert in a democracy. Initially, the expert would ride herd on the military, but over time his role could broaden to other parts of the government. On this point, Bush again agonized over his prewar allegiances. He considered the central moral achievement of the war the nation's reconciliation of the conflict between free men in a democratic society and the short-term benefits of tight control and command over a nation's resources. While insisting that democracies held inherent advantages over totalitarian nations in times of war, he based his conclusion on the presumption that free societies would embrace "a rigidity of control and a pyramidal system for the operation of the entire effort." While acknowledging that

Americans only truly prospered in an unfettered setting, he insisted that victory against future despots depended on the nation assuming "a more rigid cast for the stern task ahead." With victory now at hand, Bush wanted the "artificial controls" on Americans removed. Scientists especially must "recover that freedom of action and that healthy competitive scientific spirit." This did not imply a return to the past, but rather a new societal role for freewheeling technocrats. Only if fattened on public funds could unfettered scientists truly aid the nation.[22]

Bush's messianic sense of the scientist's part in the body politic inevitability brought him into conflict with three powerful interest groups: the military, liberal activists and the youngest generation of scientists who came of age nourished by federal spending. In opposing Bush, each group countered with its own vision of postwar America. The military, which had grown in both size and prestige during the war, saw its institutions as the most effective guardians of national security, easily capable of directing the necessary scientific expertise. The progressive remnant of Roosevelt's New Deal coalition, in awe of what science and technology had done for the military, now wanted to put these same forces in the service of social, economic and humanitarian aims. Finally, the mass of young scientists and engineers, whom Bush purported to speak for, shared little of his suspicion of military patronage, cared not for his political elitism and found his personal style remote and imperious.

The military posed the most serious challenge to Bush's vision of an apolitical, civilian technocracy. For this, Bush could partly blame himself. The very success of OSRD dramatically convinced the Army and Navy of the perils of ignoring technology—and the benefits of directly controlling its own OSRD-like outfits. Career soldiers now realized that they could not conscript the best scientists into their cause, but must allow them to retain the status and attractions of civilian life. Whereas in 1940 and 1941, military officers groused that they could achieve as much as Bush's men if only given the funds, by 1943 they largely accepted the superiority of independent technologists.

The big question for the military was how to tap expert knowhow. Under Bush's prodding, the Army formed an internal organization devoted to new weapons. Within the office of the secretary of war, meanwhile, Edward Bowles assembled a mini-OSRD consisting of academic consultants. Originally asked by Henry Stimson to help the Army deploy radar, Bowles broadened his duties until he was personally advising Army Air Chief Henry Arnold on technical issues. In sharp contrast to Bush, who seemed to delight in forcing the top brass to treat him as a big shot, Bowles brought no pretensions to his task. Viewing

Arnold as an "opportunist" with "the instincts of a lower animal," Bowles eagerly calibrated his advice to the general's mentality.

Forming a Special Bombardment Group, Bowles satisfied Arnold's appetite for expert analysis of potential air tactics. He created a group to promote the bombing of Japan "by techniques known to us but not yet fully explored or in use." Arguing for stripping B-29s of most of their defensive armament to allow for much higher bomb loads, the group shocked the plane's manufacturer and ultimately brought about a strategic shift. Defenseless bombers could operate more safely at night and from high altitudes—factors that compelled reliance on wide-area, incendiary bombing.[23]

Arnold was himself a convert to the view that technological strength determined military prowess. In June 1940, he had issued a directive to wage air war "on the continuous production of current types of airplanes." Now, four years later, he imagined future wars as brutal intercontinental aerial struggles and wanted the Army Air Forces to aggressively move beyond conventional methods. In the fall of 1944, he formed a Scientific Advisory Group, naming Theodore von Karman, a Hungarian professor at the California Institute of Technology, as its chairman. Smitten by the prospects for lethal rockets, Arnold complained that the Air Forces depended too much on "pilots, pilots and more pilots." He wanted his advisers to divine the future of air war, but offered some hints, telling von Karman, "I see a manless Air Force . . . [that] is going to be built around scientists—around mechanically minded fellows." If Bush needed an example of a soldier's transformation from technological conservative to zealot, he needed to look no farther than Arnold.[24]

But Arnold had cooled on Bush because of his skepticism of rockets. Von Karman, by contrast, was a rocket enthusiast who stood ready to do battle with Bush. The Hungarian emigré viewed Bush as a "good man," but "limited in vision." It didn't take long before von Karman saw Bush as a bureaucratic rival. When he felt a lack of support from Bush in early 1945, von Karman complained to Arnold. The general, in turn, chided Bush for showing insufficient enthusiasm for von Karman's efforts. The antagonism between the two science leaders turned on substance rather than style. Even before Bush formally declared his plan for the postwar political organization of science, von Karman decided that Bush's "concept of civilian control . . . would only *injure* progress in Air Force research."[25]

While von Karman worked the inside of the Air Force, Bowles worked the outside. In March 1945, he urged the Army (of which the Air Force was still a part) to directly hire university researchers to study new weapons and tactics. Diverging from Bush's view, Bowles advocated an approach to civil-military relations that he called "integration for national security." The concept was predicated on the belief that "the best insurance for peace is a positive attitude with respect to war," and it called for the Army to forge peacetime links with key actors in industry and academia. In contrast with Bush, who imagined that civilian scientists would keep the upper hand in these relationships, Bowles advised the Army's top brass to pull the strings. "For the Army," he wrote, "I do not believe there has yet been the opportunity for a relationship with industry and educational institutions of the sort which exists, rather spontaneously, between educational institutions and industry. With a view toward the peacetime integration I have referred to, how then can this be achieved? The power to achieve it is, I believe, well within our grasp."

For Bowles, the source of the Army's power would lie in postwar military budgets, likely to reach record levels for peacetime. With this money, the Army could buy whatever industrial and scientific services it wanted. To accomplish this, however, the Army needed "strong, highly trained men" not only to manage research, but "above all, to represent the military in dealings with the corresponding scientific body outside the Army." He recommended that West Point create a minor or a major in the application of advanced technology to military problems.[26]

Developing a cadre of technically savvy officers would take time, of course. To close the immediate gap between the Army's appetites and its expertise, Bowles proposed launching a civilian think-tank that would contract with the Air Force to plan the development and use of new weapons. By the summer of 1945, he and Donald Douglas, a principal in military supplier Douglas Aircraft Company, had concocted a plan for such an outfit with Arnold's blessing. The men were drawn together by a desire for bolder, broader scientific advice than the Air Force currently received from either OSRD or the National Advisory Committee on Aeronautics. And it helped that Arnold's son had recently married Douglas's daughter.[27]

The Navy charted a different route toward control of its technological assets. Shut out of the Manhattan Project by Bush and upstaged by OSRD in the vital field of radar, senior naval officers had suffered enough blows to their pride to realize that they must seize back some lost turf. As early as 1943, progressive officers tried to expand the scope and budget of Admiral Julius Furer's research office by attaching it to the Navy secretary's office. But Admiral Ernest King, the Navy's chief, loathed ceding responsibility to others and quashed the movement. The decision rankled Furer, who wrote in his diary that he feared that the Army—with its close ties to Bush and Bowles—"is picking up the best talent very fast. The result may be that the Navy will lose a lot of friends . . . in the long run." Furer had time and again preached to King and his fellow officers "that we are missing a bet in not recruiting some of these people. [But] the Admirals are afraid that they can't control the high-spirited talent of this kind."

Furer felt this fear was groundless. "I have had no difficulty in getting them to play ball as a team," he wrote. Unwilling to put his own career on the line for the sake of scientists, Furer held his fire. But he allowed a few younger officers, who dubbed themselves "the bird dogs," to resume their discreet agitation for a central research office under the command of the naval secretary. The bird dogs realized the Navy had neither a real research organization nor a plan for the postwar era, when they expected research to play an even bigger role in naval affairs. Since they weren't career military officers, they pressed their views despite the risk of sanctions. "We were a very small voice in the dark," recalled one bird dog, Bruce Old. "Even our own boss, Admiral Furer, was nervous about our planning. And other career officers too. They were all scared to death of Admiral King."[28]

Bush broke the logjam. In the spring of 1944, he warned the military that he planned to shut down OSRD when the war ended. "If we ever have another crisis, we can create this again," he told others. Bush's declaration shocked the services, especially the Navy, which was stumbling along with essentially the same research organization it had at the start of the war.[29]

Bush's motives for closing down OSRD were complicated. His most frequent explanation was that his researchers would bolt his organization and resume their civilian careers once Germany appeared on the verge of collapse. But there was scant evidence for such speculation, prompting suspicions that Bush was using the specter of OSRD's liquidation to prod the Army and Navy into taking postwar planning more seriously.

The gambit worked. The military saw Bush's threat to terminate as real. There was a sense among the brass that "if Bush resigned, the whole damned [OSRD] organization would fall apart."[30]

Furer quickly convened a military conference on postwar research. Held on April 26, 1944, the meeting underscored the revolution in attitudes among senior officers of the Army and Navy. Dozens of top brass attended, along with Bush, Bowles and National Academy of Sciences president Frank Jewett. A few years earlier, such a gathering would have been a grim, testy affair, with the officers scorning talk of research and deriding scientists for distracting attention from *this* war. Now it was hard to distinguish the military men from the technocrats. Everyone loved research. The group agreed to formally consider a civilian successor to OSRD, but informal comments from the soldiers revealed a desire to escape from Bush's research cocoon and a belief in the military's ultimate capacity to manage research and development. When Bush suggested that top scientists might balk at aiding the military during peacetime, the soldiers set him straight. "In my experience," one general declared, "there never has been any trouble in getting this cooperation from industry or universities if you had the money to get it."[31]

Bush recoiled at the idea of scientists being bought and sold; his high-minded view of the profession saw monetary rewards as trivial. The Navy's bird dogs hoped he was wrong. They increasingly acted as if the Navy would have a boon for researchers at the end of the war. Indeed, the service was setting aside a whopping $50 million in "leftover" war funds for research. The bird dogs hoped to spread the money around academia.

Still hiding from King, the bird dogs completed their plan in September 1944. Their proposed Office of Naval Research bore unmistakable similarities to Bush's idea for a postwar civilian agency. Imitating the OSRD's contract model, the proposed naval research office would emphasize research that had little immediate relevance to defense problems.

James Forrestal, the Navy's new secretary, rejected the bird dog proposal, though perhaps only because he planned to soon replace Furer as the Navy's top research coordinator. A former investment banker, Forrestal realized the Navy could not thrive without greater technical depth. In the summer of 1945 he explained to a visitor that the "Navy should initiate and control" its own research. The bird dog proposal for an Office of Naval Research was one means to this end, so in September Forrestal sent draft legislation for the office to Congress.

Even before receiving congressional approval, the Navy dispatched two bird dogs to canvas scientists around the country in order to gauge interest in Navy patronage. For two weeks in October and then again later in the year, two bird dogs promoted the Office of Naval Research to professors.

The reaction was predictable. Just as Bush had warned, academics feared Navy censorship and restrictions. "We visited all the major wartime research centers, Caltech, Berkeley, Columbia, MIT," recalled Bruce Old. "Everywhere we went, professors said they weren't going to contract with the Navy because we would insist that stuff be classified or not published until cleared." For the moment, they had no answer to professors' objections, so they just listened carefully. "We were on a fishing expedition in that sense," Old said. "We wanted to figure out a way to work with them."[32]

Challenged from the right by a resurgent armed services, Bush also was squeezed by the left. Imbued with New Deal faith in social engineering, liberals wanted science and technology placed in the service of social and economic ends. This was hardly a radical notion. The Depression had only been ended by World War II, and many believed that it would resume with the end of the war. Even in relatively flush times, moreover, millions of Americans remained ill-housed, poorly educated and underemployed. The return of peace, these progressives insisted, would remove from technocrats their only excuse for avoiding a direct attack on socioeconomic hardship. Given the vast power of research to alter the course of war, could scientists and engineers not do the same for the economy, they asked? And if federal funds could expand knowledge of the world, could they not be spent on gaining insights into the many social and psychological forces that mired the nation in racism and class inequities?

Harley Kilgore was the leading political proponent of this social view of research. A U.S. senator from West Virginia, Kilgore was a staunch Roosevelt backer and close friend of Vice-President Truman. An earthy small-town lawyer who carried a horse chestnut for good luck, Kilgore decidedly favored the little guy and regularly excoriated big business for robbing ordinary people of a fair chance. Elected to the Senate in 1940, he joined then-senator Truman's committee on national defense and quickly grew curious about the management of military research. He readily admitted his "utter, absolute ignorance" of science and technology, but pressed ahead gamely with ideas for organizing researchers. Impervious to rebuffs from both scientists and fellow senators, Kilgore hungered to solve problems; he epitomized the sort of sophomoric do-gooder who drove Bush nuts.

As *The New York Times* noted on Kilgore's death: "There was hardly a major national problem during his years in office for which he did not offer his own special solution. The fact that his solutions were almost unanimously ignored by his colleagues never seemed to daunt him. He always offered more."[33]

Fueling Kilgore was a populist belief that big corporations were unduly profiting from the war and not doing enough to bring innovations to market. Especially upset about the slow progress toward the production of synthetic rubber, Kilgore introduced a bill in August 1942 that would establish an office of technological mobilization. Kilgore envisioned the office as having the power to cut through conspiratorial patent pools; when he introduced a revised bill the following year, legislative hearings opened with testimony from former U.S. antitrust chief Thurman Arnold, who charged that shortages in basic materials were due to cartel control over research. Kilgore's bill would break this monopoly, he declared, and thus become the "Magna Carta of science."[34]

Kilgore's foray into research yielded no legislation during the war but certainly drew the attention of Bush, who publicly criticized the senator's views. Bush grew more concerned when in 1944 Kilgore began arguing for the creation of a postwar National Science Foundation. The senator thought that by describing his agency as a "foundation," a term with overtones of academic freedom and philanthropic enterprise, he would undercut opposition from scientists. Yet Kilgore continued to provoke opposition by complaining that monopolistic industrial agreements stifled innovation. Bush, himself a critic of patent pools and industrial cartels, considered Kilgore's complaints overdone and perhaps a smokescreen for the senator's statist economic policies.[35]

Many scientists shared Bush's unease about Kilgore; one said he hoped the senator "would drop the whole matter and leave it up to somebody else to take up." But Kilgore was nothing if not persistent. He pushed forward on a bill for a science foundation, inviting Bush in June 1944 to help draft legislation. Bush demurred, while privately conceding that Kilgore was "honestly trying to get at the root of matters" and had "certainly made some progress in the last two years."[36]

But evidently this was not enough progress to satisfy Bush. Kilgore still ignored what Bush considered a vital distinction between basic research, handled by universities, and the applied research carried out by industry. The senator also favored granting money to the social sciences, which Bush thought unworthy of federal support. Kilgore insisted on giving at least some research to universities on a geographical basis, which contradicted Bush's sense that elite schools received most government funding because they had the best people. Finally, Kilgore saw the research foundation fundamentally as an instrument of an interventionist economic policy that violated Bush's belief in the free market.

Economics drove the biggest wedge between Kilgore and Bush. The senator wanted patents arising from funded research to fall into public domain and vaguely suggested that the nation's research agenda should conform to a national goal of "economic security" and not the tastes of scientists or the whims of private corporations. If research programs were democratically controlled, he believed, "The results of federally financed research will not be perverted to private ends by monopolies and other interests at the expense of the common good." Bush, on the other hand, insisted that research contractors retain the patent rights to inventions arising from their work. This would give incentives to researchers and was, after all, OSRD policy. While accepting government funds for research, Bush rejected the idea that politicians could be arbiters of technological value. Bush presumed that useful innovations were best realized when private corporations were left free to decide how best to harvest the bounty of publicly funded research.[37]

Bush's fear that Kilgore and his fellow corporate critics would define postwar science policy fueled a desire to place his own mark on the subject. As historian Daniel Kevles has observed, Bush sought Roosevelt's blessing for a comprehensive review of research policies, "in no small part to take the initiative . . . away from Kilgore" and his allies.[38]

Besides skirmishing with Kilgore, Bush faced growing discontent in his own scientific ranks. Complaints came from two directions. Other scientific statesmen bristled at Bush's status, authority and imperious manner. Frank Jewett, president of the National Academy of Sciences, believed that his august body should play a larger role in the postwar era, a view that Bush considered suspect. Some university presidents, meanwhile, resented Bush's position as scientific kingmaker. In a telling letter to Isaiah Bowman, president of Johns Hopkins Univeristy and chair of an "Endless Frontier" committee, Bush confessed that he "ran away with the ball" during the war and had injured feelings. Given the bitterness against him it might be "highly desirable that individuals who have occupied prominent positions during the war in connection with scientific matters should stand aside when it comes to the postwar period," he wrote. "This would apply particularly to such individuals as Conant and myself." Yet Bush defended his prewar bid to forge a direct tie to Roosevelt. "I feel strongly that the path which we took was the only one possible under the pressure of the war," he wrote. "Nevertheless, there will be a better feeling among scientists generally if a somewhat new group takes over as the war comes to its end. Perhaps with this in mind we ought to look about for younger men . . . who could now pick up a good deal of the burden."[39]

The trouble was that younger scientific stars shared few of Bush's political views. While jealous of their independence, these researchers saw the federal largesse of World War II not as a long-term threat to their integrity, but as a professional liferaft. Having come of age in the Depression, these scientists had envisioned a bleak future in academia or industry—until the war came along and the money flowed. Military values were certainly alien to them, but the brass showed signs of accommodating the freewheeling style of scientists. Misty-eyed, Bush and other science leaders of his generation looked back on the 1920s and 1930s as a kind of golden age. It was hard for the younger generation to share this sentimentality. Many had no desire to turn back the clock—and none of the outsized anxiety about the future that often made Jewett and, at times, even Bush sound like party-poopers. "Coming after our education in the Depression, the war years were good: there was lots of money, high priorities, excitement," recalled Philip Abelson, an important physicist who earned his Ph.D. in the 1930s and performed military research during the war. "In the war we were immersed in the excitement. As far as we were concerned the federal money wasn't all that bad."[40]

Indeed, the very success of OSRD in unleashing the creative energies of scientists in their 20s and 30s posed a painful paradox for Bush. He saw the experience of World War II as an example to both the nation and its researchers. Yet he recoiled against the impending breakup of his tiny elite who had skillfully presented a single, coherent message about science to a perplexed public. While proud of the growth in many scientific disciplines, he worried that these special-interest groups—medicine, atomic energy, computer engineering, physics, the social sciences—would battle against one another for support and prestige. The answer, he felt, was to patch the fissures in his clubby elite, but the empowerment of so many younger scientists during the war had made it harder to protect the guardians of science from the ravages of change.

As one journalist perceptively noted:

> To the rank and file of scientists the success of OSRD is explained not so much by administration as by a kind of fruitful chaos. The organism grew so big so fast, under a steady rain of government funds, that top administrators could not hope to reach down into every laboratory to instruct an investigator to halt this line of research or take up that. Thus when a researcher had a hunch that penicillin would be useful in fatal subacute bacterial endocarditis, and his committee turned his hunch down, he just went ahead and proved that in large doses over enough time it cured many cases. The more irreverent describe OSRD as a vast mechanism that more or less ran away with Dr. Bush and his associates. It was set up to release the energies of the young, and this it did in great measure.[41]

Throughout the byzantine debates over postwar military organization and science policy, Bush kept looking for the chance to discuss atomic matters with Roosevelt. One question arose over the future of the atomic laboratory at the University of Chicago. The lab was nearly done giving aid to the Manhattan Project, but there was no firm plan for supporting it after the war. Bush thought it time for Roosevelt to plan for postwar atomic research, but the president's mind seemed elsewhere.

Another area involved the possible commercial exploitation of atomic energy after the war. Bush was skeptical of nuclear power; he had told Roosevelt in September 1944 that the technology held "great dangers . . . from a safety standpoint" and that it would be at least ten years before it made an "important" contribution to industry. But other experts were excited by the fantastic prospect of limitless, cheap energy. At Roosevelt's behest, Bush was trying to "obtain as much patent coverage in the hands of the government as possible, without taking extreme steps in this direction." He explained to Leslie Groves, chief of the Manhattan Project, in April 1945: "I have reported to the President once or twice that this was being done but have received no further instructions. It seems to me clear that the policy is a wise one." Bush admitted, however, that it might become more difficult to control the basic patents underpinning atomic energy if strong industrial companies took an interest in atomic power. "This highly difficult matter can be handled in the future much more readily," he observed, "if there are not strong industrial interests involved."[42]

It was the subject of international control of the bomb that Bush most wanted to explore with Roosevelt. He saw a chance to do so in mid-February after Russia, Great Britain and the U.S. announced plans at the Yalta conference to hold an international meeting in San Francisco in April 1945 to write a charter for a new United Nations. At Yalta, Roosevelt and Churchill stuck by their position that the bomb should remain an Anglo-American monopoly. The Allied leaders' refusal to inform Stalin of the bomb's existence greatly lessened the chances that the issue of postwar control of atomic energy would be raised with the Russians before the war's end. But it left Bush room to argue that international control of atomic weapons should fall under the proposed United Nations.[43]

Bush favored creating a "scientific section" within the UN that would oversee "communication between nations on scientific matters, particularly those involving possible military applications." It would also have enough clout to guarantee that "no aggressor, such as a reviving Japan, could prepare in secret for a war by unusual means." Bush intended to directly address the president on this matter. There was practically no one else to talk with besides Stimson, since the circle of top civilian officials who even knew of the bomb was small indeed. Bush had talked with Henry Wallace in the past, but Roosevelt had dropped him as vice-president, replacing him with a senator from Missouri, Harry S. Truman, whom Bush had never met. Frustrated with Stimson, whom he described as "so burdened that I doubt whether he will take action" on his atomic advice, Bush turned to Roosevelt.[44]

Before trying to see the president personally, Bush urged him in writing to endorse his notion of a UN science agency. In the hope of grabbing Roosevelt's attention, Bush recommended that at the UN meeting planned for April, the U.S. should back the creation of an agency that somehow made it impossible for "aggressor nations" to gain a military advantage through *future* scientific advances. Bush offered no specifics on how this would be accomplished, but his goal was clear. The agency, he wrote:

> should recommend means for policing the scientific activities of aggressor nations to ensure that they do not, in secret, provide for a new aggression by unusual methods. It should provide for full interchange between peaceloving nations on all scientific subjects which have evident military applications, to the end that no nation

shall be caused to fear the secret scientific activities of another, and on all scientific matters by which health and prosperity of nations may be enhanced. If the attempt to secure the future peace of the world by international organization progresses well, it should stand ready to recommend procedures for the interchange of information on the actual military applications of science, subject by subject, to the end that fear of secret preparations for war may be as fully dissipated as possible.

Bush recognized, of course, that his international plan ignored the many dangerous weapons technologies already developed by the U.S. and other nations. Addressing these actual threats to global peace would be more difficult, he allowed, but the UN could still try. He suggested that the agency first attempt to reach an agreement curtailing biological warfare as a kind of test of international resolve. "Further steps," tackling nuclear weapons, "should follow success with this subject," he told Roosevelt.[45]

The possible use of biological and chemical weapons had long horrified Bush, who was astonished that so far the combatants had avoided such barbarities. The memory of World War I, in which soldiers on both sides suffered grave harm from chemical weapons, certainly acted as a restraint. However, by 1943 both sides in the Pacific and European theaters had worried about and planned for biochemical attacks. Japan, which had never been the target of chemical weapons in World War I, used them occasionally in China without prompting any retaliatory attacks. For its part, the U.S. had resolved to shun biochemical weapons, concentrating instead on deterring an enemy attack. Substantial research into these weapons began only in 1942 and over the course of the war about $60 million was spent on them (compared to $2 billion on the atomic bomb).

In January 1944, U.S. restraint was weakening. Some officers in the Army Air Forces advocated gas attacks on Japanese cities. And the Army Chemical Service, the chief U.S. agency in the field, stepped up its activities in response to OSS reports that the Nazis might resort to germ warfare. (The reports were wrong. Hitler had barred offensive actions in 1939, perhaps because he had been hospitalized by a gas attack at the end of World War I. Despite pressure to change, he held to the prohibition).[46]

The sudden enthusiasm for biological and chemical weapons set Bush on edge. Roosevelt had never said whether he would retaliate against German first use of biologicals, or whether he would approve of a first strike against Japan. In the absence of a clear policy, Bush often fielded queries about U.S. options (though the OSRD had no formal responsibility for these weapons). On the heels of the OSS reports, the question came up at a February 1944 meeting of the Joint New Weapons committee. Bush appealed for caution and restraint, saying that on the subject of germ warfare, "I am always afraid of the enthusiasts in this field running wild."

Bush warned against hysterical judgments. In so doing, he showed his penchant for treating as practical choices what some saw as ethical ones. "There is as yet no clear evidence that any country is embarked on an offensive program," he told two senior officers, an Army colonel and a Navy admiral, attached to the Joint Chiefs. "We have rumors, we have hearsay, we have some evidence but not clear evidence, that Germany has a large program." Dismissing the evidence as flimsy, Bush reiterated that "there isn't any clear evidence of it yet." When the Army colonel expressed fear of a sneak gas attack from the "barbarous" Japanese, Bush allowed that it was possible but unlikely. "Probably there is no country quite as susceptible to this thing as the Japanese, and they know it," he said. "With their enormously dense population and crowded cities, they know it."

The Army officer was unpersuaded. He had traveled widely in Japan before the war and concluded that its mountainous terrain "might cause the gas [in an attack] to be blown out to sea. . . . There are a lot of geographical phases that might induce difficulties." Then the admiral interjected that the U.S. at least ought to have the capacity to retaliate, a capacity it did not seem to possess at the moment. Bush conceded that defensive preparations were needed, but pointed out the ambiguities inherent in stockpiling these weapons of mass destruction. "There is always the danger," as U.S. capabilities expanded, that Germany might mount a preemptive strike, he said. "Now I don't think that [this possibility] should be controlling. If we need to have preparations for retaliation in kind, we need to have them, and we need to take the chances, but I don't think we should prepare for it without realizing that in so doing we do take the chance of starting a race where general ignorance of the other fellow's intentions may precipitate a thing that would not otherwise occur."

This was not only a dilemma for the present war; it would be a dilemma after the war too. "I am personally somewhat terrified by what may happen on this thing in the postwar world," Bush said. "It is new. But there may be a time when it would be possible to build up by this [biological] means the kind of sudden and devastating attack that would be overwhelming, and the next dictator somewhere who has ideas of conquering the world may see this as a means, may build it secretly and get ready for it."

It was precisely because of this risk that Bush's position on biological weapons was nuanced. He thought the U.S. "ought to be pretty slow" in expanding its biological and chemical arsenals in order to avoid being accused of the moral error of "having taken the lead" in these weapons. The U.S. needed to produce such weapons as a precaution, but in doing so it ran "the distinct danger of having the enemy learn about this and regard it as an offensive program and act accordingly."[47]

By the fall of 1944, the germ warfare scare had largely subsided. U.S. preparations continued, but the likelihood of an outbreak had lessened. In October, Bush and Conant pressed Stimson on the future perils of biological warfare. In a five-page memo, the two scientists outlined the growing risks. The gist of their recommendations was that biological weapons should come under international control, preferably through the proposed United Nations. As a corollary, they proposed that the control apparatus be "divorced as far as possible from the military of each country, for quite properly military men are by training and tradition extreme nationalists in their view point." Evincing an almost messianic faith in the superiority of the scientist, they added that their professional brethren possessed an "inherent international viewpoint" that "could be used effectively in this case."

Bush and Conant also thought the very attempt to limit biological weapons would reap valuable lessons that could then be applied to the control of still-secret atomic weapons. Controlling biologicals, they noted in a cover letter to their memo, "may point the way for the consideration of still more important and highly secret matters." In the memo itself, they mounted a general case for international arrangements, which had broken down in the past but which under the threat of an atomic arms race might prove more durable. "One should not be too pessimistic about the possibility of international cooperation proving effective," they wrote.[48]

Stimson never relayed this memo to Roosevelt, so Bush made essentially the same point in his February 1945 appeal to the president. He had still not received a response when Stimson met with the president for lunch on March 15. After months of delaying, Stimson finally broached the subject of postwar control of atomic weapons with Roosevelt, describing briefly the two "schools of thought" on the A-bomb. One advocated that the U.S. take steps to maintain its atomic monopoly; the other advocated "international control based upon freedom both of science and of access." Stimson said that the question of postwar control "must be settled before the first projectile is used." Roosevelt, however, seemed more interested in discussing a curious complaint from a Washington insider who charged that Bush had "sold the president a lemon" in the form of the Manhattan Project and that the project "ought to be checked up" on before it came to a "disastrous" end. Having been warned that the president would bring up this complaint, Stimson swiftly swatted it aside.[49]

As far as Bush knew, this was the last time Roosevelt spoke of him.

Chapter 11

"After peace returns"

(1945)

There must be more—and more adequate—military research in peace-time. It is essential that civilian scientists continue in peacetime some portion of those contributions to national security which they have made so effectively during the war.

—Vannevar Bush

On April 12, Franklin Roosevelt died of a massive cerebral hemorrhage in Warm Springs, Georgia. For the nation, the loss of the president was an occasion for mourning, anxiety and soul-searching. It robbed many people, high and low, of their anchor and darkened their hopes for the future. In describing the return of the president's lifeless body to the White House, Bruce Catton, observing the crowd along Pennsylvania Avenue, captured the mood: "The people stood silent on the pavement, one hundred thousand of them, two hundred thousand, half a million, a multitude beyond counting, men and women whose lives and whose work these last few years had been shaped by this man's bidding." FDR's death, Catton insisted, created a hole in the world. "His death was about to be swallowed up in victory. But the victory had to mean something; had to be a victory of spirit, and not just a victory of the strong arm and the mailed fist. And there was in the air, on the morning of that funeral procession, the shape and the beginning of fear. The man who had made the victory was necessary, was gone. Who would interpret the victory now?"[1]

For Bush, Roosevelt's death was a professional catastrophe. It dealt a huge blow to his plans for the postwar period. It complicated his effort to build public support for technological research—on terms acceptable to him. It placed a cloud over his forthcoming science report. Finally, it reduced his influence on atomic policy.

In nearly five years of presidential service, Bush had met many times with Roosevelt. They had discussed not only the course of the Manhattan Project but atomic policy generally and the specifics of many other weapons programs. "I don't think anyone could possibly have had a better boss than Franklin D. Roosevelt," Bush later said. "He gave me a job to do, he backed me up completely, and he left me alone. You certainly can't ask for anything more than that from any man." Despite differences in outlook and experience, Bush always found Roosevelt "most cordial, and along with many another individual who differed with him on political philosophy I developed a personal loyalty and liking [for him] that was intense."[2]

The emotion was natural: Bush and Roosevelt were a good team. "On the whole, few of F.D.R.'s war administrators proved so peculiarly fitted for their special tasks or performed so smooth a job," one journalist wrote of Bush after Roosevelt's death.

Indeed, Bush owed Roosevelt a lot. Nominally director of the OSRD, he was actually an extension of the president. The personal nature of Bush's power proved problematic during Roosevelt's life and even more so after his death. The president often left no record of his instructions to Bush, even on such critical subjects as atomic policy. Often Roosevelt's laconic orders to Bush were only spoken. The president's verbal cues, meanwhile, were sometimes vague. "He would just give an implication and expect me to catch it," Bush later recalled. These elusive presidential statements "often constituted my authority for extensive activities."[3]

Roosevelt's informality and penchant for leaving decisions until the last moment added to the pressure on Bush and heightened his sense of isolation. "FDR was wonderfully supportive but he supported a lot of people," Bush recalled. "So no one ever felt they had 100 percent of FDR's support. They knew there was a competitor off somewhere else," angling for the president's blessing. By contrast, Truman seemed crisp and decisive during his first few weeks in the Oval Office. "Truman is a far better boss to work for than FDR," Bush said soon after Roosevelt's death.[4]

Bush's view was certainly shaped by the sympathy he felt for Truman, who had been put "in a very tough position" by Roosevelt's death. The former president had kept Truman in the dark on virtually every crucial war issue. Bush himself had never even met Truman until days after Roosevelt's death, when the new president called him in for a full briefing on the Manhattan Project. Accompanied by Admiral Leahy, the president's chief of staff, Bush delivered what Truman called "the scientist's version of atomic bomb." To ease Truman's anxiety, Leahy closed the briefing with a statement that reinforced Bush's contempt for the military. "This is the biggest fool thing we have ever done," he said of the Manhattan Project. "The bomb will never go off, and I speak as an expert in explosives."[5]

Truman brought a fresh eye to the atomic question. Completely in the dark about modern science and inexperienced in foreign affairs, the new president knew virtually nothing about the Manhattan Project. He wisely leaned on Henry Stimson, the war secretary. On April 25, Stimson told Truman it was now certain a bomb would be ready by August 1, 1945, in time to use against Japan. But he warned the president that the U.S. could not retain its atomic advantage indefinitely. Describing the arrival of the bomb in apocalyptic terms, Stimson said the weapon was so terrible that it could destroy civilization. The question of sharing the bomb with other nations, and on what basis, "becomes a primary question of our foreign relations," Stimson advised, echoing Bush and other scientists.

Accepting that the bomb would be used against Japan, Stimson argued that, if "proper use" of the bomb could be found, "we would have the opportunity to bring the world into a pattern in which the peace of the world and our civilization can be saved." Again echoing Bush and others, Stimson recommended Truman form an advisory group to address the bomb and postwar atomic policy.[6]

The same day, Bush met with Harvey Bundy, Stimson's aide on atomic matters, and emphasized once again the need for an advisory group to plan for international cooperation on atomic energy. "Time is very short," he told Bundy, adding that "there are certain steps which should be taken at once." The most important: the decision on when to tell the Soviet leaders about the bomb and the question of its future control.[7]

Responding to Stimson's pleas, Truman formed a study group on May 2, naming his war secretary as chairman. Two days later, Stimson named Bush to the so-called Interim Committee. Bush was elated by the assignment, writing Conant that, even though the committee was "off to a late start . . . it seems to me that on the whole we are well under way and that the job will probably get done sufficiently."[8]

Bush still worried about himself. "As the war enters its new phase," he told Stimson, "I find my burdens increasing rather than the reverse, and my support [staff] likely to decrease. There is a real danger that I now have more responsibilities than I can carry successfully, and that I may be in real trouble if the group that has aided me now tends to fall apart."[9]

The biggest question facing the eight-member Interim Committee was whether to use the bomb against Japan and, if so, whether to inform Russia and other nations in advance.

Bush never openly questioned the propriety of using the bomb. At the second meeting of the Interim Committee, on May 14, he handed out copies of the memo he had coauthored with Conant the prior September in which the two men had sounded the alarm over an impending atomic arms race. The memo read like a brief *against* using the bomb, but Bush was hardly pushing this line. To highlight this, he told the committee members that the seven-month-old memo "certainly did not indicate that we were irrevocably committed to any definite line of action but rather felt that we ought to express our ideas early." He added that "we would undoubtedly write the memorandum a little differently today."

Bush's prefatory remarks surely undermined the argument that the prospect of an arms race made the use of the bomb against Japan counterproductive. Indeed, one listener found Bush's views "very tentative," with his only distinct "idea that there should be a wide exchange of scientific information among the various nations."[10]

Two and a half weeks later, the Interim Committee recommended a surprise atomic strike against Japan. The recommendation, which resulted from meetings on May 31 and June 1, left the decision on specific targets to the military but ruled out any warning or demonstration of the bomb (on the grounds that neither would be effective). Bush endorsed this recommendation, which was opposed only by the Navy's representative, who thought that Japan should be warned. Five days later, Stimson relayed the advice to Truman, who accepted it. Now the only questions were whether the test of the bomb planned for July would come off successfully, and, if it did, when and where the bomb would be dropped.

In the decisive meeting, Bush made few remarks. The only specific advice he gave was to concur with Conant that the country's atomic programs "should be continued at their present levels until the end of the war." He also indirectly encouraged the inclination of James Byrnes, Truman's personal representative on the committee, not to share too much information with Stalin, pointing out to Byrnes that "even the British do not have any of our blueprints on [atomic] plants."[11]

Perhaps most significant, in light of later events, was Bush's contribution to a discussion about the prospects for other nations' making atomic bombs. On the specific question of how long it would take Russia to build an atomic bomb, opinion was divided. Leslie Groves, the Army's chief of the Manhattan Project, estimated 20 years. Chicago physicist Arthur Compton predicted six years. Bush figured the Russians would have a bomb in three to four years.[12]

Bush backed the use of the bomb without reservation. He offered no explicit justification at the time, perhaps because his view was entirely in character. For years he had worked to produce killing machines for the U.S. military and had turned a deaf ear to pacifists who sought to demonize certain weapons. Though he refused to allow OSRD to build devices specifically designed for assassinations, this was a curious exception to his anything-goes attitude. Indeed, throughout the war he belittled scientists who refused on moral grounds to build weapons. As early as 1941, he said, "We are living in a real and tough world. . . . It is not even wicked to work out more powerful bombs to drop on someone else." Even during the 1930s, he had thrown stones at pacifists, once saying in a speech, "Perhaps the worker on antiaircraft is more effectively a worker for peace than his brother who condemns him."[13]

The brutality of the Pacific war erased any reservations that Bush might have had about the atomic bomb. The first "firebomb" raid against Tokyo on the night of March 9, 1945, killed perhaps as many as 120,000 people, consumed nearly 16 square miles of the city and left a million people homeless. General Arnold, the Air Forces chief, was ecstatic. "Congratulations," he wired the raid's commander. "This mission shows your crews have got the guts for anything."[14]

Many firebomb raids followed. Bush never objected to any of them. He even had piqued Arnold's curiosity about firebombing in the first place by telling him about a new incendiary jelly called napalm. After hearing about the jelly, Arnold imagined the terrifying power of firebombs, fixated on one chilling image of napalm thrown in all directions and burning with such intensity that "if dropped near the entrance to a cave or a building, they caused all the air to rush out and anyone inside died from lack of oxygen."[15]

With incendiary bombing so deadly, Bush wondered why he should object to an atomic bomb that might do less harm. After all, if the bomb "brought a quick end to the war," it "would save more *Japanese* lives than it snuffed out." Besides, the support for fire raids convinced him that most Americans would welcome the use of the atomic bomb against Japan. "The moral question was hardly raised regarding the fire raids, yet that question is substantially identical in the two cases," he later wrote.[16]

Still, Bush's failure to dissent seems puzzling given his retrospective statement that Japan would have surrendered "within a matter of months even if the bomb were not used." He later confessed to being swayed by the Army's plans for a land invasion of Japan, which seemed likely to result in tens of thousands of U.S. casualties. The brutal battle in May and June to take the Japanese island of Okinawa underscored the costs of a land invasion; more than 12,000 U.S. soldiers died in the effort. One of Bush's two sons, moreover, was stationed in the Pacific. Richard, the physician, was an officer in the Army Medical Corps and worked in a base hospital on Leyte Island in the Philippines. John, Bush's younger son, was an Air Force pilot in Europe and part of a busy B-26 bomber crew. By mid-May he had flown 25 missions.[17]

Looking back on the summer of 1945 from a distance of 20 years, Bush saw no justification for an invasion. Such plans, he said, reflected a military "myth" that "no country is ever really subdued unless it's occupied [and] . . . that only the infantry on the ground" can close out a war. The U.S. insistence on invading Japan, he said, "was one of the worst decisions in recent history. And, if you please, it made the use of the A-bomb inevitable.[18]

Given his willingness to win Japan's surrender without an invasion, Bush asked, "The big question now is this: Why is it that I didn't protest vigorously against the invasion decision?" As an explanation, he cited the gulf between soldiers and scientists, suggesting that any objection would have been ignored. Yet futility had not stopped Bush from hollering at the top brass before.

At bottom, Bush did not dissent because he saw benefits to using the bomb, the chief one being that it might prompt nations to control atomic weapons in the future. The idea arose from Bush's reading of the effect of chemical weapons, which were used widely during the World War I. These toxic gases proved so horrible that the memory of them restrained World War II combatants. Deterrence arose from fear of the known, not the unknown. Bush was convinced that a similar pattern would play out if atomic weapons were used in combat.[19]

There was another consideration. In the new calculus of mass destruction, Bush ranked the bomb as among the lesser evils. He opposed most vigorously the use of biological weapons, which he considered even more deadly than atomic blasts (perhaps because he failed to appreciate fully the dire effects of blast radiation). During the war, Bush consistently argued for restricting research on biological agents to defensive measures. And he rebutted proposals to expand research, citing concern that too formidable a program might prompt rivals to launch a preemptive strike with their own biological weapons. "Great care should be taken that our [defensive] steps are not considered to be preparation for independent offensive use of biological warfare," he said in 1944.[20]

Yet Bush would not apply the same logic to atomic bombs. Perhaps this was because only the U.S. had them, while it held no advantage in biochemical weapons. This was a crucial distinction. For some indefinite period, the U.S. could use the A-bomb with impunity. Other nations were likely to build their own A-bombs eventually, but perhaps by then the U.S. would have achieved new breakthroughs in nuclear devastation. Bush, naturally, never said the U.S. would permanently maintain the lead in nuclear weapons. But his actions betrayed a certain hubris. Hearing the siren song of research, he acted as if the wizards of World War II would never falter. Indeed, he argued to the Interim Committee that funding for atomic weapons remain at present levels. He was proud of the nation's atomic scientists; he considered them the best in the world.[21]

Bush's views on atomic matters meshed with Truman's. But the OSRD director was out of step with the new president on two other key issues: the future of OSRD and plans for a postwar research foundation.

With victory in Europe imminent, Bush continued to call for the closure of OSRD at the end of the war despite intense opposition from Harold Smith. In the last weeks of Roosevelt's life, Bush thought Smith had gone on a "rampage." The budget director had even complained to FDR that Bush should not have been given authority over the forthcoming science report, which was tantamount to letting him "study his own performance." In March 1945, Smith reminded Roosevelt of the need to strike a balance between public accountability and his entrepreneurial experts. Explaining that "the whole thing was pretty dangerous," Smith asked Roosevelt to steer the Bush crowd in his direction "so that we could see that the Government was properly protected." As if he had not made his point strongly enough, Smith added a parting shot, charging without evidence "that there has been more leaf-raking and boondoggling on some of these research projects than there ever was in the" federal Works Progress Administration, the Depression-era jobs program.

As czar of the budget, Smith had an exalted view of his role. "The budget," he declared in 1944, "is at the very core of democratic government . . . because it is at the same time the most important instrument of legislative control and of executive management." Viewing himself as the unsung guardian of democracy, Smith kept close watch over Bush. In late March 1945, he railed against Bush's portrayal of OSRD as a lame-duck agency, presenting himself as a lonely defender against a scientific coup. "The battle lines are being drawn over the organization of research in the Government," Smith told Roosevelt. "This [Bush] crowd is trying to take a step in the wrong direction, and I am trying to keep them from doing so."[22]

Smith seemed to carry the day with Roosevelt, but when the president died Bush resumed his campaign. From a planning perspective, transferring

OSRD research to other military and civilian agencies made eminent sense to Bush. Few shared his view, however, and Bush paid a stiff price. He irritated Truman and Smith, angered the top brass and even upset some of his own research chiefs.

Army and Navy officers were united against Bush on this matter, insisting that OSRD operate "at full scale until the war is actually over on all fronts." While the services had grandiose plans for postwar research, they wanted to preserve OSRD through the end of the Pacific war, which planners thought could drag into 1946.[23]

By talking so heatedly about closing OSRD, Bush had caused tumult within, as well as without, his own ranks. A number of division chiefs had misunderstood his request for closure plans and had moved to shut down immediately. While Bush quickly corrected this misperception, he began to curtail NDRC's activities in May 1945. From then on, new projects could only be undertaken if the work could be completed within a year or if the Army or Navy had agreed to take over the project by June 30, 1946.[24]

The scaling-down of OSRD especially upset Lee DuBridge, director of the Radiation Laboratory, which accounted for the biggest share of OSRD spending. DuBridge pleaded with Bush in April 1945 to refrain from winding down OSRD, saying such a declaration would cause "a general exodus of personnel" from his lab. Any interruption in either funding or staffing, he insisted, would severely damage the nation's future radar capabilities. Though the end of the war approached, the Rad Lab still had 80 major projects underway. "The radar art is still in its infancy," DuBridge noted. "Important new techniques and applications are just over the horizon."[25]

DuBridge worried that Bush's desire to wind down OSRD threatened the future of his lab. This was not the first time that tensions had flared between the two men. DuBridge cherished his independence, and his Rad Lab colleagues bristled at attempts by OSRD to rein them in. As it grew, the lab did "damn well and [had] so much success that [DuBridge and colleagues] forgot they were part of OSRD," Bush later complained. "They went their own way. No one paid any attention to [me]. So one time I held up one of their checks, and they were hustling all over the place looking for [it]. Finally they landed in my office, growled about the check and I told them: 'I can tell you where that check is, right in this top drawer. Now you fellows can have it when you admit you're part of this organization and are going to follow the rules.'"[26]

Bush's rules were not always easy to swallow. His insistence on terminating OSRD flummoxed associates but sprang from his belief that the closure would give a sense of urgency to his call for a permanent successor agency. He also expected civilian life to swiftly return once the war ended. In the research realm, this meant that universities and private industry would immediately seek to regain the people they loaned to government—and risk disruption if they could not do so. The return to normalcy would require a massive demobilization, which, if mishandled, could spawn serious problems.

By April 1945, however, Bush's closest aides began to dissent. Oscar Ruebhausen, then OSRD's general counsel, argued against termination. Rather than directly confront Bush, which he considered foolish, he dropped hints and suggestions. Bush ignored them. Ruebhausen concluded he had uncovered a blind spot in his boss's ordinarily sharp vision. "Bush was destroying values that shouldn't be destroyed, but he was adamant," he recalled, adding that Bush possibly felt "guilty" about taking scientists away from their usual work and keeping them "in chains" for four years. Ruebhausen repeatedly heard Bush say of his scientists, "They've had it."[27]

Bush felt worn down too. His nerves were frazzled from the strain of preparing the A-bomb and the push for postwar research planning. He whined uncharacteristically about the burdens of maintaining the OSRD, hinting darkly that the legislative basis of his agency was suspect. He also complained about a possible revolt by his staff. "The plain fact is that I am thoroughly overloaded," he wrote Budget Director Harold Smith on April 25. "My entire organization, which has been kept to a small staff from the outset, is similarly overloaded. Moreover, a substantial number of personnel within my organization are serving on a voluntary basis without compensation, and I fear that a reaction to V-E day will deprive me of the services of some of those persons on whom I most rely."[28]

Smith had little sympathy. "I feel that it is imperative that your agency continue to function until after permanent post-war arrangements have been made to carry on," he wrote on April 27, adding that he hoped "the OSRD staff and other personnel . . . will be impelled by patriotism to refrain from dropping their responsibilities prematurely."[29]

Even as Smith implored Bush to stay on the job, however, Budget staff raised doubts about OSRD's integrity. Just eight contractors accounted for more than half of OSRD's spending, and the biggest contractor by far, MIT, had close ties to Bush and his senior deputies. "It is widely known that all the principal officials of OSRD have remained in the employ of some of the largest OSRD contractors," a Budget staffer noted dryly, then added that "little is known outside the OSRD of the reasons underlying the selection of contractors."[30]

A similar situation certainly applied to defense contractors of all sorts. Fully one-third of all war orders went to just ten companies, some of which loaned

As a boy in Chelsea, Bush had his own workshop, with shelves in a Quaker Oats box for chemicals and odd treasures in salt-cod box. On the bench is what looks like a dry cell hooked up to a clock. *(MIT)*

Bush pushing his first invention—the profile tracer—at Tufts, late 1910 or early 1911. *(MIT)*

The instructing staff of MIT's electrical engineering department, 1921. Bush, second row, second from left. Edward Bowles, third row, fourth from left. *(MIT)*

Bush as an earnest young professor. *(MIT)*

Laurence Marshall, Bush's roommate at Tufts and cofounder of Raytheon. *(Raytheon)*

Bush's mathematically adept sister, Edith, who taught at Tufts' sister school, Jackson College. *(Tufts, courtesy Fairlee Hersey)*

In the 1920s, Bush might spend a week or more on his sailboat, joined by an MIT colleague. Skippering his ketch relaxed him as few other activities did. *(Harold Hazen)*

Bush holding the Product Intergraph, a forerunner of his more powerful Differential Analyzer. *(MIT)*

Bush with his first Differential Analyzer in 1931. *(MIT)*

Bush with his mentor,
Karl Compton.
(AIP)

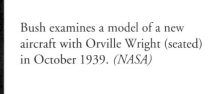

Bush examines a model of a new
aircraft with Orville Wright (seated)
in October 1939. *(NASA)*

After Pearl Harbor, newspapers discovered Bush. Here he poses in a Carnegie lab for a news service, which in March 1942 described Bush as having "access to every test tube in America and every man and woman capable of using one." *(AP/Wide World)*

One year after the first atomic explosion took place in July 1945, Bush and Conant reenacted for a "March of Time" newsreel their witnessing of the blast. *(MIT)*

The advisory council of the Office of Scientific Research and Development included almost all of Bush's significant wartime colleagues. From left to right: J.C. Hunaker, Harvey Bundy, James Conant, Bush, Admiral Julius Furer, Newton Richards, Frank Jewett, and Bush aide Carroll Wilson. *(U.S. Navy)*

Bush's role in organizing the Manhattan Project remained a secret until after Hiroshima. He, James Conant, and General Leslie Groves (from left to right) were the project's top administrators. *(Hagley)*

Actors as Bush and FDR in the 1947 film "The Beginning or the End?" *(HSTL/UPI)*

Truman pinning a medal on James Conant, with Bush standing by. *(HSTL)*

Oppenheimer, the intellectual lighting rod of the Manhattan Project, posing with his two organization sponsors, Bush and Conant. *(Courtesy, Harvard University Archives)*

Bush at the Attlee-Truman signing of the joint statement on atomic energy. *(HSTL)*

When Bush officially gave up his chairmanship of OSRD in late 1948, James Forrestal honored him, and incoming chair Karl Compton, at a supper at the Naval Gun Factory. *(U.S. Navy)*

The end of the war brought little relief as Bush pushed his twin agenda of federal support for civilian science and the unification of military research under civilian technocrats. The postwar battles were more and more fought with reports, not weapons, as illustrated by the crush of papers on Bush's desk on September 25, 1947. *(AP/Wide World)*

Harley Kilgore, the liberal Senator from West Virginia, battled with Bush over the future of science policy after the war. *(HSTL)*

Bush and his wife, Phoebe, in the doorway of their Washington home, early 1950. *(MIT)*

(Below) Promotional shot for "March of Time" documentary inspired by his book *Modern Arms and Free Men*. *(Library of Congress, courtesy Michael Dennis)*

Bush loved the sea and fancied himself a photography buff. *(courtesy Richard Hewitt)*

Bush test-driving his doomed hydrofoil boat in the early 1950s. *(courtesy Richard Hewitt)*

Vannevar and Phoebe
Bush on a 1954 vacation
at a dude ranch in
Wickenburg, Arizona.
(MIT)

Bush dictated most of his memos, but on those rare times he scribbled out his instructions
he wondered if his words were lost. He closes this letter with the apology: "Too bad you
can't read my writing. Cordially, Van." *(MIT)*

In 1956, soon after his retirement from the Carnegie Institution, Bush chats with journalist Ed Murrow in the basement workshop of his home in Belmont, Mass. *(MIT)*

Bush accepts an award from President Richard Nixon. (*Lawrence Berkeley National Laboratory*)

On October 1, 1965 at a ceremony, officiated by MIT President Julius A. Stratton (left) and Chairman James R. Killian, in which the university's Material Science Center was dedicated to and named for Bush. (*MIT*)

Bush with his Karsh photo. *(MIT)*

executives to the government. But this didn't ease Bush's sensitivity to criticism of the OSRD's clubby contracting practices. He wanted his choice of contractors insulated from outside scrutiny; the very ambiguity of this part of OSRD's charter made him want a successor agency. For several years he had tried to gain a blanket exemption for OSRD personnel from provisions in the U.S. criminal code that forbade self-dealing of the sort that routinely occurred during the war (when, for instance, Karl Compton indirectly awarded contracts to his employer, MIT). Those legislative efforts failed, leaving OSRD's top officials in possible violation of the law.[31]

Given the stakes following Roosevelt's death, Bush's worries about potential conflict-of-interest claims seemed picayune. Worse, his drive to terminate OSRD lessened the chance that Truman would embrace his plan for postwar research.

By early June, Bush was pressing the new president to take a position on his "Endless Frontier" report, which by then was circulating in near-final form. On June 1, Bush sent Samuel Rosenman, a Roosevelt confidant now serving Truman, a draft summary of his report. In a cover letter, he offered the president the choice of either endorsing the report "in whole or in part" and submitting it to Congress for legislative action, or taking no position and simply presenting the report to Congress "for study and action."

Bush expected a cool reception from Truman, partly because he wanted a big budget for the new agency, whose annual spending he expected to rise from an initial $33.5 million to $122.5 million after five years. Money wasn't the only reason for Bush's caution. He also feared the president would balk at his proposal for the director of the National Research Foundation to be selected by a board of presidential appointees rather than the president himself. Bush was torn over this point; while his letter to Rosenman made no mention of this, at the end of May he toyed with the idea of letting Truman choose the director. Many members of the "Endless Frontier" committees wanted the president to have only indirect influence over the foundation's chief, so Bush stuck with the formal recommendation. But he must have realized that Truman could perceive this as a personal slight, since he was likely to be president at the foundation's formation. The rationale for having the board choose the director was that it insulated the research agency from political interference, but there were negatives too. As Bush knew from experience, a presidential appointment helped immensely in the battles with other government agencies. Under Bush's plan for the foundation, the director might be saddled with an image problem.

Bush offered one curious concession in order to win the president's endorsement. His plan called for an independent agency, but Bush now suggested that the foundation could be lodged within the venerable Smithsonian Institution, whose mission included scientific inquiry. The idea, however, went nowhere.[32]

Even as he sought Truman's support for "Endless Frontier," Bush undercut his cause by fighting with the president over OSRD's fate.

Earlier in 1945, the War and Navy departments had agreed to form an interim research group, under the auspices of the National Academy of Sciences, to bridge the gap between the closure of the OSRD and the creation of a permanent peacetime agency. Harold Smith had fought for months against the proposed Research Board for National Security, arguing that it handed power over military research to a group outside the government. On June 8 Truman killed the research board and moved decisively to keep OSRD alive indefinitely. The agency "should not be liquidated at any early date," he instructed the War and Navy secretaries. "On the contrary it is imperative that the Office carry forward all its research work relating to the present war and that it continue to function until a suitable agency is established to take over post-war military research."[33]

Bush resisted. Rather than accept Truman's decision as final, he wrote the president on June 12, "I have given much thought to this subject and I have come to the conclusion that for [the OSRD] to undertake postwar research would be highly undesirable, for reasons which become apparent only when the matter is studied in depth." He then complained that extending the life of his agency "would reverse the understanding which I had for a long period with President Roosevelt, and with the Appropriations Committee. It would be contrary to the general principle that war agencies should not carry on into peace."

Next Bush warned that many OSRD staffers might walk off the job at the end of the war in order to protect themselves against charges of impropriety. "Many voluntary personnel of this Office are proceeding on the basis of an administrative order which I issued to avoid conflicts which seemed to me to be valid," he wrote, "but they have no guarantee that my interpretation will be sustained. They are taking a risk and they know it, in thus serving government while this cloud hangs over them. This they are willing to do in order to aid in fighting the war, but many of them would not take this risk after peace returns."

Finally, Bush described Truman's position as "a mistake," and informed the president that he would send copies of his letter to all those sent Truman's June 8 letter.

Truman never recorded his reaction to Bush's plea. But his actions conveyed plenty. On June 14, he opted not to endorse the "Endless Frontier" report.

Bush got the news from a presidential aide. He had seen Truman the same day but evidently the subject was too rough for him to broach in a 15-minute meeting. The president simply told Bush that he "liked" his report and would authorize its release.

Before leaving, Bush told Truman he was anxious to aid him in any way possible. Truman was noncommittal. Just how much he wanted Bush's help was a big question. The president's unwillingness to endorse Bush's postwar plan signaled the end of his honeymoon with the OSRD director. While not quite a rift, the nonendorsement signaled stormier times to come. Truman and Bush were quite different, after all. Bush was an elitist intellectual, capable of mixing it up in the back room but happiest manning a lathe or thinking big thoughts. Truman was a skilled politician, proud of his ordinariness and skeptical of experts. Then there were the circumstances of Truman's elevation. The new president naturally looked askance at Roosevelt's men, but he had a special reason to resent certain holdovers, Bush included, who long had kept him in the dark about the Manhattan Project. While often impatient with Roosevelt's casual way of approaching big decisions, Bush had come to see Truman's style as even worse. The president seemed like a political novice, surrounded by a clique of amateurs.[34]

Finding common ground between Truman and Bush was made all the more difficult by Harold Smith, a Roosevelt holdover who increasingly attacked Bush personally. Smith's animus confounded some colleagues. "I don't know what made Smith so suspicious of [Bush]—I can't understand it," noted Henry Wallace, the former vice-president and now commerce secretary in Truman's cabinet.[35]

Smith would not let go of his supicions of Bush. On June 14, the Budget director asked Julius Furer, the Navy's top research officer, to take a firmer hand with the civilian scientists. Bush's insistence that scientists make their own decisions on how to spend public funds struck Smith as the height of irresponsibility. Bush's attitude, he wrote:

is too much influenced by the assumption that researchers are as temperamental as a bunch of musicians, and consequently we must violate most of the tenets of democracy and good organization to adjust for their lack of emotional balance. I do not agree that they are that kind of person, in the first place, and in the second place, I can think of better ways of adjusting for emotional unbalance.

The real difficulty has been that the physical scientists are worried about government and governmental control largely because most of them—as they make rather clear to me by what they say and do not say—do not know even the first thing about the basic philosophy of democracy. This is quite an accusation, but I have seen little proof to the contrary. However, most of them have learned to accept governmental funds with ease, and I think they can adapt themselves to government organization with equal ease.[36]

Bush disagreed, feeling that Smith missed the point. Scientists must *self-organize* (and, ultimately, police themselves), no matter who paid their bills. This was the essential precondition of fruitful research. Even well-meaning politicians could unwittingly suffocate scientific freedom, Bush insisted, which was why government must adapt to the needs of science. The failure of Harley Kilgore, the liberal standard-bearer on science policy, to accept this drove Bush to block the West Virginian at virtually every turn. Yet Bush could not deny sharing a basic outlook with Kilgore. At bottom, they both wanted the government to pay for independent research, with almost no strings attached. The basic agreement between Bush and Kilgore was so striking that it alarmed the more traditional leaders in the science community. Robert Millikan, president of the California Institute of Technology, groused that Bush gave the impression that nearly "all of us scientists" favored postwar subsidies. A leader in science mobilization in World War I, Millikan certainly did not share this view, nor did "practically all of my colleagues here." More than 20 years older than Bush and more conservative, Millikan considered federal funding of research during peacetime as a move toward unwanted "collectivism."

Bush's response exposed the degree to which he considered Kilgore a fellow traveler, and objection to federal support for private activities outmoded. He replied to Millikan that he, too, opposed "stifling controls" on research by government. But Bush said of Kilgore, Millikan's nemesis, "I have at no time opposed his main thesis."[37]

Millikan was not alone in accusing Bush of misreading the possibilities for coopting liberals and mollifying the armed services. Frank Jewett, a staunch conservative and president of the National Academy of Sciences, attacked Bush's program systematically. A cofounder of the National Defense Research Committee, Jewett was a big Bush booster; he had even helped him land the Carnegie presidency in 1939. To be sure, Jewett had criticized Bush during the war on such matters as the relatively large contracts steered toward MIT. But his reaction to a draft of "Endless Frontier" overshadowed any prior differences. In a letter of June 5, 1945, Jewett chided Bush for producing an introduction that essentially expressed Bush's personal opinions. While understandable, since Bush was "in a commanding position" at the moment, it was nevertheless "extremely dangerous both for you and, more, for the success of the objective you have in mind."[38]

Jewett then warned that neutralizing critics on both his right and left flanks might prove impossible, not least because Bush had made many enemies during the war. "The whole thing is then likely to degenerate into a fight in which arrayed against your point of view are not only those who honestly disagree with you but likewise those who may not like you and the things you stand for," he wrote.

After making this prediction, Jewett ripped into the substance of Bush's postwar plan, mocking his request for "a paltry $10 million annually" in medical research, a "field where it is now easy to raise sums of nearly this amount from a willing public in connection with any one of a number of specific diseases." Then, turning the tables, he excoriated Bush for accepting without question the need for large-scale government aid, and failing to "examine *why* private funds have diminished and whether the answer may not be in the direction of revivifying that fruitful stream before plunging into the uncertain waters of the Federal tax pool."

Three days later, Jewett once more hammered Bush for forsaking the tradition of scientific independence. Of the military's response, he warned, "The Army and Navy would join in opposing parts of the report. They may differ violently on the proper way to bring top civilian science to their aid in peacetime, but both will fight any scheme which mixes control of that aid up with other non-military things." But the report should make possible a compromise with Kilgore, whom Jewett predicted "would be both delighted and disappointed" in it. While Bush fell short of embracing all of Kilgore's ideas, Jewett noted the irony that the senator would recognize "that you subscribe in part at least to his philosophy."[39]

With Bush and Kilgore close enough on the big questions about postwar research, why did they not join forces and speedily craft a permanent, civilian-run replacement for OSRD?

Important issues still divided them. Kilgore wanted the government to have patent rights on publicly funded inventions, while Bush considered this an undue restriction on entrepreneurs. Kilgore wanted the president to appoint the director of the research foundation, while Bush wanted the board to choose its own. The senator wanted to distribute some research on a geographical basis, while Bush vehemently opposed this. He insisted that the foundation have no explicit economic aims, while Kilgore wanted it to direct research into areas likely to yield socially desirable innovations, such as housing.[40]

In practice these differences were not so great; a compromise seemed possible in early 1945. Signaling his willingness to at least explore joint legislation with Kilgore, Bush asked the senator to delay any new bill and pledged to "collaborate fully" with him. After meeting with Bush in March, Kilgore decided to withhold legislation until the release of "Endless Frontier." This was a major concession on his part because the senator could have reaped political points by promptly introducing legislation on the heels of Bush's report. ("That prospect, obviously, was not appealing to Dr. Bush," an OSRD staffer recalled.) In return, Bush promised to include Kilgore in the planning for any legislation arising out of his report.[41]

Kilgore came away from the deal believing he was part of Bush's team. He wasn't. Bush had merely lulled the senator into complacency. Hoping to silence his critics, Bush moved to craft science legislation with the help of friendly congressmen. A week before the release of "Endless Frontier" Bush dispatched an aide to visit Representative Wilbur Mills of Arkansas about a bill to establish Bush's National Research Foundation. To make it easier, Bush had readied a draft bill. Mills made a slight change in the measure, then agreed to introduce it on the day of the report's release. Bush's aides arranged for Senator Warren Magnuson of Washington State to do the same in the Senate.[42]

Kilgore was livid when "Endless Frontier" was released and Bush-approved bills were introduced into both houses. The senator tried reaching Bush for an explanation, but his staff said the OSRD director was out of town and unreachable. "Senate Kilgore considers himself doublecrossed and is mad as anything," one administration staffer observed. Four days later, Kilgore introduced competing legislation. The battle was on.[43]

Why did Bush betray Kilgore? The likeliest explanation was that he feared the senator's political philosophy. He was a Yankee Republican, and Kilgore was a New Deal Democrat. A civilian research foundation, dominated by elite experts, would surely undermine Democratic calls for research aimed explicitly at social and economic goals. If too inclusive, the new agency might not only fail in its military mission but also hamper the nation's research universities. While no ideologue, Bush harbored an almost unreasoning fear that the mob—unions, the poor, the political left and sundry other "do-gooders"—would seize control of the nation's research agenda and wreck the scientific community. At bottom, the whole debate over the organization of science turned not on "whether science in this country is supported"—clearly it would be, and at an unprecedented level for peacetime—but on "whether it is also going to be controlled."[44]

For Bush, control was paramount because upon it hinged both the character of scientific knowledge and its translation into marketable products and processes. Just as scientists and engineers were the arbiters of quality research, so private corporations were the best judges of which technologies should be sold. In this vision of free minds feeding free markets, government had no role. Bush underscored this view in a letter to the president of Bell Telephone Laboratories: "I think the direct governmental aid to small industries, either in the form of aid for their research, or indeed otherwise, is likely to be rather futile. . . . I have a strong conviction that the small industrial units which are really worthwhile are decidedly able to take care of themselves. What they need is not government aid but lack of government interference in a thousand ways."[45]

Given Bush's views on political economy, his betrayal of Kilgore was a tactical blunder, not an act of bad faith. But his mistake cost him. Without Kilgore in tow, Bush could not swiftly form a replacement for OSRD on favorable terms. In the end, five years passed before the foundation law was enacted. Nearly a decade passed before the foundation's budget reached the total originally envisioned by Bush.[46]

Could the foundation have been formed in 1945? Perhaps, if Bush and Kilgore had concentrated on their common views and prevented ideological differences from setting them at odds. On paper, at least, the two men were closer than they realized. The leading scholar on the campaign to form the National Science Foundation has concluded that the rival foundation bills "actually . . . were very much alike."[47]

Even the differences between Bush and Kilgore were mainly superficial. Kilgore's zeal for government-managed technology was more style than substance. He had a limited understanding of innovation and wanted kingpin researchers such as Bush to merely make a bow toward the public good. After all, the chances of another Depression seemed high to many people as the war neared its end. Even if the government could not actually direct the fruits of research toward the general welfare to the degree it wished, politicians could at least give the people hope. The restraints on scientific latitude proposed by Kilgore were, at worst, a small burden to carry and, at best, a sensible response to certain anomalies in the awarding of research funds. Kilgore's effort, for instance, to remedy the concentration of government funds in the hands of a small number of elite schools was long overdue.

Bush, meanwhile, consistently sought to portray science and engineering as serving the interests of ordinary people. While he differed with Kilgore over the means of bringing innovations to market, he highlighted the tie between research and prosperity in nearly the same simplistic terms as the senator. "More and better scientific research is essential to the achievement of our goal of full employment," he wrote in *Science—The Endless Frontier*.[48]

Given his agreement with Kilgore on the symbols of science, Bush's betrayal of the senator was counterproductive; it jeopardized his long-range aim of reforming political decision-making. It was time, he believed, for the independent expert to take center stage, within not just the military establishment, but the whole government. In a world of growing complexity, democratic politics must accommodate the special character of expertise. Bush bet that this would occur first within the military and research areas, but would ultimately spread into other domains. His postwar plan for research was only the first phase of a long-term project to place responsibility for the nation's security in the hands of independent experts. In practical terms, this would mean putting engineers and scientists in charge of all sorts of things.

To reach this goal, Bush needed the support of diverse groups, including left-leaning enthusiasts of technology in the Kilgore mold. By finding common cause with progressives, technocrats could increase their chances of reaching the pinnacle of political power. "It seems to me that it would be much more to the point if scientists occupied Cabinet posts of various sorts," Bush told a friend in February 1945. "The proper path is not for science to aim for a Cabinet post of its own, but for scientists [and engineers] to qualify for Cabinet posts generally. This has not occurred in this country and I do not understand the reason. Part of it is due undoubtedly to the fact that scientists have not been active in the discussion of public affairs, with notable exceptions, and a great deal of it is undoubtedly due to the fact that scientists do not often have the type of skill that is necessary for the handling of such a post, nor the type of approach that is likely to result in their being selected for it." Lawyers, by contrast, gravitated toward such posts, which wasn't always for the good. "I think that a few scientists sitting as chiefs of the regular [federal] departments would make quite a difference."[49]

But a scientocracy could never be sold to the public if scientists held to an elitist style and denigrated the opinions of nonscientists. In allowing his deal with Kilgore to unravel, Bush gave in to his sense of superiority over the scientifically illiterate. In the weeks before the release of "Endless Frontier," his staunch ally, Admiral Furer, expressed the hope that despite antagonism between political camps "scientific research . . . will flourish." Yet he astutely observed that "personal feelings and antipathies between men" might stand in the way. "I am amazed," he added, "that men, who must be considered of real caliber . . . should be governed by the most primitive frailties of human nature. Pride, vanity and hurt feelings crop out in the most unexpected places and often scuttle the worthiest of enterprises."[50]

Science—The Endless Frontier arrived a few weeks past its due date. Bush and his staff had expected the report to be released publicly at the end of June, but a snag in the Government Printing Office delayed the report. It was finally released on July 19, 1945.

As the report went public, Truman, Churchill and Stalin were meeting at Potsdam to discuss the future of defeated Germany and formerly occupied Europe. In the Pacific, U.S. aircraft carriers, moving ever closer to Japan's home islands, had attacked Tokyo. Despite these events, Bush's report had the nation's capital buzzing. He had hoped "to cause a bit of a stir" by distributing about 100 advance copies of the report to Washington insiders and personally wooing some newspaper and radio columnists. The pundits responded by trumpeting "Endless Frontier" as an instant classic.[51]

Radio broadcaster Raymond Swing called the report "remarkable" and ranked it in importance with Teddy Roosevelt's call early in the century to conserve the American environment. *Business Week* called *Science—The Endless Frontier* "an epoch-making report" that is "must reading for American businessmen." *The Washington Post* applauded Bush for delivering a "thorough, careful plan for putting the needed push of the federal government behind our scientific progress and yet keeping our science independent of government control." *The Little Rock Democrat* in Arkansas gushed that "best of all" the report offers "no scheme to control the manners, morals and mores of the people."

The press response delighted Bush loyalists. One OSRD staffer termed the report "an instant smash hit" and wrote Bush, "This looks like a personal triumph as well as an outstanding report." Even critics had a hard time finding fault with "Endless Frontier." The only complaint from *The New York Times* was that the report did not call for *enough* government involvement in research. The newspaper predicted that Bush's decentralized style of doling out grants, while an improvement over past neglect, would lead to inefficiencies. Only a handful of commentators questioned Bush's basic principle that research deserved broad public funding. *The Wall Street Journal,* for example, argued that tax incentives could achieve a similar result by inducing private industry to spend sufficiently on research.[52]

Critics and fans alike viewed "Endless Frontier" as a paean to "pure" research. Closer observers also saw a milestone in the new way of thinking about national security. The nation needed a strong scientific base in order to remain the world's top military power. Research was "absolutely essential to national security" and was too important to entrust to the military, Bush concluded. "Modern war requires the use of the most advanced scientific techniques . . . [so] a professional partnership between the officers in the Services and civilian scientists is needed."

This statement of core beliefs contained a stern rebuke for the armed services. Bush wanted the military's own research limited essentially to "the improvement of existing weapons." Because the services could not foresee the weapons needed to fight the next war, "military preparedness requires a permanent, independent, civilian-controlled organization, having a close liaison with the Army and Navy, but with funds directly from Congress and with the clear power to initiate military research."[53]

By equating scientific strength with national defense, Bush raised the specter of a global research race that might determine the world's balance of power. The inadequate feeding of U.S. science, he warned, might yield dangerous fruits in the postwar era: "Basically there is no reason to believe that scientists of other countries will not in time rediscover everything we know which is held in secrecy." In particular, he fretted that the Soviet Union might challenge Anglo-American hegemony over military technology. Before Germany's defeat, Bush had viewed Nazi science as the chief rival to Anglo-American supremacy, but now he worried about Russian science. Writing to H. W. Prentis on June 9, 1945, Bush placed his report in the context of future scientific competition between the U.S. and the Soviet Union. "In the kind of world that we are likely to face in the future this country has got to have its research in science in vigorous condition or it will not be able to compete effectively with the efforts which, for example, Russia will undoubtedly devote to science."[54]

Read as a brief on the coming global rivalry in science, Bush's report illuminated a sea-change in military thought. National security was now seen to depend as much on scientific strength as on men and materiel. In a July 21 editorial on the report, *The New York Times* deftly boiled down the report: "There is no doubt that the OSRD deserves much of the credit for the success of our arms. Can we afford to relapse into our defenseless pre-war status or ignore the enormous part that science can play in maintaining efficient defense? The question answers itself."[55]

The implications of this question were far broader than the editorial writers on the *Times* allowed, however. In a 15-minute broadcast, radio commentator Martin Agronsky flushed them out. "This morning, I had a long talk with Dr. Bush," he said, "and he told me he doesn't regard his report as more than an outline of the nation's needs—and, as he pointed out, the philosophy behind it . . . is just as important and not as readily apparent."

Agronsky insisted that "Endless Frontier" described only one piece of an emerging partnership between scientists, industry and the military; this "complex"—his word—would play as central a role in preserving national security as it had in defeating Germany. Echoing Bush, the journalist explained that science, along with industry, must be integrated into military planning at the highest levels. The respective roles were clear-cut. Scientists must design new weapons and stay abreast of inventions that would make existing ones obsolete. Industry must build new weapons. Finally, the military must properly deploy the weapons that science creates and industry makes. "This complex pattern of information, invention and ultimately production—in the field of national defense—is completely interdependent," Agronsky concluded. "Unless all three—the American military, the scientists and the industrialists—keep in constant and intelligent touch, the one with the other—our country could be badly caught off base in a possible future war."[56]

The broad appeal of "Endless Frontier" lay in Bush's skillful juxtaposition of civilian aspirations with military imperatives, his equation of research with freedom and frontier values, and his paradoxical embrace of the public till on behalf of private enterprise. Making a vigorous case that "science is a proper concern of government," he wrote: "It has been the basic United States policy that Government should foster the opening of new frontiers. It opened the seas to clipper ships and furnished land for pioneers. Although these frontiers have more or less disappeared, the frontier of science remains. It is in keeping with the American tradition . . . that new frontiers shall be made accessible for development by all American citizens."[57]

But the frontier of science demanded a new politics, without which "no amount of achievement . . . can ensure our health, prosperity and security as a nation in the modern world." Bemoaning that "we have no national policy for science," Bush noted, "There is no body within the government charged with formulating or executing a national science policy. There are no standing committees of the Congress devoted to this important subject. Science has been in the wings. It should be brought to the center of the stage for in it lies much of our hope for the future."[58]

The implication was clear: the government must rely on new institutional arrangements to reconcile the scientist's requirement of absolute intellectual independence with the public's desire to know where its money went. While taxpayers deserved accountability, they could not expect it in the strict sense when it came to funding research. Bush resolved this contradiction on the side of science by insisting that taxpayers would essentially spite themselves if they failed to accept that "scientific progress results from the free play of free intellects, working on subjects of their own choice, in the manner dictated by their curiosity."

Bush's high-minded resolution of this conflict illustrated his continuing desire to play the role of "public entrepreneur" by pushing the government into unfamiliar postures. The war had shown that this could work—OSRD, after all, was run more like a sprawling private corporation than a public entity. Bush hoped to infuse any postwar research organization with this same culture. Keenly aware of the way commercial relations could shape technological outcomes, he intended to build any postwar organization around the contractual relationship, which he saw as the source of OSRD's entrepreneurial energy. Bush had thought deeply about the various meanings to both parties of the contract, yet he never questioned his basic assumption that the marketplace was the best arbiter of technological value and that scientists and engineers worked best in isolation from social and political pressures.

Some Truman allies rejected this assumption, viewing Bush's call for intellectual freedom as a cynical pretext to avoid the scrutiny that ordinarily accompanied government funds. Convinced that Bush's libertarian bent was a cover for an unprincipled raid on the public coffers, Harold Smith joked that Bush's report should be renamed, "Science—The Endless Expenditure."[59]

Chapter 12

"As we may think"

(1945)

The Rodin of the future who is inspired to make a statue of "The Thinker" may find that his models will be cogwheels, thermionic tubes and strands of wire.

—Vannevar Bush

The summer of 1945, Bush was poised to complete one of the most challenging assignments of World War II. But his future was uncertain. The liquidation of OSRD, so crucial to his postwar plans, was bogged down. His insistence on his agency's quick death had angered natural allies in the military and even among fellow scientists. The "Endless Frontier" report, while hailed as a landmark by the public, was quietly being attacked by members of the Truman administration. Finally, there was the matter of the atomic bomb. A test of the bomb was scheduled for the middle of July. Depending on the outcome, Bush could either add the Manhattan Project to his list of achievements or, as he joked, start looking for another job.

In this trying time, Bush received an unexpected reminder of his status as a technological visionary. The good news came in late June in the form of *The Atlantic Monthly,* the distinguished magazine. In its July issue, which reached newsstands on June 27, the *Atlantic* published an essay by Bush entitled, "As We May Think." In it, Bush described a futuristic machine called the "memex" that promised to "give man access to and command the inherited knowledge of the ages." He imagined the memex as a work desk with viewing screens, a keyboard and sets of buttons and levers. Printed and written material, even personal notes, would be stored on microfilm, retrieved rapidly and displayed on screen by a high-speed "selector."[1]

The memex would be more than a private library. It would be a personalized aid to memory that would remove the drudgery from human thinking. It had the capability of allowing its owner to assign and record codes that made it possible to essentially replay an entire "trail" of associations. Bush presumed that the human memory organized information by association. He imagined a process by which any two pieces of information could be linked by the memex. These links could spawn more links to create an associative trail. Rather than a person having to recall the specific connections in the trail, the memex would do so by flashing records onto a screen. For this reason, Bush considered the memex a mechanical aid to memory (hence its name). By mechanizing the process of association, the memex would extend the powers of the human mind.

This was the theory, at least. While it sounded far-fetched, thinkers had tried to find ways to sharpen memory; this was only the latest attempt. Before the birth of Christ, an unknown Roman teacher of rhetoric wrote a textbook that included techniques on improving recall. In the 16th century, an Italian, Guilo Camillo, created a "memory theater" that expanded recall by associating memories with the places and images in a building. Even the originator of the scientific method, Englishman Francis Bacon, practiced a regimen to strengthen his memory and considered such efforts part of the serious business of science. Now Bush took his place in the grand tradition of what historian Frances Yates called "the art of memory." His memex, in this context, was a modern-day "memory theater."[2]

The effect of such a machine on civilization could be immense. In the most stirring passage of "As We May Think," Bush wrote:

Wholly new forms of encyclopedias will appear, ready-made with a mesh of associative trails running through them, ready to be dropped into the memex and there amplified. The lawyer has at his touch the associated opinions and decisions of his whole experience, and of the experience of friends and authorities. The patent attorney has on call the millions of issued patents, with familiar trails to every point of his client's interest. The physician, puzzled by a patient's reactions, strikes the trail established in studying an earlier similar case, and runs rapidly through analogous case histories, with side references to the classics for the pertinent anatomy and histology. The chemist, struggling with the synthesis of an organic compound, has all the chemical literature before him in his laboratory, with trails following the analogies of compounds, and side trails to their physical and chemical behavior.

The historian, with a vast chronological account of a people, parallels it with a skip trail which stops only on the salient items, and can follow at any time contemporary trails which lead him all over civilization at a particular epoch. There is a new profession of trailblazers, those who find delight in the task of establishing useful trails through the enormous mass of the common record. The inheritance from the master becomes, not only his additions to the world's record, but for his disciples the entire scaffolding by which they were erected. Thus science may implement the ways in which man produces, stores and consults the record of the race.[3]

Bush's vision was breathtaking. His memex was no blueprint for a personal computer, but it offered something just as important: a careful description of the benefits to ordinary people of automating thought. Bush anticipated a mass market for mechanical memory aids at a time when designers could build only room-size computers and imagined that a handful of them could satisfy a nation. He also perceived more strongly than anyone else at the time that programming and human-machine interaction were the most important limiting factors to the spread of information-processing machines. Lacking the mathematical genius of a John von Neumann, he nevertheless had a practical vision about the future of computing that would serve to educate the desires of both engineers and the general public. He put into plain English a set of problems that technologists must satisfy if they were to claim victory in the drive to mechanize cognition. That he had barely a clue how to actually achieve this in no way detracts from the value of his vision: he gave a generation of inventors a target to shoot at.

For this reason, later pioneers in computing considered Bush's memex a forerunner of their own machines. When Bush declared, "I am often called the father of modern computing," he was exaggerating, but his boast contained a sliver of truth. As one perceptive observer wrote in 1985, 40 years after the publication of "As We May Think":[4]

In the pages of the *Atlantic,* Bush proposed that a certain type of device should be developed, a device to improve the quality of human thinking. . . . Bush was one of the first to see that rapid access to large collections of information could serve as much more than a simple extension of memory. Although he described it in terms of the primitive information technologies of the 1940s, the memex was functionally similar to what is now known as a personal computer—and more.

Some ideas are like seeds. Or viruses. If they are in the air at the right time, they will infect exactly those people who are most susceptible to putting their lives in the idea's service. The notion of a knowledge-extending technology was one of those ideas.[5]

To be sure, the crude technologies of Bush's day made it impossible to turn the memex idea into a functioning machine. His demand for a vast repository of data that could be instantly retrieved would not be realized for decades. Yet in 1945 the mere description of such a machine, which was really an elaborate version of the crude "rapid selector" machine Bush had designed in the late 1930s, created a sensation. The *Atlantic* likened the essay to Ralph Waldo Emerson's 1837 address, "The American Scholar." The Associated Press distributed an 800-word excerpt across the country. *The New York Times* and *Time* magazine followed with stories of their own.

Then *Life* magazine capped the surge of attention by reprinting a condensed version of the piece in September accompanied by lavish drawings of Bush's imaginary devices and a fresh interview with the OSRD director. Sketched by a staff artist, the *Life* drawings vividly conveyed Bush's brave new world of mechanized thinking and remembering. As envisioned by Bush, the devices making this possible would include a "vocoder" that would take dictation and function as a "mechanical supersecretary"; a "cyclops camera" that would be worn on the forehead and "photograph anything you see and want to record"; a "thinking machine" that would accept "premises and pass out conclusions"; and finally, the memex, described briefly as "an aid to memory" that filed material by association. "Press a key and it would run through a 'trail' of facts."

Reaching for a popular audience, *Life*'s editors also cast Bush's vision in starker terms, throwing his notions into bolder relief. With Bush's help, they punctuated his article with catchy subheadlines that underscored the revolutionary aspects of the memex:

REDUCING THE WRITTEN RECORD TO MANAGEABLE SIZE—
MICROPHOTOGRAPHY.

THE AUTHOR NEED NOT WRITE—HE COULD TALK HIS THOUGHTS TO A
MACHINE.

SIMPLE REPETITIVE THOUGHT COULD BE DONE BY MACHINE, FOLLOWING THE
LAWS OF LOGIC.

BUILDING "TRAILS" OF THOUGHT ON THE MEMEX—UNLIKE MEMORY, THEY
WOULD NEVER FADE.[6]

Editors and readers were bowled over by Bush's memex dreams, which to them seemed to spring, without precedent, from his imagination. But only

his notion of "selection by association, rather than indexing" was wholly original. Both the design for the memex and the concept of putting a whole library at a person's fingertips were in circulation well before the *Atlantic* article
appeared. In the 1920s an obscure eastern European inventor, Emanuel
Goldberg, designed and later patented a microfilm workstation strikingly
similar to the memex hardware. The prospect for mammoth encyclopedias,
meanwhile, was widely discussed in Europe between the two world wars; so
were new indexing schemes. In the late 1930s, H. G. Wells, the British futur-
ist, even wrote and lectured about what he called a "world brain." Citing microfilm as a revolutionary storage medium for the printed page, Wells insisted
that soon "the whole human memory can be . . . made accessible to every individual." Like Bush, Wells believed the new challenge for humanity was
managing its information heritage. Also like Bush, he pointed to the need for
better indexes, or coding systems, though he declared, "There is no practical
obstacle whatever now to the creation of an efficient index to *all* human
knowledge, ideas and achievements, to the creation, that is, of a complete
planetary memory for all mankind."[7]

In writing "As We May Think," Bush never acknowledged the ideas of
Wells and other European thinkers; he also had never noted the work of Gold-
berg. Bush likely was not aware of either of these European trends, despite
Goldberg having obtained a patent for his prototype selector in 1931 and
Wells publishing *World Brain* in the U.S. seven years later. None of the news-
paper or magazine articles about the memex noted any of the parallel develop-
ments in Europe, though the rhetoric seemed straight from Wells himself.
Time predicted, in its report on the memex, that microfilm "could reduce the
Encyclopaedia Britannica to the size of a matchbox [and] might even store the
whole printed record of the human race in one moving van."[8]

Bush's ability to create such a stir with "As We May Think" showed that he remained a force in the tiny field of computing and information science. His
days as an active player seemed over, however. The demands of war and his political ambition had forced him to abandon his career in computers. But as a
leader in military research he still had the chance to sponsor the work of pioneers in the design of the first digital computers.

The digital pioneers put off Bush, who worried that their plans could not
be realized in time to aid the war effort. In September 1940, MIT mathematician Norbert Wiener proposed building a computer designed on binary
principles and appealed to Bush, then chair of the National Defense Research Committee, for funding. After studying Wiener's memo and asking
pertinent questions, Bush wrote in October that he was "not yet persuaded"

of the plan. Two months later, he sounded more definite, turning Wiener down flat. "While your device would also be of aid on defense matters," Bush wrote, "it is undoubtedly of the long-range type and it appears essential that at the present time those individuals who are particularly qualified along these general lines be employed as far as possible on matters of more *immediate* promise."9

Bush's denial might have said more about Wiener's reputation as a great catalyst but weak finisher. But Bush also kept at a distance the most ambitious of the proposed digital machines, the so-called ENIAC computer. This machine promised to operate at theoretical speeds 500 times that of mechanical computers. The speed advantage seemingly justified the ENIAC's enormous size and cost, but Bush balked at funding it. The Army ordnance department instead picked up the $500,000 cost of the computer, built at the University of Pennsylvania.

Bush's refusal to fund a landmark project in digital computing partly reflected a failure of vision. A master in the idiom of analog computing, he trusted the differential analyzer because he could literally see the machine perform its function. But the ENIAC was thoroughly abstract; there were no moving parts at its core. The gadgeteer in Bush balked at this. Wiener was disappointed but not surprised. As he claimed years later, Bush had so much faith in mechanical devices that for him "gadgeteering very easily [became] a sort of religion."10

There were more pragmatic reasons for Bush to mistrust the digital advocates. He bet the ENIAC would not be ready in time to aid the war effort, and indeed it wasn't. The computer tackled its first problem in December 1945, four months after the Japanese surrender. Even then, Bush had his doubts about the ENIAC's utility. Like others at the time, he thought that failure rates on ENIAC's 18,000 vacuum tubes might make the computer impractical anyway. When the ENIAC first went into operation, the tubes failed so frequently that they were plugged into sockets in order to make them easy to replace, whereas other parts were soldered in place.11

The digital believers countered that Bush's ideas were dated and his beloved analog computers outmoded. Digital computers were more exact, more flexible and often faster at solving problems. Their success sounded the death knell for differential analyzers. Although these clanking mechanical computers would linger on after World War II, even their fans recognized the seminal shift underway. D. R. Hartree, a British physicist who had actually built copies of Bush's analyzer before the war, concluded that in computers "the main interest and activity is at present, and will probably continue to be, in the digital machines." Added mathematician Norbert Wiener, who had intimate knowledge of Bush's analog calculators, "When it comes to high speed or high accuracy, the advantage is all with the digital machine."[12]

If Bush's thinking about computer innards was passé, his ambition for the uses of computers was not. His notion of the memex was compatible with digital electronics. While he thought of the memex in the analog terms of microfilm storage, photography and mechanical retrieval, there was no reason why the nervous system of his futuristic device could not consist solely of digital circuits. In any case, whether digital or analog, the memex should hold the same appeal for people seeking a machine aid to memory and thought.

The memex stood apart from the mainstream of digital computing by virtue of its human scale. An apostle of the individual, Bush imagined a machine that could amplify the consciousness of a single person, not run an organization. The new digital computers were impersonal, designed to meet the needs of bureaucracies. They were soon dubbed "robot brains" in frank recognition of their otherworldliness. Like the Wizard of Oz, digital computers would deliver answers, perhaps even make decisions, but would never be seen by ordinary people. Instead, they would be cared for by a specialized society of computer devotees. Not unlike a new priestly class, these digital enthusiasts were already building an intellectual case for their belief that the human mind and body were essentially organic computers, reduceable to binary math. Wiener, for instance, insisted that digital computers and human brains had so much in common they should be studied by a single science. The implications of this were staggering for the concept of individual intelligence: computers might be the ideal for human beings.[13]

With his memex, Bush advanced a far different model for the future of computing. Asserting the primacy of lived experience, he sought to mold computing machines to humanity's foibles. In the rarefied intellectual climate spawned by the digital breakthroughs of the 1940s, this seemed like anthropomorphizing a revolutionary technology. Yet in the case of Bush, it actually represented his fidelity to a central but often-overlooked motive for invention. This was the desire "to help people do better what they were already doing for other reasons."[14]

This aspiration appealed to fledgling engineers not yet initiated into the secret world of giant computers. Among those smitten by the memex in the summer of 1945 was a 20-year-old American radar technician, waiting for his ship home from the Philippines.

One muggy day, Douglas Engelbart walked into a Red Cross library on the edge of the jungle on Leyte Island and found Bush's article in the *Atlantic*. Infected with memex fever, Engelbart returned to the U.S., finished his bachelor's degree in electrical engineering and set off "along a vector you had described," he later wrote Bush. Within a few years, he was harboring his own notions about how to turn intimidating digital computers into intimate amplifiers of the human mind. In the 1960s Engelbart invented some of the most durable features of modern computers, including the ubiquitous mouse, or hand-held pointing device. Still mindful of the effect of Bush's article, he confessed, "I wouldn't be surprised at all if the reading of this article . . . hadn't had a real influence upon the course of my thoughts and actions."[15]

Bush was elated by the reaction to "As We May Think" if for no other reason than that the piece had been a long time in birthing. He had tried to publish an earlier version of the essay in December 1939. This precursor to "As We May Think" was the culmination of nearly a decade of thinking by Bush about the organization of data and how human beings might devise better tools to manage their own ideas. Bush first revealed the direction of his thinking in *Technology Review,* which in January 1933 published his speculations about a futuristic machine that supplied information on demand. By punching a few keys, a user of this imaginary device would receive the exact page—of notes or printed matter—that he desired.[16]

The memex took shape in response to the looming problem of too much information. Data overload threatened individual liberty by making it harder for people to govern themselves or become broadly educated. The specialization of knowledge already overwhelmed the generalist, whom Bush counted as the backbone of the expert class. Even the synthesizer of knowledge in a single discipline, such as physics or chemistry, was hamstrung by the difficulty of quickly keeping up with the literature of the many subdisciplines. Not only academics were drowning in data. A mountain of printed material threatened to choke lawyers, government regulators, business executives and others who relied on information as a means of commerce and control. As Bush wrote an associate at the parent Carnegie Corporation in 1939, "Unless we find better ways of handling new knowledge, as it is developed, we are going to be bogged down."[17]

So for both pragmatic and philosophical reasons, Bush pursued technical fixes to information overload. In December 1939 Bush shipped a "rough" version of what would become "As We May Think" to *Fortune* magazine. Bush boasted to a friend that the article "will let the cat out of the bag on a number" of exciting possibilities, but for most of the war it sat in the files of Eric Hodgins. A writer, editor and executive at *Fortune,* Hodgins had met Bush in connection with military research and encouraged him to write about the memex for a popular magazine. While intrigued by the memex, Hodgins could not decide if Bush's piece fit *Fortune.* The delay frustrated Bush, but he was so busy during World War II that he gave only an occasional thought to his "confounded article."[18]

Certainly, the administration of the war highlighted the growing problem of information overload. From the cramped offices of the nation's capital, the war was a paperwork nightmare. So many new government agencies had sprung up, demanding so much interagency coordination, that memos in triplicate were flooding the government. Even military operations spawned massive amounts of paperwork, especially once OSRD's operations analysts tackled the task of evaluating the effectiveness of battlefield tactics. Incredibly, a federal government that had entered the war with just $650,000 worth of printing and reproducing equipment owned $50 million worth within a year. The Office of Price Administration alone produced more paper than the entire government had before the war.

As bureaucrats struggled to stay abreast of all this information, the government buckled under the demand for it. New typists were plentiful, but typewriters were not (typewriter makers having been ordered to switch to war production). By mid-1942, the government claimed to be 600,000 typewriters short.[19]

If only because there seemed no end to the burgeoning appetite for information—and no effective new ways for organizing this mountain of paper—the memex concept was worth keeping alive. With *Fortune* still undecided about publishing his article, Bush finally decided to pitch the essay elsewhere in September 1944. "I think an article of this sort, brought up to date a bit, would strike rather a desirable note in the time between the collapse of Germany and the end of the war, when people are thinking in various ways in regard to the research problems ahead," he wrote a friend.[20]

Bush first sent the essay, then entitled "Mechanization and the Record," to *The Atlantic Monthly,* which accepted it in October. Excited that "after a good deal of pulling and hauling," his memex idea would see print, Bush considered writing an entire book on the topic, but admitted to some embarrassment about spinning a web of vague predictions.[21]

To the elite Washingtonians of World War II, Bush's emergence as an American H. G. Wells seemed astonishing—especially coming in the heat of war. Politicians, the press and the public knew Bush not as an obsessive inventor but as a steely organizer of expertise for secret, military ends. The most important science administrator in an age of administration, he was a rough, practical character who fought hard for his convictions and had little time for dreamers. Few people recalled that during the 1930s Bush had been recognized by leading scientists as the world's foremost designer of analog computers. Fewer still realized that Bush remained one of the most influential figures in military computing.

In 1942, Bush's collaborator at MIT, Samuel Caldwell, completed the massive Rockefeller Analyzer (named to reflect the patronage of the Rockefeller Foundation). This new model—like earlier analyzers—used as its basic computing element a rotating disk, upon which rested a smaller disk whose rotation matched the integral of the desired mathematical function. But the new analyzer embodied improvements that shortened the time required to set up for a new problem. In earlier analyzers it could take a few days to make the necessary physical interconnections to describe a new problem. The new analyzer was programmed with punched paper tape, and Bush claimed it reduced setup time to as little as three to five minutes. Others said the machine was on a par with the first digital computers, though it worked on different principles.[22]

During World War II, various differential analyzers were military workhorses, compiling ballistic tables for antiaircraft weapons and artillery at the Army's Ballistic Research Laboratory at the Aberdeen Proving Ground in Maryland. These tables were essential: they made it possible for soldiers to aim their guns. Tables had to be produced for every possible combination of gun, shell and fuze; even the hardness of the ground, the atmospheric pressure and the rotation of the planet were accounted for. A typical firing table, which a gunner used to learn at what angle to elevate his piece to reach a certain distance, might require the calculation of 4,000 trajectories. All told a differential analyzer needed roughly 750 hours to calculate the table trajectories. The desire to obtain these calculations more quickly was "the *raison d'etre* for the first electronic digital computer." Clunky and sometimes inaccurate, Bush's machines nonetheless greatly improved the accuracy of big guns and shells.[23]

The military value of the differential analyzer was well known. Not so with another family of machines designed and inspired by Bush. This was the supersecret Comparator line, which before the war he had launched with the support of the Navy's Communications Security Group, OP-20-G. The Comparator had a troubled history, but it was used during the war to help decipher Japanese diplomatic and military codes. Codebreaking was so sensitive that Bush's involvement in U.S. cryptanalysis was probably an even more tightly guarded secret than his participation in the Manhattan Project.[24]

Bush's original work on the Comparator had ended in 1938, before he had moved from MIT to the presidency of the Carnegie Institution. But in the summer of 1940, as he plotted to win Roosevelt's endorsement for his National Defense Research Committee, officers from OP-20-G contacted Bush.

While listed on the Navy's codebreaking equipment roster in 1939, the Comparator had not been used because of performance problems.

Bush considered the machine dead, but the Navy wanted to revive the Comparator. Agreeing to try, Bush suggested that improvements in his rapid selector machine, a high-speed retrieval device under development at MIT, might be incorporated into a new version of the Comparator. Officers at OP-20-G thought the attempt would be worthwhile. On the basis of the Navy's interest, Bush arranged for the NDRC to approve a grant of $50,000 to MIT and a $20,000 grant for his own research. Now considered a vital war mobilization project, the Comparator took precedence over the rapid selector. Bush assigned three MIT engineers to the Navy and officially suspended work on the rapid selector. The NDRC informed Eastman Kodak and National Cash Register, which had agreed three years before to underwrite MIT's research on data retrieval, that the civilian project would be held "in abeyance until the emergency is over."25

Work on the new Comparator began in November 1940. Within a year, the revamped model had taken shape. Living up to its name, this microfilm-based machine promised to make at least 30,000 simple comparisons a minute, giving it ten times the power of the old model. The speed in making comparisons was crucial because codebreaking turned on the ability to find patterns in enormous numbers of messages. No machine could do the whole job, but it could reduce the drudgery involved, thus freeing cryptographers to apply their intuition to codebreaking.26

The Navy took complete control over the Comparator in 1942, directly hiring the MIT engineers and carting off all the Institute's records, drawings and prototypes. Bush's influence on the machine almost vanished, but progress did not. The Navy finished a series of Comparators, ordering about 20 during the war from National Cash Register (which had sponsored research on the rapid selector) and Gray Manufacturing Company. OP-20-G first began using the Comparator in September 1943. It was an odd-looking contraption consisting of 18 rollers for punch tape and an assortment of different-size gears, all hung from a rectangular frame. The machine was judged by OP-20-G as "extremely satisfactory for all operations it has been called upon to process." It was rated as "much faster than any hand method" at analyzing encrypted messages, but found to be "slow in comparison to some machines."27

By 1943, Bush had little sense of how his Comparator and its progeny were doing. But being shut out of the Navy's program did not stop him from promoting the civilian version of his data-retrieval device, the rapid selector. He still thought the FBI could profitably use the device to automate the task of retrieving and matching fingerprints. Other information-intensive government agencies, such as the Patent Office and the Library of Congress, might benefit from the rapid selector too, he thought. "When peace returns it ought to be applied to something," he insisted in 1944.[28]

The presumed usefulness of the rapid selector fueled Bush's faith in its ultimate success. The device "very rapidly reviews items on a roll of film, and selects out desired items in accordance with a code," he explained to the librarian of Congress. "The items are entered in succession photographically onto the roll of film, each with an identifying code in a dot pattern adjacent to it. In order to select items from a film, the roll is placed in the machine, a set of indices are placed in accordance with the code of the desired items, and the film is run through rapidly."[29]

Bush's description of the rapid selector was deceptively simple. High speeds were difficult to attain with microfilm. He wanted early versions of his Comparator, for instance, to achieve speeds that exceeded any posted by existing technologies. Data was to be read at 30 times the standard rate of telegraph equipment and 60 times faster than a movie projector. The microfilm mechanisms simply never achieved those speeds.[30]

Indexes also proved troublesome. Bush laid the blame on librarians for the growing inability to find relevant information swiftly. Control over defining index categories, he believed, had to shift from librarians to scientists. At the very least, scientists should be able to define subject terms and categories for technical literature. Rather than fixed indexes, Bush envisioned flexible ones, along the lines of the "associative trails" he described in his "As We May Think" article. These dynamic indexes would evolve on the fly, becoming personalized. The building block of this new organizing scheme was the "key word." Each document would contain at least one key word and probably more.[31]

Conceptually elegant, Bush's indexes were difficult to incorporate into hardware because they relied on a flawed coding scheme. For the rapid selector to actually locate a key word, it had to locate a specific code on a segment of microfilm. In practice, this meant that a person had to attach a generic code to each photographed item. Every system, whether it covered patents issued by the government, the FBI's fingerprint records or the photographic collection at the Library of Congress, would have hundreds, perhaps thousands of key words that corresponded with a film code. Bush presumed that coders would learn to quickly match a key word with its code, but this wasn't likely given his arcane coding scheme. He chose to use as many as 12 letters to represent the contents of a document, which resulted in such strange codes as DMUH-CORMENVS.[32]

These coding problems reflected the extent to which Bush's ambitions in data retrieval outreached his command of fledgling computer technologies. He was baffled by code, or software, which he later conceded was "the big problem on this whole affair." It was an expensive problem too. "The big cost [of creating rapid selectors] is in coding, and the problem of real usefulness is to get users to employ the code with some sense," he observed. "This is true no matter what machinery is used."[33]

For a variety of technical reasons, the rapid selector fell short of expectations. Bureaucrats also were wary of change. The keepers of government records were too swamped to think ahead, and they tended to be ignorant of the technical trends that seemed destined to bring about more efficient recordkeeping. In the summer of 1940, Bush made a restrained pitch to the FBI, presenting the rapid selector as the latest in "advanced mechanical methods for handling fingerprint files." Wary of dealing with the bureau—he told aide Carroll Wilson that "FBI people are not easy to work with"—he nevertheless saw the rapid selector as potentially an "enormous" boon to law enforcement. The FBI sorted its fingerprint files according to the index finger of a subject's right hand; if for some reason the bureau was missing the index finger and had prints of just other fingers, "they are practically lost," Bush observed.

The rapid selector promised to make it easier to match any one of a subject's ten fingerprints. Bush hoped the bureau's top officials would realize this. They did not. Six weeks after Bush formed the NDRC, J. Edgar Hoover, the bureau director, ruled out any chance of experimentation with the rapid selector.[34]

Bush ran into cautious bureaucrats elsewhere in government. Private industry also was cool toward the rapid selector. It wasn't until 1949 that a version of the machine was unveiled for the public, the result of a collaboration between the Department of Agriculture and a Minneapolis firm, Engineering Research Associates, which had on staff a prominent alumnus from the wartime selector project. Even then, the selector's inherent limitations kept it from catching on widely. More costly than a standard microfilm reader and not nearly as fast as the proliferating digital computer, the rapid selector gradually faded from view, an inspired failure that was obsolete before its core technologies could be mastered.[35]

The most curious aspect of the rapid selector was its hybrid nature; though Bush did not advertise the fact, the machine was part digital. Photographic images—microfilm images—were analog, of course, in the sense that they could contain any degree of white to gray to black tone values. But the rapid selector's selection mechanism was binary: either it made a hit or not. It was a Boolean logic system.[36]

Bush never embraced the binary, or digital, aspect of the rapid selector. He refused to accept the supremacy of the digital computer and struggled to grasp the basic technologies behind it. After so many years on the margins of the fast-moving computing field, in 1950 he wrote, "I am so confoundedly rusty on this whole business I have to hang on by my eyebrows when the discussion really gets going." While in "As We May Think" he suggested that demand for computerized aids to thinking would be insatiable ("there will always be plenty of things to compute in the detailed affairs of millions of people doing complicated things"), he gradually came to believe that advocates of digital computing were overestimating the scientific applications for their machines. In 1949, he warned Von Neumann, the world's leading computer theoretician, "After looking at the computing machine situation in the country as a whole I am rather fearful that there are some programs that have been instituted at great expense that will never go through to completion or, on the contrary, more machines will be finished than can be serviced or used."[37]

Bush's belief in the enduring value of analog computing would ultimately be borne out in the 1990s when leading-edge engineers found that analog circuits offered the most efficient means of creating electronic imitations of such complex biological organs as the eyes. While some end-of-the-century futurists even went so far as to predict the decline of digital and the triumph of analog in technologies where computer circuitry merged with bio-engineering, it was hard to deny digital had defined the electronics revolution.

Yet Bush's postwar doubts about digital fit his time. Many of his peers hardly considered the decline of analog to be predestined. As late as 1954, the computing landscape was quite heterogeneous. One survey at MIT found that of the 488 hours of weekly work performed by various machines, less than half were handled digitally. With so many hybrid machines—devices that relied on both analog and digital techniques—in circulation, Bush resisted the conclusion that the victory of digital was inevitable. "I would feel more comfortable if I thought there was a group developing analytical machinery in this country that was not obsessed with the idea of digital machines," he wrote in 1952. "One hears the statement very frequently that there are two kinds of machines, digital machines and analog machines. This indicates a blind spot which I think is quite prevalent. . . . I have felt for a long time that in the intermediate region there is an opportunity for machinery that is being overlooked."[38]

Bush's desire to retain analog devices was based on more than nostalgia for the machines of his youth. Analog devices had a powerful psychological hold on him. They linked him to a 19th-century world of gears and metal—where knowledge was a direct physical encounter of the head and the hands. An engineer could literally watch an analog calculator grind out a solution to a mathematical problem. Digital, by contrast, offered the freedom of a blank slate, but no chance to actually visualize its operations. This was an intellectual paradigm that ran against Bush's grain.

Others saw the benefits of sticking with analog. General Electric explained: "A virtue of the analog computer is that its basic design concepts are usually easy to recognize. What goes on inside is understandable since it is an analog of the real thing [whereas] the digital type computer is a product of pure logic."[39]

Pure logic could only take a person so far. Among theorists of computing at midcentury, Bush stood alone in espousing the virtues of a personal machine whose performance over time matched its owner's patterns of thought. His bold aspiration, however, drew criticism from those who accused him of "minimizing the human difficulties" in designing a memex. These principally surrounded Bush's concept of associative trails, which was his means of explaining the essence of cognition. His belief that the mind contained endless strings of associated memories was based on nothing more than a hunch about his own thought patterns. "I had the germ of an idea, I think, on search by association, but I turned it into a semi-facetious affair when I wrote the article," he confessed to a friend.

Critics asserted that only "a profound study of man's thought processes" could determine whether Bush's model of associative trails fit the brain. Bush felt no such need for empirical research. "I am sure the human brain pulls things out of its memory rapidly," he wrote, "because it has an extraordinary method of following trails of association, and it seems to me we ought to be able to do as well with machinery."[40]

Enamored of the prospect for machines that imitated the mind, Bush insisted that the very inhuman, brute-force calculations of room-size computers would not lead to the Promised Land of the digital revolution. This was a provocative notion, which ran counter to the whole course of cognitive science. But Bush's ignorance of digital computing weakened his case, and he refused to wade into this fast-paced subject. "I fear I am not going to get back into the field myself, which I regret," he promised in 1952.[41]

Bush felt he was too old to learn the fundamentals of a technology so alien to him (he showed no interest in following the new science of the mind and brain either). He would content himself with resting on his laurels. His contributions to modern computing were not limited to pregnant ideas that inspired others. The fate of his forgotten Comparator, a codebreaking computer that relied on a amalgam of analog and digital techniques, showed that Bush was not as hidebound as he sounded. Into the late 1940s, the Comparator was a standard piece of Navy equipment. Before the end of World War II, Bush learned of the Comparator's usefulness to the Navy's OP-20-G group. Confirmation of the Comparator's role in cracking Japanese codes came in curious fashion. One day, the Army rushed "me into a room, shut the door and began to talk hush-hush." As Bush later recounted with his typical wry sarcasm:

I do not think they looked under the pictures to see if there were any bugs present, but at any rate they evidently were going to bring me into something very secret. So I interrupted them to say, "Are you gentleman going to tell me about the breaking of the Japanese code?"

They went up in the air about a foot, because that was being held very confidential indeed, and quite properly so, because the fact that we had [the Japanese] code gave us an enormous advantage in the Pacific war for quite a long while. We were very careful not to use the information that came out of the intercepts in a way that would give away to the Japs that we knew their code. The military gave up some real opportunities rather than reveal this, so no wonder the people in the Army were a little disturbed. They wanted to know how I happened to know about it.

So I told them, "Well, inasmuch as I built one of the machines you're using, and inasmuch as I trained some of the young fellows that are working for you on the thing, it would be very funny if I didn't know something about it."[42]

Part Four

The New World

Chapter 13

"A carry-over from the war"

(1945–46)

I think none of us had any doubt that if we could get the bomb we ought to drop it.

—Vannevar Bush

Bush lay in the dark, ten miles from Ground Zero, clutching a piece of dark glass to protect his eyes. Exhausted from three nights of broken sleep, he pressed his face against a tarpaulin stretched over the damp New Mexico sand. James Conant, his chief deputy, lay beside him. They listened to physicist Samuel Allison counting down the seconds over a loudspeaker. Bush briefly thought to himself, "If this thing goes off with a hell of a lot more force than we've calculated, they'll have to get a new [OSRD] head."

Allison counted down to zero, ending with a scream. Then a burst of white light filled the sky, illuminating the mountain ranges in the distance. Bush covered his eyes for an instant, then peered through the dark glass at the exploding rainbow of fire. Next he felt a shock wave ripple through his body. The blast was weaker than he had expected, but only because, he later learned, it had been deflected by a rise in the ground above him. Before Bush could stand up, Conant shook his hand.

It was July 16, 1945, and the Manhattan Project was a success. The atomic age had begun.[1]

Bush's first memory of the new age was of the "great excitement" following the test. "The teams that were watching for fallout were scurrying all around the terrain," he recalled later. "They evacuated some ranches where they were afraid that there might be serious fallout due to the storm or something else."

Finally, things settled down. Bush heard that Oppenheimer, the project's scientific chief, was leaving by car for a day or two of rest. Bush walked down to the base gate and waited for Oppenheimer to pass. Five years before, he had been intensely skeptical about the prospects of an atomic bomb and doubted that it would prove to be a weapon for *this* war. He had changed his mind, persuaded Roosevelt to back a crash effort to build an A-bomb, then handed off the project to the Army in 1942. Now he credited Oppenheimer, as much as anyone, with delivering a bomb in time for use against the Japanese. Oppenheimer "had just about the most difficult job in connection with the scientific aspects of the war effort, and he handled it magnificently," Bush thought. "One would not expect a theoretical physicist with a decidedly philosophical turn of mind to be able to manage a complex affair of this sort . . . in a way that excited the admiration of everyone with whom he came in contact."

Bush saw no philosophical portents in the bomb's blast, but only the realization of an enormous gamble, at once scientific and political. In character, Oppenheimer viewed the triumph of Trinity differently. He reacted bittersweetly to the test, citing the Hindu scripture, "I am become Death, the destroyer of worlds."

When Oppenheimer arrived at the gate, his car slowed; he could see Bush, standing at attention. At the moment of recognition, Bush took off his hat to Oppenheimer. Then the car sped off.[2]

For Bush, there was little left to do in New Mexico. That afternoon, he hopped a plane back to Washington with Groves, Conant and a few other scientists. Groves recalled that the group was "still upset by what they had seen and could talk of little else." Bush remembered only feeling tremendous relief, what he later described as "about as great a release of tension as a man could possibly have."

His relief stemmed from the virtual certainty that atomic bombs would soon be dropped on Japan. Besides the bomb tested in New Mexico, there were two more in the Army's hands. Still resisting the U.S. demand for unconditional surrender, the Japanese were even holding fiercely to the island of Okinawa, forcing a bloody mopup operation that would run for six more days. The delivery of the bomb "on time," as Bush put it, might alter the conclusion of the Pacific war. If the Japanese capitulated after an atomic attack, the bomb might even save "hundreds of thousands of casualties on the beaches of Japan."[3]

The relief that came with the successful completion of the Manhattan Project was mixed with the anxiety over where atomic weapons would take the world. While Bush never spoke of his visceral fear—he never, for instance, wrote one word about his experience of the Trinity test—his actions betrayed his hidden emotions. After stopping in Washington, he immediately left for the familiar terrain of Cape Cod to join his wife at the family's 175-year-old cottage in South Dennis, some three miles from the ocean. On arriving, Bush was greeted by Phoebe and the wife of his son Richard, who had just had the first Bush grandchild (the infant had yet to see Richard, stationed on Leyte island in the Pacific).

Bush seemed happy to watch his only grandson pull himself up by the furniture upholstery and try to take his first steps. He also enjoyed the nearby Bass River and the Atlantic and the familiar surroundings of town. After Trinity, he marched around South Dennis like a hometown boy made good who, despite his hifalutin' and sometimes scary responsibilities, could still act like an ordinary Yankee. To his neighbors, he remained a "regular feller" who spoke "our language." His favorite stop in town was the post office, where he gathered with local men to exchange news and views. "All I want is to get the war over so I can show you fellows how to fish," he quipped, hoping to silence questions about the war. Postmaster Clarence Bayles, who first met Bush in the 1920s, knew enough not to probe. But others asked Bush how much longer the war would last.

A year or two more, he said.

"Are there any new weapons coming on?" someone asked him.

"No, we fired them all off at the Germans," he replied.

Bush's glib answers masked a deep unease. He later described his state of mind as "a strange state of released tension." Others considered him strange, period. "He was "terribly tense, frightfully tense," said his daughter-in-law, Catherine. "He couldn't get back to Washington fast enough." His wife, meanwhile, "didn't know what had hit" Bush, but "she knew that something had all right." He would not tell Phoebe what he had seen in New Mexico, however. Dutiful as always, she did not press her husband and remained ignorant of the bomb project.[4]

Bush soon enough would confront the bomb in the open. Truman learned of the successful test at Potsdam, where he was meeting with Churchill and Stalin. He received a complete report from Groves on July 21. The report "immensely pleased" the president, who was "tremendously pepped up by it" and imbued with "an entirely new feeling of confidence." Churchill endorsed the use of the bomb, and on July 23 the schedule for two atomic attacks was settled. The next day Truman casually told Stalin of the bomb's existence, but there was no question that the Russian leader understood the president's remarks; Stalin that evening told aides that the chief of the Soviet bomb project must "hurry things up." At the close of Potsdam, the Big Three issued a declaration that threatened the annihilation of Japan, called for the country's unconditional surrender and assured the Japanese that they could eventually pick their own government if they accepted these terms. Even retention of the country's imperial system was not ruled out.[5]

Japan ignored the Potsdam declaration, killing the offer with silence. The U.S. waited a few days to see whether Japan changed its position, then Truman ordered the attack. On August 6, at 8:15 A.M. local time, an atomic bomb fell on Hiroshima. The explosive force roughly equalled 13 kilotons of TNT. Nearly everything within a radius of about 1,600 feet was incinerated. A thick cloud of smoke mushroomed 35,000 feet into the sky. The blast killed more than 100,000 people.

Truman, returning by sea from the Potsdam conference, got the news aboard a ship. Grabbing an officer, he blurted out, "This is the greatest thing in history." In Washington, it was 7:15 P.M. on August 5 when the bomb hit Hiroshima. With the deed done, Bush could finally tell Phoebe about his secret project. She was "very much relieved" and "felt it meant surely the end of the war."[6]

The next morning at 11:00, the White House issued a presidential message, crediting the Hiroshima devastation to an atomic bomb and promising "to obliterate more rapidly and completely every productive enterprise the Japanese have above ground in any city." Calling on the Japanese to give up, Truman vowed, "If they do not now accept our terms they may expect a rain of ruin from the air, the like of which has never been seen on this earth."[7]

Bush dispatched a more somber message about the same time to his professional family at the Office of Scientific Research and Development. Of the bomb, he wrote, "It will shorten the war, and it will save the lives of many American youth. We hope that, in the future, it will become a boon and not a scourge to humanity, and toward this end we should continue to labor."[8]

It was like Bush to find hope within a peril. Many others did not. Almost immediately, thousands of intellectuals, social activists and religious people from around the world condemned the bomb and expressed bewilderment, if not outright revulsion, at the decision to use it. The outburst was not altogether predictable since the deaths of a similar number of civilians in earlier raids on Dresden in Germany and on Tokyo provoked no criticism of this intensity. Yet the most perceptive critics of the A-bomb drew attention, not to the sheer numbers of dead, but to the character of their deaths. If death gave life its ultimate meaning, after all, what meaning could be divined from the unparalleled automation of slaughter?

"We can sum it up in one sentence: our technical civilization has just reached its greatest level of savagery," Albert Camus wrote from Paris two days after Hiroshima. "We will have to choose, in the more or less near future, between collective suicide and the intelligent use of our scientific conquests. Meanwhile we think there is something indecent about celebrating a discovery whose use has caused the most formidable rage of destruction ever known to man. What will it bring to a world already given over to all the convulsions of violence, incapable of any control, indifferent to justice and the simple happiness of men—a world where science devotes itself to organized murder? No one but the most unrelenting idealists would dare to wonder."[9]

Bush was hardly an idealist and yet he dared to wonder. Within a week of Hiroshima, he set himself apart from those scientists who warned that dropping a bomb on a city would spark an arms race. While aware of this risk, Bush thought the atomic bombings had alerted Americans to the challenges posed by this new class of weapons in a way that no warning or even demonstration could have done. "We have been very fortunate indeed in timing," he wrote a friend on August 13.

"If the scientific knowledge which rendered the development of the bomb possible had come into the world, say, five years earlier we might indeed have succumbed to the Nazis. On the other hand, if it had come five years later, there would have been great danger that this country, in its peacetime easygoing ways, would not have gone into the subject in the extraordinarily expensive manner which was necessary to put it over. Yet this whole thing would most certainly have come to civilization in one way or another during our lifetime. It is fortunate that it comes now, and in the hands of a democratic, peace-loving country, but it is also essential that that country realize fully just what it is and where it stands on the matter. No amount of demonstration would ever have produced this realization in the American public. It was necessary that events happen just as they have happened in the past week, and it was very fortunate that they thus occurred."[10]

Bush's justification of Hiroshima was no passing opinion. He repeated his view to another friend later in August. "It is fortunate that the bomb arrived when it did and in a fully spectacular fashion," he wrote. Had the U.S. withheld the bomb, Bush added, civilization might "have drifted into a situation" of all-out atomic war and thus been worse off. "From here out, we at least approach [the perils] with our eyes open."[11]

But just how helpful was this wide-eyed vision? In his arguments to the Interim Committee, which had endorsed the use of the bomb, Bush had suggested that the existence of atomic weapons might promote international cooperation and new restraints on belligerent nations. There was a chance that he and likeminded political leaders could finesse the present situation, "preventing major devastation" through an international accord under the auspices of the new United Nations. Or maybe a direct appeal to Russia would work? Bush's anxieties about the Soviet Union arose from his belief that every major power would strive to build an atomic bomb once it had the means. However, perhaps Russia would slouch toward its own A-bomb, rather than mount a crash program and take three to five years to build one, as Bush expected. With the Interim Committee dragging its feet on his proposals for global cooperation, perhaps the big powers would muddle along, allowing the risks of atomic war to grow. While "not nearly as pessimistic" as such gloomy commentators on the bomb as Norman Cousins, Bush conceded "there is plenty of room for pessimism."[12]

This stubborn sense of pessimism tainted the nation's victory. It would have been easy for ordinary Americans to see the defeat of Nazi Germany and an expansionary Japan as an extraordinary proof of the superiority of the American political and economic system. Yet the manner by which the war ended robbed many Americans of a unique opportunity to feel unalloyed pride. "Indeed, the strangest thing about the entire war was our reaction to it after we had won it," observed Bruce Catton, the journalist and historian. "America did the most colossal job in history and came out of it with an inferiority complex and a deep sense of fear."

The media consensus on the bomb, hastily crafted in the aftermath of Hiroshima, saw atomic weapons as agents of ultimate doom, not victory. America had achieved victory but not security. On the evening of the Hiroshima announcement, this dour view was chillingly expressed by NBC radio commentator H. V. Kaltenborn. "For all we know, we have created a Frankenstein!" he declared. "We must assume that with the passage of only a little time, an improved form of the new weapon we use today can be turned against us." Even Harvard president James Conant was convinced that the A-bomb was in a class of its own; he recoiled at Bush's notion that it was just one of a range of emergent threats and accused him of "play[ing] down the atomic bomb too much."[13]

Bush took the fearmongers seriously. But with his usual knack for embracing contradictory moods and ideas, he took an optimistic stance. In his first major post-Hiroshima interview, Bush struck a hopeful chord, stressing the remarkable achievements of technology and its potential for good. "Dr. Bush Sees a Boundless Future for Science: The OSRD Director Holds That We

Have the Knowledge to Build a Better World," *The New York Times* headlined the story containing Bush's remarks.

Peddling an antidote for American pain put Bush in a paradoxical position. After all, he was one of the few men who had secretly decided to build the bomb and then to use it. No legislative body in the U.S. debated these momentous decisions. The public's participation was neither invited nor received. However compelling the case for secrecy, Bush's role in the atomic age invited a backlash from those who abhorred the effects of the bomb or believed that its existence mocked American democracy, perhaps even crippled it. While not alone in backing the bomb, Bush was an identifiable patron. He never felt personally at risk because of this, but those close to him wondered about his safety. After his role in the atomic project came to light, within weeks of Hiroshima, Bush's family realized their lives had been changed forever. "We worried about anything happening to Dad, or any of us," recalled his son Richard. "We worried about retaliation against his involvement."[14]

Yet Bush's role in the A-bomb's birth actually burnished his reputation. Like Truman, most Americans were thrilled by Japan's sudden surrender and the end of the war, which followed by days a second atomic attack on August 9. Rather than interrogate the leaders of the Manhattan Project, the public embraced them. Bush's reputation as a scientific seer grew; his image as an unmatched organizer of expertise solidified.[15]

For Bush, the atomic bomb capped off his five-year rise to celebrity from relative obscurity. *The New York Times,* portraying Bush as the chief bureaucrat behind the bomb, credited him with playing "an important part in bringing the war to a close." Another New York daily printed Bush's picture with the caption "Scientist and Patriot." *Business Week* plastered a drawing of a demure and stately Bush on its cover, describing him as the "spearhead of a drive to keep scientific research mobilized for peace." The March of Time, maker of newsreels, asked Bush to stage his reactions to the atomic test in New Mexico. One year after the Trinity test in July 1945, Bush and Conant visited a warehouse in Harvard Square to reenact their witnessing of the event for the benefit of posterity. The newsreel, "Atomic Power," reached theatergoers in August 1946.

Even Hollywood joined in, seeing Bush and his activities as grist for its glamor mill. Fourteen weeks after Hiroshima, Metro-Goldwyn-Mayer began planning a feature film on the bomb, envisioning a prime role for Bush's character. The studio foresaw "a great service to civilization if the right kind of film could be made." Bush wasn't so sure. He agonized over the making of the movie, nervous about "linking this sort of history with a romance à la Hollywood." But after the producer, Sam Marx, seemed to get Truman to go along, Bush reviewed the script and allowed his character to be used, signing a legal release in late 1946.

The movie, entitled *The Beginning or the End,* was released in February 1947. Bush found the actor who portrayed him "not bad," and at one point told a friend that "history was not unreasonably distorted" by the movie. A few critics were impressed. The liberal New York daily *PM* found it "reassuring," while *The New York Times* praised MGM for having "taken no obvious sides in the current atomic contentions" and for depicting the bomb as "a necessary evil."

But *The Beginning or the End* bombed at the box office. Most critics were unimpressed too. "The treatment of the moral problems exacerbated by the bomb is once-over-lightly," *Time* concluded, adding that "even as entertainment, the picture seldom rises above cheery imbecility." *The Bulletin of Atomic Scientists* was outraged by an invented scene that showed the U.S. showering warning leaflets over Hiroshima. The magazine termed this to be "the most horrible falsification of history." Indeed, the movie angered many scientists. Conant had vainly tried to halt the release of the film, and physicist Leo Szilard had even traveled to Hollywood to force changes in the scenes in which he appeared.

In the end, Bush took the narrow view, seeking merely to protect his own image from the Hollywood maw. In his dealings with MGM, he insisted that "my particular appearance does not give the public false ideas."[16]

The media's fawning attention was only one sign of Bush's importance. Scientists and engineers toasted him too. National Academy president Frank Jewett called the OSRD, Bush's brainchild, "the greatest industrial research organization the world has ever known." Arthur Compton, the leader of the Chicago atomic scientists, believed that "more than any other man, Bush was responsible for rallying the scientific strength of the United States." Julius Furer, the Navy's former research chief, insisted that when it came to convincing the military to adopt new technologies, "where Bush didn't succeed, no one else was likely to have succeeded." Rockefeller Foundation executive Warren Weaver, a leading prewar patron of natural science and an OSRD staffer during the war, thought Bush would have an even greater influence on the nation in the postwar years. Given "the experience he has had . . . Van is going to be one of our great figures in five to ten years," Weaver said. Other friends thought Bush had the right stuff to serve in the cabinet; one even said, "Bush could very well have anything he wanted, including the presidency of this country."[17]

Bush's new horizons represented more than a personal triumph. They also marked the victory of elitism in a new sphere of public enterprise. Long ignored by politicians and the military, science and technology were no longer orphans, providing fodder for fiction writers and Hollywood but little sympathy or comprehension from the wider public. Now the government found technology too valuable to leave to the vagaries of the marketplace and the contingencies of invention. The war had prompted the state to spawn revolutions in technology by applying brute force to core scientific problems. Belying the view that inventions just "happened," the OSRD's record showed that infant innovations could be force-fed into a rapid adulthood. Operations research, meanwhile, epitomized a tradition of scientific efficiency launched by Frederick Taylor, the turn-of-the-century engineer.

For prophets of technocracy, the war seemed to redeem their faith in engineers and scientists as decisive social actors. At the very least, the audacious vision promoted by Thorstein Veblen, the idiosyncratic social critic, had roots in reality. Best known for his biting observations on the nation's "leisure class," Veblen wrote a series of articles about engineers and society that were published in 1921 as *Engineers and the Price System.* The book argued that governing committees of experts and engineers should take control of the nation's industrial system from greedy financiers and wasteful entrepreneurs. Considered preposterous at the time, Veblen's "soviets" of experts seemed more plausible in the afterglow of World War II. "Veblen in his grave must now be permitting himself a sardonic smile," wrote Stuart Chase in *The Nation.* "His technicians and scientists have come roaring into their own as the acknowledged and undisputed arbiters of human destiny."[18]

In their new guise, scientific statesmen were the toast of the town. One noted: "Suddenly physicists were exhibited as lions at Washington tea parties, were invited to conventions of social scientists, where their opinions on society were respectfully listened to by lifelong experts in the field, attended conventions of religious orders, and discoursed on theology, were asked to endorse plans for world government, and to give simplified lectures on the nucleus to Congressional committees."[19]

Bush welcomed the public's new respect for technical experts, but feared it might turn into a mania. He was irked, for instance, that physicists received so much credit for making the atomic bomb. The contributions of chemists, biologists and other scientists were overlooked, while innovations in engineering and production—crucial to everything from the A-bomb to radar—were virtually ignored. "Inevitably the work of the physicists stands out in great relief, because of the somewhat fascinating and spectacular nature of their part of the work," Bush noted soon after the war's end. But the renown of the physicists "has the unfortunate effect" of obscuring the contributions of others and the fact that "physicists can go only a short distance along the path" on their own, he insisted. "As we turn toward problems of national security," he added, "it is highly desirable that the American people should have a reasonably balanced appreciation of the actual situation."[20]

Grasping the hierarchy of technologists had real consequences for the politics of expertise. Bush mainly complained about the danger that government meddling would smother innovations, but he also worried that researchers might drown in public funds now that the Manhattan Project had shown the possibilities of applying unlimited resources to a problem. The managerial lesson of the bomb seemed deceptively simple: just throw enough resources at a problem and the experts would solve it. There was the risk in the future that this handy paradigm for government intervention would endow technocrats and politicians alike with a false sense of their ability to tackle tough problems. For the moment, the attitude simply reduced in importance grand technical achievements of the past, such as the pyramids or medieval cathedrals. But Bush sensed his contemporaries falling prey to hubris, especially the physicists, whom he found since the end of the war to possess a disturbing trait of "know[ing] all the answers."

"The sudden prominence which has been given to physicists due to war progress in applications of physics has gone to the heads of quite a few people," Bush wrote an MIT professor. "They proceed to testify on subjects that they have met only for the first time and recently, which is all right, but they do so with a didacticism which is all wrong. If they do not look out they will end up by becoming discredited."

This was a rare jeremiad, however. Amid the wreckage of war and the gnawing guilt over the use of the bomb, the public's confidence in experts—physicists, especially—stood as a lonely sign that perhaps man was not obsolete after all. As Bush later wrote, "The war provided a spectacular demonstration of the power of programs. . . . It proved that things could be accomplished in a hurry, given *unlimited* funds, the intensity of war, and, most important, a background of basic scientific knowledge ready for application." Now Bush and his fellow technological wizards wondered whether they could perform the same trick for a nation at peace.

The citizenry held its collective breath.[21]

Even as he warmed to the spotlight, Bush bore the scars of having lived in the shadows for five years. Peculiarly suited to the culture of wartime secrecy, Bush now had to adapt to political life in the glare of publicity. While Hiroshima surely had ended one era and begun another, this meant different things to different people. To the masses and the media, one horrible style of warfare had given way to an even more terrible brand of killing. Having witnessed the Trinity test and received briefings on the firebomb raids on Tokyo, Bush was not stunned by the latest devastation in Japan. To him, the greatest shock of Hiroshima came from "the shattering of the little world of secrecy" in which he "had so long been confined."[22]

Bush had flourished in this secret world because he had the confidence to "run away with the ball," as he liked to say. He didn't wait for a mandate, but pushed his limits as he saw fit. "My whole philosophy on this sort of thing is very simple," he had told military men during the war. "If I have any doubt as to whether I am supposed to do a job or not, I do it, and if someone socks me, I lay off." Even when Bush socked back—and he often did—the blows were always traded in private. This way neither Bush nor his adversary had to explain to an astonished public why they had such bad tempers. Keeping the lid on these disputes, however, meant that they tended to smolder. "Van fought so many rough battles in Washington that he had an awful lot of people who were not wholly enamored to him," Weaver recalled. But to get things done, "this is the price you have to pay."[23]

Bush paid willingly. Intensely self-assured, he deferred to no one save Roosevelt and his mentor, Henry Stimson, the secretary of war. In the heat of war, his penchant for barging ahead worked wonders. The military gave more leeway to him than perhaps any other civilian in the war. Members of Congress granted his every request. "Never once did we ask for funds and fail to secure them promptly," Bush later boasted. Legislators rarely even questioned him, and when they did the exigencies of war made it possible for him to duck the tough queries anyway. He never flatly refused to satisfy a politician's curiosity, but rather dared him to comprehend the technical and military issues. Most politicos wisely kept their mouths shut. Bush kept the press quiet too, gaining the right to review some articles about him before publication and killing other pieces with noncooperation. "He was the only war administrator who could and often did refuse to see the press, gaining a reputation as a prima donna," *Fortune* noted.

The war left Bush with the impression that Congress was a malleable body, waiting to be charmed by an expert and willing to do the bidding of the executive branch. In dealing with politicians, he mixed a professorial air with a dab of vaudeville theatrics, treating them like a room full of MIT freshman. He learned before the war, during his appearances on the Hill representing the national aeronautics committee, that most members of Congress "are pretty badly bored. They sit around all day and they listen to testimony which is not particularly illuminating or exciting, and if one introduces just a bit of humor into the situation he's welcomed with open arms, because it makes a break." Sometimes the humor was unintentional. During one visit by Bush to the Senate appropriations committee, the chairman, Kenneth McKeller, mistook Bush for the head of another federal agency and directed his questions accordingly. Since McKeller and a clerk were the only ones present, "I didn't enlighten him," Bush

recalled. "And the show went on this way. He'd ask a long and involved question that had nothing to do with my agency, and I'd give him an equally long
and complicated answer which had nothing to do with my agency, and also
nothing to do with the question."

After the creation of OSRD, Bush made an annual pilgrimage to Congress in
order to gain approval of his agency's budget. This was a perfunctory appearance, because the tough questions about the budget's size were hammered out
between Bush's staff and Harold Smith's Bureau of the Budget. Bush recalled visiting a small group of congressmen "each year" during the war with Stimson to
discuss the atomic bomb project. Given that the Army's involvement began in
1942, this meant that he probably made no more than three of these visits. "In
these meetings we pulled no punches whatever," he recalled. "I did the talking."
His toughest congressional appearance was a closed-door visit in February 1944
when, joined by Army chief George Marshall and Stimson, he met privately
with House Speaker Sam Rayburn and the House majority and minority leaders.
The Army needed $1.6 billion for the Manhattan Project, and Bush, Marshall
and Stimson wanted it delivered without "a trace of evidence to show how it was
spent." Rayburn agreed, and the money was supplied with no strings.[24]

Some tactics that served Bush so well during the war were a hindrance in the
new postwar world. After the atomic explosions in Japan, ordinary citizens and
deep thinkers alike tossed about opinions on use of the bomb and plans for its
control. Roused from their slumber, members of Congress sought to influence
A-bomb policy for the first time. Even rank-and-file physicists, who had long
complained in secret about wartime regimentation, now aired their grievances
in public. While the populist vision for atomic energy had yet to emerge, Bush's
privileged position was gone. The ground had shifted under him, and he struggled to gain a new footing. Perhaps nothing revealed this so completely as the
public fight over domestic legislation to control atomic power.

For more than a year before the first A-bomb exploded, Bush had hoped to
define the shape of the nation's postwar atomic complex. In July 1944, Bush,
his executive secretary Irvin Stewart and Harvard president James Conant
completed a two-page summary of a domestic atomic energy bill. Their proposal called for a 12-member commission that would have unprecedented
peacetime powers, with the authority to control all nuclear materials, construct all bomb plants and manage any significant research. The National
Academy of Sciences would nominate five of the members; the president
would name three; the Army and Navy two each. Bush conceived of the commission as independent of politics. The commissioners were to serve on a part-
time basis and probably would not require Senate approval.[25]

For a year, little was done with the Bush-Conant plan for atomic legislation. The duo's ideas weren't taken up again with the government until a meeting of Truman's Interim Committee on July 19, three days after the Trinity test. Asked to outline postwar atomic policy, the committee was running out of time. It spent the bulk of this meeting reviewing draft legislation for domestic control by two lawyers, one an Army officer. The draft resembled the Bush-Conant proposal of the previous September; the only major change was that the unpaid commissioners would total nine, with five named by the president and four by the military. The commission's proposed powers were wide, prohibiting almost any activity not under its control. It would be unlawful, for instance, for scientists to perform atomic research without the commission's permission.[26]

If Bush was happy that the proposal tracked his first offering, he restrained himself. He now thought the commission should comprise only civilians and that its power should be narrower. He was especially concerned by what he viewed as the commission's overly broad "censorship and security" powers. He argued that the law should permit scientists to publish "information in this field which did not endanger national security." And in a position that mirrored the thinking behind his proposed National Research Foundation, Bush insisted that the commission's proposed powers over research be trimmed and that the law contain an explicit statement that the government "should normally depend on the universities to carry forward the basic research program in this field."

Bush's objections were treated merely as a starting point for compromise. Since Conant and even Groves thought a military presence on the commission nonessential, the Army's lawyers backed off a bit; they allowed that military representatives could be drawn from the ranks of retired as well as active officers. As far as the commission's powers over research, the lawyers conceded even more. They added a declaration that the commission would augment its research with efforts by "private or nonprofit institutions." In place of language giving the commission total control over atomic research, they substituted a rule banning work that released atomic energy in amounts "deemed . . . a national hazard" by the commission.

Pleased with the changes, Bush nonetheless found the proposed commission's powers "greatly overdone" and urged a complete review of the bill. But it was too late for that. Hiroshima had made it impossible to craft acceptable domestic atomic legislation in a back room.[27]

In the weeks that followed, the whole subject grew progressively messier as Truman delayed making his first major statement on atomic weapons. By the middle of September, the public was restive; *The New York Times* lamented that five weeks had passed since Hiroshima "but nothing has been done about the atomic bomb." In the meantime, scientists were expressing their discomfort. On September 1, 1945, 17 leading atomic physicists in Chicago threatened to boycott their field if the government didn't relax security restrictions. The same month, Chicago physicists formed a group aimed at educating the public on atomic power and 65 local professors signed a petition asking Truman to share the bomb with other nations to avoid an arms race.[28]

It was no longer possible for Bush to pretend that he represented the opinions of scientists on atomic matters, much less the public's interest. His own views differed markedly from those of some of the Chicago physicists, undermining his credibility as a spokesman. Even worse, the military was rapidly spreading its wings, lessening its reliance on Bush for technological advice. Freed from the daily grind of war, "The Services are running wild," Bush complained to Conant. The Army and Navy were pursuing a "very extensive and utterly uncoordinated" agenda, he added. Placing limits on their appetite for technology, possible during the war, was now much more difficult.[29]

Bush's influence on official atomic policy declined when 78-year-old Henry Stimson left the government in late September. In the final years of the war Stimson had been Bush's link to the president and his protector against attacks from the top brass. The outgoing war secretary was the only cabinet member with whom Bush shared his anxieties about international relations in the atomic age. Their relationship was more than professional. In his lonely work, Bush found Stimson "one rock stable in the current," an antidote for the confusion and nervousness that threatened to engulf him. He would miss Stimson dearly, not least because Stimson's sanity and respect for expert opinion seemed to him as rare in the Truman administration as it had been in Roosevelt's. Bush felt an uncommon affection for the secretary. "During the war you were to me as a father to a son, and my greatest satisfaction of the whole war experience came from that relationship," Bush once told him.

As a courtesy, Stimson invited Bush to attend his last cabinet meeting on September 21. They both agreed that a full cabinet meeting was not "entirely a safe place to discuss everything with complete frankness," so both "pulled punches a bit" in the discussion of atomic matters. Still, they worked in tandem to present the case for a direct appeal to the Soviet Union. Arguing that the basic scientific information behind the bomb could not be kept secret, Stimson said the U.S. had everything to gain and little to lose by sharing this information with the Soviet Union. "We do not have a secret to give away—the secret will give itself away," he claimed.

This was precisely Bush's view. After Truman cronies Fred Vinson and Tom Clark disagreed with Stimson, Bush countered that sharing information hardly meant handing Russia the secret of the bomb because the secret of the bomb lay in the details of its design and manufacture. Then, restating a point he had made to the Interim Committee two months before, Bush suggested that it was worth sharing atomic information if it improved the chances of international cooperation. The Russians would discover much on their own anyway. Bush thought they could "get to about the place we are now in five years provided they devote a very large part of their scientific and industrial effort to it."

Pleased with the debate, Truman asked everyone present for a memo since he had still not made up his mind on how to control atomic weapons. The actual decision, the president reminded his advisers, "had to be mine to make." After the meeting, Truman pulled Bush aside to say he especially wanted a memo from him.[30]

Excited by the president's encouragement, Bush composed a sweeping seven-page memo laying out the post-Hiroshima challenges. Painting in bold, clear strokes, Bush distilled his considerable insights into atomic weapons. His memo proved prophetic. In sketching alternative visions of American life under the threat of a mushroom cloud, he anticipated much of the tone that America took during the late 1940s and into the 1950s.

Bush's most striking point was that the window of opportunity to reach an atomic deal was closing. "From one standpoint it is important to make the move at once," he wrote. "If Russia becomes fully embarked on an extensive program, it will be hard to deflect her effort." (Indeed, within a month of Hiroshima, Stalin had decided to organize a high-priority atomic project. Though costly for a country with an economy ruined by war, the project was given substantial resources. Stalin told his top bomb scientist: "If a child doesn't cry, the mother doesn't know what he needs. Ask for whatever you like, you won't be refused." In early November, Molotov, Stalin's foreign minister, declared that the U.S. nuclear monopoly would not last indefinitely, warning, "We shall have atomic energy and many other things too.")

Other arresting statements hit upon the dim prospects for defense and the implications for society and industry. Bush cited three technical reasons why the peril of atomic weapons would only worsen. First, "the advent of long-range rockets" would make "interception" of any bomb "highly difficult." Second, the bombs themselves would invariably become more powerful. While not mentioning plans for hydrogen bombs, Bush clearly had something like this in mind. Third, bombs would be produced "more cheaply," not a trivial consideration at a time when only the U.S. seemed to possess the industrial resources to build bombs quickly.

These technical forces made the already slim odds of repelling an atomic attack seem even slimmer. "There is no countermeasure to atomic bombs in sight," Bush noted dryly. Absent an international weapons agreement, the U.S. and other nations before very long would have to consider extraordinary measures to better the odds of survival. "Whole nations" might have "to disperse industry and go underground," he predicted. Populations would be forced into "very unhappy" lives.

This specter made it imperative that the U.S. at least test Soviet intentions by sharing basic scientific information. Bush asked the core question: "Can we work with Russia and trust Russia? To some extent this move would enable us to find out" at little cost. He conceded that the U.S. would be "giving away the gun on our hips" even as it carried on "difficult negotiations" with Stalin, but he argued that actually this would not be "giving away anything at all." Atomic bombs were too destructive to be a useful instrument of diplomacy, and he felt the Russians knew this. "There is no powder in the gun, for it could not be drawn, and this is certainly known."[31]

Why was Bush so convinced that the atomic "gun" could not be drawn? He failed to untangle his reasons for Truman, but he had more than diplomacy in mind. Bush realized, as others would in the years to come, that the bomb represented a revolution in military strategy. A strategy of annihilation, which had been pursued against both Japan and Germany during the war with varying effect, could now be so complete as to no longer serve the aim of war unless, as historian Russell Weigley noted, the aim "was to transform the enemy's country into a desert." In their ultimate expression, atomic bombs would not serve "the rational purposes of statecraft": to bring an enemy's conduct into line.

Another consideration limiting the diplomatic value of the atomic bomb was the small number of bombs available. The attack on Japan had cleaned out the U.S. stockpile; building it up would take time. The arsenal would grow to a mere nine bombs in 1946: certainly enough to punish Russia, but probably not enough to silence her guns.

A third reason for the bomb's limited value was the effect on U.S. relations with allies if the nation dropped more A-bombs on defenseless civilians. As Bush stressed in his memo to Truman, the U.S. could gain international good will by offering to share basic atomic knowledge. "The general advantage" of sharing, he wrote, "is that this move, when it became known, would announce to the world that we wish to proceed down the path of international good will and understanding." This good will, of course, would vanish if the U.S. punished another enemy with its atomic "gun."[32]

Despite Bush's advice, Truman seemed in no rush to approach the Russians. Too many of his other key advisers favored freezing out Stalin. Though drawn toward the arguments by Bush and Stimson, Truman backtracked in public,

declaring on October 8 that the bomb was the nation's "sacred trust" and that
other nations could only "catch up . . . on their own hook, just as we did."
Truman's flip-flopping on arms control made Bush suspect that the president's
hard-line advisers, notably Secretary of State James Byrnes and Navy Secretary
James Forrestal, were not sincere about controlling the bomb.[33]

To Bush, who worshiped at the altar of rational planning, the Truman administration's haphazard behavior seemed cavalier, at best. This was consistent
with the president's waffling on civilian science policy. Refusing to endorse
Bush's plan for a national research foundation, Truman insured that the rival
plan offered by Senator Kilgore would receive a lengthy hearing. Delays on the
atomic and scientific fronts angered Bush, especially because he had devoted
the prior year to paving the way for action. He came to feel that the nation's
capital was "a bit of a madhouse. So many people have left town that about
everything that has to do with science and its relation to government descends
on me in one way or another. I hence find that I am harassed to the extent that
I do not get time to think things through and this bothers me a great deal for
there is much to think about."[34]

Indeed, the pressure seemed to warp Bush's perspective. He failed to grasp
that the reversion to politics as usual prevented the kind of fast, sure decisions
he and Roosevelt had made during the war. Blind to the storm ahead over domestic atomic regulation, he still held out hope that closed-door deals could
settle thorny issues involving private expertise and the public interest. Despite
the growing protests by scientists and the public's alarm about atomic
weapons, Bush thought that Congress might approve the domestic atomic bill
without holding public hearings. Accustomed to the supine posture of the
wartime Congress, Bush thought that if lawmakers did discuss the matter in
public, they might "at least have an explicit understanding before they start
out as to what part of the whole affair is to be aired in public and what handled
behind closed doors."

Even this degree of restraint wouldn't have been enough for Bush, who
doubted "that the doors will really be closed . . . unfortunately one cannot be
sure on that point."[35]

Bush's misreading of the prospects for atomic legislation was monumental.
Truman finally sent a message to Congress on the topic on October 3, 1945.
Representative Andrew J. May and Senator Edwin Johnson quickly introduced
the atomic energy bill, which Bush had helped to shape, into their respective
houses of Congress. Immediately the bill became a lightning rod for critics of
the government's handling of atomic matters. Scientists, meanwhile, lashed out
at Bush and Conant for signing off on the legislation. Their major objection

was to the heavy military presence on the commission. Membership, they insisted, should be a full-time job and open to civilians only. They also objected to the bill's security provisions, which seemed to interfere with nuclear research.

Behind all these objections lay a suspicion that somehow the military would hog the benefits of atomic energy, preventing the civilian economy from taking advantage of the presumed wonders of this new technology. Bush held open the possibility that civilians might tame atomic energy, but he warned that the field "is in its infancy" and that he foresaw "no immediate great commercial applications just around the corner."

Pessimism was another strike against Bush. While the advent of commercial nuclear power was a decade away, some critics of the May-Johnson bill feared that what they considered the bright side of the bomb—the potential for atomic energy to produce cheap electricity—would be kept from citizens and industries. Bush tried to dampen the enthusiasm for atomic energy. Concerned about the technology's hazards, he viewed it as almost exclusively a military asset. In a September 1945 letter to Herbert Hoover, Bush asked the former president to play down talk of atomic energy for industry, warning: "There is great danger that the country will expect too much too soon" from atomic power. The sitting president, Truman, showed no inclination to soft-pedal atomic power, however. In October 1945, he told Congress that atomic power "may someday prove to be more revolutionary in the development of human society than the invention of the wheel, the use of metals, or steam or internal combustion engines."[36]

With so much at stake, many atomic scientists felt that Bush and Conant had sold out their interests; some thought the two leaders wanted to restrict atomic power to destructive uses. Few foresaw the miasma of problems that atomic power would spawn. Fewer knew that both Bush and Conant had criticized the bill, and that Bush had pressed for an all-civilian commission. Yet neither Bush nor Conant would step outside of "channels" and air their differences with the bill.

In the Senate, the route to quick passage was blocked by opponents. But in the House Representative May tried to ram the bill through, scheduling a whirlwind hearing on October 9. Bush testified that under the law Congress would give up all control over atomic energy, an admission unlikely to delight legislators. But "rigid federal control" of atomic power was justified, he insisted, because of the risk that "some group of experimenters might set up a laboratory half a mile from my house and family and experiment on atomic energy carelessly, poison the neighborhood, or possibly blow it up."[37]

Bush's performance on behalf of the doomed May-Johnson bill appalled many who new him. Hundreds of scientists sent telegrams and letters to the

White House protesting the bill. A group of Los Alamos physicists wanted Bush and Conant, who also testified on behalf of the bill, to be replaced as scientific leaders by "men about whose views there is less doubt." Old ally Frank Jewett, meanwhile, savagely accused Bush and Conant of "doing the country and the world a great disservice" by scheming to steamroll a maimed bill through Congress. Jewett then issued a damning charge that must have stuck in Bush's craw. "If this is what the past four years has brought us to," he insisted, "then I wonder why we have been so hot and bothered about Fascism, Nazism and the Japanese."[38]

Twisting in the wind, Bush refused to defend himself publicly, but privately sought sympathy from his clubby fraternity of science administrators. He supported May-Johnson "with a clear conscience," he told Jewett, adding that he "objected rather strenuously to some parts" of the bill. To Lee DuBridge, another wartime cohort and the Rad Lab director, he tried to convey a sense of "the restraints imposed by official participation" in government policymaking. He conceded, however, that this excuse was unlikely to prevent the backlash against him from angry scientists. "I have to take the consequences," he noted, "if those to whom I cannot give a full explanation do not understand or misconstrue my motives."

While unapologetic, Bush pleaded for understanding, asking DuBridge to pass the word about that he remained "a full believer in the freedom of American science." Reminding DuBridge that politics wasn't for the pure, he observed, "I am sure that you realize that some compromise must be made if a single document is ever to emerge from such deliberations."

Bush made a good case for himself. "The scientific people in the country feel that if they have a representative with the [Truman] administration . . . [on atomic matters] I am probably that representative," he wrote. While this remained to be seen, one thing was certain: the attacks on Bush from his own camp revealed how hard it was for him to rid himself of the secrecy habit after years of meeting military needs. Now a divided person, Bush extolled freedom and individuality on the one hand, while defending the idea that regimentation and centralization were required to win not just the war, but the peace too. Expressing his misgivings about May-Johnson to someone outside Truman's inner circle, however discreetly, made him nervous. "I find myself in the highly embarrassing position that I cannot give even in a private letter the full status of the affair," he wrote.[39]

Bush was more candid, however, with his thoughts on the younger scientists leading the revolt against May-Johnson. "In the group that are making the great noise there are practically none of the scientists that I depended upon for work in important positions during the war," Bush snidely informed a member of

Congress. "The vociferous group," he added, "is largely younger men, and I fear that many of them had a bad reaction from the prominence that their efforts took in connection with publicity. The phenomenon is a natural one enough, but it is sometimes disturbing, for many of these chaps know nothing whatever about public affairs and their influence is producing bizarre results at times."

To a professor at MIT, he wrote: "Many of these youngsters are suddenly trying to become active in the political area and understand it almost not at all, with the result that they say and do quite strange things. Some of them have attacked me, for example, on the May-Johnson bill matter, but when I pin them down I find they will attack me for lack of something in the bill when it is actually in the bill and I put it there."

"I suppose they will learn after a time," he said of the younger scientists, "although the fact that they do not realize how much they have to learn makes me a little pessimistic on this score." Yet it was Bush who seemed unaware of the score. With scant resources or organizational savvy, the dissenting scientists had strangled May-Johnson—a pretty rich outcome for a bunch of amateurs going up against the military establishment of the world's greatest power.[40]

While trying to patch up the rift with the scientific community, Bush risked further isolation by courting an open breach with Truman. Five years earlier, he had realized that if he wanted to accomplish anything in Washington as an unelected expert, the president's imprimatur was essential. Now his lifeline to the White House, frayed after the death of Roosevelt, had snapped. Truman still outwardly acknowledged Bush as a technological "wise man," but he listened to the engineer less and less. Bush was torn over his growing distance from the president. A public break would give him the chance to speak his mind candidly on the Washington "madhouse." But the price would be high: adrift from the center of power, he might be ignored and, as he told a friend, "disappear from the scene."

To be sure, despite differences in style and values, Truman and Bush needed each other. As recently as August 9, the day of the Nagasaki bombing, Bush had extolled Truman as "a real leader," having left the Oval Office "walking on air" after convincing the president to issue a report on atomic-energy fundamentals. Bush's recent "no powder in the gun" memo on international atomic policy also made an impression on the president, though Bush learned this in a curious fashion. In his October 3 message, Truman sometimes sounded like Bush, saying at one point that "the hope of civilization lies in international arrangements," but "difficulties in working out such arrangements are great." In a prediction that echoed Bush's own, the president said that the failure to overcome these difficulties might mean "a desperate armament race which might well end in disaster."

Truman's credibility, meanwhile, benefited from the presence on his team of an esteemed science leader such as Bush. When on October 8 Truman revealed for the first time his decision not to share the nation's atomic knowledge with Russia and the rest of the world, he cited Bush to justify his decision. Speaking to reporters at a lakeside lodge in Tennessee, Truman distinguished between three types of atomic "secrets": basic scientific knowledge, engineering "know-how," and the "industrial capacity and resources necessary to produce the bomb." Defending his decision to withhold all three types of secrets from the rest of the world, Truman argued that since no other nation possessed America's industrial might, the first two types of information were essentially useless to other nations. So the president reasoned that it made no sense to share such information. When asked by reporters for the source of his thinking, Truman immediately said, "Vannevar Bush."[41]

Bush probably found Truman's finger-pointing disappointing, however, since he wanted the U.S. to do just the opposite. Such shortsighted decisions by Truman fueled Bush's desire to speak his mind on a range of topics: atomic control, the future of civilian science, military affairs generally and even the prospects for the entire country. Besides, Truman was beginning to rely on others for guidance about science. The president had asked John Snyder, a friend, to overhaul the stillborn May-Johnson bill; in late October 1945 Truman decided the bill no longer represented his views on the subject. Snyder didn't even invite a miffed Bush to meetings devoted to revamping the bill. Under the influence of an atomic-power zealot on his staff, Snyder believed the debate over May-Johnson—now cast in unfriendly civilian-versus-military terms—could be transformed into a benign discussion over how to bring technological progress (in the form of inexpensive electricity) to ordinary Americans. In December a freshman senator from Connecticut, Brien McMahon, introduced a rival bill on atomic energy that was written by two of Snyder's staffers. The bill by McMahon reflected the resurgent New Deal spirit. Liberal in purpose and responsive to Congress and the president, McMahon's bill placed civilians in charge of atomic power in all its manifestations.[42]

Bush thought the bill went too far to accommodate civilians, but he clung to the wartime code of staying within channels. Before stepping outside, he asked Truman to clarify "my quandary" on October 13. Bush presented a stark picture. "I cannot remain silent," he wrote. "Either my comments and advice must play an important part in the councils of your administration or I must be free to speak plainly in public on all those matters of science in which I feel that my war experience gives me a duty to speak. Hence my quandary."

Barely concealing his anger, Bush proceeded to explain that if Truman made him an "active participant" in policy discussions "having scientific aspects," he would be "anxious, as you know, to be a loyal and effective member of the team." But as a practical matter, "I am not a member of your official family, but simply a carry-over from the war."[43]

Perhaps thinking Bush simply wanted his ego stroked, Truman sent him a letter insisting he was indeed part of his official family and that he intended to "lean on [Bush] on scientific matters." Truman even had Bush to the White House for "a very frank and complete" talk after which he autographed his picture for Bush. Yet the president continued to back away from Bush, much to the latter's "distress." When on October 25 a reporter asked about Bush's role in his decision to "keep the know-how of the atomic bomb a secret," Truman pointedly said, "I was relying on my own judgment." When the newsman cited published reports "that you were relying on Bush," the president sniffed, "If that's worth anything to you."[44]

Uneasily, Bush watched his clout diminish within the Truman administration. "I am largely ignored," he complained to Jewett in early November. For the moment he made no public outcry. He clung to the belief that he could shape the new president's thinking. Vowing to endure the slights to his considerable pride, he told Jewett, "One should not let personal irritation or inconvenience stand in the way of rendering the best service possible on important matters."[45]

But it grieved him to stay silent. At the end of 1945, he saw the nation at an awesome turning point. One road led to a spiraling arms race and probably a terrible war. The other road led to an armed but prosperous truce. The nation's new responsibilities had outstripped the capacity of partisan politics, Bush felt. "Unless there are some people in a disinterested position who have the courage to speak out," he wrote Conant in late October, "the people of the country will not even have the issue before them in a way that it will become impressed upon their consciousness."

What most upset Bush was the influence of special-interest groups, especially unions, on government. With the end of the war, the whole gamut of New Deal demands for social and economic justice had resurfaced, with the added justification that the nation's economy was now strong enough to support high standards of welfare. For conservatives who saw the war as a respite from the insistent demands of the less fortunate, peace brought its own problems. It meant a resumption of class politics. Bush huffed that the country had been ruled by "a labor government . . . for twelve years."

More broadly, he wondered if the Democrats would ever lose the White

House again and if professional experts could continue to serve as a counter-weight to partisan politics. "Government by the activities of pressure groups, whatever they may be, may be inevitable," he noted, "but it is going to work only if there is a sufficient body of the citizenry that sees the point, preserves the balance and maintains the government in a position above that of merely a tool or an adjunct to the group that happens to have the ascendency at the moment."

Bush prided himself on not taking political sides, later boasting that none of the presidents he dealt with—from Roosevelt to Kennedy—ever knew his own party affiliation. In truth, he hewed to no real political line, and not even conservatism was always consistent with his brand of elitism. He favored a balance between special interests, with enough slack in the system to allow people of merit—he would later call them "a natural aristocracy"—to make significant technical decisions. "There is nothing anti-labor in my point of view, although it would promptly be interpreted as such," he insisted. "I would have been just as completely anti-capital in the same sense if I had been arguing a generation earlier."[46]

This was not the first time Bush had voiced profound concerns about the ability of democracies to deal with international crises. As early as 1939, he fretted that dictatorships "can cut rings around the democracy." During the war, he battled against recurring bouts of such anxiety, but the superiority of Allied technology and overall mobilization convinced him that democracy's supposed disadvantages were illusory. As it turned out, the U.S. and Britain had paradoxically "achieved an effective form of centralization which escaped a dictatorship." Bush had essentially made this point many times in public and private. Just a week after sounding off on labor domination of politics, he told a gathering of the Roosevelt Memorial Association that the outcome of World War II "means that, all other things being equal, a democracy can outclass any despotism in bringing to bear on the struggle the combined efforts of science, industry, and military might."[47]

Given this statement, Bush's fears about the inefficiencies of American pluralism seemed to contradict his commitment to Hoover-style conservatism. Yet his position was subtle. He was convinced that democracy had performed so well *because* of the war. Remove the prod of war and the U.S. government would revert to past form. The Truman's adminstration's "exceedingly loose" handling of arms control in late 1945 only fueled his fears.

He believed that alarmists had inflated the dangers of atomic weapons: their very existence would likely "stop great wars." Ham-handed politicians, he worried, would squander the chance to avoid great wars by striking a global atomic bargain. The quickest escape from atomic tyranny would come "by opening the doors of scientific laboratories everywhere and making the results of scientific research fully known." The road to global cooperation held perils, of course. "We have a very difficult job before us in inducing all of the best nations of the world to adopt this policy," he wrote a colleague. Some nations— he did not name Russia, but he had it in mind—might wish to close their scientific doors. "This might be the beginnings of a secret arms race, which would be appalling."[48]

Ready to break with Truman, Bush thought he had one winning card left to play at the big table. He had achieved much during the war. Perhaps no inventor since Benjamin Franklin had exerted such a direct effect on American political life. Bush expressed no guilt over the nature of his achievement: he had married science and the state, invention and destruction. To be sure, this had been done before, but had never succeeded on so grand a scale. Proud of his wartime role, Bush nonetheless wanted to leave another legacy to future generations. He wanted to contribute to the peace as much as he had to the war. His proposed research foundation, which would steer a new generation of young people into science, was one possible contribution. Another would be the control of atomic weapons.

Against the odds, Bush tried to move Truman toward global cooperation. He swam against the tide. Historic tensions between the U.S. and the Soviet Union, masked by the war, were resurfacing. The president seemed bent on monopolizing atomic technology, excluding the Russians from even basic knowledge. Since the departure of Stimson from the cabinet, Bush had lacked a sponsor for his views. But the forthcoming meeting between Truman and the British prime minister gave Bush an opening. After all, Bush had been Roosevelt's point man on Anglo-American atomic matters during the war, and he possessed more intimate knowledge of Anglo-American atomic arrangements than perhaps anyone alive.

The logical contact in the administration was Byrnes. The secretary of state, while enamored of the U.S. atomic monopoly, hoped for a better relationship with the Russians. In early November, he was preparing for a meeting with Clement Attlee, who had replaced Churchill as prime minister earlier in 1945. Almost since Hiroshima, Attlee had been seeking a meeting with Truman for the purpose of reaching a joint policy on atomic power. Such a conference would update Roosevelt's wartime promises to Churchill that the U.S. would consult Britain before using the bomb and would share in the commercial and military exploitation of atomic energy. The Truman administration considered these deals obsolete and it planned to forge a new understanding with the British in several days but had not yet settled on a plan.

Bush reached out to Byrnes. He met with him on November 3 and came away dismayed that "there was no organization for the meeting, no agenda being prepared, and no American plan in form to present." Two days later, Bush, trying to fill a void, dispatched to the secretary of state a seven-page memo spelling out an agenda for the upcoming meeting with Attlee. Byrnes welcomed Bush's help, even though Bush essentially only recast what he had told Truman six weeks before.

Reviving the notion of three "secrets" to the atomic bomb, Bush proposed a three-step plan, which had as its final stage the elimination of nuclear weapons. The first step was to invite the Russians to join with the U.S. and Britain to establish an international clearinghouse for information on atomic energy. Bush had made this demand for more than a year. "This step probably costs us nothing," he wrote, and it "will give us a chance to find out whether Russia really wants to proceed with us."

The second step called for the inspection of all laboratories and industrial plants throughout the world that dealt with the materials and processes of making A-bombs. Of course, some U.S. facilities would remain closed to international inspectors, Bush stated, "until we are assured that the inspection system is really going to work."

The final step forbade any nation from building or possessing bombs. Bush envisioned using the U.S. atomic arsenal for a series of power plants. He admitted that "we do not now know how to build such plants, but presumably we will by the time we are ready for this third step." In his mind, the elimination of atomic weapons was "many years" away.

Bush's plan was cautious. In contrast to idealists who wanted the U.S. to forswear atomic weapons now, he warned against what he called the "premature 'outlawing of the bomb.'" The U.S. approach "should be realistic at every step," he urged. This meant that the U.S. must retain atomic bombs and "be in a clear position to use them promptly, if there is any chance that our enemy has them."

Besides offering a plan, Bush used the occasion to remind Byrnes that the president needed a systematic approach to formulating atomic policy. Implicitly critical of the present setup, Bush advised that the moribund Interim Committee should be replaced by a new group that would review and shape atomic policy. Headed by Byrnes, the group should include members of Congress and should have the president's ear. "I feel it is utterly essential," he wrote, "if this administration is to present a consistent and unified point of view to the public, that there should be no statements on atomic energy from the administration until after they have been reviewed by this group."

Byrnes reacted cautiously to Bush's plan, but—in the absence of another—Truman tacitly endorsed it. Bush was then asked to help rewrite the country's atomic agreements with Britain. The assignment was a hollow victory for Bush because he hadn't convinced Byrnes that the Russians should be approached directly about the bomb. Byrnes, meanwhile, had already decided to personally initiate an overture to Russia, a decision he kept from Bush.[49]

Bush sensed that his moment as a shaper of official atomic policy had passed. Despite his public celebrity and still substantial clout in government circles, Bush stood at a precipice. He prepared to jump off, or be pushed. He anguished over which way to exit the scene. Such indecision was new to him. Bush had called his own shots throughout his life. He had abandoned people and places not to his liking, damning the consequences. Of modest means as a young man, he had walked away from a full graduate scholarship when he found his adviser close-minded. After forming Raytheon, he had warred with companies a hundred times the size of his own and boasted that his tiny lab could compete with the country's biggest. During the war, he had traded blows with Ernest King, probably the military's moodiest man, and prospered. It seemed as if he had never gone against his own instincts, but now he was torn over whether to leave Truman with a dramatic flourish or to hang on whimpering.

"Intellectually, he planned to be shunted aside" with the end of the war, recalled Oscar Ruebhausen. "He asked for it. He wanted it. He felt he was out front too long as the head of science. [Many] scientists felt he had betrayed [them] by wielding the lash much too much. He felt it was important there be new faces" at the top. But as the moment of his departure approached, Bush recoiled, "his feelings hurt." Indeed, in his showdown with Truman a few weeks before, Bush had planned on insisting that, with the closure of OSRD, his ties to the government be cut. But in the final version of his letter, he had excised this request.[50]

So now he had only himself to blame. Distant from Byrnes, Bush was "very much disturbed" by what he saw as the administration's "thoroughly chaotic" handling of atomic policy. "The actual facts are that I am not consulted," he complained to Conant. More precisely, he wasn't *obeyed.* This mystified Bush since he saw his proposal as safe and sensible. While a critic of Soviet totalitarianism, he had little patience now for those too mistrustful of the Russians to take even cautious steps toward easing tensions over the U.S. atomic monopoly.

"Whether it is possible to bring Russia along into direct participation in the family of nations on a genuine basis remains to be determined, but the attempt should certainly be made," he wrote a friend after seeing Byrnes. "One cannot jump to the end result immediately, nor should one attempt to do so, for a premature introduction of a world government before the world is prepared for it would simply not work." Warning that controlling the bomb would take "not the effort of a few weeks but the effort of a generation," Bush advised that Americans earnestly tackle the job and "settle down to a long hard pull." He added: "I have very little patience indeed for those who see brilliant solutions to be obtained overnight."[51]

In Bush's view, this meant dealing directly with the Russians. The British, however, first wanted to shore up their atomic alliance with the U.S. An Anglophobe, Bush viewed this alliance as a sideshow. Still, it was his means of influencing the "great question" of the moment. When Truman finally met Attlee in Washington on November 10, Bush stood ready to assist. The fact that he wasn't at Truman's side, however, said much about his situation. Few Americans had a greater feeling for the issue of atomic policy, yet Bush had allowed a petty squabble to distance him from the president. Byrnes had leaned toward bringing Bush to the Attlee meeting, but Bush had warned that if Samuel Rosenman, the president's speechwriter and no atomic expert, showed up as an adviser, he would walk out. So neither man joined Truman at the talks.

Despite hasty preparations, the conference opened smoothly. On Sunday, November 11, Truman and Attlee agreed "in principle" to free exchange of scientific information and international control of atomic energy. The next day, Byrnes called for Bush at his Carnegie office. Bush hurried to the State Department, where Byrnes saw him immediately and told him that Attlee and Truman had reached an agreement. Bush was astonished to learn that Truman and Attlee had agreed to Bush's own plan. Incredulous, Bush peppered Byrnes with questions, but remained confused. Then Byrnes asked him to draft an official message for the conference. Bush protested that he could not report on a conference he had not attended, but Byrnes insisted.

Bush dashed off a draft in a hour, and Byrnes accepted it. The next day, however, Byrnes summoned Bush back to the State Department and sent him to see an aide who was writing another version of the same conference message. The aide had discarded Bush's work. A furious Bush skewered the aide's "very weak" version. Turning imperious, he then declared that he was no mere technician to be used only to bail out the scientifically illiterate. "I made it clear that I did not intend to be in the position of being used to correct scientific language on a document with which I was not in agreement," Bush said. Finishing off his adversary, he added that he "had been thinking about this subject for five years with some of the best minds in the country" and that the aide had obviously not "grasped the issues."

Bush left the State Department in a huff, only to return at about 4:00 P.M. at Byrnes's request. The secretary wanted Bush's opinion of his aide's draft.

Bush saw that most of his suggestions had been taken, but still insisted his original was "better." Byrnes wasn't convinced; an hour later at a general conference attended by Bush, Truman and the British he presented his aide's draft. As talk moved to larger issues, Admiral Leahy, another Truman aide, argued for outlawing the bomb. Bush, perhaps recalling that Leahy had predicted the A-bomb would be a dud, "spoke forcibly on the dangers of any such move and received general assent," and so "disposed of any such move once and for all." While in favor of the ultimate elimination of nuclear weapons, Bush opposed "premature outlawry."

For the next two days, Bush worked and reworked drafts, finally overseeing the official copy of the message on Thursday morning, November 15. At 11:00 A.M., the two heads of state, flanked by members of Congress, released the message to the press. As Truman read aloud, Bush stood quietly to one side. The message called for an exchange of scientific information with any nation that would reciprocate, but stopped short of the sort of direct appeal to the Russians favored by Bush.

While "gratified" by what he saw as "immediate progress" on arms control, Bush despaired over the administration's poor planning. Byrnes, for instance, had done nothing to remedy the weak structure of policymaking. "One very bad slip on this whole affair," he concluded, was that the secretary of state "did not get together" the atomic advisory group Bush had sought. All in all, he wrote to Stimson, "I have never participated in anything that was so completely unorganized or so irregular. I have had experiences in the past week that would make a chapter in 'Alice in Wonderland.'" Thoroughly shocked, Bush added, "It is somewhat appalling . . . to think of this country handling many matters in such an atmosphere." The whole affair made him "very much prefer to be on the sidelines, watching the show as you are, rather than deeply involved," he told Stimson. "I fear that the things I am learning these days are principally along the lines of how not to do things."[52]

Bush got his wish. He would not stand so close to the center of atomic policy again. Byrnes followed the Attlee conference with a trip to Moscow, where he espoused some of Bush's ideas. But by then, anti-Soviet sentiment among centrist policymakers had begun to solidify, rendering much of Bush's advice moot. Liberals outside government, meanwhile, found Bush too wary of atomic cooperation. "Dr. Bush would not tell the world anything of real importance about the bomb," sniffed Waldemar Kaempffert. The esteemed science editor of *The New York Times* pronounced Bush guilty of "Yankee caution."

Perhaps because Bush was no firebrand, Byrnes invited him in January

1946 to join a "board of consultants" advising the secretary of state on atomic policy. As convinced as ever that the U.S. must try to control nuclear weapons, Bush agreed to serve on the so-called Acheson-Lilienthal committee—despite the recent Attlee debacle—because he wished "to help in any way I can no matter how chaotic the affair appears." He was overshadowed on the committee by others, however, and in the meantime the political ground had shifted.

In February, a U.S. diplomat in Moscow, George Kennan, penned an 8,000-word diatribe about Russian expansionism that electrified a Washington establishment already anxious about Stalin's moves in eastern Europe. Kennan declared, "We have here a political force fanatically committed to the belief that with the U.S. there can be no permanent modus vivendi, that it is desirable and necessary that the internal harmony of our society be disrupted, our traditional way of life be destroyed, the international authority of our state be broken, if Soviet power is to be secure." Bush had called on Truman to make a deal with the Russians on atomic controls quickly, before the window of opportunity closed. With the appearance of Kennan's influential "long telegram," the window closed.

As paranoia was becoming the rage in Washington, Bush's call for a modest overture to the Soviets seemed out of date. Even if the U.S. somehow managed to forge an arms-control deal, Bush was unlikely to play a role in it. Unlike Stimson and the first generation of A-bomb policymakers, the new crowd dominating official discussions felt no compulsion to keep a scientist nearby. That traditional politics had regained the upper hand in atomic matters was made evident by Truman's choice of arrogant financier Bernard Baruch to present the U.S. position to the UN atomic agency.

Bush considered Baruch the person "least suited" for the job and dismissed his aides as "Wall Streeters." He also saw clearly that his approach to arms control decidedly favored the U.S., which got to keep its bombs, while the Russians got punished simply for seeking parity. "We are asking a great deal indeed when we ask Russia in the first stage to lift the Iron Curtain" by allowing international inspections, Bush wrote to Conant. "In return we give them only information as to our [nuclear] supplies and knowledge of the world's supplies, a considerable fraction of [information] which they probably have already." For his part, Baruch cared not at all for the insights of technocrats. Venting his scorn, the financier told Bush in March 1946 that all he needed to know about the bomb was that it "went boom and it killed millions of people." Baruch saw arms control essentially as "an ethical and political problem" and planned to "proceed on that theory."[53]

In many ways, Baruch was right. While he cavalierly boasted about "smell[ing] his way through" the technical issues of nuclear proliferation,

Baruch understood that atomic weapons had not abolished ethics or politics. Bush had premised his involvement in atomic affairs in particular, and public policy in general, on the belief that the nation must have scientists and engineers in the highest levels of government. He saw no groundswell of popular support for this (he had hardly expected one). Even technocrats themselves differed so profoundly on basic questions as to suggest that a republic of experts might yield as chaotic and slipshod an executive branch as the one run by Truman. Bush fatally undermined his case, meanwhile, by backing legislation that reinforced the suspicion that Truman could not be trusted to select sound advisers on either atomic matters or overall research policy. Bush did not mean to insult the president, but with the return of peace and political normalcy Bush's lifelong distaste for politics had resurfaced. He told a group of scientists that "professional men" were probably right to avoid "organized political activity with all . . . of the intricate interplay of the press and radio, commentators and columnists, lines of political influence, and the practical machinery of politics."

As Truman settled into his presidency, he could not be expected to suffer patiently Bush's doubts about his competence. Truman responded by pushing independent scientists—and Bush in particular—to the periphery of atomic policy. He did the same in the area of research policy, withholding support from Bush's National Research Foundation and aiding Kilgore's bid.[54]

For Truman, Bush was no longer a useful ornament. In late November 1945, just eight days after Bush stood near Truman at the Attlee declaration—looking everything like a key adviser—the president handed his pal John Snyder the critical responsibility of overseeing legislation on all matters relating to research. Bush cried foul, ostensibly over Snyder's listing of the OSRD as an "assisting agency" on legislative matters. Actually Bush wanted this responsibility for himself. He complained that Snyder's views "have been contrary to the views which I myself expressed." Bush, for instance, "vigorously" opposed Snyder's suggested revisions to the May-Johnson bill, especially one that called for the atomic agency's real authority to be lodged in the hands of a director selected by the president. Unfazed by Bush's outburst, Snyder told him that if he could not abide by the arrangement he should "discuss [the] matter directly and promptly with the President."[55]

Bush was not about to do that. He had repeatedly asked for a more clearly defined role in the administration, to no avail. His desire for a clear charter from the president was made plain by an appeal to Byrnes in November 1945. During preparations for the meetings with Attlee, Bush explained to Byrnes "my quandary": he wanted to be "an adviser who is actually consulted or, on the other hand . . . an independent individual who could then speak without being regarded as representing administration points of view." Byrnes never gave Bush the vote of confidence he desired. In February 1946, Snyder delivered another blow to Bush's ego when he endorsed the science plan proposed by Truman's old Senate buddy Kilgore. According to Snyder, the president's thinking tracked Kilgore's bill "to a greater degree" than the Bush-backed measure from Senator Magnuson. With Truman choosing others to advise him on research issues and the OSRD a lame-duck agency, Bush no longer had an official forum with which to advance his views on two issues he cared deeply about: arms control and government's relation to science.[56]

By January 1946, Bush was out of the administration's inner circle. He had so little cachet with the White House that before leaving the capital for a month on Carnegie business, he wrote only this terse note to a Truman aide: "I am headed west next Wednesday for a somewhat extended trip . . . I trust that the President will see no need for calling on me during that interval."[57]

Bush's rapid descent from power astonished his friends. Barely a year before, Warren Weaver had considered Bush among the nation's most influential people; now he was swamped by his enemies and stymied by his own inflexibility. To Weaver's surprise, in 1940, Bush had emerged on top; now just as suddenly he fell off his perch. Weaver chalked this up to the rhythms of history; they had brought Bush to shore and now washed him back out to sea.

"Every time you have a war, you really start fresh with a new organization," Weaver said. "Although in one sense that seems wasteful, I am inclined to think this is necessary. You've got to be without precedent, without all the [past] associations and connections. You've got to start fresh with a completely agile organization and of course Van is just tough enough to do this. His toughness, his willingness to fight any kind of battle, I've always supposed that this is why Van left the world as completely as he did when the war was over. . . . He just almost dropped out."[58]

To others, "forced out" was a more accurate description. In this view, Truman's aloofness staggered Bush, and Bush's alienation from rank-and-file scientists—an unintended consequence of the May-Johnson debacle—finished him off. Heartless critics thought he had only himself to blame. "Bush misjudged postwar times and could not seem to reconvert from the convenient, authoritarian military liaisons he had so laboriously built up for victory," *Fortune* opined. "The scientists of the country, backed by democratic forces, rose in fury and smashed the May-Johnson bill—and with it went most of the confidence of the fraternity in Vannevar Bush."[59]

Chapter 14

"So doggone weary"

(1946–48)

This kind of [military] organization would not be tolerated a week in a
manufacturing concern producing bobby pins. It is utterly inadequate for
the years before us; it was made to work after a fashion during the war only
by grace of the fact that strong men are strong enough to agree during war,
to sink differences in the stress of emergency.

—Vannevar Bush

Bush stooped over a tin can, his fingers dancing over the solar collector for an
unusual pump. The midday sun scorched his hands. His feet kicked up dust
on the desert turf. An unlit briar pipe hung from his mouth.

On Bush's last trip to the desert, he had witnessed an atomic explosion.
Now in February 1946 he had less momentous reasons for visiting the American Southwest. After visiting Carnegie's California outposts, he spent a week
in a dude ranch for "some sunshine and fresh air" with his wife. It was his
longest vacation since Pearl Harbor. Once again pursuing quixotic inventions, he devoted his free time to tinkering with his model for a solar pump
and studying ways to convert the heat of the Arizona sun into power for irrigation systems.[1]

The early months of 1946 found Bush in new circumstances. As *Newsweek*
noted, he was "at the crossroads." After years of too much work and worry, he
now had too little to do. While still the president of the Carnegie Institution,

he was no longer a regular in the government's inner sanctum. The Office of Scientific Research and Development, his wartime stomping grounds, was essentially dead, awaiting only a decent burial. Bush wanted a new political vehicle, but before casting about for one he rediscovered old inventions and revived dormant hobbies.

One obvious hobby was inventing itself. At home or on holiday, he could not help but indulge his habit. *Newsweek* called him "a super-gadgeteer, who putters away in a dream world extending from such hobbies as color photography to scientific turkey raising." Most of his tinkering had no commercial value. He had recently built a bird feeder in his garden, for instance, but neglected to reckon with the pigeons, which bullied away his favorite wild birds. He outwitted the pigeons by replacing the perching rods with lighter ones too weak to hold their weight. A trick spring dumped the unsuspecting pigeons groundward, while lighter birds stayed put.[2]

Bush hoped to turn his new leisure time into some small gain. He hoped to market his solar pump and a few other obscure inventions. He took special pride in a super-sharp surgeon's knife that he had made and saw great promise in a new printing technique that he claimed would reduce eyestrain. But since he invariably signed away patent rights to MIT or other educational foundations, his motive for these concoctions was pure pleasure. "I apparently cannot help inventing," he told James Killian, an MIT administrator, after completing his western trip. "Of the things that I think of three out of four go into the waste basket as soon as I think of them. Of the remainder perhaps one in a half dozen is worth attention."[3]

With more time for himself, Bush's idiosyncratic inventions flooded his mind like fine old memories. The complete list of his postwar pursuits was startling. Aside from the solar pump, he worked on a newfangled "heat" engine, which he thought might suit an automobile, but which mystified his engineering pals. He designed "a novel sailboat" with "a fixed-wing sail." He made fishing rods and hooks. He fretted incessantly about the state of a pond he built on his New Hampshire farm. He even invented "a new technique in painting," for his "own amusement," which was fortunate since no one else seemed to appreciate it.[4]

With the end of active duty in government, Bush decided to build an elaborate "personal shop" from scratch. On his return from his western holiday, he ordered new tools: 20 drills, 17 different punches, a power hacksaw, an electric hand drill and two types of lathes. He also dusted off tools he already owned. He laid out all this gear in the 20-foot by 15-foot basement of his Washington home. To prevent this wilderness of machines from devouring him, he built benches and tables and unusual sliding plywood panels that hung on overhead tracks. The panels held the attachments for the drill press, band saw and other machines. Probably the most engrossing activity in his basement workshop was an exercise in precision production. After an untold number of experiments with pyrex glass, he made a set of optical flats: three pyrex glass discs, six inches in diameter and three-quarters of an inch thick, ground and polished to a precise flatness varying no more than one-tenth of a wave length of red light or two-millionths of an inch. "In fact it is somewhat flatter than that," Bush quipped. "It is also good to look at."[5]

When not tinkering, Bush relaxed outdoors. "You will find, of course, that I am just as soft as a man gets who sits behind a desk longer than he ought to at a stretch," he told one friend before joining him for a fishing trip in West Virginia. He shared the outdoors with his wife, of course, though he worried little about her activities. "You can be sure," he confided to this friend, "that Mrs. Bush will be quite happy loafing about anywhere she happens to be."[6]

On returning from his western vacation in March, Bush began searching for a new role in the Washington firmament. Walled off now from the policy debates in atomic weapons and civilian science, Bush saw only one realm open to him: military organization and defense research. A devotee of unification of the armed services and an apostle of expert planning, Bush had "all sorts of ideas in regard to the future of research, military organization, professional armies of occupation and similar matters," he had told Conant in October 1945. For once, his bedrock views even jibed with those of the president, who in December had called for merging the Army, Air Forces and Navy into a single administrative entity.[7]

What to do with the armed services in peacetime was a widely debated question. As usual, Bush had a knack for locating profound yet messy issues. The services ended World War II on a roll, their horizons broadened, appetites whetted and leaders immortalized. The war had transformed the U.S. military from a second-rate outfit into the world's greatest power. Naval personnel grew 20-fold from 1940 to 1945, for instance. But growth in the military's horizons was more significant. No longer complacent, the services now defined themselves in terms of potential threats abroad. As the war ended in the Pacific, the phrase "national security" came into vogue as the basic category by which to assess global hazards. The new concept, arising from an expanded sense of U.S. responsibilities abroad, underscored the importance of military power but also highlighted its relations with other factors. To maintain a strong military, scientific research and basic industries had to be linked to the nation's arsenal. So did foreign policy. The perception of Russia as a serious threat to U.S. interests closed the loop.

"The gospel of national security," historian Daniel Yergin has noted, gained "meaning, substance, and focus only when directed against another country—an external threat, a foreign danger." By the middle of 1946 virtually everyone with any influence on U.S. foreign policy—from the president to congressional leaders and members of press—had come to think the worst of Russia's aims. The problem of Russia became "linked to the future of the services in an expansive definition of national security."[8]

This meant that the peacetime demobilization would still leave the services with an expanded mission. Charting this mission was made more difficult by the pressure to unify the Army and Navy and to create a separate Air Force. Competition between the services had reached absurd levels during World War II. This necessitated two routes to Japan, a major campaign to recover the Philippines that was of questionable value, and redundant air attacks on factories that naval blockades had already crippled. "For good reason, contemporaries and historians have described the war among the American armed services as often more fierce than the war against the enemy," Michael Sherry has written.

The end of the war only heightened the potential for conflict between the Army, Navy and Air Force. The atomic bomb dramatically showed all three services how the power of technology could transform their missions. Nuclear weapons elevated the strategic importance of long-range bombers, thus putting the Air Force in a privileged position to chart the new contours of war. The service that first succeeded in creating the next decisive weapon system would likely claim control of the missions it spawned and thus enjoy a larger share of the military budget. Interservice rivalry was not born in the laboratory, but the race for new weapons made cooperation between the services more difficult.

Bush saw interservice rivalry up close. Even before Pearl Harbor, he had knocked heads to persuade the Army and Navy merely to share essential data. During the Battle of the Atlantic, he had watched in dismay as the Army and Navy jockeyed for responsibility over hunting German U-boats. Rivalries rendered his Joint New Weapons Committee impotent. The JNW, chaired by Bush, reported directly to the Joint Chiefs of Staff, which Roosevelt had formed during the war as a way to coordinate military views and present a single voice in liaisons with Britain's military chiefs. But the Chiefs proved cautious and committed to consensus. The JNW suffered as a result. With the return of peace, Bush thought the defects in the JCS would only worsen. (No less an authority than General George Marshall, the Army's wartime chief, shared VB's expectations. So did War Secretary Patterson, who complained that the JCS could not even produce an overall postwar plan. The Army and

Navy "are acting independently without integration of plans," he wrote to Bush. "We are right back where we were before Pearl Harbor.")[9]

In the JNW's first meeting in 1946, Bush spoke frankly about the military's failure to coordinate during the war, warning a group of top officers, including General Curtis LeMay of the Air Force, that things would have to change. "Now the solution of having no coordination, it is just no good," he said. When they protested that the JNW body worked well, Bush sniffed that it had only seemed that way. "In time of war almost any organization will work because in time of war men will agree; that is what has happened," he reminded them. "In times of peace, it is a very different thing."

Peace would bring less funds for military research and less urgency to get things done, Bush believed. Both military waste and inefficiency, which he considered at intolerable levels already, would only grow. The military should operate under "a provision whereby a resolution can be made and decisions reached, by one man, if necessary," he insisted. Because he doubted this would ever happen, he was "anxious to step aside" and dissolve the JNW. The generals and admirals protested, trumpeting the value of the JNW and of Bush himself. One general, Borden, said if Bush pulled out of military research, it would be "a catastrophe," adding, "I have looked on you personally as somebody who welds us together."

But Bush held to his position. He wanted a single research chief for the entire armed services whose authority could not be challenged. This struck one JNW member as tantamount to setting up a "supreme authority," or dictator. "He is not a dictator or a commander, he is a judge," Bush replied. But this military judge needed extraordinary powers. "I would put him in the position to decide for the President and in his name." Military technology was too important for either top brass or Truman to handle. The president especially, Bush said, was "far overburdened" and unprepared to settle disputes between the services.[10]

Few soldiers thought they needed a civilian czar to settle the tough questions about how, as Bush put it, "the whole practice of warfare was being revised by the laboratories." As early as April 1944, senior officers had told Bush that, in their view, weapons innovation required little more than big budgets and a penchant for duplication in order to insure that researchers pursued every logical path to a solution. Bush accepted that a degree of duplication paid dividends: the Manhattan Project had pursued four separate paths to the bomb in order to raise the chances of finishing at least one. But he resented the military's new enthusiasm for government-funded research, finding fault with the rampant belief that, with enough money, every research obstacle could be oversome. Planning helped, and of course big budgets were necessary, but neither was a sufficient condition for a weapons breakthrough. "Research cannot be conducted according to advance specifications," Bush wrote a U.S. senator in January 1946. "Moreoever, it is the ability to perform, and not the cost of research, which is the important factor."[11]

Bush's austerity message played poorly with the brass, but caught the attention of civilians. In the aftermath of the war, the Army and Navy dramatically shrank. This contraction pitted fiscal conservatives against advocates of an expanded mission for the military. In 1946, Truman ordered the budget for the Army and Navy not to exceed one-third of estimated total federal revenue, which put a projected ceiling on 1947 spending at $14 billion, one-third of the budget for the prior fiscal year. In March 1947 the Selective Service Act expired. Of the roughly 12 million U.S. troops under arms at the close of World War II, fewer than 1.6 million were in uniform at the official end of demobilization on June 30, 1947.[12]

Even amid budget cuts, however, the services scrambled to make good on the vow to deepen their research ties to industry and academia. The Office of Naval Research wooed academia with lucrative grants and formed an advisory board of civilians, including Bush pal Warren Weaver of the Rockefeller Foundation. The Air Forces, inspired by von Karman, the Caltech scientist, charted an ambitious technological future based on guided missiles and laid plans for Project Rand, a civilian think-tank. The Army forged its own research agenda, prodded by Edward Bowles, who remained a consultant to the War Department. Bowles advised the Army on such diverse topics as communications, air transport and Russia's technological prowess. Besides aiding Project Rand, Bowles penned the first and decisive draft of a landmark statement by Army chief Eisenhower on "scientific and technological resources as military assets."[13]

Issued in April 1946, Eisenhower's statement reflected just how far the military had come toward embracing the substance of Bush's philosophy, if not the form. The memo's five principles read like a blueprint for the military-industrial-academic complex:

1. The Army must have civilian assistance in military planning as well as for the production of weapons.

2. Scientists and industrialists must be given the greatest possible freedom to carry out their research.

3. The possibility of utilizing some of our industrial and technological resources as *organic parts* of our military structure in time of emergency should be carefully examined.

4. Within the Army we must separate responsibility for research and development from the functions of procurement, purchase, storage and distribution.

5. Officers of all arms and services must become fully aware of the advantages which the Army can derive from *close integration* of civilian talent with military plans and developments.[14]

Bush could hardly argue with the Bowles-Eisenhower statement, and he took pride in the Army's newfound faith in research. But he constantly carped about feckless military research, undermining his credibility with the brass. Bowles privately predicted that Bush "may antagonize the Army and Navy before he gets through." Fanning the flames, Bush gave little credit to soldiers for their newfound embrace of science and technology, instead decrying their smug confidence about the inevitability of innovations. He dwelled on the military's shortcomings in World War II. While the U.S. had "fought a good war," he reminded Truman, ". . . we nearly lost it." He also railed against the unwillingness of the three services to mesh their individual plans. This lack of coordination, which he saw as a threat to the nation, was perhaps best illustrated by the effort to build guided missiles.[15]

The U.S. had ended the war eager to build upon Germany's successes with its V-1 and V-2 rockets. But Bush undercut official enthusiasm for missiles by playing down their military potential. His skepticism was tolerable during the war, when the nation strained simply to finish the many advanced weapons under development. But after the war, the U.S. could have mounted a focused, all-out missiles program along the lines of the Manhattan Project. While many barriers prevented a unified push to produce a ballistic missile capable of reaching Russian soil, Bush did his part to doom a coherent approach by making a sweeping prediction on Capitol Hill.

"Let me say this," Bush told a Senate committee in December 1945:

There has been a great deal said about a 3,000-mile high-angle rocket. In my opinion such a thing is impossible and will be impossible *for many years.* The people who have been writing these things that annoy me have been talking about a 3,000-mile high-angle rocket shot from one continent to another carrying an atomic bomb, and so directed [by a guidance system] as to be a precise weapon which would land on a certain target such as this city [of Washington]. I say technically I don't think anybody in the world knows how to do such a thing and I feel confident it will not be done for a very long period of time to come. I think we can leave that out of our thinking. I wish the American public would leave that out of their thinking.[16]

Coming on the heels of the triumphant Manhattan Project, Bush's opinion carried great weight with Congress. His aim was to dissuade lawmakers from accepting prophetic visions of doom espoused by General Henry Arnold and other advocates of "push-button" war. To Bush, it was folly to believe that soldiers in distant command bunkers could direct lethal rockets against an enemy with the touch of a finger. The basic technologies of rocket propulsion and missile guidance were many years away from perfection, he believed. On what scientific basis he made this judgment was not clear, but he openly bemoaned that the "general public has now gotten a Buck Rogers slant on a possible warfare of the future that is going to be very hard to overcome."

As time went on, Bush's antimissile bias hardened. In January 1947, he told War Secretary Patterson: "The tendency of the American people to seize upon slogans and rationalize their way out of predicaments by facile expressions gives such a term far more force than it would have in a country that did not have these characteristics." It was "highly unfortunate," he added, that many thought the next war "would be completely an affair of dials and electronic tubes." He blamed Arnold for spreading these misconceptions, saying the general "did a great deal of disservice by his emphasis on the long-range controlled missile."[17]

The concept of "push-button" war upset Bush because he thought it embodied the peculiarly American notion of pain-free combat. "One difficulty is that the American people by reason of their general tradition and background think of wars as one-way affairs to a considerable extent," he noted. "I would be willing to bet that if the average citizen were queried about his idea of a 'push-button' war he would almost assuredly present it in terms of a group of Americans pushing buttons and would ignore the idea of an enemy pushing similar buttons at the same time."

Bush was not against studying push-button war, but wanted this done deliberately. He viewed talk of atomic missile wars as premature and irresponsible. While the technical problems were daunting, he considered military rivalries the greatest obstacle to overcome. Duplication was rife. The Army ground forces and air forces each had their own missile programs, with the work divided on arbitrary lines. The Navy had its own missile program, adding to the confusion.

The three distinct efforts sapped U.S. vitality in a young field relatively starved for funds. "Our scientists are alarmed at the prospect of our meager appropriations being dissipated in too many parallel enterprises," one journalist noted. The services seemed powerless to improve upon the situation. After reviewing the Army's programs in May 1946, a brigadier general found them hampered by "duplication of effort and inefficient utilization of available facilities." Even worse, "no successful guided missiles have as yet been developed in this country." By contrast, the Russians were well along toward achieving a unified, intense missile program.[18]

Bidding to regain a central role in military affairs, Bush tried to sort out the messy missiles program. In June, he accepted a joint appeal from Patterson and Forrestal to head a joint Army-Navy research board. Though the board had no congressional mandate, Bush launched the Joint Research and Development Board (JRDB) in the belief that it had the power to settle differences between the two services in "the whole field of research and development." Writing to the president of Johns Hopkins, Bush boasted, "I say it is authoritative and I certainly mean just that. . . . It was only when the matter was put in this form that I agreed to take the chairmanship, as I had become weary of debating societies." Nonetheless, Bush doubted his ability to tame the military, on missiles or anything else. Before accepting the job, he told Forrestal that he planned to lean heavily on a deputy because "I am exceedingly weary after war years without a real respite."[19]

The assignment proved no balm for Bush's weariness. Missiles were a mess. The technical issues were tough and divisive. Faithful to his wartime experience, Bush formed a committee to sort out the situation. The group's aim was to build support for a single, national missile test site that would serve as the focal point for research and development. But this proposal immediately set off alarms because each military branch wanted its own test site. Choosing a chair of the committee wasn't easy either. Bush's top choice was ill, so he turned to old standby Karl Compton, MIT's president. In a measure of the task's importance, Compton accepted, though only because "Van is in a jam."[20]

The jam was partly of Bush's own making. His technical sense, unerring during the war, deserted him now. Was it possible that his very success as an innovator in one technological era blinded him to the next revolutionary wave? Bush would deny such a charge, but his skepticism about long-range missiles bordered on obstinacy. Moreover, he inexplicably failed to see the military potential of satellites and related technologies. By extrapolating from existing rocketry, the trend toward satellites and space communications was clear. Bush's relative unfamiliarity with the basic technologies of rocketry was no excuse for his poor prognostications. He had known little about atomic fission when Roosevelt gave him responsibility for the "uranium" project in 1940 and yet had emerged as the chief bureaucratic driving force behind the creation of the Manhattan Project. Perhaps, as one missile enthusiast claimed, "Bush's wartime experience must have gone to his head."[21]

Bush's signal achievements during the war—and his celebrity afterward—may have dulled his improvisational edge and reinforced his arrogance. Yet his postwar skepticism toward grand technological claims also had a simpler explanation: it was rooted in his fiscal conservatism. In peacetime, he looked askance at big-ticket items. Space technologies were inherently expensive. They depended on breakthroughs across a wide range of fields: materials, electronics, propulsion and programming.

Eager to satisfy Truman's desire to hold down spending, Bush refused to take the lead in space weaponry. He even shunned the subject. In the spring of 1946, the Navy's satellite chief, who was pursuing a simple path to space using a single-stage, liquid-hydrogen rocket, cast about for funds and partners. He repeatedly tried to meet with Bush, who refused to discuss the satellite proposal. Bush's rebuff was all the more exasperating since the Navy's satellite advocate had hashed out his ideas some months before with Lloyd Berkner, Bush's executive secretary at the JRDB. Not even Berkner could budge Bush.

The Navy instead turned to the Army Air Forces, which asked Project Rand for an assessment of satellites. Rand issued a prescient report in May 1946. With "realistic engineering" and current technology, a 500-pound spacecraft could be placed in a 300-mile orbit around the earth, Rand predicted. The project would take five years and cost $150 million, but would be a bargain. A U.S. satellite, circling the globe, "would inflame the imagination of mankind and would probably produce repercussions in the world comparable to the explosion of the atomic bomb." Satellites had plenty of scientific uses, but Rand highlighted the espionage value of "an observation aircraft which cannot be brought down by an enemy who has not mastered similar techniques."

Despite such strategic potential, satellites never captivated Bush. The services, meanwhile, spent their enthusiasm on fruitless infighting. In June 1948, the Navy, squeezed by budget cuts, killed its satellite program. About the same time, the Air Force rejected a satellite proposal. Bush supported both decisions. In ignoring satellites, he cited the judgment of an RDB "technical evaluation group," which concluded that "neither the the Navy nor the [Air Force] has as yet established either a military or scientific utility commensurate with the presently expected cost of a satellite vehicle." To Bush, it seemed that a boondoggle had been avoided. He was so confident of his opinion that a year later he publicly mocked satellite boosters for peddling fantasies. "We even have the exposition of missiles fired so fast," he sneered, "that they leave the earth and proceed about indefinitely as satellites, like the moon, for some vaguely specified military purposes."[22]

With Bush defining himself as a technological conservative, the JRDB became a lightning rod for criticism. Bush took his formal charter too seriously and misinterpreted it to boot, frustrating senior officers who had no intention of relinquishing pet projects to satisfy Bush's appetite for efficiency. He viewed the board as "a judicial body [designed] to adjust differences" between the Army and Navy and as a body whose "orders become binding upon both departments." No one that mattered in the military accepted Bush's judicial metaphor, preferring to decide for themselves whether to accept or ignore its advice. A Joint Chiefs staff officer criticized the board for examining "the War and Navy Departments for strategic concepts and plans for the purpose of forming their [R&D] programs around such plans." Admiral Nimitz, chief of naval operations, convinced the Joint Chiefs to nix Bush's plan to empower his board to issue security classifications. Other military officers accused Bush of "reaching out for power and influence." Bush poured oil on the fire by asking Patterson to cut the size of funds requested by the Army for research. More attuned to the nuances of bureaucratic politics than Bush, Patterson refused.[23]

All these defeats were hard to take. "Bush opposes anyone who stands in his way," but the military "is not one to be pushed around easily," observed Bowles. Even if Bush had been more savvy in his dealings with the brass, however, he hardly stood a chance of disciplining military research. Despite the success of the centrally managed Manhattan Project, the individual services refused to place the missile effort under one organizational roof. They preferred a fragmented approach that held open the possibility that each service would retain operational control over some missiles. General Curtis LeMay typified this sectarian attitude when he reacted to a recommendation by a fellow general that the Army, Air Forces and Navy unite behind a single missile program. The architect of the firebomb raids against Japan, LeMay wanted to retain a key role for bombers in the coming missile age. The surest route to this end, he sensed, was to create the missiles themselves. "We need not go to the extreme of setting up an agency directly under the President," he wrote to air chief Spaatz in May 1946.[24]

The perception of rational planning as an "extreme" measure maddened Bush. He conceded that centralization wasn't always the answer but denied that planning was antithetical to innovation and esprit de corps (as some of the brass claimed). Using as an analogy the case of a private industry weakened by rampant competition, he made a compelling case for *containing* (not eradicating) rivalries between military branches. "We do not wish a dead level of monopoly and stagnation in our military affairs, any more than we do in our automobile industry," he wrote to the service chiefs. "But to render competition reasonable is not necessarily to kill it, and to let it run wild is dangerous." The stakes were high, he noted. "Rivalry, on a proper basis, can be an asset, but if it ever degenerates into an acrimonious struggle it can wreck our military strength."[25]

No one thought the U.S. military would self-destruct, but many believed that the level of interservice rivalry was dangerously high. In such a climate, few paid attention to Bush's calm distinction between rational and irrational competition. As new converts to scientific thinking, the services vied with each other to lure the most prestigious researchers into their camp or assemble the most star-studded advisory boards. The military had acquired the research religion, but Bush feared they worshipped their new god blindly. Instead of seeking "a professional partnership" with scientists and engineers, "military men [thought they] should learn to use scientists," Bush observed in an April 1947 speech at the Armed Forces Staff College. "I immediately ask what is meant by the word 'use.' . . . Is it intended that there should be two groups, of which one will think about military matters, and the other will think about scientific matters only? I can merely say that it will not work, or that it will not work for long. Scientists who find out that an attempt is being made to use them in this sense will merely leave. Any that remain will not be worth using."

Bush had fought against segregating scientists from strategy for the entire war, and all his postwar plans were built on the assumption that the highest levels of the military would rely on research leaders. Bowles, von Karman and other researchers were proving adept at working within existing military structures, informally advising top soldiers while taking a back seat organizationally. Their success threatened Bush and encouraged the belief of scientifically illiterate soldiers that they could indeed set research goals and plans. Boiling over, Bush told the assembled Army officers, "There is nothing more disagreeable to the truly skilled and expert man in any field than to deal with a dilettante." Too often, he sneered, a soldier presents himself "as an expert on any subject that appears, in spite of the fact that he is dead wrong or completely ignorant just as soon as one dips below the surface at any point."[26]

For Bush, the disorganization of military research was only part of a larger planning problem. Disturbed by the inconsistencies and shortcomings of military strategies and tactics, Bush preferred a more corporate style of management, in which "responsibility and authority must go together." In April, he circulated a paper entitled "Military Organization for the United States," in which he advocated a unified military command. The paper drew some barbed reactions. William Leahy, Truman's military adviser, recoiled at Bush's insistence on a single overall military chief, saying it raised the specter of "a 'Hitler' in uniform."[27]

Bush wanted to create a unified military under the command of a single chief who would report to the president. The proposal seemed too radical, disregarded the demands of the Army, Navy and Air Forces and failed to account for the willingness of most Americans to tolerate a degree of waste and duplication in government. Like Leahy, Truman also feared the emergence of a military dictator, or what the president called "a man on horseback." In July 1947, Truman approved a watered-down version of the unification model, the National Security Act of 1947. The act created a National Military Establishment, whose secretary could establish only "general" policies and exert "general" control over the individual services, which retained great autonomy (and also grew in number from two to essentially four since the Air Force (the *s* now dropped from its name) was granted equal status with the Army and Navy and the Marines were granted statutory protection). Interservice rivalries remained, however. The Joint Chiefs of Staff could hardly impose uniformity but still had great power. Though mainly advisers, the Chiefs possessed the statutory authority to prepare strategic and logistic plans; assign responsibilities among the services; establish unified commands; set policies for joint training and "review major material and personnel requirements of the military forces." The JCS, in other words, could advise the new secretary of defense and the president on the military budget.

The National Security Act of 1947 also created the National Security Council, the Central Intelligence Agency and a third agency called the Research and Development Board, which many people overlooked but Bush considered the most important. The RDB was charged with meshing the research plans of the individual services and advising the defense secretary on scientific and technical matters.[28]

Though disappointed in the narrow powers given the defense secretary, Bush improbably hoped that Truman would select him as the first secretary of defense. "Bush wants the job in the worst way," Bowles noted in his diary. Stanley Lovell, the OSS liaison with Bush during the war, urged Truman to give Bush the top defense post, and Bush himself was hopeful enough that he asked at least one colleague to work for him in the Pentagon.

Truman never considered Bush for the post. His first choice was Robert Patterson, the war secretary. When Patterson declined the job, Truman turned to Navy secretary James Forrestal. *The New York Times* listed Bush as a possible candidate only if Forrestal withdrew, but he accepted the president's offer.[29]

This left open the chairmanship of the Research and Development Board. Bush wanted this consolation prize, thinking that the National Security Act provided enough power for the RDB chairman to discipline defense research. Despite Forrestal's support, Bush was not a shoo-in. His tenure at the JRDB had been stormy and Truman's closest advisers still feared Bush's ambitions and penchant for overstepping his bounds. The president even jokingly told

Forrestal that, if he picked Bush, "he better look out for [Bush] or [he] would run away with the whole affair."

As usual, Bush walked a fine line between asserting professional independence and displaying his ego. It all added up to the same result: Bush was hard to maneuver, and everyone knew it. "The Palace Guard at the White House" opposed putting Bush in charge of the new research board precisely because, one journalist observed, he "was not amenable to political pressure in handling his work."[30]

Bush's frustrating foray into defense organization and national security fired his interest in an even more momentous topic: the future of American democracy. Bush wondered whether the U.S. system of government was strong enough to handle postwar challenges or even whether representative democracy suited the nation's new requirements. In an increasingly complex world, in which experts of all types played a growing role in public and private decisions, it seemed to him that neither the president nor Congress could meet the demands on the U.S.

First, the president was overwhelmed. "One of the great defects of the government system," Bush told a group of military men in early 1946, "is that we have burdened the president to the point where we have thrown responsibilities on him that he can't possibly carry. We have thrown powers and duties on the president to the point where it is an absurdity." Indeed, Truman personally relied on a staff of just a dozen aides, all generalists.

Bush proposed a novel solution. The president would select ten "agents," none of whom would run government departments, who would simply make decisions in the name of the president. "I would choose them of the highest caliber I could get," Bush explained, "and I would write a set of directives to these fellows, delegations of authority, and in that way take about nine-tenths of the work off [the] back" of the president. "I think that is the only hope" of achieving better executive decisions, he concluded.[31]

In December 1946, Bush actually asked Truman to assemble such a staff. With the aid of Oscar Cox, an expert on presidential authority, Bush prepared for the president a memo entitled "A Presidential and Cabinet Staff." In it, he argued that "the institution of the Presidency . . . does not have adequate staff work." His premise was straightforward: the growing complexity of public-policy issues made politicians less and less able to make informed decisions. He proposed that the president hire a "top quality" staff that would be free from "ordinary operating functions" and instead see that presidential decisions were executed; review legislation; prepare for cabinet meetings; and "resolve or arbitrate differences between departments." Cox opined that the president required no legislation to expand his staff.

In arguing for an expanded role for experts in the White House, Bush anticipated that as public policy questions became more complicated the power of traditional politicians would decline. Truman evidently had a different vision. He casually called the memo "interesting," adding, "I am happy to have it." Yet he ignored it.

Bush felt that members of Congress suffered from the same problem afflicting the president: they were too short of time and preparation to tackle the complex issues they faced. In testimony to Congress in March 1947, he bluntly drew the implications of this, declaring: "In the long run and in general, the Congress should, as representative of all the citizens, make decisions of policy. It is obvious, however, that the Congress lacks sufficient time to inform itself as to the various considerations involved in making these decisions. It must therefore delegate to others the power to make them, retaining its ultimate control through annual reports and appropriations. In my opinion, this delegation should not be to one man but should be to a group of the ablest [people]."[32]

In his critique of the presidency and Congress, Bush clearly displayed his elitism. His views jibed with those held by other establishment figures such as George Kennan. Bush was simply more frank and direct than most. He shared with other elitists a stark and not altogether distorted view of American society that pitted sober, pragmatic elites against the untutored, volatile masses. For Bush, Truman and his cronies as well as most congressional leaders clearly fell into the "masses" category. While Truman delighted in casting himself as an ordinary American, Bush—and other elite leaders—tended to view such citizens as irresponsible and sometimes irrational. The elite assumed that the mass of Americans needed patriarchal authority.[33]

In Bush's view, civilian technocrats were the solution to the inherent contradiction between the increasingly complicated problems facing government and the nation's democratic traditions. In practice, this meant that the public must pay for experts to make decisions in its name; these experts would brook little or no interference. Bush thought that changes in representative democracy were inevitable. The fight for science legislation—where researchers were asking for no-strings funding—exposed most clearly the contradiction between technocracy and democracy. But this same contradiction bedeviled debates over civilian control over the military. In the case of atomic energy, for instance, neither the civilians nor the military proved accountable to elected representatives. As Philip Morrison, a Los Alamos physicist and May-Johnson opponent, came to realize, "The civilian administration of the Atomic Energy Commission couldn't have been any more militaristic."[34]

Neither Bush's critique of democratic leaders nor his implicit call for technocracy amounted to a full-blown philosophy of government. He brought an engineer's pragmatism to these problems and valued action over conceptualizing. Since he read almost no history or philosophy, he brought little of the past to bear on his thinking. Indeed, his horizons at times seemed to go no further than his life experiences from the start of World War I to the end of World War II. And for this last conflict, Bush grew nostalgic. He yearned not for the battle, of course, but for the camaraderie with like-minded men and the way in which wartime emergency had silenced both the predictable imbecilities of national politicians and the plaintive demands of ordinary people. The war had brought cohesion to the body politic. Its end revealed the fissures obscured by war: differences in personalities, philosophies, special interests.

Bush felt uncomfortable in this new world and longed for the old. "I am still at it . . . but without the drive and enthusiasm of war," he wrote in April 1948. "I think it is neccesary to be at it, and I am glad I can still be in the affair. But times are not the same." Indeed, not. A lunch with the deputy war secretary reminded Bush of the times spent during the war with the then deputy and now secretary, Patterson. "It was a great experience, this business of fighting a war," Bush wrote him. "It had its tough moments and its rather appalling responsibilities, and of course it had its very sad side in which you and I saw friends who lost their boys and the like. But there was one part of it that was on the other side of the ledger. The privilege of working with *worthwhile* men in the common cause." To Stimson, Patterson's boss and a hero to many in Roosevelt's administration, Bush wrote simply "I miss you."[35.]

Bush's nostalgia for the war erupted in various ways. The most striking instance was his gala black-tie stag party on January 20, 1947, to commemorate the official closure of the Office of Scientific Research and Development. "The old town isn't what it used to be," Bush declared at the party, during which Admiral Nimitz pretended to nominate him for president. Joking aside, Bush had practical reasons for looking backward. He believed that only during war would Americans toe the line voluntarily. American security now depended on the links between industry, academia and the military; only through the mutual embrace of these three sectors would the U.S. outpace the Russians. Yet to achieve these links, Americans had to accept government intrusions into science, industry and other fields. In peacetime, Bush worried, citizens would reject these intrusions. "Under the pressure of war people flocked from their usual affairs to support government's effort, and all sorts of things can occur as a result," he wrote a friend, "but in times of peace it is a much tougher thing to get the sort of thinking that needs to be present in any such situation as this and under government auspices. Here I think resides the real difficulty."[36]

To solve this difficulty, Bush had advanced a vague but durable notion about the capacity for experts to govern themselves. This concept of expert self-governance was Bush's answer to the contradiction between the public's demand for accountability and the expert's need to make judgments free from outside pressure. It was a radical answer with a conservative core that stood the whole idea of public service on its head. Rather than formal accountability, it stressed results. The public interest would best be served by communities of elitists responding to public needs in accordance with the dictates of expertise as defined by that particular professional group. Bush's admittedly fuzzy concept had its roots in efforts by late-19th-century and early 20th-century progressives to upgrade government by professionalizing bureaucracies and blunting the significance of partisan politics. But it also reflected his experience in academia: after all, a university department functioned along such lines.

The failure of Bush's science legislation to win passage illustrated the entrenched resistance to having experts on top, not just on tap. As early as 1912, Woodrow Wilson, campaigning for the presidency, had voiced a common anxiety: "What I fear is a government of experts," he declared. "God forbid that in a democratic society we should resign the task and give the government over to experts. What are we for if we are to be scientifically taken care of by a small number of gentlemen who are the only men who understand the job?" The successive crises of the Depression, World War II and the emerging Cold War, however, prompted what one historian has called a "shotgun marriage . . . between professional experts and the nation's public bureaucracies." This marriage would produce a new kind of politics, with experts mobilizing public support through a variety of means rather than reacting to concerns voiced by the public and politicians.[37]

In late 1947 there were signs that Bush's elitist reformulation of representative democracy might not be needed to insure national security. The threat of push-button nuclear war—instant annihilation at the hands of a distant enemy—might so frighten citizens that they would suspend their normal suspicion of technocrats. With Russia increasingly viewed as the nation's chief adversary, the potential existed for politicians to mobilize Americans by resorting to fear. Of the prospects for this, Bush claimed ignorance. But he was curiously deaf to political shifts, noted Forrestal, the new defense secretary. "Even with both ears to the ground," Bush did "not hear the rumble of the distant drum."[38]

The postwar battle for control of civilian science policy proved to be a striking example of Bush's political deafness. Still president of the tradition-bound Carnegie Institution of Washington, Bush had a good salary and a secure position that freed him from depending on the government for a living. Unlike his attitude toward military affairs—in which he sought and definitely wanted a formal role in research organization—he had no personal designs on whatever agency would direct civilian science. But he badly wanted such an agency created. His failure to understand the contending viewpoints in the foundation debate hampered his effectiveness. Just as he was surprised by the revolt of rank-and-file scientists against the doomed May-Johnson bill, he failed to realize the extent to which older, conservative science leaders, such as Frank Jewett, had dug in their heels. They respected Bush for his war work, but vehemently opposed the continuation in peacetime of an OSRD-like agency. At the same time, a relatively small group of younger scientific activists thought Bush's approach to science legislation was too conservative. Bush took it for granted that these scientists would fall in line.

In the summer of 1945, Bush had bungled the chance for quick passage of a science measure by double-crossing Senator Harley Kilgore, the leading advocate in Congress for a populist approach to research. Kilgore wanted public ownership of patents, support for social sciences, geographical dispersal of funds to help public universities and a greater emphasis on applied research to meet purely social needs. Bush wanted patents retained by those who performed the research, a decided emphasis on the physical sciences and no explicit connection between the foundation and the marketplace. While philosophically at odds, Bush and Kilgore seemed ready to forge a compromise before Bush chose a go-for-broke strategy that cut out Kilgore, forcing him to promote his own foundation bill. By early 1946, however, supporters of Bush and Kilgore saw the pointlessness of competing bills and were poised for a compromise.

In mid-January, Bush and Kilgore met for the first time since their open break the previous summer. While ready to make concessions, Kilgore nonetheless ripped into Bush, accusing him of "bad faith" and "behind-the-back actions," of which Bush certainly was guilty. After Kilgore had vented his anger, he settled down to business, agreeing to back a single bill that was more to Bush's liking. A week later, the two men met again, with their aides, and hammered out the details of a compromise. At one point, Bush threatened to kill the talks if Kilgore refuse to concede on the foundation's administration. After much negotiation, a deal was struck. Bush's side thought "Kilgore surrendered," while Kilgore's aides and supporters in the Bureau of the Budget judged that the senator had made only minor concessions "designed to save the faces" of Bush and his allies. On February 21, 1946, both sides endorsed a compromise bill.

The Senate, however, failed to address science policy until July. By then conservative scientists had punched holes in the bill. Frank Jewett, the AT&T research executive and National Academy of Sciences president, called the Bush-Kilgore compromise "the perfect vehicle to socialize and nationalize a large and independent section of our economy." His views were echoed by a group of staunch Republicans, who decried the bill's "philosophy of centralization," describing it as a "link in the chain to bind us into the totalitarian society of the planned state." In the House, meanwhile, where Representative Wilbur Mills backed a science bill close to Bush's original conception, liberal scientists led the attack. Edward Condon, director of the National Bureau of Standards, made a veiled criticism of Bush, saying that the bill would "lead to an increasing monopolization of science by a small clique." Another activist, Howard Meyerhoff, railed against the bill in the pages of *Science* magazine and privately led what Jewett called a "vicious attack" on Bush.

The split between the scientists doomed the research bill, at least for 1946. By the summer, Bush was resigned that legislation, if enacted at all, would be handled by a new Congress following midterm elections in November. That was all right with him. While he clearly preferred the Mills bill, he found that the Kilgore compromise measure was "not at all bad considering where we started from."[39]

But once again, Bush failed to account for the shifting political terrain. During the squabbles over civilian science legislation, the military had not stood still. Bush had made a convincing argument that the government ought to support blue-sky research by civilians and stay abreast of leading-edge technical trends that might influence the future course of war. Bush had wanted these responsibilities lodged in the hands of politically insulated civilian technocrats. Convinced of the basic concept, the military saw a means of more swiftly achieving the same end. In the absence of a research foundation, the armed services would fill the breach, spawning a new breed of military technocrats that would dominate science funding for the next 20 years.

The Office of Naval Research, created by Congress in 1946, proved to be the archetypal military patron. Responsive to the rhetoric of intellectual freedom, the ONR wooed academic scientists relentlessly. Academics reacted coolly to an ONR entourage in late 1945, fearful of military control. But then the Navy's university liaison, Captain Robert Conrad, made an extraordinary package. He promised that the Navy would pay for the full cost of research, fund mostly projects suggested by scientists themselves, allow most research to be published openly and fund all fields of science for the long pull. It was an offer no leading university could refuse. None did.

In short order, the Navy became the nation's chief patron of "pure" science. By July 1946, it had agreed to spend $10 million on academic research and to set aside another $25 million for the following year. Of the first-year contracts, about 40 percent went to nuclear studies, 14 percent to electronics and about 10 percent to guided missiles. The rest of the funds were scattered among chemistry, physics and computing. Living up to its bargain, the Navy encouraged wide-ranging studies, often with little direct relevance to weaponry. Roughly half the contracts were of the no-strings type. This freewheeling approach extended not just to academics but to the naval officials managing the program. Behaving like philanthropists, they doled out money with little regard for the Navy's core mission, choosing projects based on their own intellectual judgments or those of civilian advisers. "We all had infinite freedom," one naval official said. "No one told us what to do. We decided what had to be done and proceeded."[40]

The Navy was on its way to becoming what *Newsweek* called "the Santa Claus of basic physical science." It was not hard to see why. Navy spending on university research dwarfed prewar peacetime budgets, when just a dozen universities in the entire country spent more than $1 million each for *all* of their on-campus research. The Army had no central research organization, but quickly imitated the Navy. By the fall of 1946 it had earmarked about $70 million, or one-fourth of its research funds, to universities.[41]

The military's new image among academics pleased the top brass, who dropped hints that perhaps Bush's brainchild was not needed. Soon after his retirement as chief of naval research, Harold Bowen told *Newsweek* that a civilian foundation might actually impose *more* regimentation on researchers than the military. Bowen was no friend of Bush. He had tangled with Bush before Pearl Harbor and actually been stripped of his job for not cooperating with civilian scientists. "You could move the Office of Naval Research to another building, put up a new sign on the door, reading 'National Science Foundation,' and you would have the nucleus of such an agency," he said.

As disputes further delayed the creation of a civilian foundation, the Navy adopted almost an official policy to safeguard the interests of university researchers even as it sought to further the aims of national security. "Until such a foundation is established it seems to me inevitable that the [ONR] must act in a dual capacity and shape its policies accordingly," Admiral Lee, Bowen's replacement, wrote in January 1948.[42]

Bush had mixed emotions about the sudden emergence of the military as academia's patron. A few scientists altogether refused to accept military money.

Bush's prewar pal, mathematician Norbert Wiener, wrote a researcher at a defense aviation firm in late 1946 that he planned to boycott all military research and even withhold publication of "future work of mine which may do damage in the hands of irresponsible militarists." When the *Atlantic* published Wiener's letter under the title "A Scientist Rebels," the antimilitary position gained credibility among scientists, but no more than a small minority ever joined Wiener's cause.[43]

Some scientists indignantly defended their efforts on the military's behalf, arguing in essence that their actions transcended morality because of the special character of their expertise. "The scientist can no more choose whether he works for war or for peace than the Western Electric Company can choose whether the telephone instruments it manufactures are used on domestic circuits or as Army phones on the field of battle," wrote physicist Louis Ridenour, an adviser to General Spaatz.[44]

Bush's own view was more nuanced. He favored civilian rule over the scientific estate but accepted military funding in principle. It was true, of course, that inventions could be put to evil or good uses, but the more profound question involved the extent to which the patron of basic research influenced the construction of knowledge. Bush categorically accepted the validity of any or all military research, but he insisted that the question of who sponsored research mattered. As a practical matter, he applauded the military for tiding over civilian science until Congress approved a research foundation, especially since he believed the country simply lacked the capacity (in the form of talented researchers) to absorb any more funds. For two reasons, though, he wanted military support for basic research to be a "temporary system." First, he suspected that the services might balk at being supplanted by civilians at some future date. Second, he believed support from a civilian research patron might wither if academic scientists were thriving on a diet of military funds.[45]

Whether the military would ever withdraw from basic research was an open question. It could probably only happen through an act of Congress, and that was unlikely to come. The military's funding of basic research had already reduced the sense of urgency for a civilian research foundation. There was another obstacle: Truman's inability to stake out a clear position on civilian science also clouded prospects for a foundation. Sparing Truman from criticism, Bush blamed the administration's weakness in this area on labor mediator John Steelman, who since June 1946 had overseen virtually all executive issues dealing with science and technology.

Genial and ingratiating, Steelman had been a professor of sociology and economics at a small Alabama college. Truman initially tapped him to help deal with the wave of labor unrest following World War II. In the summer of

1946, Steelman supervised the staffers assigned to shepherd McMahon's atomic energy bill through to passage. An alternative to the original May-Johnson bill, McMahon's legislation was enacted in July 1946 after concessions to the military (the biggest being the creation of a "military liaison committee" to advise the civilian commission and review its weapons plans) and the deletion of all references to the need for research on social, political and economic effects of atomic energy. While these strategic retreats vindicated Bush's confidence in May-Johnson, the resulting Atomic Energy Act discarded Bush's central organizational principles by creating a full-time civilian commission whose members were appointed by the president.[46]

With the passage of atomic energy legislation Steelman began a review of government-funded research programs in the hope of producing a report that might supersede *Science—The Endless Frontier*. He didn't like what he saw. Echoing Truman's own bottom-line scrutiny of wartime defense production, Steelman reported that "federal research activities" were now "vast," with 30 agencies spending more than $1 billion a year. "There is ample evidence that the total research program . . . is too often haphazard, lavish and uncoordinated," he told the president.[47]

In October 1946 Truman named Steelman as chairman of a new President's Science Research Board. Blunting Bush's still-formidable influence on civilian science policy ranked high among the goals of the new board, which one administration staffer admitted was a means for "coming to terms with Bush."[48]

One of Steelman's first acts as chair of the advisory board was to undertake a broad study of research in America. The report was essentially a reprise of *Science—The Endless Frontier,* but with a populist spin that emphasized the public's right to get proper value from its research money. Although a Truman aide had encouraged the president to order another study even before the release of *Science—The Endless Frontier* in 1945, Steelman's plans in 1947 for a new multivolume report seemed excessive to some colleagues. The only reason for a fresh report on research, one sniped, was to give Steelman's advisory board the chance "to show its heels to the famous Bush report."[49]

Bush disliked Steelman. He considered him uninformed about civilian science and presumed that his only qualification to opine on the subject was his personal tie with Truman. For his part, Steelman harbored a desire to put uppity scientists in their place, a trait he shared with Harold Smith (who opposed Bush at virtually every turn until his departure as budget chief in 1947). Steelman made a show of consulting Bush, whom he named to his science board. But the two men never saw eye to eye. "I'm pretty sure Steelman had no use for me," Bush said later. "I don't know how he ever got on the staff of the President and he had no standing among scientists generally."[50]

Bush tried to keep a lid on his disagreements with Steelman. One day he visited the White House with Phoebe to receive a World War II medal. The couple was waiting for Truman in an anteroom when Steelman turned up and shook hands with Bush, greeting him warmly. Bush returned the greeting. After Steelman left, Phoebe asked, "Who was that?"

"That was John Steelman," Bush replied.

Phoebe was startled to hear the name. "I thought you didn't like John Steelman," she said.

"I don't," Bush said, "and he hates my guts."

"But you greeted each other like old friends."

"That's the custom."

Bush privately trashed Steelman. "When Truman leaned on [Steelman] for his contact with science, he lost his contact with science," he told one friend. Of Steelman's multivolume report, released in late August 1947, Bush had few kind words. He grudgingly favored Steelman's request that the president consider appointing a "scientific liaison"; Bush approved of the position provided a worthy person filled it. But overall, the report, *Science and Public Policy: A Program for the Nation*, struck Bush as shallow and predictable. He labeled it Steelman's "personal report" and told an academic that he "had practically nothing to do" with it. The report "does not summarize scientific thought in this country," he bitterly wrote Forrestal. "It expresses the opinions of a small group within the government. Very few scientists were consulted during the preparation of the report and the report itself shows this clearly."[51]

Both houses of Congress passed a research bill in July 1947, calling for a "National Science Foundation." The name was Kilgore's but the law resembled Bush's original plan, especially on patents and control by part-time scientists. Bush hardly had time to celebrate, however, since Truman killed the bill on August 6. The president complained that, among other things, the bill authorized a presidentially appointed board—not the president himself—to choose the foundation's chief officer and make its key policy decisions.[52]

Truman's objection struck some as splitting hairs, especially since, in the words of a budget official, "at no time did the President indicate to Congress that this form of legislation is unacceptable to him." But the president said he felt duty-bound to reject the bill on constitutional grounds. In his official "memorandum of disapproval," he said the bill "impaired" his ability to "meet his constitutional responsibility" by vesting the power over "vital national policies [and] the expenditure of large public funds . . . in a group of individuals who would be essentially private citizens."

No judge ever ruled on the constitutionality of Bush's science plan, so Truman's objection was based on a presumption of illegality. Bush, however, considered the constitutional objections bogus. "It is almost a complete answer to the whole thing to say that the [National Advisory Committee for Aeronautics] has operated on this basis for 25 years and has been very successful," he insisted. Questions of legality aside, however, it was clear that the makeup of the science foundation collided with Truman's conception of the role of experts in public life. His veto sent a message: he wanted experts on tap, but not on top. The ambitions harbored by experts—and Bush in particular—worried Truman, especially in the arena of science. "Bush was a specialist, but Truman's attitude as President was, 'Don't let the specialists have the last word,'" recalled Don Price, an administration adviser on science policy and later an authority on the relations between science and government. The president had to walk a fine line, Price said. "He didn't want the scientific elite to dominate, but on the other hand he didn't want to keep them out either."[53]

Truman's veto destroyed Bush's hope that a research foundation would soon replace the military as the chief patron of academic science. "There will be quite a delay [in legislation], for it is very doubtful that the incoming Congress and the President can again get together," he wrote a friend shortly after the veto. "The numerous vetoes have put backs up and this is not a favorable atmosphere for a compromise of any sort." But to another friend Bush put on a brave face. He predicted (accurately, as it turned out) that "the temper of the people on the Hill" was such that science legislation in some form would rise again if for no other reason than the way "it blew up with a magnificent bang."[54]

The setback to the science foundation would give the Navy and perhaps the new Atomic Energy Commission more time to cement its ties with academia, Bush allowed. From a practical standpoint "the picture to my mind is not at all drab," since research funds were relatively plentiful. In the meantime, he would not publicly rebuke Truman. "On the whole, I think he has proven to be a pretty good President, even if he does not catch the nuances of some of these things and even if he leans on a weak group in scientific matters," he wrote John Connor, a former OSRD counsel and a member of Forrestal's staff, on September 5. As if this seemed too generous an opinion, Bush added, "He may change, and in any case the way to correct any such matter is not to howl about it, but to give him the service he needs or rather to try to get him to call for it, which may yet occur."

The "call" Bush hoped to receive was for the new Research and Development Board. Though the science veto tried his patience, he wanted to avoid any confrontation with the president that might jeopardize a chance to serve as Forrestal's research deputy at the new Department of Defense. He planned to personally discuss the position with the president in late September and hoped to smooth things over at that time. Still, he planned to hold his tongue to preserve a chance at greater power. He told Connor that he would not sacrifice his role in the military establishment in order to protest the failure of science legislation. "Truman does not have the contact with American science he ought to have," Bush noted. "He leans on people who really do not have lines of contact that are genuine and on the proper level, but he does not know this and I can hardly tell him so."[55]

Bush preferred to save his ammunition for the only aspect of government still open to him: military affairs. He had not forgotten that one impetus behind his foundation was to insure that independent civilians gained the upper hand in overseeing advanced weaponry. If he could not forge an independent role for researchers outside government, he would seek to carve out space for them within it. With the creation of the Research and Development Board under Defense Secretary Forrestal, Bush saw a fresh opportunity for influence.[56]

But Bush's energy was low. By the summer of 1947, he was worn down by the failure to create a science foundation and the frustrations of the joint research board. Nearly spent at 57 years old, he looked forward to an extended vacation. He daydreamed about spending just one more year, at most, in government service. "I am so doggone weary or bored or something that I have refused to make any dates for the fall whatever or any commitments beyond those of the Joint Board," Bush confessed to Compton in July.[57]

Escaping the stultifying heat of Washington in summer, Bush went to his New Hampshire farm, where he spent a couple of weeks in July "plain and fancy loafing." Then he traveled west, joining Hoover and other Establishment leaders at the annual Bohemian Grove gathering in California. He next visited Carnegie's outposts in Palo Alto and Mount Wilson before going fishing outside Seattle with an MIT pal. He capped off the western trip with a trek through the High Sierras that produced "a very salutary" result, he reported. A Carnegie assistant, Paul Scherer, joined Bush in the mountains. Scherer proved to be "an ideal camping companion," and Bush "had a gorgeous time."

The time away from Washington proved a partial cure for whatever ailed Bush. After nearly two months of relaxation, he returned to the capital in late August "with something of the old zest, which [had] nearly disappeared after a long drag," he told a friend. Karl Compton, visiting Carnegie, found Bush "as energetic as ever." His staff noticed his altered mood too. After a desultory summer, they were again chasing down flurry after flurry of their boss's ideas. "Bush is back," one aide said. "What hasn't already popped will shortly."[58]

What soon popped was Truman's approval of Bush's appointment as RDB chairman. Bush met the president on September 24, a week after Forrestal was sworn in as the first defense secretary, in the midst of a crisis over whether communist Yugoslavia would try to seize the Italian city of Trieste. When Forrestal stated he wished Bush to serve as his research deputy, Truman said, "So do I."

Relieved, Bush laid out his main concern. He wanted to launch the board, then "pull out in about a year." He thought the post should rotate between people "at reasonable intervals, otherwise it might come to be regarded as a one-man show." At Forrestal's behest, he agreed to stay longer if the need arose. As usual, he wished to retain his plum job as chief of the Carnegie Institution of Washington, whose trustees considered it their patriotic duty to underwrite Bush's government activities. During the war, a good deal of Carnegie's research was funded by the military but after the war Carnegie had returned to its roots in scientific research. Since Carnegie's researchers were fiercely independent and autonomous, Bush had time to pursue public service. However, he did not wish to accept the $14,000-a-year salary for serving as chairman of the RDB. Since the government wanted him accountable for his time, Bush agreed to accept compensation "for the actual time which I devote to those [RDB] duties." As a part-time official, Bush believed, he could render more impartial judgments because his livelihood did not depend on pleasing Forrestal or Truman.[59]

The two men accepted the arrangement, and Bush took the RDB job. The White House announced his appointment the next day. Washington insiders were pleased. Knowing how badly Truman's veto of the science bill had burned Bush, they had wondered whether he might simply exit government service. David Lilienthal, chairman of the Atomic Energy Commission, wrote him, "It is reassuring in the deepest sense that you are willing to continue as Scientific Chief of Staff. Would like to visit with you some day before long—perhaps lunch." Admiral Furer wrote from retirement, "I don't feel so much concerned now over the President's veto of the science foundation bill. It is very reassuring to know that the [armed] services are not going to drift around in the doldrums of research and that the taxpayer is not going to be forgotten with you at the helm. I wish you all manner of success in managing the brass hats as well as the lads in the ivory towers and in the counting houses." Many scientists in the trenches were happy too. "Had you not taken the chairmanship, all of us who are working with you would have to resign," wrote Frederick L. Hovde, president of Purdue University and Bush's chief adviser on missiles. "You have indeed a bear by the tail. I will do my best to hang on to a few tufts of hair of that tail to help you."[60]

The buoyed Bush vowed to stick with the frustrating job of rationalizing the military's research activities. Though quite a comedown from a coveted cabinet post, the relative obscurity of the RDB job appealed to him. After all, since the war domestic politics had become more partisan and Bush's role more open, which left him bloodied and defensive. "Perhaps, after all, the best thing I can do while I am still active and vigorous is that sort of odd job in Washington out of the public view without any kudos and credit," he wrote a friend. He bet the job would bring "a good deal of satisfaction if it is possible to accomplish something for national security."[61]

In no way satisfying, the RDB post was like a slow-motion automobile wreck. Bush saw the crash coming but was helpless to stop it.

His tour began badly, partly because Bush simply transferred to the new board the same techniques that had proven inadequate at the joint Army-Navy board. To tame the three military services and their overlapping technological projects, Bush had essentially relied on volunteer committees composed largely of academics. The members had neither the time to devote to their committees, nor the incentive to immerse themselves in the arcana of increasingly specialized military technologies. The board's shortcomings were most obvious when it tried to limit duplication. Bush conceived of the board's committees as judges, with the services as contending litigants. But the committees fell down in three ways. First, too many committees were formed, overlapping so obviously that even Bush admitted that the poor structure was "proving to be an oppressive drain on available manpower." Second, the committees "were not quite courageous enough" to handle the armed forces; they "lacked all the elements of independent authority that must be the backbone of any judicial system" and could not make their decisions stick. Third, the committees were inherently clumsy and slow, relying on 2,000 part-time consultants. The armed services moved so much faster in pushing projects and letting out research contracts that the board "would hardly have caught up with the parade, even if all the committee members had been eager to do so." A big part of the problem was the part-time status of committee members. "However competent the individual civilian scientists might be," one critic observed, "they simply could not compete in ideas with the competent military members who spent their full time thinking about specific weapons."[62]

Bush wounded his fledgling outfit, meanwhile, by trying to dictate terms to the services. He insisted, for instance, that the Army, Navy and Air Force representatives to his board be uniformed officers, preferably the service chiefs themselves or their research deputies. This flew in the face of a trend in the services toward greater reliance on civilian technology advisers, many of whom served as full-time employees. Bush refused to relent on this request even after the service secretaries—the civilians who directly reported to Forrestal—themselves objected. Partly because of this dispute, three months passed before the defense secretary approved the RDB's charter.

The charter proved to be Bush's undoing. He lacked the authority to realize his aims. His board could advise Forrestal on "major policy" and resolve differences between the services on "all other" matters. But the RDB could not direct or control the internal research and development activities of the individual services, which left it powerless to effect change. Finally, the charter stressed the opportunities for coordination between the RDB and the Joint Chiefs and the need to keep the chiefs apprised of new weapons. But the Chiefs naturally wanted their own staff to monitor and the charter failed to spell out the responsibilities of each body, inviting a turf battle.[63]

Bush tried in vain to enforce his will on the recalcitrant Joint Chiefs. Forrestal backed Bush's bid to unify the military's research program but organizational defects and parochial attitudes checked his progress. Bush's egotism and overbearing self-confidence sometimes blinded him to the effect of his actions. He insisted, for instance, that he gain the right to *approve,* not just review, the research budgets of the three services because they were often "hastily developed." (The request was denied.) His history of clashes with the top brass did not help his cause either. He won an agreement, for instance, to attend "all appropriate" deliberations of the Joint Chiefs, but the brass—worried about losing control over new weapons—never invited him to a meeting.

Finally, Bush hurt his credibility with the top brass by continuing to denigrate the potential of intercontinental missiles. Missile advocates imagined striking the Russian heartland from the U.S. and feared the Russians might gain the power to do likewise. Since the end of World War II, Bush had campaigned against intercontinental missiles and the related idea of push-button war. In December 1945, he flatly told the Senate, "I say technically I don't think anybody in the world knows how to do such a thing [build an accurate long-range missile] and I feel confident it will not be done for a long period of time to come." By fall 1948, such skepticism was less persuasive, yet Bush's animus toward intercontinental missiles seemed to have hardened. "I take very little stock indeed in the continent-to-continent missile, in spite of some of the unfortunate publicity" given the concept, he told Forrestal. "I think these things will be just too expensive and inaccurate to use, even if they could be built." He advised Forrestal that "we may waste some effort here if we are not careful." Bush's warnings, while of limited influence on their own, lent support to air power advocates who sought to starve missile development in favor of spending more on strategic bombers. Meanwhile, Russia charged ahead in long-range missiles, building one with a 500-mile range in 1949. The U.S. did not equal this feat until four years later.[64]

Even absent these factors, Bush had picked a bad time to try to discipline the armed forces. His effort to coordinate military research came in the midst of what one historian has called "the worst feud among the armed forces that the United States has ever known." Disputes between the services turned on roles and missions, but these differences created a poor climate to achieve consensus on research goals. Even as Bush tilted at windmills, the rift between the services widened in 1948 over control of air power, which some strategists argued could alone win a war. The Air Force saw the Navy's push for a "supercarrier" as a means of building its own bomber force and thus challenging Air Force hegemony over the skies. The worst fights, of course, came over money. Had military budgets been fat, the services might have been content to duplicate one another, but Truman's drive to squeeze military budgets stoked resentment. He limited defense spending, which in one year of the war peaked at $94 billion, to $14.4 billion in fiscal year 1947 and $11.7 billion in 1948.

Then in March 1948 a war scare erupted, triggered by a message from General Lucius Clay, U.S. commander in Germany. Previously sanguine about the prospects for a European peace, Clay cited "a subtle change in Soviet attitudes" that left him feeling war might come "with dramatic suddenness." Clay's warning—a result of Stalin's moves to tighten control over East Germany and force the Western powers from Berlin—set off alarms in Washington, where officials recognized the poor state of U.S. military readiness. Coming on the heels of the communist takeover of Czechoslovakia in late February, the war scare undermined Truman's austerity plan, boosted advocates of the military's rearmament program and added more confusion to an already thick situation.[65]

Given the global hazards and the stresses on a downsized U.S. military, it was probably impossible to achieve the one-uniform, one-service goal favored by Bush and other advocates of military unification. "It would be very simple if we had one [armed] service," he later wrote. "We would then have under the Secretary a commanding general and a general staff. . . . The Secretary would set general policies in accordance with the instructions of the President and the considerations of the Security Council. The commanding general would implement these with the advice of the general staff." Such a streamlined organization appealed to rational planners, but Bush knew that "Congress just will not accept this idea."[66]

Even for many who decried military waste and duplication, the alternative was to allow the services to create an illusion of cooperation in the hope that it might actually result in more cooperation. Bush traveled this painful road, buffeted by a battle for the soul of the Pentagon that inspired one science-fiction writer to pen a tale that quickly conveyed the realities of military culture. The story, "Project Hush," recounted an Army research mission to the moon. After the Army men set up base and explore their surroundings, they find another base, designed differently from their own. Thinking the Russians have beat the Army to the moon, one man risks capture by entering the base. The others stand by, depressed, preparing a hasty escape, when their scout returns with the worst news imaginable. The strange base, the soldier says, "is owned and operated by the Navy. The goddam United States Navy!"[67]

Bush's difficulties with the armed forces were more earthbound but just as frustrating, as his efforts to evaluate the suitability of new weapons illustrated. Deciding which new weapons suited what situations could be as technically taxing as creating the weapons in the first place. The chief lesson of World War II, Bush believed, was that it was not enough to build weapons—technocrats must insure that the armed forces use them properly. Thus, engineers, scientists and manufacturers could not make weapons in a military vacuum: they must comprehend the tactics, strategies and overall aims of soldiers. But such knowledge was not always readily available, Bush knew, because top soldiers often failed to grasp the way new technologies rendered old tactics obsolete. Moreover, conflict over the roles of individual services often was exacerbated by misunderstandings about the practical effects of new weapons.[68]

Trying to limit these conflicts and help the military realize the full potential of its new weapons, Bush proposed in early 1948 a new group, housed within the Pentagon but independent of the services, that would provide impartial evaluations of new weapons and provide a bedrock upon which tactical and strategic thinking could thrive. The group would be an outgrowth of the "operations research" technique that emerged during World War II and influenced the approach to, among other things, antisubmarine warfare. Operations research used statistical and other forms of mathematical analysis to study the effectiveness of combat tactics. It had caught on so widely that the Navy and the Air Force had permanent groups of their own and the Army had laid plans for one (the three British services had their own OR groups too). Bush felt that the service groups alone were inadequate and that the military establishment needed a top-level body that resembled Britain's Committee on Defense Research Policy.[69]

Bush's proposal, which he considered his most important organizational innovation since the end of the war, was inspired by a report from his board of "scientific advisors," formed under the JRDB. Members included physicist I. I. Rabi, a leading figure at the Radiation Lab; Alfred Loomis, the radar expert; Bell Labs physicist Wlliam Shockley; and Caryl Haskins, a close friend of Bush who would succeed him as the president of Carnegie. In the report, the advisers surveyed the whole range of offensive and defensive weapons and concluded that for the U.S. the margin between victory and defeat in "the next war" would be slim indeed. The advisers described the potential benefit of operations research in almost apocalyptic terms: "We believe the next war will so completely drain our national resources that every military plan will have to be *rigidly* examined to permit our leaders to choose the one with the minimum cost/result ratio."[70]

Forrestal liked the operations research idea, but the services did not. They suspected that the proposed Weapons Systems Evaluation Group would constrain their ability to forge war plans. Forrestal managed to overcome objections to Bush's plan, wringing a statement from the Joint Chiefs to the effect that a weapons evaluation group was "necessary and desirable." The chiefs were directed to work out the details with Bush.[71]

Bush was eager to do so. He saw the weapons group as the epitome of the professional partnership between soldiers and scientists he had tried to foster since 1940. As such, the fate of the group was a litmus test for civil-military relations in the postwar era. "This marriage of military and scientific thinking is what we are looking for; what, in fact, we must have if our planning is to be on sound ground in these days of radiological or bacteriological warfare," he insisted. With the survival of the nation perhaps in the balance, war planning now transcended age-old distinctions between the priestly and the warrior classes. "The aspects of total war are now so broad that no professional group can rightly assume ultimate intellectual authority on the whole. If military men are to continue to do the principal planning," they must break with the past, Bush argued. "Instead of professional aides, they should now seek professional partners, among the most competent of our citizenry in every field." If they failed to do so, he warned, "the alternative is not attractive." Military plans would suffer so grievously from reliance on "traditional methods" that "war planning will in fact be done elsewhere."[72]

Bush's prediction would to some degree come to pass. By the mid-1950s, civilians had taken from soldiers responsibilities for planning nuclear strategy, for instance. But the Joint Chiefs dismissed this jeremiad. Under the strain of interservice rivalries, they saw a different specter: a baldfaced power grab by their old nemesis, the wily Dr. Bush. Digging in their heels, the Chiefs pulled a fast one. When it seemed inevitable that Forrestal would force a weapons group on them, the top brass decided they wanted their own group, housed in their own organization. As one general explained, by embracing the weapons group in this way the brass "might utilize it as a basis for 'negotiation' with Dr. Bush." In any discussions, their ace in the hole would be the insistence that if Forrestal gave Bush a weapons evaluation group, he would have to give the Joint Chiefs a second one. The whole notion of dueling groups seemed so absurd as to insure compromise.[73]

This proved to be a bureaucratic masterstroke, putting Bush on the defensive by forcing him to fight for control of his own idea. Clearly no match for the brass when it came to palace intrigue, Bush won a pyrrhic victory. Forrestal ordered the creation of the weapons evaluation agency, but insisted that the Joint Chiefs inherit the outfit after its birth. The Solomonic compromise left Bush angry and humiliated. Finally boiling over after months of simmering, he wrote Forrestal: "One might conclude that JCS does not welcome the aid, that they do not wish to see it proceed unless they can control it completely, and that they fear if it were to be controlled by RDB . . . it might get out of hand and embarrass them."[74]

Bush accepted the strictures of the top brass because he had fought for the concept underlying the weapons group "ever since the war ended," and, "believing it to be of great ultimate importance, I do not like to see it die." But he knew he had lost his "own personal war" with the Joint Chiefs, telling Jewett in early August that "I seem to be getting the short end thus far" of "a first-class contest" over weapons evaluation. Bush came to view his defeat as a sad metaphor for his tenure in the Pentagon. His year at RDB "was mostly shadow boxing," he later said. "We really had no authority whatever over anything."[75]

For strong-willed Bush, the war hero, serving in the military establishment proved unbearable. In early 1948, the headaches started. Then came sleepless nights. His nerves jangled. Not even during the worst months of the war had he felt so tense. His anxiety reached such heights than he thought he must be physically ill. He asked his son Richard, the physician, for a detailed description of Parkinson's disease. As the weeks passed, his headaches, irregular heartbeat and insomnia persisted. His eyesight, never good, grew poorer. Then in July, Phoebe learned she had breast cancer and required a mastectomy. Her surgery was successful, but set Bush back. "Things have been in quite a bit of turmoil and added to the complication of the Pentagon I have been having a rather rough summer," he wrote old friend Frank Jewett. Though "very anxious to help Forrestal in any way I can," Bush admitted he wanted nothing more than to quit the RDB. "If you could tell me a good successor for my job in the military establishment it would give me the best lift I could have at the moment."

Bush would leave the Pentagon soon. The only question was whether he would walk out or be carried in a box. He suspected that his torturous headaches stemmed from a tumor in his brain. In late summer, a physician X-rayed his brain and found what looked like a tumor. Convinced he "walked pretty close to the edge," he prepared to die.[76]

The friends who knew of Bush's plight took his diagnosis of a brain tumor seriously. Merle Tuve did not. Iconoclastic and a lone wolf, Tuve had distinguished himself as the top researcher on the proximity fuze. Disturbed by the military's sway over science in the postwar era, he knew as well as anyone the depth of Bush's frustration. He also believed that Bush suffered from war guilt. Not from the atomic bomb, but from his role in aiding the ghastly firebomb raids against Japan. "For years after the war Van Bush would wake screaming in the night because . . . he burned Tokyo," Tuve later recalled. "The proximity fuze didn't bother him badly . . . even the atomic bomb didn't bother him as much as jellied gasoline [napalm]. Oh, yes, we all suffer scars you know, and I don't know how we'd help it."[77]

Bush never betrayed any war regrets; he never apologized for any of his decisions. It was possible that Tuve, far more emotional than Bush, simply projected his private pain onto his Carnegie boss. Tuve talked with Bush often, but he admitted finding Bush "hard to reach . . . detached." Still, Tuve's conjecture was plausible. Given the strain caused by war scares and interservice fighting, some people were bound to crack. Even in good times, Bush's boss, the high-strung Forrestal, was an intense workaholic who displayed insecurity. Now he was overwhelmed by his Pentagon, anguished by his shortcomings and nearing a collapse. Barely nine months after Bush's painful summer, Forrestal would crack up and kill himself. Air Secretary Stuart Symington, another Bush colleague, acted frenetically at times, perhaps because of stress and high blood pressure. Indeed, within the Truman administration as a whole there was so much free-floating anxiety that not long after Bush hit upon his brain tumor explanation *The Nation* proclaimed hypertension "the Great Man's disease of 1949."[78]

Bush wished to avoid medical treatment. He was a terrible patient. He hated hospitals. He didn't like anyone telling him what to do. Especially doctors. Since some brain tumors were inoperable anyway, Bush chose to ride out the storm. His life would play out according to its own logic. Congenitally optimistic, he tended to take his blows rather than duck them. Sickly as a child, he had grumbled throughout his 58 years about various aches, pains, colds and fevers. He had succumbed to short breakdowns twice since 1940: once not long after FDR tapped him to launch the NDRC and the second time right after the first atomic test in 1945. On those occasions, he had bounced back immediately, but not this time. Now he feared the loss of his intellectual keenness, his wit, his gift for handling gadgets. Stuck in the doldrums, recoiling from the possibility of surgery, Bush saw only one way to survive: leave government for good.[79]

Bush had never thought much about his own death. While close to death during the war, he had never been personally in danger. Despite a lifetime of complaining about aches and pains, he sought no divine cures for his ailments.

He was not religious and never attended church. He kept his opinions about death, and the possibility of an afterlife, to himself. His idea of a final ritual was to travel west with his close friend Caryl Haskins, a scientist of independent means who was one of Bush's RDB advisers.

Haskins knew of Bush's illness. Asked on short notice to make the trip, he cleared his calendar, packed his fishing rods and joined Bush. They flew to Billings, Montana, on August 15. A professional guide met them in town. The guide arranged for horses and a pack train and led them them into the wilderness. They would ride for five or six miles a day, then pitch tents. The guide would tend to the horses, rustling them up in the morning. The riding was hard but great fun. Haskins had not ridden since college, when he served in a horse-drawn artillery division. Bush, who never rode much either, did not complain. He stayed quiet for long spells. Since they rode single file on trails, it was hard to talk much anyway. Still, Haskins thought Bush preferred the silence and that "he was indeed recuperating, way down deep."

The two men and their guide at some point reached a wide river. They rode along the banks of the river, stopping each day to fish for trout. Bush enjoyed fishing and he was awfully good at it, so that they caught too many fish to eat themselves. They built a smokehouse and smoked the part of their catch they could not eat. They went at the fishing and the smoking with zest. At the end of the two-week trip, they divided up the trout. There was plenty. Haskins's share alone was enough to satisfy him for a whole year.[80]

Chapter 15

"The grim world"

(1949–54)

We must be so strong that the free world will not be attacked. We must be strong in every way, and in all places, for the assault upon us has sought our soft spots, geographically and psychologically, and will continue to do so. We must have friends, and increase their strength. And we must accomplish all these things under the recognized disadvantages of a democratic system, without wrecking it in the process, and without breaking down our economic health on which all else depends.
—Vannevar Bush

Bush returned home from Montana with new priorities. His resignation from the government's Research and Development Board was now a mere formality. As a parting shot, he complained again to Defense Secretary James Forrestal about the Joint Chiefs of Staff. He cited new evidence of the disregard for his board by the top brass, reprising earlier criticisms of their weakness—this time in shriller tones. The new provocation, and Bush's tirade, provided a neat excuse for him to resign as RDB chairman. "There has now been a clear invasion by the [JCS] into the affairs of the Board which has seriously injured its prestige and standing," he told Forrestal. "There is now a general impression about that, since the Joint Chiefs can invade the affairs of the Board without notice, the latter is in some manner secondary or subsidiary. I find my ability to serve you in connection with the responsibilities you have placed in my hands seriously impaired."[1]

Word of Bush's impending departure quickly leaked out. *Newsweek* reported that Forrestal planned a "reorganization" of the RDB, which the magazine said "has been cumbersome and slow moving at coordinating big service scientific projects." The article drew no reaction from Bush, whose main concern now was his health.[2]

Two weeks in the wilds of Montana had boosted his morale and made him more patient with others. On returning to work, he quipped, "I noticed that the human race had changed appreciably while I was away." But his headaches had not let up; he still felt near death. So for the first time, he took decisive steps to uncover whatever medical mysteries lay behind his condition. Newton Richards, the wartime chair of OSRD's medical committee, recommended Bush see a specialist at Presbyterian Hospital in New York City named Robert Loeb. On September 19, he wrote Loeb, who agreed to evaluate him.

Bush held off from visiting Loeb immediately, but set aside the entire week of October 11 "to jump through any hoops that you and your colleagues may have set up." Bush did not look forward to Loeb's medical tests, but he had confidence in the doctor, whom he hoped would settle—one way or another—the brain tumor question. Eager for an answer, he peppered Loeb with recollections of his ailments, telling him that earlier in the year he had suffered from "violent fits of coughing just after arising." Before Loeb could reply with an explanation, Bush wrote the doctor again and pointedly asked for "a tentative guess as to what may be wrong."[3]

Loeb played coy. Bush, meanwhile, officially left the Pentagon. The White House announced his resignation on October 5, releasing a letter from Bush that suggested he had planned all along to step down after a year atop the RDB. Truman and Forrestal issued statements in praise of Bush, who declined to give interviews. The next day, both *The New York Times* and *The Washington Post* devoted as much attention in their stories to the RDB's new chief, MIT's Karl Compton, as to Bush. Bush had no complaint. His resignation relieved him of an enormous weight. He told Loeb that he had felt "better in the last four days than I have for three months."[4]

Though Bush kept tight-lipped about his personal situation, journalists sensed that his departure signaled the end of an era in military research and in science policy generally. In the years since the war he had gained a growing reputation for conservatism and for a penchant for shooting down the high-tech dreams of military enthusiasts. Bush would never again formally advise any president or cabinet secretary. Bolder, younger research leaders were now consulted. These successors walked in Bush's footsteps to a degree, and many actually trained under him. "There is hardly a field in the government's scientific front, from the Atomic Energy Commission down, where a Bush-trained man is not taking a prominent part," wrote one journalist. "The same is true of many scientific research efforts in universities and private industry."[5]

The nation paid its respects to Bush's contribution. "President Truman spoke for the whole country when he paid warm tribute to Dr. Vannevar Bush," *The Cincinnati Post* wrote. *The New York Times* added, "The President was right in saying that his regret [at Bush's departure] will be shared by scientists and higher military officers." *The Kansas City Star* celebrated the man "who has been chief of staff for the technological side of our defense efforts since 1940."

The press was respectful, even reverential toward Bush as they were toward nearly all American leaders at the time. No newspaper or magazine supplied even a hint that Bush left government a broken spirit, bereft of influence, repudiated by the military leaders whom he wished more than anything else to stand among. Indeed, just a few months before his health failed, one national magazine enthusiastically profiled Bush, calling him "chief of U.S. brain power" and "one of the most important men in America."[6]

The image of Bush as a tragic hero, cut down by his refusal to bend to the military's will, was not a story the press could either comprehend or nicely package. The sharp decline in Bush's clout among scientists and the Washington establishment, hinted at two years before by *Fortune,* simply did not square with the durable celebrity awarded heroes of World War II. Paradoxically, Bush was more celebrated now, when he lived on the margins of power, than he had been in 1944, the high-water mark of his influence on government. Few charted Bush's rapid descent. Only his circle of close associates— men such as Warren Weaver, Jewett, Conant and Compton—realized that Bush's audacious campaign to bring government under the full sway of experts had come to an end.[7]

Bush's defeat and expulsion from power were predictable from one standpoint. It was not just Truman's aloofness from Bush that cost him his clout. Neither did the National Security Act of 1947, despite giving too much autonomy to the services and too little authority to department officials, doom Bush. His problem was personal. "There isn't much middle ground about the way people react to Bush," a journalist wrote in mid-1948. "Either you think he's wonderful . . . or you hate him violently." For a time, Bush's friends counted more than his foes. Then the tables turned. He had stoked the military's resentments during the war, and these resentments came home to roost. Too many military men and civilian scientists disliked him, if not for what he stood for then for his sometimes-imperious manner. Whereas other civilians, such as Edward Bowles, adapted to the military's existing institutional arrangements, Bush demanded "coequal" status for civilians and a level of interservice cooperation that seemed unthinkable to the top brass. Neither could be achieved without a revolution in civil-military relations. During a deepening Cold War, this was not about to happen. Both soldiers and civilians were too stressed by immediate demands to overhaul the military's structure.[8]

But there was a larger problem too. Bush's style was ideal for wartime, with its crises, secrecy and snap decisions. A master at sizing up situations, he worked best when interfered with the least. But the whole texture of American life changed with the return of peace, leaving Bush nostalgic for the past and defensive about his old-fashioned values. He could not accustom himself to the state's enlarged role in science and society, defense and world affairs. He sought to reconstitute a prewar voluntarist model, in which associations of private citizens and people of merit carried the burden of furthering national aims. Bush sensed that the world had grown too complex—and U.S. obligations too large and intense—for voluntarism to work in the form expressed in the 1920s by Herbert Hoover. But his reformulation of the Hooverist ideal failed to captivate either traditional conservatives or pragmatic centrists who shared Bush's goals but not his means.

Caught between two worlds, Bush struck one later observer "as a Moses who pointed the way to a promised land, remained behind while others moved into the new domain and realized to his dismay that the new promised land did not match his vision."[9]

Though dead politically, Bush refused to exit the stage. The results of his physical tests proved extraordinarily good. In Bush's few days in the hospital, Loeb found no evidence of a brain tumor. Indeed, on October 22 he flatly told Bush that his headaches, sleeplessness, anxiety and general depression had no organic cause. Elated yet embarrassed, Bush confessed to friends that he had fallen prey to what he called "a psychosomatic difficulty." It was a "reaction to the letdown" following the war, his son said. Others chalked it up to the stress of his political defeats. "He'd been under a lot of strain and pressure for so long that it was probably psychosomatic," said John Connor, a friend.

To Bush, the whole painful experience still seemed like a close call. "I evidently walked pretty close to the edge and I guess it is fortunate I did not walk over the brink," he wrote Compton. He told Laurence Marshall, his Tufts roommate and Raytheon cofounder, that his tests "worked out very well indeed. They went over me decidedly thoroughly and the outcome is that I have no organic difficulty of the nature I feared." To Lilienthal, the atomic energy chairman, he confessed, "The medical threat which then looked rather serious has now evaporated under closer examination and I feel sure that I will be in good shape again after I have a bit of respite."

The prognosis strengthened Bush's resolve to stay clear of government for good. Loeb supported his patient's inclinations. "Don't get into the government service again," he ordered. Bush's insomnia still lingered, which made him think he could suffer a relapse. Resigned to a slower pace, he told a friend, "The docs say I have to relax, but that will be OK."[10]

Bush was not planning to drop out of sight altogether, however. His post at Carnegie would give him a window on broad scientific advances, and he planned to devote his now ample spare time to speaking and writing about national security matters. Since the end of the war, he had wanted to write a book on the topic in order to show that "the way in which we handle our future security depends upon the way in which we manage our democratic processes." In 1949, he published it under the title *Modern Arms and Free Men: A Discussion of the Role of Science in Preserving Democracy*.

A treatise on military organization and readiness, *Modern Arms* restated for public consumption views on military weapons, tactics and strategy that he had often advanced within the Truman administration. As in his private discussions with senior soldiers, Bush debunked simplistic technological visions, playing down the chances of a push-button war involving intercontinental missiles armed with atomic bombs. He also rejected the argument of airpower advocates that the atomic bomb was an absolute weapon and held out the hope that in the long run defensive measures could limit the damage done by atomic attacks. Sounding characteristically optimistic, he wrote that "the technological future is far less dreadful and frightening than many of us have been led to believe, and that the hopeful aspects of modern applied science outweigh by a heavy margin its threat to our civilization."[11]

Published in late November 1949, *Modern Arms* was a main selection of the Book of the Month Club and a best-seller. *Life* magazine published a long excerpt in two parts. The book was widely reviewed, drawing as much ink as "the latest work of a celebrated writer of fiction," one critic noted. "This is not surprising," he added, since Bush "is a leading contemporary figure in science" and devotes a good portion of the book to "a persistent fascination for our unhappy period . . . war and weapons."[12]

Designed to calm an anxious public, *Modern Arms* put Bush back in the news. Within weeks of the book's publication the Associated Press named him "man of the year in science," calling him "a powerful leader of the science forces . . . [whose] speeches are still top news." Walter Millis, a noted military analyst, was so impressed with the value of Bush's book in "preparing this nation for its real military responsibilities" that he thought it was "probably worth a couple of divisions or a whole group of B-36s." The book also earned a front-page notice in *The New York Times Book Review,* prompted the editorial board of *The Washington Post* to discuss its virtues and provided the basis for an hour-long March of Time documentary. Dozens of other reviewers wrote enthusiastically about the book in the U.S. and England. The *Post* called Bush's analysis "refreshing because it combines scientific insight with philosophical reflection" and complimented him for talking "some hard-headed sense which debunks both the wild tales of the scaremongers and some of the pet one-weapon theories of the strategists." Even Bush's critics praised the book. The *Bulletin of Atomic Scientists,* which had emerged out of the rebellion against the May-Johnson bill, found the book "well done" and thanked Bush for delivering it "in this time of need."13

Indeed, the timing of *Modern Arms* was superb. On August 29, 1949, the Soviet Union secretly tested its first atomic bomb on the steppes of Kazakhstan. Several days later, a U.S. reconnaissance plane on routine patrol from Japan to Alaska found that a filter paper, exposed for three hours 18,000 feet over the North Pacific, gave a radioactive reading slightly higher than the amount that called for an official alert. The Air Force, which had been testing air samples as part of a program aimed at detecting a Russian atomic explosion, dispatched a few planes to gather more evidence; one detected a record amount of radioactive particulate matter. After several days of study, military officials, suspecting a Soviet atomic test, summoned Carroll Wilson, Bush's wartime alter ego and now the general manager of the Atomic Energy Commission. A deputy defense secretary showed the data to Wilson and explained that the news of a Russian test would be so emotionally charged that Americans might not believe it. In the deepening Cold War with Russia, the U.S. took comfort in its presumed technological superiority. The news of a Russian bomb would shake American confidence in the country's ability to contain communism. Before declaring proof of a Soviet bomb, the Defense Department wanted an independent scientific panel to bless the findings. Asked to assemble the panel, Wilson immediately thought that Bush, his old boss, would make a fine chair.

Both the Atomic Energy Commission and the military found Bush a natural choice to lead the panel. He agreed to take the assignment. On September 19, a Monday morning, he convened a meeting of the panel in an Air Force office. Joining Bush were physicist Robert Oppenheimer, a dozen other scientists, high-ranking Air Force officers and a British delegation. By lunch, it was clear from the hundreds of air samples taken from a broad swathe of the Northern Hemisphere that the Russians had exploded an atomic bomb. Although the evidence made it hard to pinpoint the exact time of the explosion, Oppenheimer estimated that the Russian test had occurred sometime between the 26th and the 29th of August.

With confirmation in hand, the Truman administration considered how best to break the news. On September 24 the White House announced the Russian test in a terse statement. It shocked the nation. "The prospect of a two-sided atomic war," said Raymond Moley in *Newsweek,* meant a "towering change in the world outlook."[14]

The news came just before *Modern Arms* went to press. To take the Russian achievement into account, Bush hastily added a brief foreword and revised a few words "here and there" throughout the manuscript. But he basically left the book unchanged. An entire chapter on the A-bomb made no mention of the Russian test, and the book retained such statements as, "It is a far cry indeed from the time when the enemy has a bomb." Bush insisted, however, that the Russian bomb did not fundamentally alter his outlook: since he always assumed that the U.S. atomic monopoly would not last, he never thought it afforded much of a strategic advantage in the first place. His soothing advice echoed statements he had made following the atomic bombings of Japan. Four years later, atomic bombs remained scarce, he argued, so with the arrival of Russia's bomb war had not suddenly become a "completely different order of magnitude." Other than his surprise at the timing, he wrote Lilienthal at the Atomic Energy Commission, the Russian test "does not change the way I look at the whole problem, for I am concerned with the long pull rather than the immediate." Critics still counted Bush's surprise against him. Along with blithely ignoring satellites, intercontinental missiles and other weapons of the sky, he could now be tarred with underestimating the Soviets too.[15]

This last charge was unfair. Before Hiroshima, Bush was almost alone in warning that other nations would rather quickly build their own atomic bombs. In September 1945 he predicted that Russia could have a bomb in five years if it devoted enough resources to the task. More recently, Bush had grown complacent about Soviet atomic progress. But he was hardly alone in this. The nation's intelligence sources were caught off guard too. Less than two months before the Soviet test, the head of the CIA had told Truman it was "remotely possible" for the Russians to build a bomb by mid-1950, but thought "the most probable date" was mid-1953.

Part of the reason for the misjudgments by Bush and others was their ignorance of the extent of Soviet espionage. As the Russians were setting up their crash program after the war, British physicist Klaus Fuchs delivered Moscow a detailed description of the U.S. plutonium bomb. This information surely quickened Soviet progress; Fuchs himself estimated that his clandestine aid shaved "one year at least" from the project.[16]

In Bush's new role as a private citizen, he sought to counter the frenzy over Russia's atomic test by reminding Americans that their nation's true edge was its commitment to individual freedom and an open society. In the first pages of *Modern Arms,* he stressed that one of his book's "chief conclusions" was that "the democratic process is itself an asset" in the competition with communism. "If we can find the enthusiasm and the skill to use [democracy] and the faith to make it strong, we can build a world in which all men can live in prosperity and peace."[17]

Like many Americans, Bush wanted to take important decisions out of politics. World War II had created a false sense of unity. The postwar era returned to the fore the fissures in American life—over wealth, race and foreign policy. The bitter rivalry among the armed services was only one example of this. The political arena seemed increasingly fragmented. Lacking the cohesive presence of Roosevelt, the Democratic party split in three parts before the 1948 election, with southerners bolting to the right, some liberals following Henry Wallace to the left and Truman desperately trying to hold party loyalists together. The high level of partisanship in the late 1940s—whether the topic was military strategy, the national science foundation, atomic weapons or the danger of communists in American society—only reinforced Bush's worry that a democracy might fail to meet the challenge of Cold War.

Never questioning the need to contain Soviet influence throughout the world, Bush worried that Russia might defeat the U.S. without firing a shot because too many of his fellow citizens "want to turn the country into a wishy-washy imitation of totalitarianism." He was especially concerned about attacks on big business. The war had greatly concentrated industry, and its aftermath brought a surge in federal antitrust cases. Unions, meanwhile, were on the rise. After the war strikes broke out in almost every industry, making 1946 the most strike-torn year in U.S. history. Congress retaliated the following year by passing the Taft-Hartley Act, which curtailed union activities by outlawing secondary boycotts, holding unions liable for contract breaches and legalizing injunctions against strikes that threatened the nation's health and safety. While Truman had angered labor by crushing railroad and mining unions, he seemed a better friend to the working man than Thomas Dewey, his Republican challenger. In the 1948 presidential election, Truman carried the nation's 13 largest cities as wage earners recalled the president's opposition to Taft-Hartley. When he learned of his upset victory, Truman said, "Labor did it."[18]

Militant unions were not the only stain on America, according to Bush. The government also encouraged the belief that large industries required either tighter regulation or state aid. The growing dependence on government contracts by such big industries as aviation averted the possibility of nationalization, but it did reflect a new relationship between the state and business. To Bush, the economy seemed headed toward socialism by another name. "Where are the Socialists in this country?" he asked Lloyd Cutler, a savvy Washington attorney. "They are all through the machinery and they are pulling strings everywhere."

But he conceded that there was no easy solution to the problem of bigness. The "tendency in modern society toward . . . monopoly" distressed him. "Anything that knocks out the little industry certainly moves us toward assembly of great industrial aggregations, and this moves inevitably toward statism," he told Cutler. While agreeing that antitrust laws and regulation could act as a brake on monopoly power, he thought the government ignored the equally valuable safeguard of market forces. The public must "see to it that the big fellows . . . have plenty of competition which is real," he wrote. This could often be "best brought about by the advent of small new industrial units, for if these latter have half a chance they can cut rings around the great stodgy concern."

Two trends, however, worked in favor of harmful monopolies. First, many progressives attacked "mere size," forcing the regulation of some industries that were not natural monopolies. Railroads were so tightly regulated as to "become in effect a tool of government, and steel has gone in that direction quite a distance," Bush wrote. Second, the government's "tendency to crack down on private initiative of any sort . . . is the most insidious part of the whole business." He cited the tax system as "the most effective tool for this purpose." While taxation hardly equaled the horrors of "totalitarian control such as there is in Russia," it could certainly render "the creative instinct . . . relatively sterile."[19]

In a widely reported speech at MIT, just days after the publication of *Modern Arms and Free Men,* Bush picked up on some of these themes, attacking the government's growing tendency to aid specific pressure groups through financial subsidies. To meet the problems of the postwar economy, the government generally ran at a deficit in the late 1940s and Truman backed subsidies for public and private housing, the unemployed, farmers and the medically needy. He also proposed raising the minimum wage and social security payments. To Bush, this seemed like a headlong plunge toward the "nanny" state, especially since it came at a time of labor militancy and growth in the power of unions. Opposed to restrictions on capital as much as on individual freedom, Bush complained that "every man's hand is out for pabulum and virile creativeness has given place to the patronizing favor of swollen bureaucracy."

Bush's complaint echoed the pleas by conservatives such as Senator Robert Taft to resist "creeping socialism" spawned by Truman's "fair deal." Only if the nation continued to reward merit and eschew government handouts, Bush declared, would democracies outpace dictatorships. Fearing the American people lacked discipline, he railed against the specter of "a people bent on a soft security, surrendering their birthright of individual self-reliance for favors, voting themselves into Eden from a supposedly inexhaustible public purse, supporting everyone by soaking a fast disappearing rich, scrambling for subsidy, learning the arts of political log-rolling and forgetting the rugged virtues of the pioneer."[20]

Bush sounded these same themes in *Modern Arms,* linking the Cold War and the drive for more advanced military weaponry with the strains on American democracy created by an engorged state. No armchair political philosopher, Bush's motives for thinking about democracy were practical. "We often do not see democracy at its best," he complained. Impressed with the potential efficiencies of dictatorships, he fretted about the weakness of democracies. "Someday we may be able to go a long way toward a system in which voluntary conformity to group opinion is fully controlling," he wrote. "In the meantime we need to be sure that just because we have free speech, we do not create soft or wishy-washy organizations." Locked in struggle with Russia, Americans could not afford "to turn their liberty into license," but rather must embrace an ethos of sobriety and self-discipline, falling into line behind their leaders following an elite decision. "Free expression of opinion before the event [cannot lead] to malingering after the decision is made," he wrote.[21]

Bush unfortunately never defined what he meant by "democracy," but he certainly shared the belief of his wartime cohort James Conant, who thought more systematically about the subject. In a 1940 essay, Conant had argued that the essence of democracy was the drive to replace an aristocracy of wealth with one of talent. In Conant's view, democracy did not depend, as Jefferson thought, on widespread property ownership, the marriage of mental and manual labor or the the exercise of citizenship. Of "paramount importance" to a democracy, Conant believed, was that "careers [were] open to all through higher education." Conant had his critics. Christopher Lasch called this view "paltry" and accused Conant of seeing "in democracy nothing more than a system for recruiting leaders," a way of guaranteeing the "circulation of elites."

That same criticism could be leveled against Bush, who also saw the top universities as breeding grounds for future government leaders. More frank in his elitism than Conant, Bush perhaps made a stronger case for it. Even Lasch conceded that "the idea that democracy is incompatible with excellence, that high standards are inherently elitist . . . has always been the best argument against" majority-rule government. This was the point of Bush's complaint in *Modern Arms* about the tendency of democracy to yield waste, "confusion" and "government by committee." "Most of the important actions, the executive acts, need a man with authority who can make decisions and carry them out," he wrote, sweeping aside possible benefits of public participation in government decisions and ignoring his own penchant for relying on comittees. As early as the 1920s, Walter Lippmann had argued that public opinion was an unreliable guide for a complex industrial society and that the public should be content to leave government to experts—provided they brought home the bacon. Bush's experience in World War II made Lippmann's then-controversial view seem axiomatic.

Yet acceptance of the primacy of elite judgments made democracy no less messy. Political factionalism, of the sort that had rent the military establishment since the war and the Democratic party in the 1948 election, clashed with Bush's technocratic sensibility. "The central question is whether there is order and control, whether orders are carried out promptly and cheerfully, in the government itself and in the organizations that are subordinate to it," he wrote in *Modern Arms*. While admitting that his language "has a connotation of conformity," he asked Americans not to fear the emergence of dictatorship at home. Bush only wanted the "best" people in charge, steering the nation through its crises by dint of their superior expertise and good sense.

Many Americans willingly accepted rule by experts, whether their expertise was military, scientific or social. So Bush's defense of expertise was consistent with the times. He joined a chorus. By the late 1940s, Michael Sherry has written, "Americans were told, perhaps more than at any other time in their history, to trust objective experts to solve all sorts of problems, from the riddles of nuclear strategy to juvenile pathology and family conflict."[22]

As the years wore on, the prevailing respect for experts emboldened Bush to advocate technocracy even more unabashedly. In the mid-1950s, he made headlines by calling for a "natural aristocracy" that would govern the "the climate of opinion" in the country out of which politics and values arise. Denying accusations that he favored a "fascistic intelligentsia," Bush claimed at a "Democracy Workshop" for youth that it was the duty of a minority of perhaps a few million Americans to "establish the climate of opinion that will determine the actions even of the great men in places of power." Shrugging aside complaints that democracy demanded mass participation, he frankly admitted that talent was too unequally distributed to run a great nation on such a mass basis. "In one way or another," he said, "there must be those who lead and those who follow, and there always will be."²³

The elevation of technocrats to new heights invited a backlash. Some Americans had lost faith in the power of rationality to solve problems and the inevitability of human progress. These twin pillars of expertise were increasingly seen as obsolete in the aftermath of a murderous war that left a legacy of horrible weapons and a chilling Cold War. "Since the war years, the optimistic rational faith has obviously been losing out in competition with more tragic views of political and personal life," C. Wright Mills observed in 1951. These "tragic" views implied a return to mass politics and a yearning for the Jeffersonian generalist.

Bush's notion of faith in the superior individual now competed for cultural supremacy with the more tragic view of life arising from the ashes of total war. Bush tried to quiet this rising sense of a doomed humanity by calling for discipline in the face of what he considered an unreasoning fear of annihilation. But he found his own advice problematic. The very logic of expertise diminished the individual's capacity for self-governance and thus his humanity: a specialist could not offer a general solution and this undermined the authority of expertise. For Bush, the expansion of the welfare state was the most egregious example of a trend toward furnishing "paternalistic care" for the less prosperous or emotionally troubled. Even among elite institutions, he foresaw the decline of old-fashion virtues of self-reliance, discipline and the capacity to work with both the head and the hands. The great sin of liberalism was not hubris (say, about the efficacy of economic planning, or the potential to truly improve the lot of the disadvantaged), but its corrosive influence on individual character. "There has been altogether too much emphasis at times on the privileges of citizenship," Bush insisted. "It is time we emphasized more of the responsibilities that go with freedom."²⁴

By the early 1950s, Bush's conversion from insider to outsider was complete. No longer constrained by serving within government, he often spoke his mind. Whether the subject was nuclear weapons, military affairs or the wonders of science, Bush struck a tough-minded but balanced pose, setting himself apart from alarmists on the right and the left. To an anxious public, he presented a sober message from the ultimate technological wise man.

In his heyday, Bush had traded blows at a moment's notice and inspired as intense a loathing in his critics as he did admiration in his pals. A rough backroom brawler, he inspired public support for pure research and helped to create some of the most terrible weapons ever known. The paradox of his career left him seeking a more benign, even avuncular image. Now the press helped him to construct one. This was no act of generosity; it was difficult for ordinary people to square how the finest among them could be both visionaries and killers, fiercely independent themselves and yet demanding of conformity in others.

Coronet magazine best captured the spirit of Bush's transformation in a 1952 issue that sought to banish the ambiguities of his career by juxtaposing his bureaucratic achievements with his folksy authenticity. In a full-page photo, Bush sported a dreamy look, one hand grasping his chin and his squinty eyes gazing out into the distance. "It is interesting that Mister Science looks so much like Mr. America," the magazine observed in a pithy three-paragraph portrait. "If this is the face of the hydrogen bomb, it is also the face of a boy who pumped the organ in his father's church . . . and whose favorite invention is a gadget that protects songbirds from pigeons, not by killing the latter but by dumping them good humoredly off the perch."

What manner of man could incinerate a city yet spare a pigeon? *Coronet* announced: "If you lost your wallet in a strange town and had to borrow a dollar from a passer-by, you might choose a man like this. For this is the face of a kindly, comfortable, and approachable man, who has probably lost a wallet himself at some time while fishing in his pockets for a tobacco pouch. He reminds you of somebody—Will Rogers? Uncle Sam? Anyway, you've seen this face before, and it belonged to a man you liked."[25]

Coronet's portrait exaggerated Bush's folksiness, but there was a lot to like (especially if one did not *oppose* him). Bush could not say the same about his countrymen. He liked little about America in the 1950s, and it showed. A critic from the right, he occupied a curious position in the nation's cultural firmament. No longer the confidant of warriors and kings, he still issued his jeremiads. Nostalgic for the past, he decried what he saw as the slipshod way in which the country faced the future. He was by no means alone in looking to the past for the means to solve tomorrow's problems. Compared to his years in government, he now had a far bigger audience than ever for his views. Yet his proclamations meant less and less to the people who mattered most to him.

The problem of national security troubled Bush. The Korean War, which broke out in June 1950, confirmed both his doubts about the shortcomings of military planning and his opinion that the country must maintain its capacity to fight conventional wars. Invading South Korea, North Korea overran its neighbor, prompting the U.S. to retaliate by evicting the invaders and penetrating into the north. But the war turned into a stalemate after China—a World War II ally under communist rule since late 1948—sided with North Korea and Truman chose neither to withdraw nor to escalate. Mired in an

Asian land war, the nation rallied behind Truman's call for a huge defense buildup. Money poured into military research programs, ending years of relative austerity. Overall spending on defense and foreign affairs tripled in the fiscal year that began three months after North Korea's invasion.[26]

In this climate, Bush wrote a 25-page essay, arguing that the U.S. must produce more advanced weapons on "a few quick" basis. While prompted by the steep rise in research funds and the belated introduction of more effective weapons in Korea, Bush's aim was not to prescribe a winning strategy for the Korean War. Rather he wished to highlight the necessity of improving weapons even during a conflict—and doing so economically.

Bush circulated his paper widely in late 1951, sending copies to senior military officials, including Joint Chiefs chairman Omar Bradley and Pentagon chief Robert Lovett. Drawing on his World War II experiences, Bush cited numerous examples of how his wartime researchers had actually built small batches of weapons and equipment, based on little more than prototypes, in order to rush these advances into battle. Bush called this concept "a few quick," because it contrasted so sharply with ordinary large-scale production. The latter approach required designs to be frozen well in advance and modified only to meet the limitations of material supplies. It guaranteed production of the *most* weapons, but not the *best* ones. "We are faced with the old quandary," Bush insisted. "Do we get masses of mediocre weapons, or do we get really advanced equipment . . . that will give us striking advantage."

Bush argued that the leaders of mass production would never give enough freedom to the advocates of "few quick," so that the latter needed their own stream of funds and "should of course be backed up very thoroughly indeed, protected against predators." World War II had demonstrated the conflict between mass producers and innovators. As one historian has noted, "Established mass production methods, such as those pioneered by the automobile industry, typically proved too rigid for the high level of uncertainty and rapid pace of innovation imposed by the war economy."

The solution was not to revive OSRD, however. "The bureaucratic controls of Washington would smother any small independent outfit doing queer things," Bush insisted. "We have to face reality." He favored instead a fluid, ad hoc organization composed of soldiers and civilians, so that there would be no "danger of decision by weight of stars rather than by weight of evidence." He could not say what these research rebels would build, but he urged that they be given the chance—and receive no penalties for honest failures. "There are plenty of keen officers of this type," he allowed, "but they don't always get the chance to operate without the heavy hand of authority on them." The trick was to not allow a bureaucrat to hem them in. "More initiative has been balked by arbitrary military requirements than by any other device invented by the mind of man," he warned.[27]

Bush's "few quick" credo made sense, and the military welcomed it. But in typical fashion, each service was urged to launch its own "few quick" unit. Such top-down reactions masked the growing possibility that research innovations would bubble up from below. To be sure, the military still squandered the talents of many people simply by squashing them. It also had a bias toward spending large sums of money on known quantities. These failures were inevitable in any large organization, and the military was the largest in the U.S. By 1953, the Department of Defense consumed half the federal budget. It spent more money in a year than did the 12 largest U.S. corporations. Within this behemoth, waste and conformity reigned, but pockets of entrepreneurialism thrived, though Bush overlooked them.[28]

The most prominent example of the "few quick" spirit involved Hyman Rickover. An insurgent naval captain, Rickover shared with Bush a single-minded, no-nonsense attitude. Rickover was obsessed with the idea of a nuclear-powered submarine. Because of limitations in power supplies, subs spent too much time either on the surface or near it. Atomic power promised far longer periods of submersion and at greater depths. But the path toward building a nuclear-powered sub was crooked. No reactor had yet operated on a ship, much less on a submarine. Safety issues loomed; at a minimum, the crew had to be shielded from hazardous levels of radiation.

Technical issues were surmountable. But government bureacracy stood in the way, posing greater difficulties. As a result of the Atomic Energy Act of 1946, the armed forces were in the curious position of having to beseech a civilian body to obtain nuclear materials and weapons. Shut out of the Manhattan Project by Bush, the Navy was understandably keen to jump on the atomic bandwagon after Hiroshima. Rickover wanted to form a nuclear propulsion project, but his superiors gave the AEC a chance to tackle the job. However, they allowed him to learn about nuclear technology at an AEC plant in Oak Ridge, Tennessee. By 1947, Rickover had come to see the nuclear submarine as "merely" an engineering problem. The following year he convinced the Navy to back his dream.

Rickover embodied the burgeoning military-industrial-academic complex. He held a master's degree in electrical engineering from Columbia University, worked alongside academic scientists, divined the military aims for his sub and even went so far as to force management changes in his industrial contractors to insure his project's success. To gain support of his superiors, meanwhile, he used a trick perfected by Bush: writing letters for his bosses to sign.

Bush probably never met Rickover, but he must have heard about him from his protégé Carroll Wilson, the AEC's general manager, who had dealings with the Navy officer. While skeptical about the commercial prospects for nuclear power, Bush gave Rickover's efforts a nudge when his Research and Development Board asked the AEC to collaborate with the Navy on all aspects of the nuclear sub. This made good sense because, as Rickover's boss noted in an AEC conference in April 1948, just 1 percent of the design required for a submarine reaction had been completed. The Navy blamed the AEC for the lack of progress. After much foot-dragging, the AEC agreed in 1949 to appoint Rickover as its reactor chief, a job he held concurrently with his naval post. Two years later, he gained approval to build the first nuclear-powered submarine.

Rickover's drive to deploy an atomic submarine unfolded slowly, but otherwise reflected the "few quick" philosophy. Not only did Rickover master a difficult and dangerous technology, he translated his knowledge into a working prototype. He disregarded bureaucratic boundaries and overcame the sort of opposition that Bush predicted could befall innovators in military technology. He even gained the backing of members of Congress, enhancing his clout in somewhat the same way Bush had used the backing of President Roosevelt. In 1955, Rickover's pushing paid off. The first nuclear submarine, named *Nautilus,* steamed 1,300 miles submerged, a distance ten times longer than the norm. Her speed was unmatched.[29]

Bush's next broadside against the military came in September 1952. It was directed at the Joint Chiefs of Staff. The Chiefs were Bush's favorite whipping boys; he traced almost every defense lapse to their failures. It wasn't a question of finding the right Chiefs; for Bush the problem was structural. The Chiefs reported both to the president and to the secretary of defense. Each chief, moreover, naturally felt the tug of loyalty toward his own service. These conflicting allegiances made a single, cohesive military plan impossible.

The time was right to raise the issue. Truman, his popularity at a low ebb, had chosen not to run for reelection. The presidential race, pitting Eisenhower against Stevenson, was in many ways a referendum on the government's conduct of the Cold War. Hoping to influence the defense organization of the next administration, Bush lashed out against the Joint Chiefs in a speech at the Mayo Clinic in Minnesota. The present defense setup, he said, "is not effective" and the country "is not in a position to do its military planning adequately and well."

Citing the possibility that "an all-out war could indeed come," only to find the U.S. poorly prepared, Bush proposed that the Joint Chiefs be limited to an advisory role and that armed services report directly to the secretary of defense.

To make his case, he dissected the weaknesses of the Joint Chiefs, finding the staff too thin to craft proper plans, incapable of resolving differences between the services and an obstacle to fuller civilian control of the military. "The system is wrong. . . . We have confusion," Bush concluded. "Our top planning agency does not operate well for its intended purpose. It dips into matters it should avoid, it fails to bring well considered resolution to our most important military problems, and it fritters away its energy on minutiae."[30]

The speech drew wide notice. Dozens of newspapers reported on it. Walter Lippmann, the syndicated columnist, called the speech "extraordinary" and said "no one, in or out of uniform, is better qualified" to "speak about military planning in the Pentagon." *The Christian Science Monitor* opined that "most experts . . . would welcome the broad fundamentals" espoused by Bush, though they would oppose "giving so much detailed military command responsibility" to civilians. *The Minneapolis Star* urged its readers to discuss Bush's speech, asking, "Is our national defense paralyzed from above?" *The Washington Post* editorialized in Bush's favor, agreeing that even the best military chiefs "can be gravely handicapped by poor organization."[31]

The response to the Mayo speech lifted Bush's spirits. Still carrying the scars of his defeat at the military's hands in 1948, he desperately wanted his notions on defense reform to take hold. Asked to lobby Congress for higher spending on civilian research, Bush took a pass, telling Lee DuBridge, president of the California Institute of Technology, "I am getting too deeply involved with my private war with the Joint Chiefs of Staff" to spare the time. To James Killian, MIT's president, Bush wrote: "My object at the present time is to stir things up if possible in the direction of getting action in Congress when it convenes, and for this purpose it is not essential that there be agreement on all points, in fact it probably would be better if I brought up forcibly points on which I was sure there would be disagreement."[32]

The Mayo speech was an opening salvo in a flurry of statements by Bush on military leaders and the inability of Congress or the president to improve the nation's defense without sweeping reforms. He followed with another address two weeks later at Tufts College, his alma mater. Then he capped off the offensive by publishing a lengthy article in the popular weekly *Colliers* entitled, "What's Wrong at the Pentagon?" Bush's magazine piece laid out the stakes raised by his critique more incisively than had his more polite speeches. In short, poor military planning put the nation's very survival at risk.

"We may be in serious trouble soon unless a change is made," he insisted, then ticked off a list of ignored defense needs. The most notable: the failure to build an air defense system capable of stopping a Russian atomic attack. Each of the three services, he wrote, was involved in defense against air attack, yet no effective system existed because of the lack of "a single unified, integrated plan." Even with better overall planning, he admitted, "we cannot hope to build a system capable of stopping every bomber that attacks. But we can build, in a few years, a far more effective system than we now have or now plan."

Taken together, Bush's speeches and writings constituted a primer on how to increase national security without busting the federal budget. "The questions he posed really go to the heart of many of the greatest problems of war, peace and fiscal policy," Walter Millis wrote in *The New York Herald-Tribune*. While alienating himself from the top brass and the outgoing administration, Bush showed by his critiques that he "probably knows more about the nature and workings of [the military] system than any other man in civilian clothes."[33]

Having been applauded for pasting the Joint Chiefs, Bush felt emboldened. He tackled a more difficult issue: the hydrogen bomb.

The H-bomb promised destruction on a new order of magnitude. Its explosiveness could be increased many times by adding fuel. It could devastate one hundred square miles and shower radioactive fallout for hundreds more miles. It insured that any nuclear war would likely become a planetary disaster.

The idea of the H-bomb dated back to the Manhattan Project. Technical difficulties—chiefly the problem of containing the bomb's potential force until detonation—convinced J. Robert Oppenheimer to set aside this possibility. After the war H-bomb enthusiasts pressed for government support, but failed to convince key scientists. With the Russian A-bomb test in 1949, the H-bomb lobby seized the opportunity to win support for what they called the "Super," a hydrogen bomb with an atomic trigger. Seven of the eight scientists advising the AEC, including Oppenheimer, advised against pursuing the Super in October 1949. The advisers cited both technical and moral grounds. The AEC sustained the advice on a split vote, but Truman reversed the decision in January 1950, ordering work on the Super to continue. Six weeks later, the president approved an all-out effort. His decision stemmed in part from the arrest and confession of Klaus Fuchs. The revelation of his espionage underscored the possibility that the Soviet Union was much further along in nuclear weapons than the U.S. realized. In opting for an H-bomb, Truman listened to military leaders and a minority within the atomic establishment. This hard-line faction was led by AEC commissioner Lewis Strauss, who told the president that a "government of atheists [the Soviet Union] is not likely to be dissuaded from producing the weapon on 'moral' grounds" (Strauss was right: the Russians had the basic concept for a workable H-bomb by the end of 1948 and began their construction effort about November 1, 1949).[34]

In less than two years, the U.S. was ready to test its first H-bomb. The date was set for November 1, 1952. This occasioned a renewed debate over U.S. disarmament policy and whether an approach should be made to the Soviets about banning the H-bomb. In April 1952, Secretary of State Dean Acheson appointed a five-person committee, including Bush, Oppenheimer and CIA chief Allen Dulles, to advise the government on arms control.

As a member of the panel, Bush learned of the forthcoming test and grew alarmed. "I was very much moved [to action] at the time," he later said. He thought the U.S. ought to cancel the test and open talks with the Soviet Union on banning all H-bomb detonations. No such talks had occurred before Truman's 1950 order. Bush now feared that a first test would doom any chance to control this new class of nuclear weapons. The risks posed by these weapons was so great that Bush proposed at the panel's second meeting, in early May, that both countries simply pledge not to test H-bombs. No inspections were needed, he insisted, because evidence of an H-bomb test could not be hidden. The agreement would be "self-policing in the sense that if it was violated, the violation would be immediately known."

Bush's opinion represented an extraordinary turnabout. He had long accepted the inevitability of nuclear weapons and had been the most influential adviser in Roosevelt's 1941 decision to approve an all-out project. He had never expressed guilt about his role in making the atomic bomb and even had argued that the use of the weapon had forestalled the development of horrible biological and chemical weapons. Yet in the 1950s, his views had shifted subtly. Now he wanted more to slow the spread of nuclear weapons. While still voicing no regrets about either the Manhattan Project or Hiroshima, he began to show remorse for his atomic past, his son Richard believed, by trying to invent heart valves and other devices that might help to save human lives.

Stopping the first H-bomb test would be a more direct way for Bush to overcome his past. In the spring of 1952, the State Department arms panel—in a top secret, unsigned memo—urged that the H-bomb test be delayed. Bush wanted to make a similar, but personal, appeal to Acheson. Oppenheimer and the panel's secretary, McGeorge Bundy, encouraged him to do so. Bundy, whose father, Harvey, had worked closely with Bush during World War II, called Bush as an ideal emissary since he was "very clear" on the issues and unafraid of going against the prevailing mood.

Bush realized the Soviets might simply ignore a U.S. appeal to ban H-bomb tests, but he thought it was worth a try. He also had two other objections to the test. First, the test was scheduled to take place just three days before the presidential election. Bush felt it was only fair to allow the new president, whom he assumed would be Eisenhower, to decide whether to test an H-bomb. However, given Bush's distance from Truman, this argument looked like a slap against the sitting president. Never long on politeness, Bush thought the circumstance too serious to spare the president's feelings.

Bush had a second and in some ways more powerful argument against the test. He believed that the fallout from a hydrogen explosion would tell the Russians (not to mention scientists from other countries) much about how the U.S. actually made its H-bomb. While this was a reasonable technical concern, Bush could not quantify just how much help an analysis of fallout would give the Russians.

Bush made no record of his conversation with Acheson, but Bundy later described the gist of the talk: "Bush made his case, and Acheson was not persuaded."

Bush went no further. It was hopeless to visit the president, who no longer listened to him anyway. Truman subsequently considered a delay in the test after complaints from an AEC commissioner about the proximity of the test to election day. But the president decided to go ahead when he learned that weather conditions would cause a delay in the test of at least a month if it did not occur in early November. The test, codenamed "Mike," went off flawlessly on schedule. The blast, roughly a thousand times more powerful than that of the Hiroshima bomb, obliterated an island in the Pacific one mile in diameter and left a huge crater in the ocean floor.[35]

Truman called the explosion "a dramatic success." Bush saw it as a setback. Two days after the test, he replied to a query from Bundy, suggesting cynically that, after the election and the departure of Acheson as secretary of state, no one would want advice from the arms panel. Since the purpose of the Mike test leaked to the press—it was to have been disguised—Bush felt free to voice his discontent in public. The featured guest on Walter Cronkite's *Man of the Week* show, Bush explained on November 23 that cheering the test was misguided. "We are in a hydrogen bomb race," he said gravely. "We must assume that if we can make the H-bomb the Russians can, too."[36]

Since the test was still officially a secret, Bush would not publicly say he had tried to cancel it. He contained his anger, allowing it to boil over only in early 1954. Two events stirred up talk about the nation's nuclear program and set Bush afire: the second H-bomb test and the charge against Oppenheimer of disloyalty, lodged chiefly because of his principled disagreements over H-bomb research. In late March, Bush told a friend that the public was finally grasping the ramifications of an H-bomb race. Looking back remorsefully at his failed attempt to halt the first test, he noted, "I feel that an opportunity was missed and that history will probably record that it was."[37]

Speaking in defense of Oppenheimer in April 1954, Bush went further. He damned those who had pushed so relentlessly for a test without even making a show of forging a pact with the Russians. "That test ended the possibility of the only type of agreement that I thought was possible with Russia at that time, namely, an agreement to make no more tests," he said. Then he added: "I still think that we made a grave error in conducting that test at that time, and not attempting to make that type of simple agreement with Russia. I think history will show that was a turning point when we entered into the grim world that we are entering right now. That those who pushed that thing through to a conclusion without making that attempt have a great deal to answer for."[38]

Bush's words reverberated for many years. Aboveground tests of H-bombs caused irreversible damage to islanders in the Pacific and, because of radioactive fallout, raised the incidence of illnesses worldwide. In 1963, the U.S. and the Soviets agreed to conduct only underground tests. These did not cease until the 1990s.

With Eisenhower's election in November 1952, Bush looked forward to more enlightened management of the armed forces. If anyone could tame the military, Ike could, Bush said. If he "appoints a good man" as secretary of defense, some Pentagon problems "will take care of" themselves, he wrote a friend. Of his own efforts to reform military organization, he told Karl Compton, "I really think there is a chance something positive may be accomplished."[39]

For a time, it looked that way. On the losing end of too many political battles, Bush relished the chance to influence official policy. His broadside against the Joint Chiefs and the Pentagon had coincided with a trenchant review of the military establishment by outgoing defense secretary Robert Lovett, who shared many of Bush's views. Soon after the election, Lovett told Truman that the next secretary needed more authority over the three military services and a clearer relationship with the Joint Chiefs. He recommended that all boards in the Office of the Secretary, including the Research and Development Board, be abolished, with the responsibilities transferred to the secretary.

While the individual services naturally balked at the sort of centralization urged by both Lovett and Bush, no small number of civilians also defended the current system of fragmented defense responsibilities and conflicting missions. Robert W. Johnson, of the Johnson and Johnson company, warned that "centralized military management does well at the start but loses its wars," arguing that in "modern management . . . there is merit in centralization and equal or perhaps greater merit in decentralization." Johnson told one administration official that Lovett and Bush were "committed to the German [military] system which captured civilian authority, lost two wars and ruined Germany."

Sweeping aside critics of centralization, Eisenhower moved to reorganize the Defense Department. In February 1953, his choice as defense secretary, General Motors executive Charles Wilson, asked Nelson Rockefeller to chair a committee to review the options. Bush was asked to serve on the group, as was Lovett. The choice was a natural one since Bush's views on defense reform jibed with those of Lovett, who also joined the committee.[40]

The Rockefeller committee made quick work of its task, issuing a 25-page report on April 11. It backed Lovett's call to abolish the Pentagon's boards, which would be replaced by an array of assistant secretaries who would serve as the equivalent of senior executives in a functionally organized big corporation (one would oversee military research). It also called for "a single channel of command or administrative responsibility" within the Pentagon, more authority for the defense secretary, and more power for the chairman of the Joint Chiefs, who under the current setup had no formal vote in the body. Eisenhower accepted the committee's advice. In late April, he told Congress of his plans to reorganize the Defense Department.[41]

The reforms took effect June 30, 1953. By Bush's standards they were modest but still a vindication of his outspoken stance. "Of course it is not exactly the result that I would personally have chosen," he observed, "but I am sure that it is going to be a result that I can personally support with enthusiasm." The military reorganization, one analyst wrote, "wiped out the last vestiges" of the 1947 concept of three semiautonomous departments coordinated by a secretary of defense and, in giving more power to the Joint Chiefs chairman, enabled the Pentagon to take "an enormous step toward a single military commander over all the armed forces." While certainly not creating "an all-powerful military chief"—a post basically favored by Bush—"it clearly moves toward a supreme military chief with his own general staff."[42]

Despite the new administration's acceptance of some of Bush's ideas on defense reform, his honeymoon with Eisenhower proved short. He had actually never been enthusiastic about Ike's candidacy, just relieved that 20 years of what he called "Labor" government had come to an end. Within two weeks of Ike's election he had felt "rather blue" about defense affairs because he felt it was "presumptuous" to give the new president advice on military matters. But at the same time "this discourages me quite a bit. . . . What I fear will happen is that Ike will choose a Secretary soon, and then a number of Department Secretaries and Under secretaries will be appointed for various reasons without a planned attempt to form a coherent and harmonious group and the whole thing will get frozen pretty much in the present pattern."[43]

To Bush, this was pretty much what happened. Some of his dour assessment was just sour grapes, of course, since he had no real contact with Eisenhower or his administration other than through the short-lived Rockefeller committee. The president's national security adviser, Robert Cutler, occasionally exchanged ideas with Bush, but that hardly counted as systematic advice. In time, Bush grew disappointed with the president. Like many others who neither knew him well nor observed him up close, Bush saw Eisenhower as an absentee executive, timid in his dealings with the services and unwilling to expand the nation's conventional defenses. He "does not get to the bottom of things in time to give us the strong leadership we need so badly on so many matters," Bush told Nelson Rockefeller in May 1954. A few weeks later, he confessed to Conant: "The difficulty with Ike as I see it is that he regards himself as a sort of ceremonial President. He apparently does not read much and the people around him tell him very little. So I doubt whether he really grasps what is going on."[44]

In the area of nuclear strategy, Bush had no doubts: Eisenhower did not get it. The president seemed bent on using the nation's nuclear arsenal as a way to insure security on the cheap. While a fiscal conservative, Bush wanted more emphasis on defense against nuclear attack. In early 1953, the State Department panel on arms control, on which Bush served with Oppenheimer, issued its final report. It reflected Bush's twin priorities in the area of nuclear strategy. First, the report called on the government to adopt a policy of candor toward the American people, by revealing all of the dangers posed by the arms race with the Soviet Union. This advice was not inspired by any democratic impulse, but by the elite view that Americans, terrified by the full panorama of nuclear horrors, would back higher taxes to pay for possible remedies. Bush's favored remedy was described in the panel's second important recommendation. It urged that the military intensify its defensive efforts to protect the nation against a surprise knockout attack by the Russians.[45]

The panel's advice was thrown into bold relief in August when the Soviet Union tested its first hydrogen bomb. The Soviet test put Eisenhower in a quandary, since it was now clear, as Oppenheimer vividly noted in a July article in *Foreign Affairs,* that the U.S. and Russia were like "two scorpions in a bottle, each capable of killing the other, but only at the risk of his own life." Reiterating the call made in secret by the State Department arms panel, Oppenheimer insisted that the government candidly describe its nuclear arsenal.[46]

By November 1953, Eisenhower had still not embraced the "candor" tactic or endorsed an all-out effort to build a continental defense against nuclear attack. At a White House dinner, Bush, who rarely saw the president, was ready to pounce on the topic of arms policy. When Bush arrived late for the predinner cocktails, Eisenhower lightened his mood by personally getting him a drink. "He might have called a waiter, but that wasn't Ike's way," Bush thought. After dinner, Eisenhower baited Bush by commenting on "people's seeming reluctance to recognize the threat of the Hydrogen bomb." Bush "immediately took up the case for scaring the people into a big tax program to build bomb defenses." But the prospect of a hysterical public, throwing fiscal caution to the wind, was precisely why Eisenhower saw risks in telling citizens too much. Security must not be bought at the price of American values. "We must not create a nation mighty in arms that is lacking in liberty and bankrupt in resources," he had told the Congress in April 1953.[47]

Instead of candor and beefier defenses of home turf, Eisenhower opted in 1954 for what he called the "New Look," which aimed to contain Soviet expansion by threatening "massive retaliation" through nuclear attack. The hope was that the costs of defense could be contained through new technologies, which by threatening nuclear annihilation would discourage limited wars. The strategy had the benefit, in the president's view, of holding down defense costs, then consuming 70 percent of federal spending. Eisenhower was convinced that cuts in military positions around the world were possible because, as his secretary of state put it, the nation now had the "great capacity to retaliate by means and at places of our own choosing."[48]

The New Look was certainly inexpensive, but it represented an extraordinary gamble. If Russia ever called the U.S. bluff—say, by invading Western Europe or lauching a first strike on North America—the U.S. could only really respond by making nuclear war. If the U.S. failed to flatten Russia, it would likely be hit by Russian bombs. The exchange could well devastate a good part of the world. That Eisenhower failed to appreciate these possibilities came through rather chillingly in his comment, "Where these things [nuclear weapons] can be used on strictly military targets and for strictly military purposes, I see no reason why they shouldn't be used just exactly as you would use a bullet or anything else."[49]

Bush joined many Americans in rejecting the New Look. He did so privately but forcibly. Uncomfortable voicing his complaints directly to Ike or Secretary of State John Foster Dulles, Bush sent a long letter in February 1954 to Dulles's brother Allen, the CIA chief, whom he had come to know during their stint on the arms panel. "The Secretary [of state] may be right," Bush wrote. "Any war may be bound to become an atomic war, and to leave ourselves and our enemy prostrate. But I am not ready to accept this yet."

In a perceptive critique of the New Look, Bush echoed the views of a raft of scientists who were starting to show the technical feasibility of defensive measures against nuclear attack. But he also deftly exposed the contradictions in the administration's approach to security at home and abroad.

Certainly I do not subscribe to the idea that the hordes can be held only by a strategic atomic air force of great power [Bush told Dulles]. We say in effect that if the enemy starts to hit us in a vital spot, we will demolish him with A- and H-bombs, for that is what will be assumed retaliatory power will mean. I trust we could get through to the Russian centers with our bombing fleets. Larry Norstad says we could, and he ought to know. I would feel better personally if I had convincing evidence that Russia has not built and cannot build in a reasonable time a defense system that could stop us. But usually, when one says—you move and I'll sock you—he has a pretty good defense of his own. *We are still wide open.* Our defense system is only now being pushed forward, and we have lost several years. Even now it is not organized to be one of our main efforts. . . . If we were going to say—any war and it is an atomic war—I think we might look very hard at our defenses first.

It could work the other way. Suppose Russia takes a similar position. How will our people stand up if we are confronted with an ultimatum or a move, and it is probable that if we meet it, we will have 20 million casualties within a few days? It won't make much difference what the Russian people think. In such a game it seems to me we are on the short end.

At the end of the war our A-bombs were a deterrent. With a few of them in sight we necessarily considered strategic use. When A-bombs multiplied we might have altered our approach. Europe would have felt better if we had then emphasized their use to stop armies. But the strategic idea has had enormous momentum, and some of its support has been emotional rather than logical. We are still going down this alley, and, it seems to me, somewhat blindly.

Bush's point was that the nation had bet too heavily on its nuclear card. He wanted greater conventional military power than Eisenhower and his advisers thought necessary. He wanted a wider range of military options, from "a powerful strategic striking force" to "an air defense system far tighter than that of Russia" to all the "things that are modern for land warfare." While a fantastic wish list, this was a precondition for a philosophy of security that eschewed brinksmanship. Only then could the U.S. act rationally on the world stage. "Then I think we would take the position," Bush concluded, "that, if the enemy moves, we will destroy his military forces."[50]

Bush's discontent with the nation's military posture mirrored his attitudes toward many aspects of American society and culture in the 1950s. Unsure of the nation's direction, he played the part of prod and gadfly, steadfastly reassuring Americans that no catastrophe lay around the next curve on the road but relentlessly reminding them of their weakness and lack of foresight. His cultural criticism, however, often seemed confused. He fretted publicly that

Russia was "training more scientists than we are," yet complained that the U.S. government's free-spending ways would lead to a glut of scientists. He feared the quality of research, force-fed on scholarships and grants, was declining. He congratulated the Russians for screening their "country for talented kids more than we do," yet chastised the Supreme Court for striking down in 1954 the "separate but equal" educational doctrine that consigned African-American schoolchildren in the South and elsewhere to inferior schools. He denied that "to segregate Negroes placed a stigma on them" and failed to realize the extent to which racial discrimination constricted the national pool of scientific talent.[51]

On the subject of civilian research, an area of expertise, his complaints hit the mark. In the spring of 1950, Congress passed, and the president finally signed, the law creating the National Science Foundation. Truman called the new foundation "a major landmark in the history of science in the United States." For Bush, it was too little, too late. His friend James Conant agreed. "Truman signed [the NSF law] . . . too late," he wrote. "The armed forces had taken over."

The NSF's charter was related to the one Bush envisioned in his 1945 report, *Science—The Endless Frontier,* but many important features had been lost in the five years of wrangling. The foundation had no military division, which in many ways had been the linchpin in Bush's plan. There was no medical division either; no provision for the foundation "to develop and promote a national policy for scientific research"; no real power in the hands of the foundation's board, since the director was appointed by the president and made the key operational decisions. Congress also imposed a $15-million ceiling on the foundation's annual budget and approved just $250,000 for its first year. "Rather than subsuming federal science policy under a single agency, as [Bush] originally envisioned, the NSF only added yet another contestant to an already crowded field."[52]

Still widely seen as the foundation's parent, Bush felt he had given birth to an orphan and had little to do with it. He even refused to allow researchers from the Carnegie Institution, which he still headed, to accept foundation grants. Of the foundation's first chief, meanwhile, Bush had little positive to say. Alan Waterman had been considered a safe choice to run the foundation. The former chief scientist of the Office of Naval Research, Waterman was a low-key administrator with exceedingly modest ambitions for his new agency. Bush felt that with the birth of the foundation, "It is completely improper for military agencies . . . to contract with universities for basic research." Waterman, however, refused to infringe on the military's sponsorship of academic research, content to fill in the gaps.

Waterman was cautious. "That was his character. Some men grab power and others just don't," said Emanuel Piore, who replaced Waterman as the top civilian at the Office of Naval Research. As the type of man who grabbed power, Bush had little sympathy for Waterman, who so frustrated him that he finally threw down the gauntlet. In a March 1954 note to Waterman, Bush warned that "the Foundation will now either take a position of strong leadership, or it will cease to exist as an effective agency. I do not know, of course, how the Board will go about its affairs on this matter, but it certainly has to move positively and promptly. The job can not be done without antagonizing some of the other agencies, at least in minor ways, and an attempt to do it by general agreement and compromise, would, I think, now fail."

The tirade had no discernible effect on Waterman. Averse to conflict, he exercised his authority so quietly and cautiously that even the foundation's official historian called it "an insignificant agency" as late as 1957. Upon Waterman's retirement as NSF director in 1963, Bush was asked to address a gathering in his honor. Bush had never publicly voiced his disappointment in Waterman. He would not do so then either, but his speech betrayed his feelings. After reading ten single-spaced pages on the foundation's history and his own exploits, Bush finally arrived at the subject of Waterman—and devoted all of one paragraph to him, lamely hailing him for shaping "the course of science in this country in a salutary manner."53

Bush always resented the NSF's subsidiary role. Still insistent that independent civilians should hold sway over the military's hired technologists, Bush ignored the practical difficulties facing any civilian agency that competed with the military for funds. He also offered contradictory advice, at times emphasizing the need for civilian leadership and at other times warning against profligacy. Still uneasy about the state's role in research, Bush often complained that funds for research now flowed too swiftly. At a foundation seminar in 1953, his chief point was the "danger" to the foundation posed by its core practice of "giving away money." Sounding more like archconservative Frank Jewett than the author of *Science—The Endless Frontier,* Bush told an audience of foundation supporters to be ever vigilant against pork-barrel spending on research. "Concealed or disguised subsidy invites lobbying, puts a premium on scientific salesmanship—[and] might eventually prostitute the universities," he said. Still fearing that second-rate colleges and bogus research projects would grab federal grants, he wondered if top-flight scientists could achieve "freedom from political influence." Given "the present state of maturity of this republic," this was "a great deal to ask."54

Bush had long worried about political interference in research. However, the Red Scare of the late 1940s and the early 1950s revealed a new and troubling dimension to the question of scientific independence and relations between the research community and the government.

As a moderate conservative of unquestioned integrity, Bush was never accused of disloyalty. He accepted the need for security checks and took pride in the OSRD's unblemished record of keeping secrets. But his experience during World War II had taught him that even unorthodox researchers could be unstintingly loyal and so deserved toleration.

The end of the war posed new insecurities. Like many Americans, Bush at first accepted the heightened level of concern about suspected disloyalty of goverment employees and consultants. He hoped any issues of actual treason could be dealt with discreetly and by men of good will, removed from the partisan politics that lay barely below the surface of many security allegations. The war had shown beyond any doubt that even scientists with stridently liberal views could keep defense secrets. This reality made it imperative that the country not reflexively place every critic of capitalism under suspicion. He wrote War Secretary Patterson in March 1947:

> There is an enormous difference between the liberal, who may have, from the philosophic standpoint, a yearning for some new system other than the present capitalistic system, but who dislikes totalitarianism in any form and specifically dislikes the Russian form and is a good American on the one hand, and the individual who *claims* he is a liberal but who is really working for the success of Russia's attempt to impose its absolutism on the world. Our problem of distinguishing is a difficult one and I hope we do not do foolish things in this country. Above all things, I hope we can continue to utilize the services of those men who were far out in their liberalism before the war but who have seen the light and are now as intensely loyal Americans as anyone else, sometimes more so because of their experience.[55]

Bush was right to worry that the nation might act foolishly in its zeal to root out subversives. A year later the first celebrated disloyalty case involving a scientist arose. The target of the House Committee on Un-American Activities, which led the drive against purported subversives, was physicist Edward Condon, director of the National Bureau of Standards. The House committee on March 1, 1948, labeled Condon "one of the weakest links in our atomic security," accusing him of ties to suspected communist-front groups, notably the American-Soviet Science Society.

There was scant evidence against Condon. The most serious claim was that he associated with an alleged Soviet spy. But even if this were true, which it probably was not, Condon stood accused of guilt by association. The aggrieved congressmen never suggested his actions had harmed U.S. security. Condon defended himself unhesitatingly, even poking fun at his accusers. "If it is true that I am one of the weakest links in atomic security that is very gratifying and the country can feel absolutely safe," he said, "for I am completely reliable, loyal, conscientious and devoted to the interests of my country, as my whole career and life clearly reveal." Vouching for his loyalty was his employer, the Commerce Department, which had cleared him of suspicion the preceding month.

Still, the charges against Condon split the scientific community. Individual scientists and liberal scientific groups such as the Federation of American Scientists defended Condon forthrightly. Older, established groups were more restrained. The National Academy of Sciences, for instance, decided against issuing a ringing endorsement of Condon, settling for a tepid statement that voiced concern about the process of judging Condon but noted that House committee members had assured the Academy the physicist would be treated with "complete fairness."[56]

The Academy's response—so meek that no newspaper featured it—came despite the desire of an overwhelming majority of its members to take a stronger stand in Condon's favor. The Academy's president, A. N. Richards, rejected this path on the advice of, among others, Bush, who raised serious doubts about Condon's conduct. Richards had served as chair of OSRD's medical committee and knew Bush well from the war. Bush's relations with Condon, meanwhile, were strained, because after the war Condon had spoken publicly against Bush's legislative plans for both atomic energy and civilian science.[57]

Bush did not object to the Academy issuing a statement on Condon's behalf, especially if it made clear that "scientists will serve the country through hell and high water but they don't enjoy" the type of treatment accorded Condon. Bush, however, wanted the Academy to speak softly because he thought Condon might have acted improperly. "Underneath [Condon's protests] there is in all probability a record of lack of proper care in the type of remarks he has made and the type of associates he has sometimes had," he insisted, without citing any evidence.

While mistrusting Condon, Bush worried about the implications of attacking scientists or any other citizen. But he saw this as necessary. "The system is wrong of course," he wrote. "It violates the bill of rights, and makes us fear for our liberties. But anyone who criticizes [the loyalty program] needs to show either that it is unnecessary, or that there is a better way. There is [a better way], of course, but it consists in getting more fairmindedness in Congress and that is hard and takes time." Bush thought there was not even "the slightest chance" of that happening, chiefly because of the public's attitudes and the strange logic surrounding the whole subject of disloyalty. As he observed:

Specifically, I think the American people feel:

1. that Commie infiltration constitutes a genuine menace in this country

2. that the atomic energy matter is a field where we must be especially careful

3. that any government official handling secrets in the field should be rigorously careful of his associates and employees

4. that Condon has not been that careful

5. that quite apart from loyalty, those who serve him, and he himself, have much to explain . . .

6. that even though they may deprecate the methods used by [Congress] some such move was needed to uncover a situation that was dangerous

7. that those involved and their friends would do well to improve their ways rather than protest

8. that people who protest loudly either do not understand or have ulterior motives for protesting.

I believe all this in their minds will submerge the process of fairness.[58]

The Condon case revealed the depth of Bush's sympathy for the witch-hunters. For a time, he looked askance at those who decried the mania for security, insisting in 1949 that the mania had not and would not make scientists reluctant to aid the government "for fear of smear." The same year, Bush's frank desire to guard against spies prompted the FBI to consider him as a speaker at a graduate ceremony for new agents. In 1949, he also told Conant that while he would "make a place" for "honest liberalism of any stripe," he wanted the right to "call a traitor a traitor."[59]

Yet even as Bush stayed alive to possible subversion, the FBI opened a "preliminary investigation" into Bush's loyalty in May 1948. The secret inquiry, which came just months after he turned his back on Condon, was prompted by a report to the bureau that Bush had received an invitation to attend a forum by a left-wing scientific group. The inquiry turned up "no derogatory information whatever," and the bureau closed the case.[60]

By the early 1950s, scrutiny of scientists was high; in the 22-month period beginning June 1952, one federal department, Health, Education and Welfare, canceled nearly 30 grants to researchers for "security reasons." Senator Joe McCarthy, meanwhile, made a mockery of legitimate security procedures by fabricating allegations against innocent people in order to reap political gain. Bush began to lose his patience with the witchhunters. Then in late 1953, the Red Scare hit close to home. He learned that the loyalty of Robert Oppenheimer, the scientific chief of the Manhattan Project, was in doubt.[61]

Bush held Oppenheimer in high regard. Hours after the test at Trinity, in 1945, he had stood at attention to salute as the physicist's car passed him on the road. In 1947, he had helped stop an attempt to deny Oppenheimer a role in the new Atomic Energy Commission based on his sporadic contacts with Communist party members. In 1952, they had served on the State Department panel that sought a delay in testing the H-bomb. Bush was not inclined to look for government conspiracies, but he knew Oppenheimer had made many enemies by questioning the technical, strategic and moral dimensions of the H-bomb. It did not take Bush long to conclude that if there was ever a trumped-up, phony security case, this was it.[62]

The case against Oppenheimer was indeed an attempt to settle old scores. After the Soviet Union first tested its H-bomb in August 1953, an Oppenheimer critic sent FBI chief Hoover a rehash of the old claims about the physicist's leftist ties, adding the incendiary charge that he had opposed development of the H-bomb in 1949 in order to harm the U.S. The attack on Oppenheimer found a receptive audience, especially in the chairman of the Atomic Energy Commission, Lewis Strauss, who had rejected Oppenheimer's advice five years before and convinced Truman to go ahead with the H-bomb.

In December 1953, Eisenhower secretly stripped Oppenheimer of his access to any atomic secrets. Bush learned of the order soon afterward, telling a stunned James Conant "the grim news" on a visit to Germany, where Conant was serving as U.S. ambassador. Oppenheimer declined an offer to resign as an AEC adviser in return for dropping the charges; he planned to defend himself at a private AEC security hearing. Bush agreed to testify on his behalf. In March 1954 he told Oppenheimer's attorney, Lloyd Garrison, that he would not only vouch for the physicist's loyalty, but also defend his 1949 decision to eschew development of the H-bomb on technical grounds.[63]

Bush found the Oppenheimer imbroglio lurid, and not just because the physicist was being put through "an ordeal which may break him." He also felt that for the first time the loyalty craze might seriously damage the intricate relations between researchers and government. The intimate connection between these two realms was Bush's most durable legacy; now he acted as if his favorite child were in distress. A few days before testifying at Oppenheimer's hearing in April 1954, he wrote to Lewis Strauss, the AEC chairman. Bush had never liked Strauss, whom he had met during the war, when he had succeeded in keeping the then-naval officer at a distance from OSRD matters. Strauss epitomized for Bush the perils of placing politically motivated dilettantes in positions of influence over technological affairs. Masking his disdain, Bush told Strauss that he feared the "disintegration of morale" among scientists in the wake of Oppenheimer's humiliation. "My whole thought, in these trying days," he wrote, "is to attempt to further effective relations between scientists and their government, which I labored to create during the war years, and which now are in jeopardy." The risk was that the U.S. would come to resemble the "one thing we object to in the Russian system. . . . When the top oligarchy has decided upon the party line, if a citizen then even expresses doubts his career is destroyed."

Bush was "deeply troubled" by the situation, but as so often during the postwar years he seemed paralyzed by conflicting ideals. He valued individual freedom above all else, yet accepted the goverment's need to control information and dissent on matters of military technology. At a loss for a solution, he told Strauss that he considered making an appeal to President Eisenhower, but was "baffled" about how "he could act to solve" the problem. His patience was nearly exhausted. "The hunt for subversives," he wrote, "has done great harm to our national reputation. . . . I have looked to the President to lead us out of this morass, to insist on decency with vigor, and now I am confronted with this."[64]

Bush assumed Eisenhower had not clearly looked over the Oppenheimer case, but as usual he underestimated the president. Ike had spent nearly three days on the matter, from April 9 through 11, and come to believe that Oppenheimer had inexcusably tried to slow the project after Truman's decision to pursue an H-bomb. He wanted Oppenheimer removed from the AEC so he did not have the chance to spread doubts among nuclear scientists. His reasoning reflected very much the notion that the physicist should toe the government's line once a decision had been made. Eisenhower made no apologies for the move against Oppenheimer, nor did he see any parallels to the Soviet system. He wished to keep the proceedings secret, he told an aide, "so that all our scientists are not made out to be Reds."[65]

That proved impossible. On April 6, Senator McCarthy declared in a televised address that a procommunist faction in government had conspired against the H-bomb, delaying its completion by 18 months. Who caused the delay, he asked? "Was it loyal Americans or was it traitors in our government?" McCarthy knew about the case against Oppenheimer, but kept quiet about it. On April 13, the charges against the physicist erupted in the press, illuminating McCarthy's veiled attack.[66]

The day before, Oppenheimer's security hearing had opened in secret before a three-person panel. Eleven days later, on Friday, April 23, Bush arrived at room 2022 in AEC building T-3 at 2:00 P.M. In 50 minutes of testimony he showed all the outrage he could muster. It was plenty. Not only did he defend Oppenheimer, he attacked those who had pushed pell-mell for an H-bomb test. "History will show," he said, that the nation "made a grave error." He questioned the authority of the AEC to bring a case against Oppenheimer, saying it had "made a mistake . . . a serious one" in "placing a man on trial because he held opinions." This was "quite contrary to the American system" and "a terrible thing" because it left the impression that Oppenheimer "is being pilloried because he had strong opinions, and had the temerity to express them." This dishonor to Oppenheimer had "deeply stirred" the scientific community and made Bush wonder whether "the republic is in danger because we have been slipping backward in our maintenance of the Bill of Rights." He then alluded to his own opposition to testing the H-bomb, implicitly asking who would be consumed next by the witchhunt. He sought to expose the fact that by the committee's twisted logic many decent Americans could be judged guilty of imaginary crimes for dissenting against consensus opinion. Indeed, good government was dependent on the willingness of people to dissent when they saw fit. The very reason independent advisers could aid government, he believed, was their willingness to disregard fashionable opinions:

> I think this board or no board should ever sit on a question in this country of whether a man should serve his country or not because he expressed strong opinions. If you want to try that case, you can try me. I have expressed strong opinions many times, and I intend to do so. They have been unpopular opinions at times. When a man is pilloried for doing that, this country is in a severe state.[67]

Bush sat tight after his testimony, fearful of the outcome but still hopeful that Strauss or Eisenhower might come to their senses. "Of course, I have not talked to the press, and I have urged others not to, while the Board considers the matter," he wrote Strauss on April 28. Two days later, he explained to a professor at Columbia that he refused to defend Oppenheimer publicly because his "effectiveness, if I have any, is best preserved" by advocating for the physicist through official channels. But he admitted it was hard to stay silent in public. "This whole thing has stirred me so deeply that it is difficult indeed for me to be normal when any aspect of it is mentioned," he wrote. In holding his fire, he hoped he was "taking the right course."[68]

As it turned out, Bush's restraint went for naught. Oppenheimer's hearing ended on May 6, after 19 days of testimony. Three weeks later, the security board voted two-to-one to strip the scientist of his clearance. Bush was livid. He no longer saw any reason to stay within channels. Off went his gloves. On

June 11, he kicked out of his office an FBI agent who was doing a routine security check for the AEC on a civilian scientist. When the agent began asking
questions, an irritated Bush blew up. "I don't have time to answer your questions," he said. "You can go back and tell your boss that." About the same
time, Bush gave an impassioned talk at the private St. Botolph Club in Boston,
telling his audience that the Oppenheimer case represented "thought control"
and warning that "democracy has walked up to [the] brink of chasm."[69]

Then on June 13, Bush delivered in *The New York Times* what one historian
has called "the most devastating broadside" yet against Oppenheimer's accusers. No citizen should "suppress his honest opinions in order to slavishly
follow a policy arbitrarily laid down," he wrote. A grateful David Lilienthal, a
liberal who had chaired the AEC when Oppenheimer had dissented against
the H-bomb, thanked Bush for his pronouncement. "Your article . . . may
prove to be among the most important of your many contributions to your
country," Lilienthal wrote. "Common sense and decency are having a bit of a
rugged time these days—your article will help get us back on the track—and
certainly make most people realize how far off we're getting."[70]

Indeed, many Americans were fed up with the scourge of anticommunism,
or at least its worst excesses. Earlier in June, McCarthy finally received a stinging rebuke, when in a hearing on his conduct, a soft-spoken lawyer named
Robert Welch destroyed him with the simple question, "Have you no sense of
decency, sir, at last?" (In December, McCarthy would be censured by the Senate.) Just as the McCarthy hearings came to an end, Bush wrote Conant on
June 17, "On balance I think that we probably have reached the peak of absurdity and that the tide will now begin to turn. I do not of course mean that
I think the country will become sane over night, but we may see less of excess
in the months to come."[71]

Bush's caution was soon justified. Despite the outrage over Oppenheimer's
treatment, the full Atomic Energy Commission, led by Strauss, confirmed the
security board's verdict on June 29. It found "proof of fundamental defects in
his character" and judged his associations to be "far beyond" the tolerable limits of prudence and self-restraint.

Oppenheimer had lost. He would never again work on U.S. nuclear
weapons or influence arms policy. To Bush, the case was a personal tragedy but
also an indictment of the government. "Our internal security system has run
wild," he told *Newsweek*. The AEC's verdict spoke volumes about the politi-
cization of expertise but nothing about Oppenheimer. The commission's decision, he said, "does not affect my complete confidence in Dr. Oppenheimer's
loyalty and deep devotion to the security of the U.S."[72]

The defense of Oppenheimer pained Bush but carried a measure of satisfaction for him. In the immediate years after World War II, many scientists had dismissed Bush as an autocratic conservative who had lost touch with the well-springs of intellectual freedom. They found him too closely allied with the military, a charge that seemed ironic after his intense battles with the armed services. Bush's transformation from the ultimate insider into a strident out-sider was established beyond doubt by the record of the Oppenheimer hear-ing, published as a book in July 1954. Reviewing the thousand-page tome, Alfred Friendly of *The Washington Post* listed Bush as a "leading defender" of Oppenheimer and among the few involved in the security hearing to emerge with a burnished image. To those scientists who had once turned against Bush, Friendly offered a sweet postscript. In siding with Oppenheimer, Bush had proven himself to be, after all, "the Grand Old Man of American science."[73]

Chapter 16

"Crying in the wilderness"

(1955–70)

Now scientists are regarded as supermen. They can do anything, given enough money. If America wants to put a man on the moon, which is really a tough engineering job, they just gather enough thousands of scientists, pour in the money, and the man will get there. He may even get back.

—Vannevar Bush

Not every person has his time. Some are born too soon, others too late. Some seize their moment, others bungle their opportunities and then pine for a second chance. Some overthrow conventions, others worship them. Some wear masks through the spectacle of life and don a new mask with each new scene. Others stay their course, realizing their moment as fully as if they had been bred for it. Successful because of their substance, these people are either broken by their times or stand fast, taking the blows and inspiring allies and opponents alike.

Bush was one of these people. In a world at war, he seemed so essential that few imagined the nation surviving without him. Emerging from obscurity, he mined the rich veins of expertise and helped to win the war. Beholden to no single group, he borrowed from academia, industry and the military. In a crisis, he crafted a system of innovation that would endure long after he exhausted his own utility as an administrative and technological seer.

379

Then the world changed. The pace of innovation quickened. Governmental arrangements that one generation thought of as temporary were considered indispensable by the next. War turned cold. A respite from politics as usual gave way to an orgy of partisanship. While its economy still ebbed and flowed, America's affluence loomed over the rest of the world. Without anyone really knowing why, Bush's practical know-how lost its relevance and it was no longer natural to follow his lead.

Perhaps it was vanity that made Bush among the last to realize that his moment had passed. The Oppenheimer affair certainly highlighted the gulf between him and the national-security elite. Having been "present at the creation," as Dean Acheson put it, Bush now watched a new American nation journey in what he saw as the wrong direction. Public life had sustained him since his withdrawal from active research in 1940. Now he was too old to resume his academic career and too important to play on a small stage. A hero without a cause, he seemed to be against *everything*. He had helped to unleash technological, military and governmental forces that now eluded not only his grasp but his comprehension. "We are living in a world that is altering very rapidly," he wrote in early 1955. "It is difficult indeed to visualize what this country may be like in another generation, if indeed it continues to exist in its present form."[1]

Yet he had too much dignity to strike a forlorn pose or join an organized opposition. Because of the public's sentimental affection for him (as for other World War II leaders), he had the luxury of mounting a lonely crusade for his brand of decency. As one journalist noted, "There are some qualities in this man, which cry out for a park bench, a special table at a restaurant or club, a fireside armchair or a well-stocked library. He needs a forum."[2]

In the summer of 1954, a few months after the Oppenheimer hearing, Bush decided to retire from the Carnegie Institution of Washington, effective at the end of 1955. Nearly 65 years old, he wished to return to his native Massachusetts. He launched a search for his replacement, put his house up for sale and ordered a new one built in Belmont, a haven for MIT faculty near Boston. In November 1955, his retirement was marked by a series of low-key dinners. At the most memorable one, the president of the California Institute of Technology, Lee DuBridge, former chief of the wartime Radiation Lab, shared a "hunch that the particular Yankee who has been playing king is not yet dead." If "Van is looking forward" to retirement, "he must have some devilment up his sleeve whose nature and extent no one will dare to imagine."[3]

The next month, Bush left the nation's capital for good. "Washington will not be the same without you," Nelson Rockefeller told him. Bush made a clean break, resigning from a National Science Foundation advisory board and refusing to serve even as CIW trustee. He did "not wish," he said, "to get in the way of [the] incoming president," his pal Caryl Haskins.[4]

Bush welcomed retirement. He had many friends, many interests, many connections. He had no financial worries, due to Carnegie's generosity and his role in forming a few successful companies. His net worth exceeded $500,000. And he had plenty of available cash from the sale, in September 1955, of a few hundred thousand dollars worth of stock in Metals & Controls, a government contractor that had bought a company he had cofounded.[5]

While he missed his old Washington comrades, Bush substituted the pomposity of corporate life for the "hurly-burly of government affairs." He served on the boards of two major corporations: American Telephone and Telegraph Company and Merck, a leading pharmaceuticals company. He kept a watchful eye on a third company, Millipore Filter Corporation. His younger son, John, formed Millipore in 1954 with Bush's encouragement and aid. John, a pilot in the war, had worked for a chemical company owned by Stanley Lovell, the spy who had served as the liaison to Bush for the Office of Strategic Services. After the war, the Army contracted with Lovell to make prototypes of a sophisticated filter taken from a captured German laboratory. The filter had the potential to screen bacteria and other microscopic particles from water. Lovell made a small number for evalution and to show it was possible to mass-produce them. He then offered to sell John Bush, who had worked on the project, the rights to the filter technology for $300,000.

After John Bush raised most of this amount on his own, his father agreed to "promptly pick up enough stock" in the venture to guarantee that his son had sufficient funds to close the deal. Bush essentially gave his son a short-term loan and moral support. "He encouraged me to go forward, but afterwards made an issue of not getting involved," John recalled. "He let me wing it."

John repaid his father's trust. Millipore "made money from the outset," and its filter became part of the standard method of testing drinking water in the U.S. In the 1960s, the company moved into water purification and other businesses.[6]

In retirement, Bush kept busy. He raised turkeys—some 800 of them—at his farm in New Hampshire. He lavished time on his many hobbies and inventions. He made about a dozen pipes a year from briar blocks and built furniture. He experimented with pendulums and gravity clocks. He analyzed the action of bird wings and imagined assembling a car from scratch. Three of his projects qualified as serious affairs. He designed a new type of car engine, a military boat powered by a hydrofoil, and tiny medical valves for use in the treatment of brain and heart ailments.

Bush hoped to commercialize these inventions. The new engine garnered some enthusiastic press reports but was shunned by the auto industry and died. This prompted Bush to write later of Detroit's leaders: "I think the men who manage the industry are dumb." He came closer to success with his hydrofoil boat, forming a company, investing some of his own money, attracting other backers and doing extensive tests over a decade. A hydrofoil is a flat or curved finlike device that raises the hull of a boat above the water's surface. It increases a boat's speed by reducing the drag caused by contact with water. Besides offering faster speeds, hydrofoil boats promised greater stability. Bush was convinced that hydrofoils "are going to be important in military affairs."

Hydrofoils were not a new concept; Germany had used them during World War II and Sweden since the war. But the U.S. Navy had never relied on the technology. Bush won a research contract from the Office of Naval Research. In 1953, he tested a 35-foot-long hydrofoil, weighing 20,000 pounds, on the Chesapeake Bay. At one point in the test, the manual controls failed and Bush and his team were nearly tossed into the water. The boat proved too slow to win the Navy's favor, however. Bush told a friend that development "costs have run us ragged." Citing the unexpected expenses, the Navy cut off funding— soon after a Navy crane operator accidentally dropped a hook through Bush's craft and sank it.

Bush was bitter about the Navy's rebuff. The service "really need[s]" the hydrofoil, he told Nelson Rockefeller, complaining that "cutting off support in an economy wave" makes no sense "just as results are being produced." Bush's basic concepts were considered sound, and the Navy continued to study hydrofoils—but not ones designed by Bush's company.[7]

Of his serious inventions, making minute medical valves gave Bush the most satisfaction. His son Richard, a surgeon, thought Bush enjoyed trying to help save lives after his years of aiding the military. Bush built prototypes of the valves in his home workshop, designing them in concert with local surgeons. He made heart valves from gold or rubber and cut pieces of nylon to patch faulty arteries. He visited hospitals where he saw experimental valves implanted into dogs. One of his valves—designed to relieve fluid pressure on the brain of a child—was about an inch long and the diameter of a darning needle. A magnetized rod controlled a tiny gold ball that fitted over the tube entrance. While useful experiments all, according to Richard, the valves were never mass-produced. This was partly because Bush knew a lot about materials but not much about the human body. Replying to a letter from his father about the difficulties of inserting into the body valves made of rubber, Lucite and other materials, Richard wrote: "Unfortunately the wall of the heart does not tolerate irritation from foreign bodies very well." Other physicians were more direct. One told him repeatedly, "Bush, you don't know anything about the blood system."[8]

Bush's private pursuits left him time to address the wider public. Keenly aware of the interplay between science, technology and national security, he wondered whether men could "live without war." Now that "the glamour of war is gone," he asked whether the kind of direct combat "that once had a real appeal for the red-blooded man" was obsolete. Others had noted that modern technology had made war impersonal and that the "virile attributes" of war, which enlivened societies in the past, would have to arise from another source. But Bush's romantic yearning for an earlier stage of combat seemed peculiar given his role in exploiting the very technologies that further dehumanized war. "Where do the virtues of war lie today?" he mused. "Is courage needed to watch a radar screen or adjust a guided missile? Where is the daring of the soldier when the folks at home encounter equal risks? When one man guides a plane that can destroy a city, what becomes of the infectious influence of comradeship, the sense of being engaged with many others in a common hazardous campaign?"

His meditations on the nature of war led him into the fuzzier territory of religion, leadership and spirituality. For an engineer who instinctively sought solid ground, this was a strange turn. It led to confusing attempts on his part to rebut mechanistic explanations of consciousness, show the limitations of objective knowledge, describe the "art of management" and assert that scientists, despite their surface arrogance, actually felt humbled in the face of the great cosmic mysteries. Humanity should "follow science where it leads," he advised, "but not where it cannot lead." A lifelong pragmatist, he praised "a second kind of culture that has nothing whatever to do with utility."

Some dismissed his wordy essays, which bore cryptic titles such as "For Man to Know" and "Science Pauses," as sheer pontification (even though they conceded that Bush had earned the right to pontificate). At the core of his windy verbiage, however, lay the concern that Americans had lost their way. He did not blame any specific cause for the country's lack of direction. He even played down the chief culprit, nuclear proliferation, postulating that nuclear war was impossible because the world's major powers had achieved an "atomic stalemate."

Bush's anxiety had a different source: the global instability wrought by rising affluence and the spread of new technology. Worried about rapid population growth and the depletion of natural resources, Bush called for new sources of energy, food and materials. He predicted "distress, famine and disintegration" unless world leaders backed alternative energy sources (including solar power), population controls and techniques for harvesting food from the ocean. "Wars aside, man is still . . . headed for catastrophe unless he mends his ways and takes thought for the morrow," he wrote in December 1955.[9]

Bush still enjoyed speaking to the public, but he increasingly seemed too strident and obsessed with weighty topics. During a live interview from Bush's Belmont home in June 1956, the journalist Edward R. Murrow listened intently as Bush declared in his corncrake voice, "All-out wars are obsolete as a means of settling differences." Rather than debating this big thought, Murrow pressed for a tour of Bush's elaborate workshop (transferred whole from Washington) and chatted idly with Phoebe, who predictably cheered on her husband.[10]

What isolated Bush from mainstream America was not his conservatism, but his contrarianism. While he defended the nation's right to stockpile nuclear arms—and even use them against an enemy—his belief that nuclear strikes would serve little military or diplomatic purpose indirectly aided a U.S. peace movement that in the late 1950s was awakening from two decades of retreat. Bush had no use for pacifism, but his objections to the H-bomb led logically to an observation made by Martin Luther King, Jr., who said in 1960 that the choice facing Americans "is no longer between violence and nonviolence. It is either nonviolence or non-existence."[11]

In the small round holes of 1950s culture, Bush was a square peg. Given his penchant for angry defiance, he seemed closer to a boardroom rebel than a defender of the elite. He had long harbored doubts about large corporations, bristling at their tendency to smother competition and ignore innovations that might benefit customers but upset their operations. Yet Bush had contributed to the expansion of big business during World War II by relying mainly on large, established industrial contractors as opposed to upstarts. This was consistent with the times. After the war, the corporation emerged as the archetypal American institution. The so-called "organization man," described by William H. Whyte in his 1956 book by the same name, was now the backbone of corporate life. The individual was in retreat. "Man exists as a unit of society," Whyte wrote. "Of himself he is isolated, meaningless." Describing the corporate philosophy, he added, "Society's needs and the needs of the individual are one and the same."[12]

While a friend of corporations, Bush refused to sacrifice the individual on the altar of team play. He found the postwar world, despite its embrace of procedures and bureaucracies, still dependent on the power of one man to make himself heard. He promoted old-fashioned character traits of self-reliance and personal vision. "Define [your] objective: don't follow the crowd," he told a group of MIT freshmen in 1963. Make money, he insisted, but do so "constructively," by building new products and processes, and not by being "parasitic." "Be sure you welcome responsibility. Only a few . . . really do." He also urged students to strike "a reasonable balance" between private and professional interests. In too many corporations, he saw the "sad sight" of a "business man with [his] nose to [the] grindstone."

While hoping "many" students would "aim to be business executives," Bush wished some would consider launching their own ventures. His stress on entrepreneurship during an age of corporate battleships set him apart from mainstream economic thinkers. In his 1952 study, *American Capitalism,* John Kenneth Galbraith conceded that competition was imperfect and would never limit private power (as it should under classic economic theory). Galbraith proposed that the "countervailing power" of society's three behemoths—labor, capital and government—would curtail excesses. Galbraith thought that the ordinary man need not worry about the size and scale of the fundamental American institutions, but such advice was cold comfort to Bush. He saw "bigness" as a problem, not a solution. He found an escape from social conformity (whether enforced by democratic or totalitarian systems) in the nation's 19th-century tradition of self-reliance and its 20th-century embrace of technology and change. His own thinking about the role of the individual in economic life married the sensibility of New England idealist Ralph Waldo Emerson with the visionary economics of Joseph Schumpeter. Emerson's faith in individual conscience ("Whoso would be a man, must be a nonconformist") fit neatly with Schumpeter's view of "creative destruction" as the engine of capitalism.[13]

Bush's own life certainly epitomized this curious combination of new and old. He never worked a day in his life for a corporation except as a consultant and often he owned shares in the company and had a say in its management. Yet in 1957, he became chairman of Merck's board of directors (by dint of his relationship with the company's chief executive, Jack Connor, who had worked for Bush during the war). The top post at Merck was largely ceremonial, but it placed Bush in the clubby top tier of American management and gave him a new forum in which to espouse his views. He did not share the aversion to turmoil and dissent that froze many corporations at the time. After less than a year on Merck's board, Bush bluntly told then-chief George Merck that the company's organization was "atrocious," its "lines of authority blurred" and its board of directors too meddlesome in company affairs. Bush even criticized George Merck for leading by consensus and not taking enough personal responsibility for corporate decisions. "You now have a management committee. It is a contradiction of terms," he insisted. "A committee cannot manage."[14]

George Merck was not used to being flayed by his directors, but he restrained himself. He had met Bush through his work on biological weapons during World War II and considered him a special case. Merck politely told Bush that he "appreciated very much" his comments and was "delighted that you want to spread your interest [in Merck] beyond the area of research and development." Bush probably would not have backed away from a fight if Merck had been offended by his comments. His harsh evaluation was perhaps aimed at testing Merck's tolerance. The freedom to voice his views forcefully was essential. His concept of a successful career involved the clash of professional judgments; the resolution of differences came not through the homogenization of judgment but through the triumph of one view over another. As he told a friend in 1965, a person with a good job "can look his boss in the eye and tell him to go to hell. That's my chap."[15]

Consumerism bothered Bush as much as conformity. Like many commentators on American life in the 1950s, he seemed pulled in two directions: between the allure of future technology and nostalgia for the style of the small town. While he celebrated technological innovation and the free market in the abstract, he disliked some prime examples of the American way: television, cars, even movie stars.

"There must be something wrong with my genetic constitution," he wrote a friend. "I see lots of girls that have better figures, better features, more graceful movements, than these popular wenches." Like other social critics, he blamed advertising for corrupting American tastes. "Madison Avenue believes that if you tell the public something absurd, but do it enough times, the public will ultimately register it in its stock of accepted verities." He viewed the power of advertising as another sign that the masses could not govern themselves. He suspected that some kind of group hysteria, akin to the religious outbursts and speculative bubbles of bygone days, lay behind the peculiar power of advertisers to persuade consumers. "When I talk to individuals I find them far from gullible . . . yet in mass they buy fairytales and extravagances."

The American attitude toward automobiles exemplified Bush's new existential dictum: "In mass we do not seem to make much sense." Advertisements for the new 1965 car models reinforced his sense that the auto industry—the emblem of U.S. industrial might—had ossified, and car buyers had lost their heads. The ads, he wrote:

> are all alike, and so are the automobiles. The ads start out with rhapsodies and claims that this is the year of great changes. Then one reads the ads, and there is not a change noted more important than moving the tail light six inches. I wonder where the auto industry is headed. It can no longer play on style changes. Every car now has about the same lines, and departures such as fins or bulges have caused enough grief to those

who have introduced them to discourage any further departures from the norm. And any really radical change is barred by the cost of introducing it. What are they now going to sell? Nothing but hogwash, and the public seems to fall for it. . . . As long as the public is gullible, and satisfied with hokum, why move? My wonder is that the public seems to have unlimited capacity to swallow fantasy.

Bush was convinced that U.S. carmakers had no real desire to improve their products and that the industry's concentration enabled them to ignore innovations that would make cars safer and better. No U.S. auto company had accepted his design for a more fuel-efficient engine. The rejections struck Bush as collusion, which he knew how to recognize (his first experience with industrial subterfuge had come in the 1920s when the cartel of radio patent holders tried to destroy his Raytheon venture). Heightening Bush's suspicion was a letter from General Motors that contained "a queer slip." GM told Bush that his engine was not an improvement over its own, but that "even if it were a better engine, [GM] would not be interested in it."

Monopoly power had robbed the auto industry of its passion for technical advance. "No American company will stick its neck out, the cost and risk are too great," Bush insisted. The solution was competition. But where would it come from? He had no clue. Blind to the possibility that the trickle of auto imports would turn into a torrent, Bush prematurely concluded in 1964: "No foreign company has the strength to really overturn the American market."[16]

Bush's distance from the American mainstream was best measured by his reaction to the Soviet launch of *Sputnik* and the subsequent race to put a man on the moon. The Russians launched the satellite on October 4, 1957, stunning U.S. scientists and provoking a public outcry. Besides highlighting Soviet scientific capabilities, *Sputnik* heightened fears that Russian nuclear missiles could reach the American heartland. *Time* and *Newsweek* saw *Sputnik* as the "greatest technological triumph since the atom bomb." *The New Republic* compared the satellite to "the discovery by Columbus of America" and cited the spacecraft as "proof . . . the Soviet Union has gained a commanding lead" in vital technologies. Physicist Edward Teller made the most damning comparison, declaring on television that the U.S. had lost "a battle more important and greater than Pearl Harbor." James Killian, the president of MIT, believed that *Sputnik* created "near hysteria" in the U.S. and "apprehension throughout the free world."[17]

Not everyone viewed the Russian breakthrough with alarm, however. Many scientists, frightened by the public's response, noted that the U.S. had the capability to lob a similar piece of metal into orbit—but had not wished to do so. That was the problem, according to those who saw the Russian achievement mainly in cultural terms. Americans were too obsessed with material gain and unwilling to sacrifice for the good of the state, they said. To these critics, *Sputnik* represented a useful prod. The military analyst Walter Millis, for instance, thought the satellite justified "a fresh national appraisal" of national security policies. Yet some agreed with sociologist C. Wright Mills, who dismissed *Sputnik* as "a lot of malarky." Warning against a space race with the Soviets, Mills asked, "Who wants to go to the moon anyway?"

The *Sputnik* debate excited Bush, who took a perverse glee in the Russian advance. "If it wakes us up, I'm damn glad the Russians shot their satellite," he told *Newsweek*. "We are altogether too smug in this country."

Yet like Mills, Bush made no grand claims for *Sputnik,* saying that the satellite did not mean the Russians could put a nuclear missile "on our doorstep." Like Millis, he thought the satellite exposed a weakness in U.S. military organization, but not the nation's technological base. In Bush's mind, the problem was that the Army, Navy and Air Force still pursued separate research programs, insuring duplication of effort and a fight for funds. This hampered the U.S. in its race with Russia. Bush's solution was simple: "Unify our military planning. Without it all else is futile."[18]

Given his stature, Bush easily joined the national conversation about *Sputnik,* which culminated in congressional hearings on U.S. satellite and missile programs. In late November 1957, Senator Lyndon Johnson summoned a list of luminaries to Capitol Hill, including Edward Teller, the "father of the H-bomb," and James Doolittle, who led the 1942 air raid on Tokyo. In a sign of his cachet, Bush was asked to appear before Johnson's Armed Services subcommittee on the first day of its inquiry.

After Bush swore to tell the whole truth, Johnson fawned over him. "Dr. Bush, for many years Americans have been in the habit of turning to you for good advice and good counsel. It has been a wise habit, and we members of this committee turn to you once again in time of crisis."

Johnson wanted guidance, and Bush obliged. Repeating the themes sounded in his *Newsweek* interview, Bush counseled against rash actions. He opened by striking a somber note, declaring that *Sputnik* was "far more than merely a problem of an advance in weapons."

Raising the specter that Russia could "devastate us" with a rain of nuclear missiles "and we could not reply," Bush grimly said, "this country now faces definitely a situation where it must prevent *at all costs* being in the position where it can be overcome without the possibility of answering."

Overall, Bush's message was considerably brighter than his initial doomsday reference. Rapid gains in space technology were possible, but the U.S. must end "contests between the services, which have been damaging in the past and which sometimes, in my opinion, have been disgraceful." Having had a "wake-up" call from Moscow, Bush believed, Americans now would realize "we are in a tough competitive race where we have got to do a lot of good, tough work."

Besides offering scholarships to lure more talented but needy students into science, the nation need not push the panic button. "If we had really integrated central planning in this country . . . to be sure that the emphasis is sufficiently intense," he said, we would "at the same time find, I am sure, many ways in which money can be saved, so that in the long run . . . I am not sure that our military expenditure needs to go up enormously." Spending more, he warned, might help, but it was not the key to achieving every technological feat. "You do not get results simply by pouring money," he insisted. "If you put ten times as much money into basic research today, you would not have ten times as many scientists."

During his cross-examination, Bush escaped questioning on his own role—a decade earlier—in the nation's confused planning for space weapons. After the war, "I was exceedingly skeptical whether that very tough problem [of missile guidance] could be solved," he explained. But obviously the Russians had done so by shooting a rocket into space and then guiding it back to Earth. Later in the hearings, Bush was attacked for his poor forecast, acceptance of cuts in missile spending after the war and failure to grasp the military potential of satellites. One witness, trying to explain Bush's lack of foresight, testified that "there are people who are highly creative at certain periods of time and then at other times conclude that creative things can't be done." The implication was clear: having led one technological revolution, Bush had slammed the door on the next—possibly out of psychological exhaustion.[19]

Bush denied this damning charge. He defended his record of picking winning innovations, saying he had played down the potential of satellites and missiles only because enthusiasts had irresponsibly exaggerated the speed with which these technologies would emerge and greatly underestimated the technical difficulties in mastering space weapons (the biggest being that a space weapon "could not then be counted on to hit anything worth hitting, with sufficient assurance to make it a practical device").[20]

Despite his protests, Bush undeniably had grown more cautious since the peak of his power in 1945. His desire for a deliberate response to *Sputnik* illustrated this. This made sense and mirrored Eisenhower's reaction. The president had tried to proceed at first as if little of significance had occurred, but he quickly succumbed to the pressure for bold action. Within a few months, he named his first science adviser and increased by 20-fold the amount spent by the government on space-related research. He created an agency within the Defense Department to supervise all space-related research by the armed services and then steered a bill through Congress calling for a civilian agency (the

National Aeronautics and Space Administration) to manage future space ventures. The efforts paid off. By the early 1960s, the U.S. had overtaken the Russians in both space and missile technology.[21]

The frenzy of activity sparked by *Sputnik* vastly increased government spending on scientific research and forced the Pentagon to accept more responsibility for coordinating initiatives in military technology. But despite these gains, Bush excoriated the space program for raising false hopes and substituting the pursuit of an engineering spectacle for genuine innovations. While some critics relented as the space program grew, Bush kept on complaining. In March 1960, he sharply criticized the space program before a House committee, then followed with a strongly worded letter to T. Keith Glennan, NASA chief. Warning that the first attempts at space flight "will result in killing some nice young chaps," Bush could "see the time coming when results are few and far between, and simply bore the public, which is after all fickle." In the June issue of *Technology Review,* Bush insisted that "putting a man in space is a stunt; the man can do no more than an instrument, in fact can do less." The effort, even if it manages to "kill some promising youngsters in the process," would not move the nation closer to space colonization or travel to other planets. Debunking the claims of space boosters, he noted that "the days when men will be in space for long periods and for varied purposes are so far off." Many of the claims for space exploration were "simple unadulterated absurdity."[22]

After the first U.S. manned space flight in early 1961—Alan Shephard's 15-minute suborbital flight—the stakes in space grew. John F. Kennedy, the new president, raised the bar on American ambitions. In the presidential campaign, Kennedy had cited the "missile gap" as evidence of U.S. technological failure. Now he found that manned space missions "had the dash and drama" that "fit perfectly with the spirit" of his administration. Kennedy's desire for grand achievement in space conflicted with the opinions of his science adviser, Jerome Wiesner. Also a former MIT administrator, Wiesner opposed challenging the Russians with "stunts." But the U.S. had little choice. Based on their lead, the Russians were likely to be the first to put crews of two and three men into orbit, launch a space station and orbit the moon. If the U.S. wanted to win the technological and propaganda battle over space, its astronauts must be the first to land on the moon.[23]

The moon goal radically changed the space debate. Even before Kennedy asked the nation to back a moon program in the spring of 1961, the elite had closed ranks behind him. Criticizing U.S. ambition in space suddenly seemed un-American. For his continued attacks on NASA, Bush earned a rare tongue-lashing from *The New York Times* editorial page. Calling the space race "decidedly worth while," The *Times* castigated Bush and implied that his criticism was unprincipled. "Dr. Bush knows very well that many of the greatest scientific discoveries were made by . . . serendipity . . . in other words chance." To add insult to injury, the newspaper cited nuclear fission as one example of this principle.[24]

Bush's earlier pessimism about guided missiles and satellites made him vulnerable to attack. Trying to shake his reputation as a space naysayer, he declared that he had always supported "fundamental work on guidance systems, propulsion systems and the like" and had only opposed "premature excursions into making hardware I knew would not operate." But technical issues were of secondary importance. Bush's chief objection was that the space program would never justify its costs. "The scientific information to be obtained is not worth the money." This legitimate objection was lost on a nation blinded by its abundance and hostile to the ideal of fiscal stringency.[25]

Bush failed to realize the futility of his position. Once more misjudging the drift of opinion, Bush tried to persuade Kennedy to slow down on space in 1963. Writing to James Webb, then chief of NASA, Bush skewered the moon project and asked that his views be given to the president. Webb, who had befriended Bush while working in the Truman administration, passed on the engineer's message. Kennedy ignored it.[26]

In November 1963, Bush again publicly complained about the moon program, publishing a pithy letter in *The New York Times* in which he ghoulishly predicted that a bungled mission would leave "one or two young astronauts . . . caught in a vehicle" and "left to starve, or to commit suicide" in their space capsule. Bush thought this a stiff price to pay for technological spectacle. "To put a man on the moon is folly, engendered by childish enthusiasm," he noted. "It will backfire on those who drive it ahead."[27]

When Kennedy was assassinated a week after Bush's letter, the nation's resolve to land astronauts on the moon stiffened. If this was their dead president's desire, why not?

The unpopularity of Bush's outlook on space prompted a rare period of self-examination. "I wonder myself whether I preach too much, whether I might get an entirely different reaction from the one I would aim at," he wrote Carroll Wilson.[28]

After years in the limelight, Bush now got the cold shoulder. "It is well known that I have opposed the moon shot program . . . ," he told his banker. "I found out long ago that, if one talks a great deal about such a subject, he is soon ignored." Bush liked nothing less than to be ignored, but he had only himself to blame. NASA had so far avoided the death of an astronaut. Rather than congratulate the agency, Bush attributed its performance to "great luck" and stuck to his predictions of gloom. This only drove him further to the margins of public discourse. On the "moon program," he wrote a friend, "I have been a voice crying in the wilderness for there are so many advocates that a lone voice of protest does not get too far."

The intensity of Bush's opposition to the space program reflected his growing sense that federal support for research had foundered. The profligacy of the 1960s—fiscal, military and technological—ran counter to his belief in a limited charter for public enterprise. To be sure, conservatism shaped his objections to the space race, but he also felt betrayed by a support system for science and technology that he had hoped would stay closer to the core values of national security and scientific competency. Instead, the government oversold technology in the 1960s, planting the seeds of a backlash.[29]

By 1970, when he reached the age of 80, Bush was such an anachronistic figure that about the only people still keen to see him were historians. The nation's foremost historians of science and technology—Gerald Holton, I. Bernard Cohen, Daniel Kevles—sought his views and recollections. Bush helped these scholars only sparingly, partly because he disliked the inexact methods of history and partly because he wished to discourage anyone from writing a book about him. "I hope nobody'll ever write a biography of me, because I think it probably would be terrible." As long as he lived, he believed he could "fend it off," but he worried that a biographer would corner him after his death. "The feeling that some of my friends over the years might live long enough to read it kind of jars me," he admitted.[30]

To forestall a biography and satisfy an old promise to a friend in publishing, Bush cobbled together a book, *Pieces of the Action*. Part essay and part memoir, the book, published in the fall of 1970, was a poor substitute for a coherent account of Bush's life and thought. Written with the aid of two professional writers, the book had its origins in the early 1960s, but Bush's mixed emotions had kept the project in limbo for many years. He first considered writing his memoirs in 1946, but the idea repelled him. "My general impression of memoirs by people who have been mixed up in all sorts of affairs is certainly sad, so that I have rather come to the conclusion that it would be a mistake to add to the total," he told a friend. "Moreover, I am not one who keeps a diary, nor do I write down the little interesting sidelights, and I presume that this is because I do not take myself too seriously and never feel that anyone later will be much interested . . . I believe . . . the importance of memoirs is often much exaggerated, particularly in the mind of the chap who writes them."[31]

When finally completed, *Pieces* was a rambling book that contained a dollop of Bush's biting personality but a good many saccharine insights into invention, leadership, organization and government. The book was no autobiography; it contained no revelations about his stormy public career and barely mentioned his role in the efforts to build and control nuclear weapons. A sanitized account, the book was essentially a pep talk for Americans who wondered—at the end of the divisive 1960s—whether the Establishment still deserved to lead the nation.

Pieces fell flat: no excerpts in major magazines, no big news stories, no paperback edition. Still, the book received respectable notices. It "is never dull," opined *Technology Review,* "and from it emerges an engaging, forceful, puckish, and occasionally ruthless personality." Writing in *The New York Times,* historian Elting Morison compared Bush to Benjamin Franklin and noted that his ideas "on how to order human affairs in a technological universe are extremely useful and should be pondered and understood by all who share his concerns." *The Los Angeles Times* was less kind, finding Bush's advice on how to handle the current crop of social problems too cryptic. "Working within the system is what Bush advocates, a system he played a large role in fashioning," the reviewer noted. "But the clues he offers us as to how to do this most effectively are frustratingly concealed." Probably the most perceptive critique of *Pieces* came from the pen of William Gilman, writing in the *Saturday Review.* Gilman correctly labeled Bush "a technocrat" and chided him for looking askance at ordinary folks "who meddled with a military technology beyond their comprehension." But Gilman was nonetheless forced to conclude that Bush still had much to teach, noting: "If our troubled democracy, facing a seeming myriad evils created by or ignored by science, summons technology to the rescue, the results need not be the mechanistic gamble that humanists mistrust—*if* technology can offer enough managerial men like Bush. He has a reassuring nonmechanistic streak in him. In 1962 he wrote in praise of 'a second kind of culture that has nothing to do with utility.' And today there remains a dissonant note in his admiration for the computer and its electronic myrmidons."

The mixed reaction disappointed Bush, who recalled the strong response to his 1949 volume, *Modern Arms and Free Men.* Of the new book, he confessed, "It went over but not nearly as well as I hoped it would."[32]

Even if *Pieces* had been less a pastiche and more a true memoir, Bush still could not have kept others from shaping his image and giving their own interpretations to his life. In his old age, myths about him thrived, and his legacy became contested ground. Even those who cast Bush as a hero often did so with specific aims in mind. His protean career—inventor, entrepreneur, administrator, military thinker and, finally, author and pundit—made him uniquely suited to serve the political and cultural ends of virtually any movement that sought legitimacy for its views on the interactions between technology, capitalism and the state. In this terrain, Bush embodied three great themes: the move of the civilian expert to the center of military affairs; the marriage between intellectuals and government; the emergence of the computer as the most profound technology of the 20th century.

Bush's image as the archetypal military technologist was the most durable. In the first memoir by an atomic scientist, a 1956 volume by Chicago physicist Arthur Compton was faithful to this image. Since Hiroshima the roles of Oppenheimer and Groves had been glorified—these striking men stood for all high-energy physicists and hard-nosed soldiers—with Bush relegated to a bit part. But Compton carved out a place in the pantheon for the pragmatic engineer and wily organizer of expertise. In *Atomic Quest,* Compton wrote of Bush:

> His understanding of men and science and his boldness, courage and perseverance put in the hands of American soldiers weapons that won battles and saved lives. But his great achievement was that of persuading a government, ignorant of how science works and unfamiliar with the strange ways of scientists, to make effective use of the power of their knowledge. Books have been written about the remarkable wartime accomplishments of the Office of Scientific Research and Development of which he was head; but the American people will never know how great a debt we owe to this hard-thinking, energetic engineer. He could talk straight to generals and cabinet officers and the President, and make them take it.[33]

In the mythic rendering of his life, Bush's actions in World War II grabbed center stage, blotting out his later disappointments. Admirers thought he epitomized the marriage between science and the state; they ignored his failure to arrange this marriage on his terms. They played down his inability after the war to quickly form a civilian research agency, curb the trend toward fragmentation in government research or halt the military's domination of science. Finally, they forgave his inability to grasp the importance of space technology, which rightly or wrongly became the measure of the nation's military and industrial strength in the 1960s. In this popular view, the 25 years following 1945 counted as "the Vannevar Bush era" because of the wide acceptance of his ideal of an unfettered scientific estate, funded by the body politic and yet free to pursue its own ends.[34]

In the 1960s, technocrats suddenly faced an increasingly restive public that no longer trusted the elite to deliver scientific supremacy without prodding or advice. By the early 1970s, the politics of science had moved far from Bush's technocratic conception. Research became more partisan, subject to the passing whims of each president (and, increasingly, to Congress and professional activists). Kennedy wanted a man on the moon. Johnson wanted to end social decay. Nixon wanted a cure for cancer. These three presidents promoted two insidious practices: first, they treated technology as spectacle, as an exercise in national prestige cut off from a mooring in commerce and national security; second, they viewed research dollars as akin to old-fashioned pork, opening research centers in geographical backwaters and bolstering the budgets of universities in the jurisdictions of political friends. As cynicism about the research enterprise grew, scientists became more aware of the political ramifications of their work and raised more questions about "the comfortable notion that science is ethically neutral and that the scientist must necessarily be absolved of responsibility for what is done with his work."[35]

Given the breakdown of consensus on aims and values of elite research, Bush's ideal of no-strings funding became the mythic prime directive for scientific leaders. Among this crowd, his famous report, *Science—The Endless Frontier,* stood as Holy Writ and Bush as the Great Father in a tall tale about postwar history. It went as follows: the government would have abandoned science without Bush's foresight. U.S. global economic dominance from 1945 to 1970 stemmed largely from the innovations spawned by government-funded scientists and engineers (and not from the collapse of the European and Asian economic powers after the war). Economic stimulation flowed equally from military or civilian research. Even research mandated by politicians was welcomed by the research community provided recipients could pursue their studies (useful or not) free from political meddling.

This Bush myth was an exaggeration, but a popular one. Lloyd Berkner, a Bush colleague at both Carnegie and the Research and Development Board, gave classic expression to the myth in his 1964 book, *The Scientific Age.* Berkner argued that postwar American prosperity (and the federal bureaucracy that nourished it) sprang wholly from the mind of his mentor. "No other nation enjoys equal scientific and technological support," Berkner wrote. "Indeed our transition from an economy of scarcity to an economy of plenty is *directly* traceable to the prescience of Vannevar Bush."[36]

Bush's identification with prosperity would prove a double-edged sword, however. During World War II, the convergence of civilian production methods with military needs reached its zenith. Technocrats could say without qualification, as did Frederick Terman, father of Silicon Valley, "War research which [is] now secret will be basis [for] postwar industrial expansion." For two decades, military innovations indeed strengthened U.S. competitiveness, with the nation's staggering lead in computers and electronics the most tangible benefit of government spending on research.[37]

But there were signs of a looming crisis, no matter how distant. Specialization not only undermined the ultimate authority of experts by promoting fragmentation of views, it also spurred the military to make increasingly "baroque" demands on technology. This divergence from commercial practice eroded manufacturing competence, which was apparent as early as 1949 when a startled Pentagon official told Karl Compton, "From a few sources we have been receiving information which might lead us . . . to suspect that current design of military equipment is so elaborate and varies so materially from commercial standards that industrial concerns are becoming apprehensive that the manufacture of equipment of this type will require so much machinery and new tooling along with special skills that its manufacture . . . will be materially retarded."[38]

By the early 1970s, German and Japanese industry had emerged as tough competitors and U.S. industry seemed to lose vitality as military and civilian technology diverged. Bush's warnings about the inherent sluggishness of big companies gained new credibility as oligarchic control of key industries (in steel and autos, notably) left them vulnerable to foreign penetration. The dependence of U.S. industry on military contracts, meanwhile, warped managerial priorities in the crucial manufacturing sector. This went unanalyzed but not unnoticed. In 1961, Eisenhower had complained about the dangers of a military-industrial-academic complex, spawning much soul-searching. But in economic terms, the ensuing debate never moved beyond a vague disquiet over the domination of the economy by big institutions.[39]

Bush shared Eisenhower's unease about the alliance between academia, the military and industry. But he never identified the most serious problem with this mutual embrace: the growing divergence between industrial and military technologies and the willingness of the government to shield big corporations from the market effects of poor technological choices. The great defect of *Science— The Endless Frontier* was its neglect of industrial innovation. "The implicit message of the Bush report," Harvey Brooks has noted, "seemed to be that technology was essentially the application of leading-edge science and that, if the country created and sustained a first-class science establishment based primarily in the universities, new technology for national security, economic growth, job creation and social welfare would be generated *almost automatically.*"

Bush's faith in the ability of private corporations to wisely harvest basic science—without any interference or "aid" from government planners—seemed justified in the 20 years following the end of World War II. In that period, U.S. industry dominated the world. But by the mid-1960s, there was growing evidence that Bush—in his zeal to avoid interfering with private-sector exploitation of government-funded research—had ignored the possibility that corporations might consistently fail to convert basic breakthroughs into competitive products. This flaw in *Science—The Endless Frontier* arose from a deeper conceptual confusion. In his report, Bush had embraced uncritically the linear notion of technological progress. In this view, advances in science lead to advances in technology, and not the reverse. Even as he promoted the report, Bush knew the linear model was an outmoded framework for describing the relations between science and technology and the sources of commercial innovation. But he had no ready conceptual alternative. Instead of searching for one, perhaps vainly, he took the convenient course of celebrating science as the font of all well-being and passing off commercial engineering and invention as subsidiary concerns. Indeed, the popular press reinforced the image of technology as a handmaiden to science.

While trading on the prestige of science proved a successful tactic for winning public funds, it carried a bill that came due in the 1970s. By then the connection between research, whether technical or scientific, and the marketplace had been severed in many key areas. But thanks to those who had treated *Science—The Endless Frontier* as gospel, U.S. industry found itself in a paradoxical situation: awash in theoretical knowledge, it was starved for the basic processes and products that lead to victories in commercial contests.

On one level, the root problem was frustratingly simple: defense and space research concentrated on *performance* improvements, regardless of cost, while the best industrial companies emphasized technical innovations that could be mass-produced relatively inexpensively and became less costly to produce over time. The one-sided incentives for producing exotic breakthroughs that required gold-plated manufacturing left industry woefully deficient in the very skills that determined commercial success in globalizing markets. The weakening of U.S. manufacturing repudiated the central premise of *Science—The Endless Frontier* that strong research produced a strong economy, which in turn bolstered national security.[40]

Even defense suffered from distorted priorities and the growing bureaucratization of research. For perhaps 20 years after World War II, military weapons and infrastructure retained their cutting-edge qualities, "but then military R&D ran away with itself," recalled Frank Press, President Jimmy Carter's science adviser. "It became so bureaucratic, specialized and expensive. And the procurement times took longer and longer." Far from achieving Bush's ideal of "a few quick" weapons, so much time would pass between conception and delivery "that by the time a weapon arrived it was 15 years out of date." It was hardly fair to blame Bush for the failure of the military and industry to achieve optimum performance, especially since he fought against the ossification of his paradigm. But with the onslaught of economic competition, and slowing growth in the 1970s, "We started to have doubts," Press concluded. "With all this great scientific leadership, people asked, how come the economy isn't great? That's when people began to blame the scientific community and think it was a mistake to have channeled all these resources into it."[41]

The third battle for control over Bush's image raged between two obscure technical camps. While little noticed, each was on its way toward becoming a major force in commercial and scientific life. These were the computer pioneers. Bush's active role in computer research had ended during World War II; he made no technical contribution to modern computers. But the publication of his essay "As We May Think" in 1945 made him for decades the best-known advocate in the U.S. for information-retrieval systems that both responded to and expanded on human inquiries. Even as late as the mid-1960s, when corporations widened their use of digital computers and his earlier analog machines were thoroughly discarded, shapers of popular opinion continued to cast Bush as one of the founding fathers of the whole field. As an editor at *Time* magazine told him in 1966, "We think of you as the man who envisioned information technology. . . . The field which you and only a few others built and nourished seems at last to have come into its own."[42]

By then Bush's vision of the memex—a personal information processor small enough to fit in a desk and flexible enough to create personalized "trails" through forests of data—stood well outside the mainstream of the field. Most computer designers concentrated on finding ever-faster ways for large, centralized machines to perform mathematical calculations. In contrast to Bush's insistence on placing machine intelligence in the service of the basic human desire to manage information, the sponsors of early digital computers saw them as the engines for large-scale, impersonal defense and corporate systems. Whether these computers tracked incoming missiles or business orders, people were expected to mold their behavior to the system's demands. These machines fundamentally served the needs of institutions, not individuals.

By the mid-1960s a new generation of computer designers had begun searching for ways to find a place for the individual in the digital world. Their Holy Grail was a *personal* computer, a machine easy enough to use that it could take its place alongside three other great consumer technologies of the century: the telephone, the radio and the television. The problem, of course, was that the requirements of processing power and memory seemed to demand that computers stay large and difficult to use. Mainstream designers only grudgingly accepted that computers could display information visually on terminals or screens. But proponents of personal computers realized that it was only a matter of time before parts became smaller, cheaper and more flexible.

For these digital rebels, Bush's memex seemed to offer a mirror onto which they could project their own hopes and dreams. To them, Bush was a useful cypher; his stature as a Cold War visionary promised to lend legitimacy to their own efforts to push computing in a new direction. With the dominant mode of mainframe computing seemingly gaining strength by the day, insurgent designers paid homage to Bush, partly out of genuine admiration for his vision and partly as a ploy to gain credibility for their infant machines.[43]

The inventor Douglas Engelbart, whose experiments in the early 1960s anticipated the advent of both the personal computer and the online computer community, tried to strike up a friendship with Bush, but was ignored. He adopted Bush as his patron saint anyway. Another influential designer of alternative computers, MIT engineer J. C. Licklider, also proudly noted the connections between Bush's vision of machine-augmented intelligence and his own research into human-computer interactions through visual symbols (rather than text, which was the only form in which conventional computers displayed information). Writing in the introduction to his 1965 book, *Libraries of the Future,* Licklider credited "As We May Think" as the "main external influence" on his ideas and dedicated the book to Bush. Yet he had never read the essay until *after* he finished his most important experiments and he thought so little of Bush personally that he never even sent him a copy of the book.[44]

The cult of "As We May Think" and Bush's status in the computing pantheon grew enormously after Theodor Nelson, a rebel computer designer with affinities to the 1960s counterculture, embraced Bush as his technological guru. A clever conceptualist, Nelson transmuted Bush's analog notion of "associative trails," or the personalized links between different stored records, into "hypertext," its digital cousin. In the early 1970s, Nelson concocted an alluring synthesis that mixed the ideal of personal computers with the counterculture desire for complete liberation from social strictures. The linchpin in Nelson's metaphysics was "hypertext," and "As We May Think" was his Bible. Once again, Bush's actual predictions, and the problem of information overload that inspired him, never mattered much to Nelson. Fighting an uphill battle to gain respect for an approach to computing shunned by large corporations and the military, he found Bush's imprimatur useful, indeed invaluable, to conveying the impression that the triumph of his brand of personal computer was somehow foreordained.[45]

Bush never minded the various uses to which his memex concept was put. After all, he was flattered. "Recognition and admiration by his peers was very important to him," an old friend said. Marginalized politically, he took solace in his growing reputation—deserved or not—as a computer visionary. In the 1960s, "As We May Think" was lionized by a variety of scholars, whose descriptions of the article included "the manifesto for information science"; "the earliest, and perhaps the most important single description of the potential uses of computers for information processing"; "the start of modern documentation"; and "an inspiration to many later workers in the field."[46]

Despite the acclaim, Bush never endorsed any particular reading of "As We May Think" and remained uncomfortable with digital computers. Binary electronics robbed him of the chance to watch a machine grinding out an answer. His analog machines, after all, literally recreated in metal the physical dimensions of the problem to be solved. Though the digital rebels saw "associative trails" as a possible style of writing code, or software, for computers, Bush was baffled by code. He persisted in believing that to really tame a thinking machine he had to have a wrench in his hand and see the gears turn. Thus he stood apart even from the insurgents who embraced his vision and celebrated his name.

Even though Bush never returned to the design or construction of computers after World War II, for the rest of his life he thought about the big questions of the digital age. Though he failed utterly to anticipate the flexibility of personal computers, he insisted that digital enthusiasts paid too little attention to the gulf between machines and people. "Both races appear to be prospering and proliferating," Bush told an MIT seminar on computers in 1961. "The only difficulty seems to be that the rapport between them is not all that might be desired."[47]

Bush still thought the memex might be the means of escape from the paralysis of information overload and thus lead to a revival of the generalist, the broadly knowledgeable person now under siege from specialized knowledge and the pace of scientific change. "Our libraries are filled to overflowing, and their growth is exponential," he told a meeting of engineers in 1955. "Yet in this vast and ever-increasing store of knowledge we still hunt for particular items with horse and buggy methods. As a result there is much duplication and repetition of research. *We are being smothered in our own product.* While we record with great care the work of thousands of able and devoted men, full of significance of timeliness to others, a large and increasing fraction of their work is, for all essential purposes, lost simply because we do not know how to find a pertinent item of information after it has become embedded in the mass."[48]

In an unpublished essay dated August 1959, Bush laid out his most fantastic vision of the memex, describing a mind-amplifier that would transcend the man-machine interface altogether. Rather than being controlled by a person's voice, this memex would sense "the activity of the brain without interfering with its action." These brain waves would then be translated into a form recognized by the memex. Alas, Bush allowed that such a machine "will have to wait until the psychologists and neurologists know far more than they do now."

In 1965, on the 20th anniversary of "As We May Think," Bush charted the practical progress toward his ideal information retriever and was pleased with what he found. "Now man takes a new step," he wrote. "He builds machines to do his thinking for him. These are still in their infancy, but their significance is great. It is one thing to supplement muscles and sense. It is a far more profound thing to supplement intellectual power. We are now in the early stages of doing just this."

The potential was enormous. "The personal machine," he added, will deliver "a new form of inheritance, not merely of genes, but of intimate thought processes. The son will inherit from his father the trails his father followed as *his* thoughts matured, with his father's comments and criticisms along the way. The son will select those that are fruitful, exchange with his colleagues, and further refine for the next generation." The computer, then, even promised a measure of immortality for everyone, and relief from the ravages of time. "No longer, when [a person] is old, will he forget," he predicted.[49]

"Now is all this a dream?" Bush asked in "Memex Revisited," a 1967 essay. "It certainly was, two decades ago. It is still a dream, but one that is now attainable. To create an actual memex will be expensive, and will demand initiative, ingenuity, patience, and engineering skill of the highest order. But it can be done."[50]

Seven years later, the first personal computers were built. While falling far short of the theoretical memex, these were machines that Bush would have recognized, if not as his own, then at least as kindred souls.

Bush flirted with heretics in the fields of computing and space exploration, but his challenge to the dominant culture only went so far. He remained wedded to the Establishment on the biggest issue of his time: the prosecution of the Cold War. Since the late 1940s, he had accepted the need to contain the Soviets abroad and maintain leading military technology at home. Nothing more clearly illustrated his deep attachment to the Cold War's frayed logic than his reaction in the late 1960s to the worsening war in Southeast Asia.

"The difficulty in Vietnam as I see it is, first, that we cannot win and, second, that we cannot get out," he wrote a friend in 1967. He saw "no solution in escalating" or gathering "our troops into a perimeter around a selected part" of South Vietnam. An invasion of North Vietnam, meanwhile, would not "bring China into the war formally," but "would indeed render the war a far more serious one, more expensive in money and casualties," and not make victory likely. Yet U.S. withdrawal from South Vietnam was out of the question since it "would mean all of our friends in South Vietnam would be promptly murdered. If we deserted our friends in this way our influence throughout the world would rather largely disappear, and deservedly so."

The dilemma clouded Bush's senses. He admitted to having no clue about how to end the Vietnam War, but he attacked the peace movement anyway. "Most of what we hear consists merely of statements that we must stop the war, without any suggestions as to how to do it," he said. He hoped that "some wise individual somewhere" might offer "a reasonable solution" to the war, but at the same time he confessed that just talking about finding a "dignified escape" from "the mess" in Southeast Asia damaged U.S. credibility in the world.[51]

In 1969, the Vietnam War came home to the Massachusetts Institute of Technology, where Bush kept an office and was honorary chairman of the board. MIT had long served the ends of national security. Thanks to Bush and Karl Compton, MIT had been the biggest academic defense contractor during World War II. In the ensuing years the Institute remained on top, a testament to the durability of wartime patterns and the strength of the engineers and scientists gathered along the Charles River. In the 1968 fiscal year, the Pentagon paid MIT $119 million, or about double the amount it paid to Johns Hopkins, its second-largest university contractor. MIT's close ties to the Pentagon had provoked only scattered criticism in the past, but the Vietnam War inspired a massive peace movement that even included some MIT faculty. In January 1969, dissident professors called for a strike, to begin on March 4. They were unhappy with the militarization of U.S. science and university support for technologies used against the North Vietnamese.

Because critics dubbed MIT "Pentagon East," the March strike drew wide attention even though most students and professors spent the day in classrooms or laboratories. An estimated 1,400 people jammed an Institute auditorium to hear discussions on military conversion, arms control and academic-government relations. "What we're concerned about is the misuse of talents and resources," one organizer said.

The following April, about 50 MIT students protested against missile guidance research performed on campus. The Institute responded by agreeing to review the role of two of its military labs. Five months later, hundreds of students struck the school and picketed one of the military labs. Waving Viet Cong flags, the protesters shouted "Shut it down!" They were routed by police, but vowed to fight military research at MIT "until the cost of keeping it is higher than the cost of ending it."[52]

The rebellion at MIT mortified Bush. When such a bulwark of the Establishment fractured, surely the society that he had done so much to preserve was deeply troubled. To be sure, Bush's dark mood was exacerbated by the death of his wife, Phoebe. After years of living with hypertension, Phoebe awoke one July morning in 1969 with what seemed to be symptoms of a stroke. About noon she suffered a massive coronary; she never recovered and died on July 24. Relieved that she did not linger in a paralyzed condition, Bush still missed her awfully. He lived alone now with a housekeeper, nursing his own ailments. A gall bladder operation a few years before had impaired him for months. He marked his 78th birthday on March 11, 1968, by writing to a friend in England, "I creak at the joints and I feel old and decrepit." Two years later, he railed against old age for robbing him of the joy of tinkering: "I find I am just as busy as before. But I can't putter about in the shop now: One eye gone and no binocular vision."[53]

Whether because of Phoebe's death, his own decaying physical state or the rebellious times in America, Bush recoiled against what he saw as the degeneration of his country. Fighting off negativity, he rose to defend his country from the barbarians at the gates. Just as he found years before that the atomic bomb had not doomed humanity, so he believed now that inner-city riots and youth rebellion did not mark the irreversible decline of the U.S. "We have lost our bright spirit, and with it our perspective," he wrote in 1970. "The times are tough, but let us look back and see if we can recover balance."

Compared to the first world war or the second, the campus protests and the street riots of the 1960s were unremarkable, he believed. What if Hitler had been first to build an atomic bomb? "Suppose he had. Washington, London, Moscow would soon have been rubble," he answered. "We would now be living in a very different world, democracy would have vanished not to return for generations. Yet we look about us today and become discouraged, many of our citizens spend their time baying at the moon. We are in trouble. But we have seen far worse."[54]

Few were reassured by Bush's words. But then, none of the elite leaders of the World War II generation had managed to retain the public's full esteem. This was only partly due to the disillusionment brought on by the failure of the "best and brightest" to handle Vietnam. Even without the war, the elite shapers of mass opinion would have suffered a setback. The terrain of expert opinion and administrative governance had become so contentious by 1970 that decisions covering every aspect of American life—from education to energy and weapons to child-rearing—were beyond the control of a cozy elite. The backlash against experts undermined the authority of all the professions, but none suffered more than scientists. The proliferation of nuclear weapons, the rise of environmental hazards and the evident political partisanship of many scientists—all combined to engender a cynicism in the public about the aims and evidence of science. The cynicism was one not of principle but of practice: experts were giving bad advice because they were insensitive to the global systems in which their microjudgments existed. As a result, what in 1965 was Ralph Lapp's controversial view—that "a new priesthood of scientists may usurp the traditional roles of democratic decision-making"—had by 1970 become conventional wisdom.[55]

Bush rued the breakdown of expert authority. He remained convinced of the inability of democratic institutions to effectively tackle the sort of complex, technological problems that increasingly dominated the world. Nonpartisan technocrats were the best answer; their very effectiveness depended as much on their detachment from politics as on their expertise. As Bush boasted in 1967, "Good lord, I worked with Hoover, Truman, Eisenhower, Roosevelt, Kennedy, and I don't think any of them ever knew what my political philosophy was or were in any way interested in it."[56]

Yet there was an element of false bravado in Bush's idea of the detached expert. He sensed that the very logic of expert hegemony undermined its durability. Elite reliance on experts, as a shield from mass interference, was built on the power of scientific knowledge and the promise of delivering the goods. The specialization of expertise had advanced knowledge but also expanded the possibilities for legitimate disputes about its import. As with every other postmodern conundrum, Bush saw no solution and only barely grasped the problem. He failed to grasp the abstract political dilemma, seeing instead a personal challenge in the form of the question, what made for "the full man"?

> The graduate student [he wrote the president of Tufts in 1967], when one tells him that he should study broadly, will reply, that, if he does, he will not be able to compete in his own field of specialization with its intense competition and its hectic pace. And there is something in it.
>
> So my approach is a bit different. I admit that specialization is inevitable and intense. I admit the difficulty of being broad in the present world without at the same time becoming superficial. Benjamin Franklin could be, in a sense, a master in a wide variety of fields, but it cannot be done by a graduate student today. I see the sadness of knowing more and more about less and less. But I would not turn to knowing about everything, and knowing about nothing.[57]

Postscript

"Earlier than we think"

When I get downcast I read history. . . . I go back and read about Tippeca-
noe and Tyler too—slogans, mudslinging, torchlight processions, and not
a policy or principle in sight. We may yet be crude but we have come far.
                                                                    —Vannevar Bush

New Year's Eve, 1973. Bush recovered slowly at home from a prostate operation. A live-in housekeeper and a team of nurses looked after him round the
clock. His weight had dropped to 117 pounds earlier in the year, but in the
past few months had risen steadily so that he was now only six pounds shy of
his normal 150. But at 83 years old he was worn down, and he knew it. His
eyes were failing and his hands were no longer steady. He could not work in his
shop and found it tiresome to put on paper the messy scrawl that had long
passed for his handwriting. He had shrunk an inch, and now stood barely five
feet, ten inches tall. Even worse, he apparently had lost his fierceness. He
missed his wife, Phoebe, and often just stayed in bed. "Maybe his age is against
him now, he just wants to lie there," his housekeeper said. "We all are in hopes
that he will get mad enough some day and get up and get the show on the road
again."[1]

   On this, the last day of the year, Bush sat in a chair, ready for visitors. Old
habits dying hard, he smoked a pipe ("I could choke doing it but I don't stop,"
he cracked). At midmorning, the house bell rang. His nurse opened the door.
Two men stepped inside. Bush noticed right away that one carried a tape
recorder. He checked his impulse to send the men away. Tape recorders made
him freeze up. He routinely refused to be recorded by researchers and re-

porters, even turning down requests from sentimental friends who wished a record of his voice. Careful about his public comments after years in the spotlight, he fretted that one of his vulgar outbursts might be caught on tape.[2]

He allowed the men to stay. Their visit had been carefully planned and he wanted the company. One of the men, Harold Edgerton, ranked among MIT's most successful professors, a pioneer in stroboscopic photography. Bush knew the other man too. Karl Wildes, also an MIT professor, was writing a history of the Institute's electrical-engineering department and wanted to interview Bush about bygone days. After showing him a "new model" tape recorder—startling Bush with the news that it was made by a Japanese company, Sony—Wildes pressed the record button and said, "So you're on."

The talk covered familiar terrain. Bush took charge, first asking Wildes if he had looked at his files and then talking about his wartime experiences, puffing on his pipe all the while. Edgerton lifted Bush's spirits; his presence allowed Wildes to proceed more casually. Finally, he managed to steer Bush onto the subject of MIT, and then the men regaled themselves with obscure tales about spats and misunderstandings within the department of electrical engineering. (Of one colleague, Bush snidely recalled: "He didn't know anything about electrical chemistry or anything else as near as I could make out. He came from the proper part of Brookline—that was his principal asset.") Wildes had carefully reviewed Bush's records from the department, studying the construction of Bush's four analog computers and even reading over his 1915 application to graduate school ("I'll be darned," Bush said).

After 45 minutes of reminiscing, Bush felt melancholy. Compared to his illustrious past, his present life meant almost nothing. "I'd like to do a little more writing for the simple reason there's nothing else I can do," he said glumly. "It isn't the writing. I can write after a fashion but I can't concentrate. I don't know if I'll ever get back to where I can write." Wildes, uncomfortable, began drawing the interview to a close, saying he did not wish to wear out his host. Edgerton, a good ten years younger than Bush and still active professionally, rose to leave and politely told Bush he looked fine. "Yeah, I'm doing all right," Bush sighed. Then, as if changing his mind, he said, "It's awful slow, Harold."[3]

Five months later, Bush suffered a cerebral hemorrhage. He then succumbed to pneumonia and never recovered. He died in his home on June 30, 1974.

His death drew wide notice. *The New York Times* published a front-page obituary with the simple headline: "Dr. Vannevar Bush Is Dead at 84." Neatly summing up his life, the *Times* wrote: "A master craftsman at steering around stubborn obstacles, whether they were technical or political or bull-headed generals and admirals, Vannevar Bush was the paradigm of the engineer—a man who got things done." *The Washington Post* recalled Bush's role in the creation of the atomic bomb. In its syndicated dispatch, the Associated Press stressed his service to President Roosevelt in "organizing American science and technology during World War II."[4]

MIT arranged a private funeral service for Bush and held a formal memorial in the fall. At this later gathering Jerome Wiesner, MIT's president and science adviser to President Kennedy, paid Bush the ultimate compliment by declaring, "No American has had greater influence in the growth of science and technology than Vannevar Bush, and the 20th century may yet not produce his equal."

Most of the crowd—studded with World War II–generation scientific leaders such as Harvard's James Conant—wholly agreed with Wiesner. Bush's influence crossed narrow disciplines, shaping the panorama of American experience. His analog computers foreshadowed the emergence of the digital computer, the most far-reaching tool ever devised. Not since Benjamin Franklin had an inventor played so large a role in government. Bush's management of atomic weapons research, despite initial caution, was a model for later "big science" projects. His unstinting support for federal funding of science and engineering after World War II altered the face of higher education and guaranteed U.S. supremacy in military and civilian technology. His repeated call for military planning and coordination, at a time when defense spending devoured the lion's share of the federal budget, provided a beacon for reformers.[5]

Yet Bush suffered many disappointments. These setbacks were too large to ignore, even for those who respected and admired him. After World War II, his technical vision grew blurry. He no longer brought uncanny judgment and a sense of urgency to military research. His rigidity held up the birth of the National Science Foundation, stunting its life and hastening the rise of a military-industrial-academic complex. His acceptance of the logic of arms proliferation—that nuclear weapons insured that wars would be limited and the Cold War a stalemate—made him cling to nuclear weapons even as he tried to contain them. His elitism—indeed, his hostility to participatory democracy—meant that he never built mass support for his axiom that researchers worked best in isolation from political and social concerns. When a restive population demanded control over the nation's research agenda, as it began to do in the late 1960s, he had nothing to offer but platitudes. His preference for centralization over pluralism reflected his faith in efficiency and order. But it also highlighted his lack of appreciation for the benefits of competition between public entrepreneurs and the value of countervailing institutions in both civilian and military spheres.

In hindsight, how does one judge Vannevar Bush? Guilty or innocent? Right or wrong? Good or bad? Success or failure?

Such questions certainly would strike Bush as absurd. He almost never graded, judged or even examined himself. His was a life not of looking back, but of charging ahead. Yet no one had to tell him that his finest hours came in World War II. He knew it. His appetite for quick decisions, camaraderie and the awesome responsibilities of the war years proved this convincingly. If after the war he lost his sure touch politically and technologically, then he was not alone. Many lost their way; whole nations did. If he erred in judging democracy too harshly and in tying the destiny of the state to the competence of an expert class, he at least did so out of a commitment to excellence and integrity that reinforced his belief in the power of one person to make a difference. If he helped to strengthen forces that made his country more bureaucratic, more militaristic, more dominated by big institutions, then at least he had the character and clarity of mind to face the contradictions of this new world and plainly state his objections to them.

Bush never claimed to be a prophet. He had an engineer's respect for failure, and the unanticipated opportunities arising from failure. He had a simple faith, again born of his scientific bent, that men and women, while not perfectible, were not doomed to repeat their mistakes either. He believed that his fellow Americans should neither judge his generation too soon, nor conclude that the nuclear cloud over humanity would never lift.

Late in life he adopted a phrase that stood as his answer to the many Americans who doubted their moral and material worth. "It is earlier than we think," he liked to say.[6]

List of Abbreviations

AEC:	Atomic Energy Commission
AIP:	American Institute of Physics
B-C:	Bush-Conant correspondence, S-1 Files, Office of Scientific Research and Development, Record Group 227, National Archives and Records Administration
CIW:	Carnegie Institution of Washington
DOD:	Department of Defense
DOE:	Department of Energy
EE:	Electrical Engineering
FDRL:	Franklin D. Roosevelt Library
FOIA:	Freedom of Information Act
GR:	General Records
HD:	History Document
HO:	History Office
HSTL:	Harry S. Truman Library
JBC:	James B. Conant
JCS:	Joint Chiefs of Staff
JFP:	Julius Furer Papers
JNW:	Joint New Weapons Committee
JVNP:	John von Neumann Papers
LOC:	Library of Congress
MIT:	Massachusetts Institute of Technology
NARA:	National Archives and Records Administration
NAS:	National Academy of Sciences
NASA:	National Aeronautics and Space Administration
NDRC:	National Defense Research Committee
NER:	New England Region
NSA:	National Security Agency
NSF:	National Science Foundation
NWP:	Norbert Wiener Papers
NYT:	*New York Times*
OF:	Official Files

OH:	Oral History
OSD:	Office of the Secretary of Defense
OSRD:	Office of Scientific Research and Development
POA:	*Pieces of the Action*
PSF:	President's Secretary's Files
RAC:	Rockefeller Archives Center
RAMP:	Robert A. Millikan Papers
RBNS:	Research Board for National Security
RDB:	Research and Development Board, Department of Defense
RG:	Record Group
RPB:	R. Perry Bush
RS:	Rapid Selector
VB:	Vannevar Bush
VB OH:	Vannevar Bush Oral History. This is an 800-page transcript, housed at the Carnegie Institution of Washington, of tape recordings made by Bush from July 1964 to May 1965. A slightly different version of the transcript is kept by MIT.
VBP:	Vannevar Bush Papers

Notes

Prologue: "Call it a war"

Epigraph: VB, "A Few Quick," Nov. 5, 1951 (LOC, VBP, 139).

1. Accounts of OSRD farewell party drawn from: VB correspondence (LOC, VBP, 125; VB POA, 136); interview, John Connor, a former OSRD attorney who attended the party; Hershberg, 310.

2. On VB's stature among 20th-century Americans: A. Chandler to author; interview, Frank Press; Ratcliff, J. D., "Brains," *Colliers,* Jan. 17, 1942. On VB's resemblance to Franklin: in 1970, historian Elting Morison wrote that VB "shares many of the qualities" possessed by Franklin (NYT, Sept. 27, 1970).

3. VB testimony, "In the Matter of J. Robert Oppenheimer," Transcript of Hearing before Personnel Security Board, 562.

4. "important": Shalett, Sidney, "Chief of U.S. Brain Power," *The American Magazine,* June 1948.

5. VB, "The Case for Biological Engineering," *Scientists Face the World* (1942), 33–45.

6. VB to L. Cutler, Oct. 4, 1949 (LOC, VBP, 30/677).

7. Warner, 183; Meeting of the Joint Committee on New Weapons and Equipment, May 26, 1942 (NARA, RG 218, 15).

Many people attested to VB's resemblance to Rogers, including the photographer Yousuf Karsh, Stanley Lovell and John Connor. VB's physical mannerisms and personal style contributed to this comparison, which could reach embarrassing heights, as when *The Boston Herald* called VB the "Will Rogers of the Atom Bomb."

In his memoir, *Pieces of the Action,* VB wrote: "I sorely miss Will Rogers, who could remind us of our absurdities and do so without rancor. One new Will Rogers would do us more good than a dozen economics professors lecturing us on our sins. I have been looking for him, and have not found him. I shall continue the search. I miss his touch" (25).

"Sock" quotation: Meeting of the Joint Committee on New Weapons and Equipment, May 26, 1942 (NARA, RG 218, 15).

8. Interview, Walter Wriston.

9. VB OH, 99—101; VB to JBC, March 20, 1967 (MIT, VBP).

"I have a horror of autobiographies and memoirs," he once wrote. Another time, he explained, "This is not only because [the subject] can't look at himself, but it's also because he's bound to try to play down something and play up something else."

10. Though classed as a conservative, Bush failed to meet English political philosopher Michael Oakeshott's concise definition of conservatism. To Oakeshott, conservatism was a matter of "disposition" rather than "doctrine," where one takes "delight in what is present rather than what was or may be." While Bush relished the pace and personality of small-town society, he never shied away from the implications of technology. He welcomed material change and saw no reason why government, acting pragmatically, could not help to steer such change in desired directions.

11. "Natural Aristocracy Asked by Dr. Bush," *Washington Post,* Feb. 20, 1955; Wiener, 112.

12. VB to A. W. Horton, Jr., May 3, 1938 (LOC, VBP, 120/2942).

Chapter One: *"The sea was all around"*

Epigraph: VB to K. Roberts, Sept. 23, 1948 (LOC, VBP, 98).

1. McCollester, Lee S., "R. Perry Bush," *The Christian Leader,* May 1, 1926; *Chelsea Evening Record,* April 3–5, 1926; Horace S. Ford, remarks at George Putnam Fund Annual Meeting, March 6, 1956 (LOC, VBP).

Besides officiating at funerals, Perry conducted an estimated 5,000 weddings in his Chelsea tenure. He "knew everybody, was welcome everywhere," Ford recalled.

2. "The Three Hundredth Anniversary of the Landing of the Pilgrims," official Program of the Celebration, Aug. 29–Sept. 6, 1920, issued by Provincetown Tercentenary Committee; Smith, Nancy P., "Provincetown: 200th Anniversary," 1927 (courtesy James Theriault).

3. McCollester, Lee S., "R. Perry Bush," *The Christian Leader,* May 1, 1926; VB OH 1; "In Memoriam: R. P. Bush: 1828–1903," Oct. 5, 1903 (King Hiram's Lodge, Provincetown, courtesy James Theriault).

4. Jennings, H., *Provincetown* (1890), 23–25; VB POA, 238–39; Dec. 27, 1912, Abstract of Proceedings of the Grand Lodge of Massachusetts.

5. VB OH, 4–6; address to "The Friday Club," Marion Jones, April 1949 (LOC, VBP).

6. "friends": June 9, 1926, Abstract of Proceedings of the Grand Lodge of Massachusetts, 234.

"If universalists are true to the spirit of faith," wrote Clarence Skinner, a leading Universalist thinker in 1915, "they pledge themselves to free humanity from the economic degradation which fetters it, body, mind and soul, in the twentieth century. The logic is relentless, the implication clear. Universalism, by its very genius, is led into the great social maelstrom, because it is essentially a battle for the freedom of the common person."

7. Ruby, Lorraine, "A Brief History of Chelsea" (undated); Gillespie, Charles B., *The City of Chelsea* (1908) (City of Chelsea Archives).

8. Warner, 182; VB POA, 239; Hazen, Harold L., *Memoirs: An Informal Story of My Life and Work* (unpublished, 1976), 2–69.

Perry Bush was such a good pool player, VB recalled, that "he had to quit, for he began to attract an audience and he feared his fame might reach some of his more rigid parishioners." On beating a boy at billiards: Harold Hazen remembered VB telling him that the wayward youth challenged Perry to a game, thinking "he knew how to put that minister in his place. . . . By luck the boy drew the first shot to break the set-up of balls whereupon Minister Bush proceeded to clear the entire board." The boy challenged Perry to another game, "offering again to break the cluster. Bush repeated his first performance."

9. *Chelsea Evening Record,* April 3, 1926; VB to Edith Bannon, Nov. 21, 1958 (MIT, VBP).

10. Bush, R. Perry, *Poems,* n.d., private printing.

11. RPB speech reprinted in Carpenter, E. J., *The Pilgrims and Their Monument* (1911), 140–44.

Roosevelt did not acknowledge the toast; he reportedly left "during one of the bursts of applause that greeted" Perry's remarks.

12. VB to Helen King, Aug. 23, 1966 (MIT, VBP); Dec. 27, 1912, Abstract of Proceedings of the Grand Lodge of Massachusetts.

13. Dec. 27, 1894, Abstract of Proceedings of the Grand Lodge of Massachusetts, 306–7.

14. VB POA, 240–41.

15. Interviews, Catherine Bush, Elmer Colcord; Hopkins, Anne G., "Manuscript: Village Directory," 1897 (Provincetown Public Library); address to "The Friday Club," Marion Jones, April 1949 (LOC, VBP).

16. Address to "The Friday Club," Marion Jones, April 1949 (LOC, VBP); Hal, F. O., "Perry Bush," *The Christian Leader,* May 1, 1926.

17. Interview, John Bush.

18. VB OH, 1—3; interview, Richard Bush; VB to Lucinda Boss, Sept. 7, 1939 (LOC, VBP, 13).

19. VB POA, 72.

20. VB OH, 85–86 (CIW).

21. Owens, "Straight-Thinking," 103–4; VB to Frederic Delano, Sept. 16, 1944 (LOC,VBP, 31/718). "High seas": VB to R. Ivelaw-Chapman, May 18, 1948 (LOC, VBP, 23/511).

22. POA, 238; VB OH, 5–6.

23. Interviews, John Bush, Elmer Colcord; president to E. Bush, May 10, 1921; "Activity Plus: The Word for Jackson's Dean," *Tufts Weekly,* July 29, 1943; "Jackson's Retiring Dean Bush Sincere Admirer of Youth," *Boston Sunday Herald,* April 27, 1952 (Tufts Archives, E. Bush File).

24. Chelsea School Committee report of 1910 (Chelsea Archives).

25. *New Scientist,* June 2, 1960; VB photo and caption (MIT Museum).

26. "Tidal wave": Hughes, 14; Bilstein, 8.

27. "Scientific Son of a Minister Named Vice President At Tech," *Boston Globe,* March 13, 1932.

28. Douglas, 193–207.

Chapter Two: "The man I wanted to be"

Epigraph: VB OH, 1–1A (CIW).

1. VB OH, 7 (CIW); Horace S. Ford remarks, March 6, 1956 (LOC, VBP).

2. VB to Elihu Root, Dec. 2, 1955 (LOC, VBP, 126).

3. Panama Canal brochure, VB Tufts Scrapbook (courtesy Richard Bush).

4. Interview, John Bush; Owens, "Straight-Thinking," 110; Undated dance card, VB Tufts Scrapbook (courtesy Richard Bush).

5. Owens, "Straight-Thinking," 112–13.

6. Tufts began the examination practice in 1899 and ended it in 1925. In the first 14 years, just 23 students finished the combined program.

7. VB POA, 251.

8. *Boston Globe,* Feb. 2, 1964.

9. VB biography (Tufts, VBP).

10. VB POA, 155–56.

11. VB, "An Automatic Instrument for Recording Terrestrial Profiles," Tufts Master's Thesis, Jan. 1913.

12. Jefferson quotation: *The Story of the American Patent System* (1940, Byron Miller Papers).

13. H. Wellman to VB, April 18, 1939 (LOC, VBP, 118/2823); Roy, William G., *Socializing Capital: The Rise of the Large Industrial Corporation in America* (1997), 1–20.

14. VB POA, 157.

15. VB POA, 242–43.

16. VB POA, 244.

17. Arts and Sciences Dean to H. C. Bumpus, Feb. 17, 1915 (Tufts, VBP).

18. "Hermon Carey Bumpus," *Science,* Jan. 14, 1944.

19. H. Bumpus to G. S. Hall, Feb. 23, 1915 (Tufts, VBP).

20. VB to E. Hodgins, Nov. 23, 1964 (MIT, VBP).

21. VB to H. Bumpus, Sept. 7, 1915 (Tufts, VBP).

22. VB interview, Dec. 31, 1971 (courtesy Mrs. Harold Edgerton).

23. VB POA, 212.

24. Interview, Richard Bush.

25. VB POA, 253; VB OH, 31–33.

26. Wildes and Lindgren, 84–85.

27. Interview, Truman S. Gray; VB, "A Tribute to Dugald C. Jackson," Electrical Engineering, Dec. 1951.

Gray was a graduate student under VB in the 1920s and later became an MIT professor.

The most detailed account of Jackson's philosophy is Carlson, W. Bernard, "Academic Entrepreneurship and Engineering Education: Dugald C. Jackson and the MIT-GE Cooperative Engineering Course, 1907–1932," Technology and Culture, July 1988.

28. Noble, *America,* 137–40; VB, "A Tribute to Dugald C. Jackson," *Electrical Engineering,* Dec. 1951.

29. VB to H. Bumpus, March 11, 1916 (Tufts, VBP).

30. VB OH, 557.

31. Miller, Russell E., *Light on the Hill: A History of Tufts College, 1852–1952* (1966) 391–95; VB POA, 163; Barnow, 27–37.

WGI's broadcasts continued every two weeks (and later weekly) until 1917, resuming in 1919.

32. VB OH, 792 (MIT).

33. VB POA, 163; Scott, 6.

34. VB, "Research on the Hill: The Story of the S Tube," March 1921, *Tufts College Graduate* (CIW VBP); VB OH, 558.

VB added: "University laboratories are studying the proper way to mix concrete. Commercial laboratories are busy dissecting the atom. Every large, progressive manufacturing organization numbers in its staff, chemists and physicists who strive to extend the field of human knowledge."

As historian Edwin Layton, Jr., aptly described the situation in *The Revolt of the Engineers* (1971): "The engineer is both a scientist and a businessman. Engineering is a scientific profession, yet the test of the engineer's work lies not in the laboratory but in the marketplace. The claims of science and business have pulled the engineer, at times, in opposing directions."

35. VB POA, 245–47; VB to E. Hodgins, Nov. 23, 1964 (MIT, VBP).

36. Conant, James, *Modern Science and Modern Man* (1952), 9; Hughes, 118–26; Franklin, 68–75.

37. Kevles, *The Physicists,* 105–26.

38. VB POA, 71.

By late 1917 the most important detection research was performed in New London, Connecticut, at a lab under Millikan's direction.

39. VB to H. Bumpus, May 8, 1917 (Tufts, VBP); VB POA, 72.

40. VB POA, 72–73.

41. H. Bumpus to N. Rush, May 28, 1917 (Tufts).

42. VB POA, 73.

43. VB POA, 74.

44. Bumpus speech to the Manufacturers Dinner Conference, "The Demand of the Government for Efficient Men," Feb. 28, 1918 (Tufts).

Chapter Three: "Blow for blow"

Epigraph: VB OH, Columbia, 1967.

1. VB interview, Dec. 31, 1973 (courtesy Mrs. Harold Edgerton).

2. Etzkowitz, Henry, "The Making of an Entrepreneurial University: The Traffic Among MIT, Industry and the Military," *Science, Technology and the Military* (1988), 515–40; Wildes and Lindgren, 62–63; "immediate success": Noble, *America,* 141–44.

3. "Dazzled by the prosperity of the time and by the endless stream of new gadgets, the American people raised business in the 1920s to a national religion and paid respectful homage to the businessman as the prophet of heaven on earth," William Leuchtenberg has written. "As government looked only to the single interest of business, so society gave to the business-man social pre-eminence. There was no social class in America to challenge the business class. To call a scientist or a preacher or a professor or a doctor a good businessman was to pay him the most fulsome of compliments, for the chief index of a man's worth was his income. 'Brains,' declared Coolidge, 'are wealth and wealth is the chief end of man'" (Leuchtenberg, 187–88).

4. Wilson, Joan Hoff, *Herbert Hoover: Forgotten Progressive* (1975), 12–17; Baritz, Loren, ed., *The Culture of the Twenties* (1970), 385–99.

5. J. D. Ratliff, "War Brains," *Colliers,* Jan. 17, 1942; F. Fassett to VB, Feb. 22, 1945 (LOC, VBP, 37/886).

Fassett recalled that in one such lecture VB "stood in [Room] 10-250 shaking a Stillson wrench at 700 awed freshmen."

6. Hazen, Harold L., "Memoirs: An Informal Story of My Life and Work" (unpublished, 1976), 2–69.

7. VB POA, 164–67; Warner, 186–87.

8. VB POA, 168; VB to Power, Feb. 8, 1922 (Raytheon Archives); Scott, 19–20; VB POA, 164–67.

9. VB's Research Department reports, Nov. 1920, March 10, 1921 (Raytheon Archives).

10. A 1992 analysis of Smith's design by Arden Steinbach of MIT confirmed VB's conclusion. Smith's "original theoretical analysis was wrong and the refrigerator could not have worked in the manner [Smith] thought it would" (Norman Krim to author, Dec. 11, 1992).

11. Krim, Norman, "Vannevar Bush and the Early Days of Raytheon," unpublished paper.

12. Ibid.; VB, "Record of Tests on Gaseous Rectifier Tube," Aug. 7, 1924 (Raytheon Archives); Scott, 23–28, 32.

13. *The Reader's Companion to American History* (1991), 903–6

14. Undated issue, circa 1920s, of *The Saturday Evening Post,* 163 (Raytheon Archives).

15. VB POA, 199; Scott, 39–40.

16. Wiesner, Jerome, "Vannevar Bush," *Biographical Memoirs,* vol. 50 (1979); Wildes and Lindgren, 68; Owens, Larry, "MIT and the Federal 'Angel,'" *Isis,* 81 (1990), 188–213.

17. VB interview, Dec. 31, 1973 (courtesy Mrs. Harold Edgerton).

18. Hazen, Harold L., "Memoirs: An Informal Story of My Life and Work" (unpublished, 1976), 2–72.

19. VB to D. C. Jackson, Oct. 1, 1924; VB to M. Koes, Oct. 1, 1924; Parry Moon to VB, June 30, 1925; VB to Parry Moon, July 6, 1925 (MIT, EE).

20. McCollester, Lee S., "R. Perry Bush," *The Christian Leader,* May 1, 1926.

21. *Chelsea Evening Record,* April 3, 1926; April 5, 1926.

22. *Boston Globe,* March 13, 1932 (MIT Musuem, VB File).

23. Bush, Richard, "Recollections of Vannevar Bush" (n.d., unpublished).

24. Owens, Larry, "Vannevar Bush and the Differential Analyzer," *Technology and Culture,* Jan. 1986, 78.

25. Hazen, Harold L., "Memoirs: An Informal Story of My Life and Work" (unpublished,

1976), 2–75; interview with Russell Coile; VB to W. Weaver, Jan. 6, 1933 (RAC); H. Hazen, "Predifferential Analyzer Work," July 19, 1967 (MIT, 106, 6, 165).

26. Bowles, Mark D., "U.S. Technological Enthusiasm and British Technological Skepticism in the Age of the Analog Brain," *IEEE Annals of the History of Computing,* 18:4 (1996), 5; VB OH, 266–67; Hazen, Harold L., "Memoirs: An Informal Story of My Life and Work" (unpublished, 1976), 2–74 (courtesy of Katherine Hazen); Berkeley, Edmund C., *Giant Brains: Or, Machines That Think* (1949), 72; Wildes and Lindgren, 87; Shurkin, 76–77.

27. VB POA, 181–82.

28. Bowles, Mark D., "U.S. Technological Enthusiasm and British Technological Skepticism in the Age of the Analog Brain," *IEEE Annals of the History of Computing,* 18:4 (1996), 6–7; VB POA, 183.

VB held contradictory opinions on the importance of Kelvin's work in the history of analog computers, saying that the Englishman "had practically all of the basic ideas that were involved in my machine." There was also the fact that one of VB's graduate students who worked on the differential analyzer had learned of Kelvin's ideas in the 1920s, but discarded them as unsuitable.

29. Interviews, Claude Shannon, Truman Gray; Morse, 121; Shurkin, 78.

30. Interview, Arthur Porter.

31. Hazen quotation: *Technology Review,* May 1932.

32. Weaver Diaries, Nov. 21, 1932 (RAC).

33. VB, "Mechanical Solutions of Engineering Problems," *Tech Engineering News* (1928), cited in Owens, Larry, "Vannevar Bush and the Differential Analyzer," *Technology and Culture,* Jan. 1986, 85–86; VB POA, 262.

34. Morse, 121.

35. VB to Weaver, Jan. 6, 1933; VB to Weaver, March 1, 1933 (MIT, Office of the Vice President, 1932–38, 1/16).

36. Weaver Diaries, Nov. 21, 1932 (RAC).

37. Wiener, 139–40; VB OH, 50–50a.

38. Julius Stratton to M. J. Kelly, Oct. 4, 1944, cited in Pang, Alex, "Edward Bowles and Radio Engineering at MIT, 1920–1940," *Historical Studies in the Physical Sciences,* 20:2 (1990).

39. VB interview. Dec. 31, 1973 (courtesy Mrs. Harold Edgerton).

40. Wildes and Lindgren, 110.

41. Pang, Alex, "Edward Bowles and Radio Engineering at MIT, 1920–1940," *Historical Studies in the Physical Sciences,* 20:2 (1990); Owens, Larry, "MIT and the Federal 'Angel,'" *Isis,* 81 (1990), 188–213.

42. VB to Thomas Barbour, May 20, 1941 (LOC, VBP, 9/194).

43. Schweber, S. S., "Big Science in Context," *Big Science: The Growth of Large-Scale Research,* Galison, Peter, and Hevly, Bruce, eds. (1992), 160; "odor of shop": Owens, Larry, "Vannevar Bush and the Differential Analyzer: The Text and Context of an Early Computer," *Technology and Culture,* Jan. 1986, 93–94; Pfeiffer, John E., "Our Top Man in American Science," NYT, Oct. 17, 1948.

44. VB OH, 50–52.

VB once wrote of Compton: "No man in academic life was ever more genuinely loved" (VB POA, 32).

45. Robley D. Evans OH, 70 (American Institute of Physics, 1978).

46. Wildes and Lindgren, 109.

47. Interview, Edward Bowles, April 13–17, 1987 (conducted by Donald R. Baucom), 26.

48. Interview, Richard Bush.

49. Interview, Edna Haskins; John Bush to author, July 3, 1991.

50. Interview, Richard Bush.

51. Interview, Richard Bush; train and archery anecdotes: Bush, Richard, "Recollections of Vannevar Bush," undated (author's possession).

52. Interview, Richard Bush.

53. Interview, John Connor.

54. Interview, Edna Haskins.

55. Interview, Caryl Haskins.

56. Interview, Katherine Hazen; K. Hazen to author, June 1, 1991; interviews, Edna Haskins, Richard Bush.

57. *Boston Globe,* Oct. 1, 1945.

58. Interviews, Katherine Hazen, Richard Bush.

Chapter Four: "Versatile, not superficial"

Epigraph: VB speech, "The Place of Science in the World Today," March 30, 1935 (CIW).

1. Henry C. Johnson to VB, Sept. 7, 1962 (MIT, VBP, 1).

2. VB OH, Columbia (1967), 8.

3. Leuchtenberg, 187.

4. Ibid, 260.

5. Kevles, *The Physicists,* 238.

6. VB, "Stability in Changing Times," speech to Kilwinning Lodge, Lowell, Mass., Nov. 30, 1934 (CIW, VBP).

7. VB to N. Wiener, Oct. 30, 1935 (MIT, NWP, 1/43).

8. "The Place of Science in the World Today," handwritten notes of VB address to Twentieth Century Association, March 30, 1935; "The Stimulation of New Products and New Industries by the Depression," VB address to Cambridge, Mass., Society of Industrial Engineers, Nov. 22, 1934; "Stability in These Changing Times," VB address to Kilwinning Lodge, Nov. 30, 1934; address to Norwich University, Oct. 22, 1934; notes for May 7, 1935, talk; VB to N. Wiener, Oct. 30, 1935 (MIT, NWP, 1/43); "The Place of Science in the World Today," March 30, 1935 (CIW).

9. Cochrane, 352, 378.

10. Greenberg, 63.

11. Greenberg, 65; Auerbach, Lewis E., "Scientists in the New Deal," *Minerva,* Summer 1965, 466.

In a canny assessment of the SAB, Auerbach added: "The Science Advisory Board was not a failure. The brevity of its existence, however, might be interpreted as a failure in adaptation. It was the failure of the bearers of a tradition of pure and disinterested science, grown up in an environment in which government had played a relatively small role, to adapt themselves to a new situation in which government sought their assistance and offered them its support. Both possibilities disturbed them."

12. K. Compton to FDR, Oct. 23, 1936 (MIT, 77-9, 1).

13. Interview, Katherine Hazen.

14. Kevles, *The Physicists,* 260–61.

15. March 22, 1938 (RAC, Weaver diaries).

16. Cited in Owens, Larry, "MIT and the Federal 'Angel,'" *Isis,* 81 (1990), 209.

17. Owens, *Straight-Thinking,* 291–97.

For primary documents, see David Lilienthal Papers, 88, Princeton University, especially VB to A. Morgan, Dec. 7, 1933, and VB to W. Sutherland, May 3, 1934.

18. VB to Wiener, Oct. 30, 1935 (MIT).

19. VB, speech to New England Gas Association, Feb. 9, 1934; "Relation of Patent System to Stimulation of Industries," Report of the Science Advisory Board, April 1, 1935 (CIW, VBP).

20. VB, address to Norwich, June 8, 1935 or 1936; "Pursuit of an Objective," notes for VB talk to sophomores, Feb. 1934; VB "Talk to Freshman Class," April 3, 1934 or 1935 (CIW, VBP).

21. VB, "Critical Analysis of the Examination System of American Engineering Schools," *Journal of Engineering Education,* Jan. 1933.

22. VB, "In Honor of Professor Elihu Thomson," *Science,* May 5, 1933.

23. Steel, 218–19.

24. Cited in Reingold, "Bush's New Deal," *Historical Studies in the Physical and Biological Sciences,* vol. 17, part 2, 307.

25. VB, speech delivered June 22, 1937, in Milwaukee, Wisc., and printed in "The Engineer and His Relation to Government," *Science,* July 30, 1937.

26. Hoover, *Industrial Explorers;* Owens, *Straight-Thinking,* 306, 334.

27. Reingold, "Bush's New Deal," 307.

28. Hearings before the Temporary National Economic Committee, Concentration of Economic Power Hearings, 1938–39, 871–72.

VB's testimony came on Jan. 17, 1939, after which he wrote to L. Marshall that the committee "kept me on the stand practically all day, and I think I accomplished a few things" (LOC, VB 95/2191).

29. *Boston Globe,* Oct. 1, 1945.

30. VB to C. E. K. Mees, Dec. 7, 1951 (LOC, VBP, 72).

31. "Power from the Sun," VB memo, Sept. 19, 1936 (MIT, AC4, Office of the President, 1930–58, 43, Cabot folder); VB to J. Killian, March 25, 1946 (LOC, VBP, 62/1471).

32. VB to Jewett, June 14, 1937 (LOC, VBP, 55).

33. Owens, Larry, "Vannevar Bush and the Differential Analyzer," 78–79; VB to Warren Weaver, March 10, 1934; Weaver diary, March 29, 1935 (RAC).

34. VB to Weaver, March 17, 1936 (RAC, 224D, 1.1 Projects, folder 23).

35. Owens, "Vannevar Bush and the Differential Analyzer," 79–80.

36. Davis visit: Farkas-Conn, 16–20; Davis, W., "Memorandum of Visit with Dr. Vannevar Bush," Nov. 15, 1932, cited in Burke, *Information and Secrecy,* 116–17. Burke concludes that "little came of [VB's] meeting" because Davis was seeking financial aid, which VB wasn't in a position to offer. VB also may have considered Davis a competitor for foundation grants (179).

37. Even five years after their meeting, VB was referring people interested in his rapid selector, and the underlying concepts of automated retrieval of information, to Watson Davis, "who," VB wrote to R. E. Freeman on May 1, 1937, "seems to be in the center of the matter at the present time" (MIT, VBP). From 1939 to 1942, Davis and VB corresponded a few times regarding science clubs and radio reports sponsored by Science Service (LOC, VBP, 31, Davis).

Davis searched widely for patrons of his "Bibliofilm" service, but he failed to garner big support. He did succeed in convincing libraries in the U.S. Department of Agriculture, the Army, the Census Bureau and the National Archives to employ microfilm equipment (RAC, Warren Weaver Diary, Oct. 19, 1937; American Documentation Institute—Bibliofilm Service, RG1.1, S 200D, 124).

38. VB, "The Inscrutable Past," *Technology Review,* Jan. 1933.

VB's article was reprinted in a collection of VB's essays, *Endless Horizons* (1946).

39. Within a few years, Hoover might have wished he had viewed VB's offer more seriously. By the start of World War II, the piles of FBI fingerprint files reached to the ceilings, and the bureau couldn't find enough clerks to work down the backlog (D. Brinkley, 108).

40. VB to W. Weaver, April 14, 1937 (RAC, 224D, 23, 1936–37).

41. After World War II, VB undoubtedly heard about Goldberg, who visited Washington in 1949. By then Goldberg was living in Tel Aviv. John Green, an official in the Commerce Department, met with Goldberg and informed VB, whom he described as the "parent of the selector." Green attached a copy of a 1931 U.S. patent held by Goldberg for a technique that anticipated some aspects of VB's selector. J. Green to VB, Sep. 22, 1949 (LOC, VBP, 44).

42. VB POA, 187–89; Burke, *Information and Secrecy,* 180–85.

From what VB termed the "somewhat strange history" of the rapid selector, Burke identifies four distinct versions built in the late 1930s.

43. Buckland, Michael, "Emanuel Goldberg, Electronic Document Retrieval, and Vannevar Bush's Memex," *Journal of the American Society for Information Science,* 43-4 (1992), 284–94; Shaw, Ralph, "Machines and the Bibliographical Problems of the Twentieth Century," Department of Agriculture, March 15, 1951 (LOC, VBP, 95/2186); interview with Russell C. Coile; Burke, *Information and Secrecy,* 94–95.

44. VB OH, 111; VB POA, 192; Flamm, 35–36.

45. VB POA, 192.

46. VB OH, 112; VB POA, 193.

47. Burke, *Information and Secrecy,* 69–71, 129, 137, 142–48.

48. VB POA, 192–94; Burke, Colin, "The Other Memex," unpublished paper, 8–9; Burke, *Information and Secrecy,* 8–9, 53, 59, 151, 168. VB to Admiral C. C. Courtney, Dec. 11, 1937 (LOC, VBP, 118).

49. L. F. Safford to VB, Dec. 10, 1938; VB to Safford, Dec. 13, 1938 (LOC, VBP, 67/1665); Burke, *Information and Secrecy,* 156, 167–69, 173–74.

50. VB, "Outline of Scope of Activities," adopted Nov. 19, 1937 (NAS, Scientific Aids). From the minutes of the meeting: "The Committee unanimously adopted with slight modifications the proposed outline of the scope of its activities submitted by Dean Bush on May 20, 1937." VB's outline closes with an appeal to seek industry participation: "Recognizing that large commercial interests are involved in connection with many important devices, the Committee expects to cooperate with industry. . . . In subsidizing a commercial device it would aid to introduce it into uses where the commercial interest would not warrant development, and where advancement of scholarship would result."

51. Kevles, *The Physicists,* 288: Conant OH (Columbia); "Lunch at the Century," By Bethuel M. Webster, June 1948; VB to B. Webster, June 3, 1949 (VB, 117, LOC).

52. Cochrane, 368.

53. VB, *Modern Arms,* 19.

54. Leuchtenberg, 284; Millis, *Arms and Men,* 268.

55. Biddle, Wayne, *Barons of the Sky* (1991), 272; Perret, Geoffrey, *Days of Sadness, Years of Triumph,* 29–30; Dupree, 367; Weigley, *Army,* 413–17.

56. VB Phi Beta Kappa address at Tufts, Nov. 1, 1935 (CIW).

57. Cochrane, 369–81.

58. Transcript, Industrial Research Institute (NAS, E & IR file: IRI, 1938).

59. Interview, Caryl Haskins; quoted in Henry F. Pringle, draft of "The Great Harmonizer" (LOC, Pringle Papers, 1).

60. VB to F. Jewett, May 4, 1937 (VB, LOC, 55/1375).

61. Purcell, Carroll, "Science Agencies in World War II: The OSRD And Its Challengers," *The Sciences in the American Context,* Nathan Reingold, ed. (1979); "Empire & Emperor," *Time,* Jan. 1, 1940.

62. VB to H. King, Aug. 23, 1966 (MIT, VBP); Charles Dollard OH (Columbia, 45–46). VB later said that the Carnegie Corporation, which oversaw CIW, had been interested in hiring him for some time. Carnegie's smaller Pittsburgh institution wooed him to take its presidency earlier in the 1930s, but VB firmly turned them down (VB OH, Columbia, 1967).

63. Interview, Caryl Haskins.

64. F. Walcott telegram to J. Merriam, May 27, 1938 (CIW); Delano to VB, June 6, 1938 (VB 21, LOC); minutes of CIW board meeting of June 2, 1938 (CIW); Wiesner, Jerome, "Vannevar Bush," *Biographical Memoirs,* vol. 50 (1979).

65. Emphasis added. Warren Weaver diaries: Sept. 7, Oct. 28, 1938 (RAC); June 3, 1938, NYT; *Boston Globe,* June 5, 1938.

66. MIT had a respectable program in aerospace engineering and its own airplane for test purposes. As vice-president, Bush monitored the program. He also took an interest in the vig-

orous student gliding club, mainly to make sure no one got killed (VB to Allan W. Rowe, April 24, 1933; MIT, Office of Vice President, 1932–38, 1/16).

67. Bilstein, Roger, *Orders of Magnitude: A History of the NACA and NASA* (1989), 4.

68. Memo by Secretary of State, Sept. 22, 1938 (LOC, H. H. Arnold Papers, 16, Lindbergh file).

69. Bilstein, Roger, *Orders of Magnitude: A History of the NACA and NASA* (1989); Bush to W. S. MacDonald, Oct. 14, 1938 (LOC, VBP, 67); C. Lindbergh to Ames, Nov. 4, 1938 (NASA HO).

70. Dec. 7, 1938, probably from *Boston Transcript* (Boston Globe archives).

Chapter Five: The minor miracles

Epigraph: VB to Eric Hodgins, April 10, 1941 (LOC, VBP, 50).

1. Hughes, 182; Owens, Patents; Temporary National Economic Committee, Concentration of Economic Power Hearings, 1938–39.

2. *Washington Star*, Jan. 20, 1939; NYT, Jan. 19, 1939 (LOC, VBP, 24).

3. Interview, Edna Haskins; Brinkley, D., 4.

4. VB to Wardman, May 13, 1939 (LOC, VLB, 116).

5. VB to CIW trustees, Oct. 20, 1939 (FDRL, Delano papers, 2).

6. VB to Osborn, Oct. 2, 1939 (CIW).

7. VB POA, 274.

8. W. Weaver diary, May 24, 1939 (RAC).

9. Kevles, Daniel J., *In the Name of Eugenics: Genetics and the Uses of Human Heredity* (1985), 102–18.

10. The handling of Laughlin by VB and CIW is recounted in Singer, Maxine, "Vannevar Bush as Statesman for Biology," unpublished paper (CIW); A. Kidder to J. Merriam, June 21, 1935; Report on Eugenics Record Office, A. Kidder, June 28, 1935 (CIW).

11. J. Merriam to A. Kidder, June 24, 1935; J. Merriam to H. Laughlin, Dec. 31, 1938; VB to H. Laughlin, Jan. 4, 1939; H. Laughlin to VB, Jan. 6, 1936; VB to H. Laughlin, April 21, 1939; VB to L. Weed, May 3, 1939; VB to H. Laughlin, May 4, 1939; VB to A. Blakeslee, June 8, 1939; VB to H. Laughlin, June 8, 1939 (CIW).

On May 3, VB decided there was "no real solution" except to retire Laughlin. The following day he wrote him and suggested that if he wished to continue his "personal objectives" in the area of public policy, he should "drop out" of CIW.

12. VB to A. Blakeslee, June 23, 1939; N. Wiener to VB, June 20, 1939 (LOC, VBP, 119); R. Reynolds to F. Delano, Dec. 8, 1939; F. Delano to VB, Dec. 8, 1939 (CIW).

13. VB to W. Forbes, Dec. 11, 1938; "Memorandum Re Retirement of H. H. Laughlin," Dec. 5, 1939 (CIW).

14. W. Weaver diary, March 17, 1939 (RAC).

15. Interview, Stacey French.

16. Weaver diary, May 24, 1939, Oct. 17, 1939 (RAC); I. Bernard Cohen, "A First Interview with Vannevar Bush," circa 1945.

17. VB OH, Columbia (1967), 44.

18. Weaver diary, March 6, 1940 (RAC); I. Bernard Cohen, "A First Interview with Vannevar Bush," circa 1945, author's possession.

19. VB to J. H. McGraw, Jr., Oct. 9, 1939 (LOC, VBP, 67).

20. VB to F. Keppel, March 3, 1939 (LOC, VBP, 61/1462).

On the pre–World War II roots of the military-industrial-university complex: Koistinen, 23–61; Schwarz, Jordan, "Baruch, the New Deal, and the Origins of the Military-Industrial Complex," *Arms, Politics and the Economy*, Robert Higgs, ed. (1990); Hooks, 42–124.

21. British quotation: Larrabee, 17.

22. VB to Jewett, March 23, 1939; VB to KTC, May 4, 1939 (MIT, AC4, Office of the President, 1930–58, 42).

23. VB to H. Hoover, April 10, 1939 (LOC, VBP, 51); Huie, W. B., "The Backwardness of the Navy Brass," *The American Mercury,* June 1946.

24. VB to W. Whitman, March 23, 1939 (LOC, VBP, 119).

25. VB POA, 129; VB, OH, 697; VB, NSF speech, June 21, 1963 (MIT, VBP, 11).

"The first time I appeared before a [congressional] committee, I was scared stiff," VB recalled in 1963.

26. Hartman, Edwin P., *Adventures in Research: A History of Ames Research Center, 1940–1965* (1970), 5–17.

27. Bilstein, Roger, *Orders of Magnitude: A History of the NACA and NASA* (1989), 25–28.

28. Ibid.; VB to Mead, Dec. 20, 1939 (LOC, VBP, 72/1737).

29. VB to R. E. Doherty, Feb. 24, 1940 (LOC, VBP, 33).

30. Mosley, Leonard, *Lindbergh: A Biography* (1976), 259–60; VB to Lindbergh, Dec. 6, 1939 (LOC, VBP, 65/1572).

31. Sherwood, 152.

32. VB to Gordon Rentschler, Jan. 5, 1940 (LOC, VBP, 95/2206).

Aware of the need for military allies, VB had already deemed the Army his most likely backer, adding that his plans "cannot go ahead of course unless in one way or another they are given the knowledge and support from the Army that they need in order to get somewhere."

33. VB POA, 33.

34. Coffey, 154–61.

35. Ibid., 207; Bilstein, 128.

36. Coffey, 207–8.

37. VB to members of Division of Engineering and Industrial Research, Sept. 7, 1939; VB to W. L. Blatt, incoming chair of division, Dec. 21, 1939 (LOC, VBP, 84); VB to F. Jewett, Oct. 31, 1939 (LOC, VBP, 55).

38. Jewett quotation: Cochrane, 389–90.

39. VB to John Victory, April 11, 1940 (NASA HO).

40. VB to F. Jewett, April 15, 1940 (LOC, VBP, 55).

41. Hershberg, 115–26.

42. Guerlac, Henry, "Conversation with Alfred Loomis," May 21, 1943 (National Archives for New England Region, Rad Lab, Loomis folder, 49A).

43. Baxter, 14.

44. Dupree, A. Hunter, "The Great Instauration of 1940: The Organization of Scientific Research for War," *The Twentieth Century Sciences,* ed. Gerald Holton (1972), 445.

Dupree wrote about the committee apparently without the benefit of its internal records. He rightly finds that its members "felt that no existing institution could effectively reorganize science for the looming emergency."

45. VB OH, Columbia (1967), 23.

46. Conant sets the date of the lunch in *My Several Lives,* 234; in "Lunch at the Century," June 1948 (LOC, VBP, 117), Bethuel Webster describes the mood. VB to George R. Harrison, March 2, 1955; March 10, 1955 (LOC, VBP, 46).

47. VB to C. L. Wilson, Feb. 10, 1966 (MIT); VB POA, 33; VB to George Harrison, March 10, 1955 (LOC, VBP, 46).

On Compton's hopes in the summer of 1940, VB wrote: "Karl was principally following up the possiblity that some of the machinery of the Academy, such as the Science Advisory Board on which he had been active as you know, could be utilized."

48. Greenberg, 76.

49. VB quotation: Kevles, *Physicists,* 301.

50. Graham, Otis L., Jr., *Toward a Planned Society* (1976), 52–56; Conkin, Paul K., *The New Deal* (1945), 46.

The National Resources Planning Board was killed off by Congress in 1943.

51. "Delano's note to Early, Watson's reaction and VB's memo: President's Official Files, 4010 (FDR Library); memo also cited: Dupree, A. Hunter, "The Great Instauration of 1940: The Organization of Scientific Research for War," *The Twentieth Century Sciences,* Gerald Holton, ed. (1972), 450–51.

52. VB speech, "Aeronautical Research, a Vital Link in Our National Defense" (NASA HO).

53. VB POA, 35; Sherwood, 154; Adams, Henry H., *Harry Hopkins: A Biography* (1977), 164–66; VB to George Harrison, March 2, 1955 (LOC, VBP, 46); Greenberg, 76.

54. VB to F. Keppel, June 5, 1940 (LOC, VBP, 62); VB to Loomis, June 6, 1940 (LOC, VBP, 66/1598).

55. quotation: Larrabee, 65.

56. The date of VB's meeting with FDR, from president's appointment records (FDRL, document courtesy of Robert Buderi); Sherwood, 154; Kevles, *The Physicists,* 297; Records of the Advisory Commission to the Council of National Defense (FDRL, NDRC).

A one-page memo by Bush, dated June 12, 1940, is in the Roosevelt Library. The document doesn't have "OK—FDR" written on it, so may not be the memo carried by VB on June 12, but it otherwise conforms to VB's recollections.

Regarding the length of VB's meeting with FDR, Sherwood describes it as "a few moments." VB told a government historian in 1944 that the meeting lasted 15 minutes ("Interview with Dr. Vannevar Bush," Henry Guerlac, NARA, NER, Rad Lab, 49).

Kevles sets the length of the interview at ten minutes, citing his own interview with Bush. In Kevles's account, Roosevelt says, "Put 'O.K., FDR' on it," indicating that perhaps Bush or Hopkins actually signed the memo.

57. VB to F. Jewett, March 17, 1947: Owens, "Counterproductive Management," 518; Kevles, *The Physicists,* 295; Hart, David. M., "Competing Conceptions of the Liberal State and the Governance of Technological Innovation in the U.S., 1933–1953" (unpublished dissertation, 1995), 308.

58. VB, who often fought back against critics who said he handed contracts to pals in academia and industry, pointedly noted that technically he didn't qualify as a "dollar-a-year" man, since "I didn't even get a dollar" as chief of NDRC or its successor, the Office of Scientific Research and Development (VB OH, Columbia, 1967, 23).

59. O'Neill, *Democracy at War,* 78; Larrabee, 1–2, 32, 53; Burns, James MacGregor, *Roosevelt: The Lion and The Fox, 1882–1940* (1956), 403.

60. FDR's Complete Presidential Press Conferences, vol 15. The next day, June 15, 1940, Roosevelt signed a letter of authorization to Bush (which Bush himself had written with the aid of Hopkins), which described the hopes for the NDRC:

"The function of your Committee is of great importance in these times of national stress. The methods and mechanisms of warfare have altered radically in recent times and they will alter still further in the future. The country is singularly fitted, by reason of the ingenuity of its people, the knowledge and skill of its scientists, the flexibility of its industrial structure, to excel in the arts of peace, and to excel in the arts of war if that be necessary. The scientists and engineers of the country, under the guidance of your Committee, and in close collaboration with the armed services, can be of substantial aid in the task which lies before us. I assure you, as you proceed, that you will have my continuing interest in your undertakings" (Sherwood, 155).

61. Interview, Bruce S. Old, who heard the story from the researcher, Jerome Hunsaker.

62. VB to George Harrison, March 10, 1955 (LOC, VB, 46).

63. Baxter, 16.

64. Stewart, 191.

65. Conant, 236.

66. Price, *Government,* 65–94; VB to F. Fassett, Aug. 3, 1969: Owens, Counterproductive Management," 525.

67. Paul Forman suggests that this did occur over the next two decades under the weight of the overwhelming support for scientific research by military sponsors. Assaying the effect of "patronage" on "the direction of research," Forman wonders "what *kind* of science" resulted from military sponsorship? He concludes: In the case of academic science we have been too ready to accept . . . that this system of patronage aimed chiefly at "the advancement of the best possible physics," too ready to ignore the real rationale for the support from the military. "Behind Quantum Electronics: National Security as Basis for Physical Research in the United States, 1940–1960," *Historical Studies in the Physical and Biological Sciences,* 18: 1 (1987), 200–201.

68. VB POA, 31–32.

69. VB to Keppel, Aug. 1, 1940 (LOC, VBP 62 1462); Pringle notes of VB interview (LOC, Pringle Papers, 27); VB to Hopkins, Aug. 1, 1942 (LOC, VBP, 51/1269).

VB wasn't back to full strength, though, writing Hopkins on Aug. 7, 1940, that his "physical situation doesn't seem to be clear yet, so I may be away a bit for a few more tests" (LOC, VBP, 51/1269).

Chapter Six: *"Don't let the bastards get you down"*

Epigraph: "Yankee Scientist,". *Time,* April 3, 1944, 57.

1. Woolf, S. J., "Chief of Staff on the Science Front," NYT, Jan. 23, 1944.

2. Lovell, 15.

3. VB OH, Columbia (1967), 44.

4. VB, Report of the President of CIW, Dec. 1940 (CIW).

5. The description of CIW's building comes from the pamphlet, "The Administration Building" (CIW); interview, Oscar Ruebhausen.

At the first meeting of the NDRC, Bush agreed to take space in Carnegie. Before choosing it as the NDRC's home, he briefly considered housing the committee in the offices of the National Academy of Sciences, which at the time were located across the street from the War and Navy departments. Carnegie was chosen chiefly to satisfy VB, who "would be able to devote more time to the committee's business than if the offices were situated elsewhere." By the end of World War II, the OSRD, the NDRC's successor agency, had outgrown Carnegie. While VB and his staff remained, the Committee on Medical Research was housed at the National Academy. OSRD operations also were housed at Dumbarton Oaks, an estate in the heart of D.C. that had been donated to Harvard University (NDRC minutes, June 18, 1940, NARA, RG227, Office of Historian, Subject Files 7; Stewart, 32).

6. VB POA, 334–35, Eleanor J. Poirier to S. Callaway, April 29, 1948 (LOC, VBP, 32/740).

Callaway was so valued that, after the war, VB arranged for him to receive a Presidential Certificate for Merit for his contribution to the war effort. He had been the secretary for VB's predecessor and outlasted VB at Carnegie, retiring three years after him in 1958.

7. The "teas" were paid for out of donations from regular participants. Interview with H. Guy Stever; C. L. Wilson memo, Oct. 7, 1944 (LOC, Alan Waterman Papers, 31).

8. Interviews, John Connor, H. Guy Stever, Oscar Ruebhausen, Caryl Haskins; Guerlac, Henry, "Conversation with Alfred Loomis," May 21, 1943 (NARA, NER, Rad Lab, 49–49A).

Of the sign on VB's desk, Ruebhausen recalls: "Someone gave it to him and he liked it." Even professional writers were impressed by VB's writing and speaking. Journalist and popular historian Henry Pringle credited him with "a literacy rare among engineers or scientists" (*Esquire,* April 1946).

9. Interviews, John Connor, David Ginsberg.

10. VB to G. Pyke, May 29, 1942 (LOC, VBP).

11. Interview, John Connor; VB, "A Few Quick," Nov. 5, 1951 (LOC, VBP, 139); VB to R. Major, Nov. 29, 1951 (LOC, VBP, 69).

VB's emphasis on men was deliberate. Not a single NDRC division chief was a woman. None of his top aides were women either. The "good men about" idea endured as the pillar of VB's approach to management. After the war, he cited this principle as crucial to corporate executives.

12. Catton, 13.

13. Interviews, David Langmuir, Oscar Ruebhausen, Bruce S. Old; Baxter, 16; VB to H. S. Morgan, June 9, 1945 (LOC, VBP, 78).

Wilson's importance in VB's life was clear. "Wilson was the way to get to Bush" said Bruce Old, a naval officer who dealt extensively with the NDRC.

14. H. S. Pringle memo (LOC, Pringle Papers, 27).

There is no biography of Tuve. The best source on Tuve's pre–World War II activities is Thomas D. Cornell's dissertation, "Merle A. Tuve and His Program of Nuclear Studies at the Department of Terrestrial Magnetism" (1986, Johns Hopkins University).

15. Merle Tuve Oral History (AIP, 1967).

16. Proximity Fuze," *Life*, Oct. 22, 1945; interviews, David Langmuir, P. Abelson; Baxter, 226; "Final Report on the Development of the Radio and Other Proximity-Fuzes," March 31, 1944 (CIW, DTM).

Tuve struck many people as a blend of appealing and unappealing traits and the epitome of a lone wolf. "He was idealistic and a lover of humanity but he was also a driver and he drove himself and others," recalled Phil Abelson, a Carnegie colleague.

17. Interview, David Langmuir.

18. Interview, Caryl Haskins.

19. Greenberg, 79.

20. Hewlett, Richard G., "Mobilizing Technology for War: Thomas A. Edison and Vannevar Bush" (1979, unpublished).

21. "A Technological High Command," *Fortune*, April 1942, 67.

22. Interview, Lee Anna (Embrey) Blick; VB to J. Furer, Feb. 5, 1942; B. Old and R. Krause to J. Furer, Feb. 4, 1942 (NARA, RG298, Naval Coordinator of R&D, General Corresp., 1941–45, 20).

Removing some of the barriers to Army-Navy communication took time. In early 1942, VB wrote to Furer: "To date no Army liaison has been given access to the Navy project control record, nor has any Navy liaison been given access to the Army project control record. However, it has recently been suggested that some advantage to the armed services might be gained if such references are permitted. Unless you see some reason to the contrary, I plan to open each of these books to the other service." Not only did Furer not object to VB's proposal, his own staff called the lack of communication "an unhealthy situation." VB's stimulus led Furer to meet with the War Department to establish a new procedure.

23. "Interview with Dr. Vannevar Bush," Henry Guerlac, Aug. 20, 1944 (NARA, NER, Rad Lab, 49); Bowen, 178.

Harvey Bundy, special assistant to Secretary of War Stimson, asserted that "Stimson's interest in the scientific . . . was stimulated very much by" Loomis, who was keen on developing radar as both an offensive and defensive weapon. Stimson caught the radar bug from Loomis during the winter of 1941, when Britain's survival hung in the balance. At the time, Bundy recalled, "The Army didn't know radar from a hole in the ground. . . . But Stimson really brought this seriously to the attention of a great many at the top of the Army." The secretary even watched a demonstration of radar tracking at the Radiation Lab in Cambridge, Massachusetts.

24. H. Bowen to F. Knox, Jan. 29, 1941; H. Bowen to F. Knox, Feb. 20, 1941 (NARA, RG80, 410, 33–41); H. Bowen to F. Knox, March 17, 1941: Sapolsky, 17.

Bowen was the Navy's representative on the NDRC, so his comments had special weight, but other Navy officers felt threatened by VB's outfit. The growth of the NDRC "looks very objectionable," a Captain G. L. Schuyler wrote on March 12, 1941. He added: "The eventual status of the NDRC admittedly presents a problem. It is building itself up rapidly in a way that perhaps few in the Navy have realized" (NARA, RG80, 410, 1933–41).

25. Sapolsky, 16–18; Gannon, 26.

In a few months Bush would choose Hunsaker to replace him as chair of the National Advisory Committee for Aeronautics, a position Hunsaker held for 15 years.

26. VB, NBC interview, June 5, 1965 (MIT, VBP).

27. "Organization of Defense Research," VB to Hopkins, n.d., with annotation "sent to Harry Hopkins 3/3/41" (LOC, VB, 51/1269).

28. Kevles, *The Physicists,* 300; Harvey Bundy OH, Columbia (1961), 172.

29. Bundy OH, 170–72; VB POA, 147.

30. Bundy OH, 169–74; VB to R. G. Leffingwell, Feb. 16, 1945 (LOC, VBP, 20).

VB appreciated Bundy, probably more than the self-effacing Stimson aide realized. Near the end of the war, he freely recommended Bundy to foundations and universities seeking top executives, extolling Bundy's virtues but frankly admitting, "He is not at all a forceful individual and he always obtains his results by patience and care, rather than by aggressiveness." Bundy's personal traits, VB conceded, made their relationship "somewhat difficult from my standpoint, for he is my official contact with [Stimson's] office, and if I do not get action through that contact I am in a quandary. However, he does not become disturbed if I get out of channels and this has helped greatly." Since VB often acted "out of channels," Bundy's traits were fortunate indeed.

31. VB recalled in POA (82): "It was important that the word 'Development' appeared in the title of O.S.R.D. and that the Appropriations Committee of Congress . . . saw the point and supported the idea that the laboratory could proceed beyond research and build hardware to demonstrate its accomplishments."

32. Baxter, 124, 31; VB POA, 45–47.

The NDRC's first $15.5 million had come directly from Roosevelt's emergency fund via the Council of National Defense, an arm of the executive.

33. VB to George Merck, Oct. 29, 1947 (LOC, VBP, 72/1751); VB POA, 47; VB interview, Dec. 31, 1973 (courtesy, Mrs. Harold Edgerton).

Never as central to the OSRD as the NDRC, the CMR nonetheless made important advances in treatment of malaria, insecticides, transfusions and penicillin. It spent $24 million on about 600 contracts with 133 universities, foundations and firms (Baxter, 300).

34. Interviews with Lee Anna (Embrey) Blick, John Connor. NDRC meeting, Jan. 2, 1942 (NARA, OSRD, RG 227, NDRC Minutes, 2); Owens, "Counterproductive Management," 533, 555–56.

35. Pringle, H. S., "Science, War and Vannevar Bush," *Esquire,* April 1946.

Larry Owens, after reviewing OSRD contract records, has concluded that administration, rather than the activities of NDRC's operating divisions, was "the solid foundation of OSRD's power." Owens finds that Bush's administrative chief, Irvin Stewart, ran "an empire whose assets comprised administrative offices, legal documents, and memoranda, and whose front-line troops were not physicists or engineers but rather lawyers, accountants, office managers, and filing clerks." This judgment seems to place an undue importance on administration, which at best was a means to an end. It seems clear that Bush's clout stemmed from his ability to deliver new weapons and equipment.

36. "Report of the NDRC for the First Year of Operation, June 27, 1940, to June 28, 1941." VB signed the report, July 16, 1941 (FDRL, PSF, Safe File, Bush folder 2).

37. *Radar: A Report on Science At War* (1945), 8–12; Buderi, 2; Kevles, *The Physicists,* 303.

Alfred Loomis believed the delivery of the magnetron advanced U.S. radar development by two years.

38. Ibid.

39. Buderi, 15.

40. Larry Owens has concluded that the siting of the Rad Lab had "enormous consequences for the siting of postwar R&D." Even during the war, it made MIT "a significant regional patron," through subcontracts with local suppliers. Owens, "Counterproductive Management" 554–55. Guerlac, Henry, "Interview with Dr. Vannevar Bush," Aug. 20, 1944 (NARA, NER, Rad Lab, 49).

41. Genuth, Joel, "Microwave Radar, the Atomic Bomb, and the Background to U.S. Research Priorities in World War II," *Science, Technology & Human Values,* Vol. 13, Nos. 3 & 4, Summer & Autumn 1988, 280–82; VB to K. Compton, Jan. 5, 1940 (LOC, VBP, 26).

42. Guerlac, Henry, "Interview with Dr. Vannevar Bush," Aug. 20, 1994 (NARA, NER, Rad Lab, 49); Edward Bowles interview, July 14, 1987 (courtesy Martin Collins and Michael Dennis), 19–21.

43. Rigden, 133–45.

44. Karsh, Yousuf, *Portraits of Greatness* (undated).

45. VB memo, NDRC No. 20, Dec. 11, 1940 (CalTech, Von Karman Papers, 84.7), VB speech, Oct. 24, 1941, printed in *Journal of Applied Physics,* Dec. 1941.

46. "A Technological High Command," *Fortune,* April 1942; *The Wall Street Journal,* June 17, 1943, 1.

47. J. Furer diary, April 10, 1942 (LOC, JFP).

48. VB to F. Jewett, March 11, 1941 (LOC, VBP, 55).

49. Beaumont, Roger A., "Quantum Increase: The MIC in the Second World War," *War, Business, And American Society,* 127; Baxter, 456.

50. Wayne Coy to VB, Aug. 21, 1941 (FDRL, Delano Papers, 2); Stewart, 280, 186, 13.

Coy's ruling was based on a four-page opinion by Oscar Cox, a Treasury Department attorney who had informally handled various legal matters for Bush and the NDRC during its first six months of life. On Dec. 9, 1940, the NDRC hired a permanent counsel, but it continued to consult Cox until late in 1942.

Cox, an intimate of Harry Hopkins, held a variety of governmental posts from 1938 to 1945, including posts in the Justice Department and the Office for Emergency Management. He was best-known for authoring the Lend-Lease Act, which allowed the U.S. to materially aid Britain against Germany before the formal declaration of war. In the summer of 1940 Cox wrote the legal interpretations that enabled Roosevelt to initially fund NDRC out of his discretionary funds. Cox, an MIT graduate, would serve as Bush's personal attorney following World War II (FDR Library, Cox Papers; NYT, "Oscar Cox Dies," Oct. 6, 1966).

51. VB to Marshall, Sept. 18, 1941 (LOC, VBP, 69/1701); VB to T. G. Brown, Sept. 18, 1941 (LOC, VBP, 123, 1941); VB to T. G. Brown, Sept. 30, 1941 (LOC, VBP, 123, 1941); Scott, 91.

In his letter of resignation on June 3, 1938, VB cited his new post in Washington D.C., as the reason for his departure from the board. He added, "Needless to say I will maintain interest even in the absence of formal connections."

52. Jewett cited in Owens, "Counterproductive Management," 552–53; Saxenian, 14–15.

Charges of self-dealing against VB never stuck; more credence was given claims that VB had usurped the authority of the traditional leaders of the science community, such as the National Academy of Sciences. By 1945 VB preempted critics by freely admitting he "ran away with the ball" and played a new game by his rules. VB to I. Bowman, Feb. 6, 1945 (LOC, VBP, 13).

Raytheon's boom ended with the war. The company's sales, which topped $170 million in 1945, fell to $105 million in 1946. The company posted a loss of $333,000 for the year, compared to a profit in the prior year of $3.4 million. By 1948, the company's sales had fallen to $53 million (Scott, 97, 182).

53. VB speech, Oct. 24, 1941, *Journal of Applied Physics,* Dec. 1941; NYT, Oct. 25, 1941.

54. VB to E. Hodgins, April 10, 1941 (LOC, VBP, 50).

55. Caryl Haskins, a close associate during the war, recalled that VB "was very conscious of burnout dangers, for himself and the people who worked for him." On the mission to Britain: VB to FDR, Feb. 1, 1941; FDR to VB, Feb. 4, 1941: "I am delighted at the prospect of Dr. Conant going to England as a member of a Mission to interchange technical information with the British. He will do a grand job" (FDRL, Official File 4010, NDRC).

56. Woolf, S. J., "Chief of Staff on the Science Front," NYT, Jan. 23, 1944; *Saturday Review,* Aug. 1, 1959; Henry Pringle, notes of VB interview, Sept. 12, 1945 (LOC, Pringle Papers, 27); VB to R. Major, Nov. 29, 1951 (LOC, VBP, 69/l686).

VB mainly let off steam through tinkering. For some of his gadgets, however, he held hopes of reaping commercial gains after the war. Writing to Norbert Wiener, the mathematician, he noted that his "very strange and bizarre" solar pump could "make the desert bloom," adding: "I puzzle away at something which has nothing to do with the war, and which gets me away from the shuffling of papers . . . to the thing I like to do." As the war went on, VB's inability to pursue idle technical questions frustrated him further. In November 1942 VB wrote a friend, "If there wasn't a war on, I'd be delighted to attempt to collaborate with [another researcher] in handling the problem of the distribution in depth of the photographic image, and it would be a lot of pleasure to do so. As it is I can hardly expect to do serious work of any sort." VB to NW, April 19, 1943 (LOC, VBP, 119); VB to A. C. Hardy, Nov. 9, 1942 (LOC, VBP, 46).

57. Henry Pringle, notes of VB interview, Sept. 12, 1945 (LOC, Pringle Papers, 27).

Pringle's notes contained a more blunt assessment of Bush's emotional state than the resulting profile in *Esquire.* But even the published article noted mysteriously that "Bush did not crack up while supervising all these activities."

58. Interviews, David Langmuir, Edna Haskins.

59. Interviews, Dorothy McDonald, Edna Haskins.

60. Interviews, John Connor, Edna Haskins.

61. Interview, Lee Anna (Embrey) Blick.

62. VB to H. R. Shepley, April 8, 1941; VB to New England Trust, Aug. 29, 1941 (LOC, VBP, 123, 1941).

63. The Jaffrey farm was purchased in late 1937 or early 1938 for $19,500. VB recalled locating the farm late in 1937 but not actually taking title to the property until April 1, 1938. VB to L. B. King, Sept. 29, 1943 (LOC, VBP, 63); VB to PB, July 17, 1941 (LOC, VBP, 123, F441); *Boston Globe,* undated, Nov. 1943 (MIT). VB to E. L. Gillett, April 29, 1942 (LOC, VBP, 42).

64. VB, Oct. 4, 1941 (LOC, VBP, 123, 424).

65. VB to John French, June 3, 1939 (LOC, VBP, 19); VB to R. D. Merson, June 20, 1942 (LOC, VBP, 76/1757); VB to Archibald MacIntosh, Feb. 24, 1942, (LOC, VBP, 48/1176)

66. The Lend-Lease Act was fiercely opposed by isolationists, who considered it virtually a declaration of war against Germany. Roosevelt protested that the U.S. remained neutral in the European war, but the act's author, a government attorney named Oscar Cox who aided Bush in legal matters facing the OSRD and later became his personal attorney, essentially admitted that the isolationists were right, on the third anniversary of the legislation. "At long last, the public and Congress is seeing the effectiveness of lend lease as a weapon of war," Cox wrote (FDRL, Cox diary, March 11, 1944).

67. VB commencement address, June 19, 1941 (CIW, VBP).

68. VB to M. Sloss, Dec. 15, 1941 (LOC, VBP, 108).

Chapter Seven: "The man who may win or lose the war"

Epigraph: Meeting of the Joint Committee on New Weapons and Equipment, May 26, 1942 (NARA, RG 218, 15).

1. VB to Paul Klopsteg, July 10, 1941 (LOC, VBP, 63).
2. VB to George Mead, April 4, 1941 (LOC, VBP, 72/1737).
3. VB to Paul Klopsteg, Oct. 24, 1942 (LOC, VBP, 63); Greenberg, 93; *Saturday Review,* Aug. 1, 1959.

As late as the mid-1950s, Klopsteg's designs were still kept secret by the military, which cited concerns that teenagers or criminals might use them.

4. *Boston Globe,* Oct. 1, 1945.

Given the way VB hectored the Army and Navy about not being open enough to new weapons, VB's passion for the relatively backward bow and arrow suggested hypocrisy to some military officers. However, the more perceptive realized that VB was underscoring how important it was to find a technology that suited the problem at hand. Even when fighting an electronic war, as VB was, he instinctively knew that mechanical technologies still might win out in certain situations. There was also the paradoxical and as yet barely appreciated problem of higher failure risks associated with more complex equipment. A perceptive Navy liaison, Captain Lybrand Smith, indicated his understanding of this rather elusive point, writing VB on April 5, 1943, about a passage he'd read in a new book on the history of weaponry: "Opposition to the change from bow to musket was powerful. As an example, Colonel Sir John Smyth wrote to the Privy Council in 1591: 'The bow is a simple weapon, firearms are very complicated things which get out of order in many ways . . . a very heavy weapon and tires out soldiers on the march. Whereas also a bowman can let off six aimed shots a minute, a musketeer can discharge but one in two minutes.' Many other John Smiths felt the same distrust of the new arms" (LOC, VBP).

5. This comment, by Eric Hodgins, is on p. 178-A of a lengthy Oral History with VB that Hodgins conducted from July 1964 through May 1965. Hodgins, a writer for *Fortune* magazine who had met Bush in World War II, sprinkled the transcript of VB's statements with his own observations. Slightly different versions of the transcript can be found at CIW and MIT.

6. Ratcliff, J. D., "Brains," *Colliers,* Jan. 17, 1942.

The burst of recognition was galling for Bush's critics in the Navy, who complained that the *Colliers* piece contained too many "revelations." Furer, who fielded the complaints, considered them "not terribly serious, although they might easily have been made somewhat more innocuous by the deletion of only a few words." Furer agreed to bring the matter up with Bush and perhaps suggest to him that journalists writing about the OSRD submit their articles "for censorship before publication" to the War or Navy departments in addition to Bush's own office. Furer warned the dissenting naval officers not to raise their hopes. "I had better take the matter up with Dr. Bush diplomatically, as there is no need of going to the mat over the matter." Just five months later—as Bush's positive press notices mounted—Furer expressed a less-generous view in his diary. "Bush is, of course, in a perfectly human way, looking for praise for his work, to which he is actually entitled in my opinion, as the scientists are doing a fine job, although Bush himself does not exactly have a passion for anonymity" (LOC, Furer Diary, Jan. 12, 1942; May 26, 1942, 1).

7. Weigley, *American Way,* 175.

8. Graebner, 66.

In *Public Entrepreneurship: Toward a Theory of Bureaucratic Power* (1980), Lewis examined the careers of J. Edgar Hoover, Hyman Rickover and Robert Moses to illustrate his theory. VB's actions during World War II meet Lewis's criteria. Exploiting the contradictions between laissez-faire and state management, VB borrowed from both to create an amalgam that at once expanded state power while enriching private institutions and broadening their scope.

9. VB OH, 219.

10. Interview, H. Guy Stever.

11. Interview, R. Bolt (R. D. Glasow, 1980), cited in Gannon, 58–59.

12. Polenberg, 54, 176; Karl, 201.

13. VB saw the need for deferment for researchers as early as the fall of 1940, saying at a meeting of the NDRC that "there is no provision in the law for deferment of men who may, at a subsequent time, become necessary to a research program," NDRC minutes, Oct. 25, 1940 (NARA, OSRD, RG 227, NDRC Minutes, 1); Adams, 76–79.

14. VB to FDR, Jan. 9, 1942 (NARA, RG 51, 39.27, 19, "Civilian Scientific Corps"); H. Smith to FDR, Jan. 16, 1942 (FDRL, Smith Papers, 3); Childs, Marquis W., "Superman of the Budget," *Washington Star,* May 3, 1942; "Man Who Holds the Purse Strings," *Nation's Business,* Nov. 1944; Smith entry, *Current Biography,* 1943.

15. VB to H. Smith, Smith reply attached, Jan. 30, 1942 (NARA, RG 51, 39.27, 82).

16. VB to Furer, Feb. 6, 1942 (LOC, JF Diary, 1); Kevles, *The Physicists,* 321; Baxter, 127–29.

17. Greenberg, 86; VB to Harley Kilgore, Aug. 27, 1943, (MIT, AC4, Office of the President, 1930–58, 42, VB, 1941–42); VB POA, 288.

18. Ranelagh, John, *The Agency: The Rise and Decline of the CIA* (1986), 37–39; Brown, Anthony Cave, *The Last Hero: Wild Bill Donovan* (1982), 147–85, 235–38.

19. The sentence in Baxter (125) is: "Division 19, 'Miscellaneous Weapons,' was created April 1943, to meet certain needs of the Office of Strategic Services."

20. VB OH, 684–85; Meetings of the Joint Committee on New Weapons and Equipment, Sept. 29, 1942, June 3, 1943 (NARA, RG 218, 15); VB to W. Weaver, Feb. 26, 1946 (LOC, VBP, 117).

21. VB OH, 684; H. M. Chadwell to VB, Oct. 27, 1942; VB to JBC, Oct. 29, 1942 (RG 227, contents of VB's MIT safe).

Chadwell officially served as VB's "special representative" and chair of OSRD's new subcommittee on "special government agencies." VB grew to like Chadwell so much that when Carroll Wilson left as his chief aide after the war VB hired Chadwell to do the job.

22. H. M. Chadwell to VB, Nov. 6, 1942; VB to H. M. Chadwell, Nov. 11, 1942 (RG 227, contents of VB's MIT safe).

23. Baxter, 265; *Summary Technical Report of NDRC: Miscellaneous Weapons,* vol. 1, 1946 (NARA, RG 227); Lovell, 39–42.

24. VB to Parry Moon, Aug. 14, 1943 (LOC, VBP, 78/1806); VB to H. Hoover, April 19, 1943 (LOC, VBP, 119).

Writing to former president Hoover four months before, VB had insisted, "We have to have force in the world, the whole problem is how to have it responsive to the best will of all the people."

25. VB OH, 685.

In 1963, Lovell published a memoir in which he stated his admiration for Bush and detailed some of the Division 19 research for OSS.

26. Dawidoff, Nicholas, *The Catcher Was a Spy: The Mysterious Life of Moe Berg* (1994), 148; Lovell, 35–45.

27. Lovell, 35–45.

While Lovell is the sole source on numerous aspects of OSRD's relations to OSS, VB privately endorsed Lovell's account in his OH, which was conducted shortly after the publication of Lovell's book. VB doubted whether he should describe his relations with the OSS in his own memoir (he ultimately didn't), but he said the issue was no longer one of security because "Lovell spilled the whole affair" in his own book. "I couldn't really object because he got it cleared in the regular form but I didn't like the idea of a lot of this stuff being published" (VB OH, 684).

28. VB OH, 685–86.

On balance VB considered the contributions made by OSRD to OSS "very valuable" and found "there was no difficulty with security." He credited H. M. Chadwell, director of Division 19, for his "not merely effective but pleasant" handling of matters (VB to W. Weaver, Feb. 26, 1946, LOC, VBP, 117).

29. Ibid., 59–61.

30. Meeting of Joint Committee on New Weapons and Equipment, June 23, 1942 (RG 218, JNW, 15–17); Lovell, 61–62.

31. R. C. Tolman to VB, Nov. 20, 1940; VB to C. Eastman, Nov. 20, 1940; Hoover to VB, May 13, 1943; VB to Hoover, May 18, 1943 (NARA, RG 227, OSRD Dir., GR, 37).

32. Kaempffert, Waldemar, "War and Technology," *American Journal of Sociology,* Jan. 1941, 431–44.

33. VB to J. P. Baxter, Sept. 16, 1944 (NARA, RG 227, History Office, 7, Org: NDRC-OSRD). VB to Roosevelt (FDRL, PSF, safe file, Bush folder); VB OH, 176, 425.

34. Buderi, 93.

35. Furer diary, March 20, 1942 (LOC, JFP, 1).

36. Stimson and Bundy, 465.

War Secretary Stimson, and his aide Harvey Bundy, helped VB make his case to Roosevelt.

37. Larrabee, 153–57.

38. Buderi, 94; Furer diary, May 26, 1942 (LOC, JFP, 1); Meigs, *Submarines,* 44–45.

39. POA, 91; Meigs, *Submarines;* Furer diary, March 21, 1942 (LOC, JFP, 1).

40. "Interview with Dr. Vannevar Bush," Henry Guerlac, Aug. 20, 1944 (NARA, NER, Rad Lab, 49); Buderi, 96; VB OH, 760–61.

On why Bowles left the Rad Lab: VB explained that Bowles wanted to emphasize engineering methods more than the lab's physicist leadership thought was wise. "It was Bowles against the field—they pasted the hell out of him," VB noted. By early 1942, Bowles was "reduced to escorting distinguished visitors around the lab."

On Stimson's interest in radar: VB said Stimson "was unhappy at the way radar was being used or not sufficiently used in the Army." The secretary "independently" came up with the idea of hiring a radar expert and Bundy, Stimson's aide, asked VB for names of candidates. VB ranked Bowles the top choice.

VB respected Bowles despite their tendency to butt heads. He often recommended Bowles to officials seeking to improve the military's radar capabilities. As far back as 1939, VB had given the name of Bowles to Charles Lindbergh when he inquired about radio detection (VB to C. Lindbergh, May 4, 1939, LOC, VBP, 65/1572).

41. VB to H. S. Morgan, June 9, 1945 (LOC, VBP, 78). VB's comments on Bowles came in a letter describing Bowles's suitability for a postwar executive position.

42. Kevles, *The Physicists,* 319.

The Navy was following the same path as the Army, albeit more gradually. In August 1942, the legendary mathematician John von Neumann severed his ties with NDRC, hiring on directly with Navy, joining the Bureau of Ordnance's Mine Warfare Section. JVN to VB, Aug. 31, 1942 (Princeton, JVNP, 16); Aspray, 26.

43. Collins, Martin, "Integrating Vertically: Edward L. Bowles and His Corporate Vision of the National Security State," unpublished paper delivered at the annual meeting of the History of Science Society (1991).

44. Stimson and Bundy, 514.

45. E. Bowles to Major General D. Olmstead, Aug. 5, 1942 (NARA, RG 165, New Dev. Div., 334, 471.61).

46. Emphasis added. VB to H. Hoover, April 19, 1943 (LOC, VBP, 119).

47. Baxter wrote (31) after the war that the JNW "proved only a partial solution to bringing the civilian scientist in at the planning level." But even this appeared a generous assessment as Baxter concedes, "JNW never was brought in contact with the [military's] staff planners." Then,

undoubtedly parroting VB, he concluded, "The difficult problem of the relationships of scientists to the military at the strategic level awaits a solution."

48. VB to McNarney, Sept. 30, 1943 (NARA, RG 107, war secretary's top-secret files, 10).

49. Marshall order, Oct. 25, 1943 (NARA, RG 227, Office of Historian, 9).

50. "History of NDD," Sept. 1, 1945 (NARA); Furer diary, April 9, 1943, Oct. 14, 1943 (LOC, JFP, 1); VB to James Conant, Oct. 19, 1943 (NARA, OSRD, Reports to President, 5).

Chapter Eight: "A race between techniques"

Epigraph: VB OH, 744.

1. Overy, 53–57; VB POA, 88.

2. Buderi, 104–7.

3. King's "Escort is" quotation in Larrabee, 177; King's "remainder of" quotation in Meigs, *Slide Rules,* 91.

The quotation comes from an address by King to a special Anglo-American conference on antisubmarine war. He also said, "I see no profit in searching the ocean, or even any but a limited area, such as a focal area. All else puts to shame the proverbial search for a needle in a haystack."

4. VB to W. R. Purnell, Feb. 9, 1943 (NARA, RG 218, JNW, 1).

5. VB to FDR, Feb. 17, 1943 (NARA, RG 227, OSRD Reports to President, 5).

6. Morison, *Turmoil and Tradition,* 464–75; Kevles, *The Physicists,* 314–15; quotation from Stimson diary, July 23, 1942: Meigs, *Slide Rules,* 69.

7. Larrabee, 179–80; Morison, *Turmoil and Tradition,* 475–76; Meigs, *Slide Rules,* 91.

8. VB to Marvin McIntyre, March 31, 1943 (FDRL, PSF, Subject File, 97); VB to Frank Jewett, April 14, 1943 (NA, RG 227, NRC-Navy Cooperation, 34, E-13).

9. Furer diary, April 9, 1943, Oct. 9, 1944 (LOC, JFP, 1).

10. VB to King, April 12, 1943 (NARA, RG 227, NRC-Navy Cooperation, 34, E-13).

On the difficulty of seeing beyond World War II, VB confessed in Feb. 1942, "One cannot look ahead to the end of the war with any accuracy." VB to Sloss, Feb. 4, 1942 (LOC, VBP, 108).

11. VB to Frank Jewett, April 16, 1943 (NARA, RG 218, JNW, 2).

12. Baxter, 32; VB memorandum on conference with King, April 19, 1943 (NARA, RG 227, NRC-Navy, 34, E-13); Buderi, 110.

13. Baxter, 32; Meigs, *Slide Rules,* 97; VB to King, April 20, 1943 (NARA, RG 227, NRC-Navy, 34, E-13). As Low's three radar advisers, VB chose three NDRC division chiefs, John Tate, Hartley Rowe and Alfred Loomis, Stimson's cousin.

14. Larrabee, 181; Stimson quotation in Morison, *Turmoil and Tradition,* 478.

15. Quotation by Admiral Karl Doenitz in Rigden, 160; Kevles, *The Physicists,* 315.

16. Kevles, *The Physicists,* 311.

The accuracy of reports of combat actions was the Achilles heel of operations research, and part of the motivation to have trained observers not only analyze but collect raw data. Harvey Bundy, a War Department aide, later recalled that the Air Force, for instance, had a tendency to report that "they destroyed everything in sight. That became very doubtful when it appeared later that the same things they destroyed were operating a week later or a month later. So that there was a danger of impulsiveness and enthusiasm of the reports of battle operations." The hope was that scientists, "who were accustomed to dealing objectively with facts," would provide greater accuracy (Bundy OH, 193).

17. Meigs, *Slide Rules,* 51–58; *Radar: A Report on Science at War* (1945), 24.

18. Morse, 177–87.

19. Baxter, 410–14; Greenberg, 94; Kevles, *The Physicists*, 315; *Fortune*, "The Great Science Debate," June 1946, 117; Thiesmeyer and Burchard, 9–12, 28–29; Furer diary, Dec. 14, 1943 (LOC, JFP, 1).

VB left the decisions on aid in the field to those closer to the action, but he crafted a set of principles, released Aug. 14, 1943, that were meant to guide scientists in their choice of field assignments and alert them to the pitfalls they might face. "Experience has shown," Bush declared, "that successful use of such personnel . . . requires (a) that the officer to whom they are detailed definitely wants them; (b) that they be allowed access to such information as they may need for their work; (c) that they be allowed reasonable freedom as to the way in which they do their work; and (d) that they be responsible to the Commanding Officer and make their reports and recommendations to him."

20. Buderi, 158–59; VB POA, 106–10.

21. Baxter, 221–22.

22. VB POA, 107; Baxter, 223–32; Buderi, 159; "Final Report on the Development of the Radio and Other Proximity-Fuzes," March 31, 1944 (CIW).

23. VB POA, 109–10; Baldwin, 245; JNW meeting, Nov. 3, 1943 (NARA, RG 218, JNW).

While VB generally accepted the restriction on proximity fuzes, he argued that certain fuzes would take longer for the enemy to copy than others. Fuzes in shells, for instance, he felt were harder to make and would take the Germans at least a year to copy. Of the Japanese, he said, "I'm not at all afraid that the Japanese will copy our shell fuzes." Bomb fuzes, however, were easier to copy, he believed. VB raised this distinction with the Navy and Army members of the JNW because he felt the Joint Chiefs were ignoring this distinction.

24. JNW meeting, Feb. 16, 1944, undated meeting number 56, probably in May or June (NARA, RG 218, JNW).

25. VB to James Conant, May 6, 1944 (Harvard, JBC Papers, Special Subject File, 6).

26. VB POA, 307; Powers, 353; Henry Pringle, notes of VB interview, Sept. 12, 1945 (LOC, Pringle Papers, 27).

Concerned about German rockets armed with biological toxins or poison gas, VB formed a committee to study the problem under the auspices of the Joint New Weapons committee. The JNW secretary, Burton L. Lucas, sent the Joint Chiefs a memo on Jan. 6, 1944, advising that "the growing mass of intelligence regarding an enemy secret weapon undoubtedly has a serious basis in fact. It centers on long range rockets of two types, directed principally at London and Bristol; and also about biological warfare" (NARA, RG 218, JNW, May 1942–45, 1).

27. Roland, *Model Research*, 189; O'Neill, *Democracy at War*, 120, 410–11.

In the cases of both the jet engine and the torpedo, the U.S. deficit could be blamed mainly on its lackadaisical attitude toward military technology before mid-1940. As Roland frankly concludes from his study of the case of jet engines, "The United States was, quite simply, egregiously late in appreciating and developing jet propulsion for aircraft."

For the transformation of U.S. torpedoes, see *Hellions of the Deep: The Development of American Torpedoes in World War II* (1996). Author Robert Gannon found that "American torpedoes at the beginning of World War II were, at best, mediocre. Compared with those at war's end— and with those of the adversaries in 1941—they were primitive dullards with the intelligence of a garden hose. Those at the end of the war, however, were sophisticated mechanisms crammed with technological innovations, outfitted with organs of voice and hearing, reliable, trustworthy, awesome in their capability, and exceedingly important in the history of weaponry" (12).

28. Guerlac, Henry, "Conversation with Alfred Loomis," May 21, 1943 (NARA, NER, Rad Lab, 49–49A).

29. Keegan, *Second World War*, 582; Neufeld, 13–23; Bergaust, Erik, *Wernher von Braun* (1976), 82–85.

Von Braun's biographer states that von Braun only imagined designing rockets that reached the U.S., saying his plans "never passed the paper study stage."

30. Overy, 5.

31. VB to H. H. Arnold, Jan. 24, 1943, G. E. Sratemeyer, Jan. 29, 1943 (NARA, RG 227, Dir. GR 60); JNW meetings, Sept. 29, 1942, Sept. 1, 1943 (NARA, RG 218, JNW); VB to General S. G. Henry, May 9, 1944 (NARA, RG 218, JNW, Subject File, 2); Hacker, Barton C., "Robert H. Goddard and the Origins of Space Flight," in Pursell, *Technology,* 263–75; Bilstein, 122; von Karman, Theodore, *The Wind and Beyond* (1967), 243; McDougall, 77; J. Furer diary, August 2, 1944 (LOC, JFP, 1).

In the Sept. 29, 1942, meeting of the Joint Committee on New Weapons and Equipment, VB fumed about the slow pace of advances in rocketry and missiles, blaming the services for not sharing enough information with civilian researchers.

A missile enthusiast at Caltech, von Karman recalled that VB had expressed his skepticism to colleague Robert Millikan. Indirectly, von Karman confirms that Bush's attitude was widely held: the word "rocket," von Karman recalled, had such an amateurish connotation that "for practical reasons we decided to drop it from our early reports and even our vocabulary."

VB apparently never blamed himself or the military for the failure to harvest innovations by Goddard, whose research in the 1920s and 1930s anticipated (though it did not realize) the chief features of the German V-1. The OSRD's official history, *Scientists Against Time,* makes just one passing reference to Goddard, who played no significant role in the agency's own scattered rocketry efforts. Goddard's attempts to work directly on behalf of the armed services, meanwhile, proved frustrating. In 1938, Goddard complained to Charles Lindbergh, a friend and patron, that even though "the first practical use for [his] liquid-propelled rockets is likely to be for military purposes," the military was treating him with skepticism. This was partly because Goddard's rockets seemed costly and inaccurate, but with the outbreak of World War II cost concerns diminished considerably. In May 1940, just as Germany moved on France, Goddard's desire to develop large rocket missiles for the military was ignored at a major armed services conference. In the end, all the Army wanted from him was a rocket device to aid aircraft. While VB did not cite Goddard's rough experience as a justification for forming the NDRC when he did, he certainly could have. The rebuffs to Goddard, a physicist, typified the military's attitude toward scientists before the advent of NDRC.

By contrast, the German Army launched its program to build liquid-propellant rockets in 1930 and soon surpassed Goddard's achievements.

To be sure, Goddard's brilliance was elusive. Like many inspired but isolated inventors, he squandered his advantages and may even have traveled badly off course. Von Karman raised this possibility in his memoir when he asserted, "There is no direct line from Goddard to present-day rocketry. He is on a branch that died. He was an inventive man and had a good scientific foundation, but he was not a creator of science, and he took himself too seriously. If he had taken others into his confidence, I think he would have developed workable high-altitude rockets and his achievements would have been greater than they were." Perhaps sensing these same flaws in Goddard, VB chose to ignore him. But VB did respect famed aviator Charles Lindbergh, whom he had served with on the National Advisory Committee on Aeronautics. Lindbergh knew of Goddard's prowess and had aided him before the war.

Many were frustrated by VB's dour view of missiles. In August 1944, he flatly told Furer that missiles were "not a development which can be brought about for this war." VB's pessimism tested the mild-mannered Furer, who privately accused VB of "consistently" holding back on missile research.

32. VB to DeLany, Dec. 4, 1943 (NARA, RG 218, JCS, JNW, May 1942–45, 2).

33. Keegan, *Second World War,* 581–82.

34. VB OH, 165; VB to H. Stimson, April 21, 1948 (LOC, VBP, 109).

Four years after the attack, VB wrote Stimson, recalling "our relief when we realized that the V-Bombs carried only conventional war heads."

35. Powers, 352–55.

36. VB POA, 110.

37. Baxter, 233; Buderi, 163.

38. Baxter, 236; VB POA, 110–11.

39. VB POA, 111.

40. VB OH, 451–52.

41. Patton quoted in Baxter, 236, and VB to H. Bundy, March 13, 1945 (NARA, RG 227, Office of Chairman NDRC & Director OSRD, GR, 53); Parker, 212.

VB delighted in Patton's letter, which he felt vividly dispelled any lingering doubts about the value of the OSRD. In March 1945, VB considered sending excerpts from the letter to a congressman who openly mused at a public hearing that OSRD had simply copied captured German weapons. Such comments, VB wrote Bundy, "make it quite evident that he believes we in this country have done no new development whatever but simply copied the Germans and have been generally behind. It is of course impossible to correct this point of view in an open hearing." In private, however, VB was quite willing to disabuse the congressman of this notion.

42. Woolf, S. J., "Chief of Staff on the Science Front," NYT, Jan. 23, 1944.

By the end of the war, radar-aided attackers were sinking nearly one U-boat a day. Proximity fuzes were so widely used that 10,000 U.S. production workers were assembling rugged tubes, essential to the fuze. At peak production 2 million fuzes were built a month, drawing on the resources of 300 different companies (*Radar: A Report on Science at War* [1945], 27; Baxter, 232–33).

43. "Science Mobilizes a Test-Tube Army," NYT, Jan. 3, 1943; "War Brains," J. D. Ratliff, *Colliers,* Jan. 17, 1942; Karl, 221.

This swing from civilian to military priorities resulted in an image makeover for many U.S. leaders, notably the president himself. By 1943, Roosevelt told the press, "Dr. Win the War" had replaced "Dr. New Deal."

44. *Time,* June 3, 1944, 1, 52–57.

45. VB to Albert Fisher, March 31, 1944 (NAS, Jewett file). Some of the people VB feared the cover story might offend were gracious in response to his apologies. Frank Jewett, the president of the National Academy and an original member of the NDRC, was ignored altogether by *Time,* but he dismissed Bush's complaints. "Such articles, quite properly, have to center around the leader not merely because he is the leader but because so much of the success of any great undertaking stems directly from his statesmanship," Jewett wrote VB on April 3. "So far as I am concerned, old top, you can forget that my name or that of the Academy and Council was omitted from the article. I didn't need your letter to Fisher to let me know how you felt."

46. VB to James Conant, May 6, 1944 (JBC Papers, Special Subject File, 6, Pusey Library).

VB valued discretion. "Our external relations with other agencies and organizations have sometimes been lurid," he noted, "but they have been kept as internal affairs and we have never been part of the Washington strife as viewed by the public."

47. Guerlac, Henry, "Interview with Dr. Vannevar Bush," Aug. 20, 1944 (NARA, NRE, Rad Lab, 49–49A).

48. Kevles, *The Physicists,* 321–22; Polenberg, 21–22.

49. Baxter, 131–32; Lee DuBridge OH, Sept. 21, 1965 (courtesy Daniel Greenberg).

DuBridge, director of the Rad Lab, recalled "a few cases of individuals drafted," but said that draft status of most researchers was resolved in the lab's favor on an individual basis. "We found it difficult to convey to [local draft boards] the importance of keeping these people at work," DuBridge said. He called this "a continual problem but not a serious deterrent to our work."

50. Stimson to Roosevelt, March 4, 1944 (FDRL, PSF, Confidential File, OSRD, 8); VB to Stimson, April 3, 1944 (NARA, RG 227, Office of Historian, 8); Bundy OH, 193; Baxter, 133–35; Henry Pringle, notes of VB interview, Sept. 12, 1945 (LOC, Pringle Papers, 27).

51. VB to Stimson, April 3, 1944 (NARA, RG 227, Office of Historian, 8). What VB didn't tell Stimson was something the war secretary probably knew: that the weeding out of specialists weakened the morale of the infantry. It also thinned the potential ranks. Eisenhower was so desperate for foot soldiers in December 1944 that he offered the thousands of GIs inside Army prison stockades in England and Europe a pardon and clean record if they would pick up rifles and go into battle (Parker, 176; Ambrose, *Eisenhower: Soldier, General,* 370).

52. "Rape of the Laboratories," *Time,* April 24, 1944; Baxter, 133–35; Kevles, *The Physicists,* 322–23.

53. VB to C. Williams, Sept. 13, 1944 (NARA, RG227, OSRD, GR, 53).

Two months earlier, VB seemed slightly more certain of his desire to recede from view, writing Williams: "I do think, however, that by the time the war ends I will be ready to quit and let someone else carry the ball" (VBP, LOC).

54. VB recounted his "tax tangle" in a March 31, 1944, memo (LOC, VBP, 63).

55. VB to C. H. Bradley, March 23, 1942 (LOC, VBP, 14).

56. "Dr. Bush Wins Domicile Row Over Commissioner Long," *Boston Globe,* Feb. 22, 1944; "Bay State Disputes Dr. Bush's Residence Claim in Tax Case," *Boston Globe,* undated (MIT); VB to L. B. King, Feb. 22, 1944 (LOC, VBP, 63).

Chapter Nine: "This uranium headache!"

Epigraph: *Pic* magazine, June 7, 1941 (courtesy William Lanouette).

1. Rhodes, 269–73; POA, 57; The Manhattan Project—US DOE (1990).

2. Thornton, Winthrop, "Science Discovers Real Frankenstein," *Boston Herald,* June 4, 1939; VB to W. C. Forbes, June 8, 1939 (CIW).

3. VB to F. Jewett, May 2, 1940 (CIW).

4. VB to F. Keppel (LOC, VBP, 62).

5. VB POA, 58–59.

6. Conant, 278–79.

7. Rhodes, 360.

8. K. Compton to VB, March 17, 1941 (B-C, 1/1).

9. VB to L. Briggs, March 27, 1941 (B-C, 35); VB to F. Jewett, June 7, 1941 (B-C, 1/1).

10. K. Compton to VB, March 17, 1941 (B-C, 1/1); VB to F. Jewett, June 7, 1941 (B-C, 1/1); Rhodes, 143–45, 360–62; VB to E. Lawrence, July 14, 1941 (University of California at Berkeley, Lawrence Papers).

11. VB to L. Briggs, March 27, 1941 (NARA, RG 227, S-1, Briggs files, 2).

12. E. O. Lawrence to VB, July 10, 1941; VB to E. O. Lawrence, July 14, 1941 (University of California at Berkeley, Lawrence Papers).

13. VB to Jewett, April 15, 1941 (B-C, 1/1).

14. Goldberg, 440.

15. VB to F. Jewett, June 7, 1941 (B-C, 4).

16. VB to F. Jewett, July 9, 1941 (B-C, 9/91).

17. VB POA, 56; Powers, 481; H. Pringle, notes from VB interview, Sept. 12, 1945 (LOC, Pringle Papers, 27); VB speech, June 14, 1949 (AIP, Goudsmit papers).

Four years after the end of the war, in a commencement address at Johns Hopkins University, VB gave his fullest account of his thinking about Nazi progress toward an atomic bomb and why, as it turned out, "the Nazis had not even reached first base."

18. VB to F. Jewett, July 9, 1941 (B-C, 9/91).

19. As late as July 18, Jewett thought an atomic bomb to be a "very long range program and very expensive" (NARA, OSRD, RG 227, NDRC Minutes, 2). For Szilard's criticism, see his May 26, 1942, letter in *Leo Szilard: His Version of Facts* (1978), 151. In *Atomic Quest,*

Arthur Compton recalled (49) worrying about a complete withdrawal of government support that summer.

20. Goldberg, 444–49; VB POA, 59.

After the war, VB indicated that the Nazi threat was the foremost reason for a U.S. bomb-building effort, but he left no record of this sentiment at the time. Perhaps because after the war revulsion for the Nazis was so great—and remorse over the U.S. role in beginning the atomic arms race was so great too—VB chose to characterize the anti-German motivation as dominant rather than as one of several factors leading to a decision to back an all-out effort.

21. Wallace OH, Columbia.

22. "Report of the NDRC for the First Year of Operation," June 27, 1940, to June 28, 1941" (FDR Library, PSF, safe file, VB folder, 2). VB signed the report, July 16, 1941.

23. Wallace OH, Columbia.

24. Conant, 279; VB to J. Conant, July 25, 1941 (B-C).

25. Wallace OH, 3978, Columbia.

26. VB to Conant, Oct. 9, 1941 (B-C, 1/1); VB to Roosevelt, March 9, 1942 (B-C).

27. Wallace OH, 1199–1200, 4372–75, Columbia; Goldberg, 499.

28. Bundy, 45–48; Hewlett and Anderson, 46.

29. F. Jewett to VB, Oct. 6, 1941 (B-C, 4/24); Goldberg, 446–47.

In its third and final report, the Academy pegged the cost of a bomb at $133 million.

30. VB to F. Jewett, Nov. 4, 1941 (B-C, 1/1).

31. VB to FDR, Nov. 27, 1941; VB to Murphee, *Dec. 4, 1941* (B-C).

32. Stanley Weintraub, *Long Day's Journey Into War: Dec. 7, 1941* (1991), 62-63.

33. VB to J. Conant, Dec. 16, 1941 (B-C, 2); VB to C. Darwin, Dec. 13, 1941 (B-C).

34. Italics added. VB to FDR, March 9, 1942 (NARA, Harrison-Bundy, 58); FDR to VB, March 11, 1942 (FDR, PSF, 97, VB folder).

35. VB to FDR, June 17, 1942 (NARA, Harrison-Bundy, 58).

36. VB to Bundy, Aug. 29, 1942 (NARA, Harrison-Bundy, 58).

37. J. Conant, Sept. 17, 1942; VB to J. Conant, Sept. 21, 1942 (Harvard University, Conant Papers, Special Subject File, 6); Groves, Leslie, *Now It Can Be Told* (1962), 20–21; VB OH, 466; Goldberg, Stanley, "Groves Takes the Reins," *Bulletin of the Atomic Scientists,* Dec. 1992.

38. Meigs, "Managing Uncertainty," 80–81; Helwett and Anderson, 80–83; VB POA, 60–61; VB interview, NBC, June 5, 1965 (MIT, VBP); Rhodes, 378–79.

The two military members on the committee were Admiral William R. Purnell of the Navy and General W. D. Styer of the Army. Conant was VB's alternate. The two administrators divided their duties. "My task was relations with Congress, the President and with other agencies of government," VB later said. "And [Conant] handled nearly all the relations with the scientists themselves."

39. VB believed that Groves never put him under surveillance again. After the war, VB and Groves made a show of unanimity, saying they always saw eye to eye. "General Groves was a strange individual and still is," VB later said. "I got along with him excellently after the first encounter." Groves wrote more affectionately of VB in his memoir, recalling that relations "were always most pleasant and we soon became, and remain, fast friends. . . . Never once throughout the whole project were we in disagreement. He was a pillar of strength upon whom I could always rely" (Groves, Leslie, *Now It Can Be Told* [1962], 20–21; VB OH, 466).

40. VB to Conant, Dec. 16, 1941 (B-C, 2).

41. Bowen, 189; Sherwin, 45; Burke, Colin, "The Other Memex: The Tangled Career of Vannever Bush's Information Machine, the Rapid Selector" (unpublished manuscript).

42. VB to Styer, Feb. 13, 1943 (B-C, 35).

43. VB to F. Aydelotte, Dec. 15, 1941 (BC-15); VB to F. Aydelotte, Dec. 30, 1941 (B-C, 72). Bush consistently distanced himself from Einstein, though he never directly questioned the

physicist's loyalty. Near the end of the war his writing assistant inserted a quotation from Einstein into the draft of a speech. Bush sent it back with the quotation stricken out and a note on the margin, "I will not quote Einstein" (interview, Lee Anna [Embrey] Blick).

44. Bush to Conant, Feb. 24, 1942; Bush to Thayer, Feb. 28, 1942 (B-C, 15); Meigs, "Managing Uncertainty," 69.

In *Atomic Quest,* Compton blames his security problems on his association with "Communist promoters" who "had used my name in connection with their organizations completely without authorization" (120).

45. VB to Charles Darwin, Sept. 20, 1941 (B-C).

46. VB to Jewett, Nov. 4, 1941 (B-C, 1/1); Rhodes, 378.

47. Hewlett and Anderson, 82.

48. Lanouette, William, *Genius in the Shadows: A Biography of Leo Szilard* (1992), 196–213.

49. L. Szilard to VB, May 26, 1942; *Leo Szilard: His Version of the Facts* (1978), 151–52.

50. Wallace OH, Dec. 4, 1943, Columbia; Powers, 190–92, 353, 522–23.

Bethe and Weisskopf exhaustively discussed their plan on Oct. 28, 1942 (Dawidoff, Nicholas, *The Catcher Was a Spy: The Mysterious Life of Moe Berg* [1994], 201). The next day Weisskopf wrote Oppenheimer about their plan, concluding, "It is evident that the kidnapping is by far the most effective and safest thing to do." Recalling the episode nearly a half-century later, Weisskopf noted in his memoir, *The Joy of Insight,* "Today I have a hard time understanding how I could have proposed such a harebrained idea." As Powers recounts in lush detail in *Heisenberg's War,* the U.S. spent two years studying the possible kidnaping or assassination of Heisenberg, finally dispatching an agent of the Office of Strategic Services to a Heisenberg lecture in Zurich in Dec. 1944. The agent, former baseball player Moe Berg, was instructed to shoot Heisenberg in the lecture hall if anything the physicist said convinced him the Nazis were close to a bomb. As Powers recounted, at the lecture "Berg did nothing."

51. J. Conant to L. Groves, Dec. 21, 1942 (B-C, 16); Powers, 209; VB to G. V. Strong, Sept. 21, 1942 (B-C, 35).

52. VB to FDR, Dec. 16, 1942 (B-C, AEC HD, 121); Henry Wallace, Dec. 4, 1943, Columbia OH, 2884–86.

53. Bundy, 45; VB POA, 277–79.

54. Powers, 209–13.

55. Wigner memo, "On Relation of Scientific and Industrial Organizations in the Uranium Project," Nov. 20, 1942 (NARA, RG 227, contents of VB's MIT safe).

56. Bundy, 46.

57. Hewlett and Anderson, 52.

58. Goldberg, 450.

In 1992, Stanley Goldberg made waves in the small community of nuclear historians when he asserted that VB was the decisive actor in the transformation of official attitudes toward the building of an atomic bomb. Though critics claim that Goldberg overstated VB's role, a few of his contemporaries (though not any atomic physicists) recognized his contribution. No less a figure than Lucius Clay, an Army general who had criticized the A-bomb project "on the grounds that it took too much manpower and critical materials," said VB "was really responsible for keeping it underway." Clay was deputy director of the Office of War Mobilization and Reconversion in 1944 before serving in the U.S. military government in Germany. In a 1951 interview with a government consultant, Clay "praised Vannevar Bush very highly, calling him a most worthy adversary" (HSTL, William T. Golden memos, Jan. 26, 1951).

59. Hewlett and Anderson, 279.

"There remains little doubt," VB wrote on Dec. 15, 1942, "that man has available a new and exceedingly potent source of energy, in a form which can be practically utilized. In the form now in sight, the cost of power obtained on this basis might not be exceedingly low, for the high plant cost offsets the lack of fuel bill. The first applications may hence come in areas where power is needed, but where there is neither coal, oil, nor water power available. However, this is the beginning of a new art, possibly the beginning of an entirely new phase of man's control of the processes of nature, and it is impossible to know where it may lead" (B-C).

60. VB to Jewett, April 15, 1941 (B-C, 1/1).

61. VB, "Churchill and the Scientists," *Atlantic Monthly,* March 1965: Sherwin, 73, 83.

VB arrived in England on July 7 and departed July 29. The official "notes" on his mission contain no references to meetings with Churchill (NARA, RG 227, OSRD, RG 59).

62. Sherwin, 85–89; Hewlett and Anderson, 256-88.

63. VB POA, 115; Powers, 284–85, 355–57, 481; VB speech at Johns Hopkins University, June 14, 1949 (AIP, Goudsmit Papers); Groves, Leslie, *Now It Can Be Told* (1962), 190–91; Keegan, *Second World War,* 582.

Though at first shocked by the German atomic failure, VB in time saw it as part of the Nazis' overall inability to exploit science and technology, ticking off a list of U.S. innovations unmatched by Germany: the proximity fuze, advanced radar and dozens of other useful devices.

VB blamed Germany's failure in atomic science on its regimented "totalitarian system" and a botched war organization, which made it impossible for scientists to pursue their instincts. In 1970, he elaborated on the reasons for German failure, citing Hitler's persecution of the Jews, which drove top physicists from Germany, and the lack of "genuine collaboration between scientists and military men." Two studies of Germany's effort provide a sharper picture. In *German National Socialism and the Quest for Nuclear Power, 1939–1949* (1989), Mark Walker surveys a range of flawed decisions, including the apparent failure of German physicists to understand the basic path to a bomb. In *Heisenberg's War,* Thomas Powers recounts the story in spellbinding detail, concluding that "German efforts to build a bomb were killed by the technical pessimism of leading German scientists who had no desire to make a bomb for Hitler." While calling VB's version of the German effort a "cartoon history" that "has been heavily amended by historians," he nonetheless credits him with getting right the basic "outlines" of the story (481–82).

64. VB interview, NBC News, Jan. 5, 1965 (MIT).

65. O'Neill, *Democracy at War,* 422; VB interview, NBC News, Jan. 5, 1965 (MIT).

William O'Neill bluntly makes a political case for using the bomb: "In a democracy the existence of the Bomb compelled its use."

VB recalled that FDR only told him the bomb should be finished during the war, and definitively shown to either work or not. "He felt that it was utterly essential that we get it in the period of the war," VB said. "Now, of course, it's easy enough to see why that was very important to him . . . because he had backed an idea that never could be tested on a small scale, that could only be tested full-scale."

66. Sherwin, 209–10; Hewlett and Anderson, 253; Weinberg, 885.

Groves told Stimson in April 1945 that "the target is and was always expected to be Japan."

67. VB to JBC, April 17, 1944 (B-C, 37); VB to JBC, Sept. 26, 1944 (B-C).

68. VB to JBC, Sept. 25, 1944 (B-C).

Unbeknownst to VB, Russia had formed its own nuclear project in April 1943. Only three months before, it had requested ten kilograms of uranium metal from the U.S. through the Lend-Lease Administration. General Groves approved the request out of fear his denial might draw attention. Russia also effectively spied on Anglo-American atomic efforts; Stalin knew about the Manhattan Project and possibly had advance notice on the first atomic test (Holloway, 96, 101, 116–17).

69. VB memo, Sept. 22, 1944; VB to JBC, Sept. 25, 1944 (B-C).

FDR created a situation "in which there was no arrangement for the orderly and timely consideration of questions that went beyond making the bomb as fast as possible" (Bundy, 47).

70. VB-JC to Stimson, Sept. 30, 1944, in Sherwin, 286–88.

71. Sherwin, 126–27; VB to H. Stimson, Sept. 30, 1944 (B-C, 19).

72. VB to JBC, Oct. 24, 1944 (B-C, 37); Hewlett and Anderson, 334–35; Sherwin, 129–34.

Stimson shared VB's worry about a potential arms race, but he was absorbed with the task of devising a postwar settlement. He advised FDR on Dec. 31, 1944, that it was not yet time to tell Stalin about the bomb. FDR took his advice.

Chapter Ten: "The endless frontier"

Epigraph: "At 80, Scientist Bush Looks Back at Eventful Years," *Boston Globe,* undated, probably Fall 1970 (Boston Globe library).

1. "Vannevar Bush: General of Physics," *Time,* April 3, 1944.

2. Guerlac, Henry, "Conversation With Alfred Loomis," May 21, 1943 (NARA, NER, Rad Lab, 49–49A).

3. Bush aide Irvin Stewart, in recalling his wartime experience, observed in the official OSRD history, *Organizing Research for War* (321): "In practice research tended to merge into development."

The blurring of lines between science and engineering, experimentation and application, was exacerbated by the pressure of war. As early as the end of 1943, OSRD researchers were actually spending most of their time helping the military solve practical problems. Such were the demands on technical experts that "many groups of investigators now find themselves not concerned primarily with research and development but rather with providing expert knowledge on week-to-week demands by the various Services," a VB aide informed a Joint Chiefs officer (L. Chalkley to B. L. Lucas, Dec. 23, 1943, NARA, RG 227, OSRD, RG 54).

4. Because of the importance of the report, the origin of the Roosevelt letter has piqued the curiosity of scholars. The distinguished historian of science I. Bernard Cohen once asked VB about the provenance of the letter. Without hesitation he replied that the idea for the letter had been his. Cohen recounts, "He then addressed my question of who wrote the letter. He looked me right in the eye, smiled again, and then stated unequivocally, 'I did. I wrote the letter.'" Nothing in the documentary record contradicts VB's declaration. However, an OSRD attorney, who worked closely with VB on the drafting of the letter, credits Oscar Cox with the idea for the letter. The attorney, Oscar Ruebhausen, recalled that VB came up with three of the letter's four questions about the relation of government to science; James Conant came up with the fourth. Ruebhausen claimed to have actually written the draft himself (letter from I. B. Cohen; interview with Oscar Ruebhausen).

5. FDR to VB (draft), Oct. 27, 1944; O. Cox to H. Hopkins, Nov. 9, 1944 (FDRL, Cox diary).

6. H. Hopkins to O. Cox, Nov. 10, 1944 (FDRL, Cox diary).

The draft "had a lot of praising rhetoric for OSRD. Bush probably knew perfectly well it would be cut out," Oscar Ruebhausen recalled (author interview).

7. O. Cox to H. Hopkins, Nov. 11, 1944 (FDRL, Cox diary); H. Hopkins to S. I. Rosenman, Nov. 15, 1944 (FDRL, President's Official File 4482, OSRD); S. I. Rosenman to FDR, Nov. 17, 1944 (FDRL, President's Official File 4482, OSRD); interview, Oscar Ruebhausen.

Rosenman made about a dozen alterations to the final letter. Other than adding the final paragraph, all were minor changes in wording.

Though released on Nov. 20, 1944, the letter still carried a date three days earlier.

8. Americans worried about a new Depression even as quickening war production ended joblessness. Using findings from an opinion survey, the poet Archibald MacLeish, a presidential appointee, informed Roosevelt in May 1942 that seven out of ten Americans "expect to be personally worse off after the war." So Roosevelt knew well that scientists could not expect much public support in peace if the economy fell apart (Blum, 29).

NYT, Nov. 21, 1944: "Roosevelt Urges Peace Science Plan: Asks Dr. Bush, Head of Military Research Unit, to Study Postwar Projection of Program."

9. McDougall, 79, Cowan, Ruth S., *A Social History of American Technology* (1997), 310–18.

The entire debate over postwar research centered on its *form;* virtually no one questioned its *function.* At least publicly, there was unanimity that the OSRD model achieved the best results. VB extolled the OSRD's achievements, and leading publications, such as *The New York Times* and *Fortune,* celebrated it. A few scientists, however, privately groused that, from a bird's-eye view, OSRD was disorganized and inefficient. Joel H. Hildebrand, a professor at UC Berkeley, criticized the amateurish organization of OSRD's chief operating unit, the NDRC, saying it was "organized on lines that betrayed the academic mind. One possible war gas was developed in Division 9, because it was organic, another in Division 10, because it was inorganic." Hildebrand added, "I do not want to appear unappreciative of the vast achievement of OSRD," but "I do not believe that OSRD should continue into the future in anything like its present form" (J. H. Hildebrand to F. Jewett, Nov. 30, 1944, NAS). After the war, a group of researchers in Minnesota complained, "When the true record is written the waste, inefficiency, ignorance and obtuseness in utilizing scientific knowledge in the recent war will be apparent to all" (Senate testimony: Greenberg, 94).

Another set of criticisms, rarely directly aimed at OSRD, centered on VB's skillful avoidance of the War Production Board. The miasma of problems associated with scarce materials such as aluminum and rubber periodically made headlines and prompted pleas for VB's aid. At the urging of Frank Jewett, VB basically turned a blind eye toward the WPB (Jewett warned him: "I'd shun it [WPB] like I would the seven-year itch"). VB's refusal to research problems of production had the benefit of freeing OSRD to concentrate on investigating and designing weapons and their countermeasures. But to advocates of the WPB, this was also seen as shirking a tough job. Partly to quiet such complaints (and respond to an appeal from WPB chief Donald Nelson), in the summer of 1942 VB dispatched heavyweights James Conant and Karl Compton to study U.S. rubber research (Conant, 305–28; Pursell, *Science Agencies,* 370–72).

10. Even Frank Jewett accused VB of narrowness in choosing committee members. In disregarding citizen opinions, VB mainly sought to wall off his report from advocates of the view that scientific priorities must reflect social needs and diversity. He also slammed the door on rank-and-file scientists or activists with frankly left-leaning sympathies. VB's exchange with the national secretary of the American Association of Scientific Workers—a group affiliated with the militant CIO union—was instructive in this regard. "I am mindful of the statement in your letter to me that you hope American scientists will come forward to submit their opinions and discussions in regard to the program which you are working out," wrote the national secretary, Harry Grundfest, in late December 1944. "It has been my past experience that scientists, perhaps because they are mostly divested from the job of administering their own work, are an unusually silent group of people, who don't often express their opinions, at least in the field of social policy. It would be a pity, however, to miss having the views of this intelligent trained body. . . . Please allow me, therefore, to suggest that you bring them out of their shells, perhaps by publishing a statement to the effect that you would welcome suggestions and discussion, both in communications to you and in the scientific press."

VB replied two weeks later, politely saying Grundfest's suggestion about inviting opinion "is an excellent one." He promised that a press release would be issued soon naming the committee members assisting him. At that time, "I shall take that opportunity to invite comment and suggestion from all those who may be interested in giving their views." VB evidently was thinking of ordinary citizens relaying their views, because he closed by bluntly reminding the activist that the scientific rank and file needed no special input into the process because "I am a medium through which the considered judgment of American scientists on these important matters can be expressed" (VB to H. Grundfest, Dec. 26, 1944; H. Grundfest to VB, Jan. 1, 1945; VB to H. Grundfest, Jan. 9, 1945, NARA, RG 227, Director OSRD, Reports to President, 2–3).

Watson Davis was the only member of the "Endless Frontier" committees who could remotely be considered a lay person. A science journalist, Davis was active in many efforts to improve communication between scientists and the public. Davis, who years before had sparked VB's interest in microfilm as a medium for the rapid retrieval of data, served on the committee pondering ways to identify talented youth.

Of the 49 committee members, certifiable cronies of VB numbered at least 14, including such obvious pals as Carroll Wilson, James Conant, Karl Compton, attorney Oscar Cox, Carnegie researcher Merle Tuve and Irvin Stewart, OSRD's chief administrative officer. VB knew none of the medical committee's nine members well, and it showed. The committee resisted his suggestions.

11. VB to I. Bowman, Jan. 10, 1945 (NARA, RG 227, Director OSRD, Reports to President, 3).

VB went so far as to read minutes of the meetings held by Bowman, which he used as the basis for further guidance (VB to I. Bowman, Feb. 23, 1945, NARA, RG 227, Reports to President, 3).

12. Carroll Wilson, acting on VB's behalf, tried to push the chair of the medical committee into withdrawing the endorsement. Calling the position "a preliminary judgment," Wilson exhorted the chair to reconsider, writing, "Frankly I think it is unrealistic to expect the creation of another independent agency such as your memorandum proposes." Then Wilson raised the grim possibility that if the report called for the creation of two new federal research agencies it "might well result in failure to get even one" (C. Wilson to H. Smith, March 14, 1945, NARA, RG 227, OSRD Reports to the President, 2).

In the final report, VB overrode the endorsement of the medical committee, recommending that all civilian research be handled by a single foundation. Before the release of the report, he killed an extended discussion of patent reform, a subject of special interest to VB but not specifically mentioned in the Roosevelt letter. In late April, Henry Wallace, the secretary of the Commerce Department, which housed the U.S. patent office, advised VB to limit his references to the patent system. Wallace's order "entirely changes my relationship to the study of the patent system," VB wrote to an executive of Standard Oil, whose chairman served on the Bowman committee. "I am nevertheless going to continue to revise the draft . . . for it is just as important as it was before that I should crystallize my thinking on this subject" (VB to B. Brown, April 26, 1945, NARA, RG 227, Director OSRD, Reports to President, 3).

In VB's 34-page summary of the final report, patent policy rated just one paragraph, containing no recommendations for reform of patent law but arguing that the proposed foundation should have the power to set its own patent arrangements and that "there should certainly *not* be any absolute requirement" that "discoveries" funded by the foundation "be assigned to the government." The Bowman committee's section of the report briefly referred to patents as "the life of research" and noted that "no study on the aids to research . . . would be complete without an inquiry" into the nation's patent laws. It explained the absence of any "detailed recommendations on the patent aspects of research" by citing a study of the subject that VB was making "independently."

Of the four committee reports, the one on publication of scientific information was submitted on Jan. 9, 1945. This report was essentially a policy statement on the best way to lift secrecy restrictions. Every member of the committee had a direct connection to OSRD and its chair was Irvin Stewart. Two other reports, on medical research and government aid to research, were submitted in April. The report on the last question, the encouragement of scientific talent, was submitted on June 4, 1945.

13. VB sensed that the benign bubble around science was about to burst, at least for some of its members, and that scientists themselves might promote the antiscience sentiment. "The physicists are very likely to have a reaction against intense devotion to ingenious ways of killing the enemy," VB presciently commented to a friend in August 1944, a full year before the atomic bombs were dropped in Japan (VB to T. Barbour, Aug. 22, 1944, LOC, VBP, 9/194).

VB was so serious about an elegant introduction to *Endless Frontier* that he considered hiring a professional writer for the job. Among the possibilities suggested by an aide were Edmund Wilson, the esteemed literary critic, and James Reston of *The New York Times*. VB ultimately decided against an outsider, writing the summary himself from a draft supplied by Carroll Wilson, his chief aide (O. Ruebhausen to VB, March 10, 1945 [NARA, RG 227, Director OSRD, Reports to the President, 1]).

14. VB, "The Engineer and His Relation to Government," *Electrical Engineering*, Aug. 1937.

The popularity of Frederick Jackson Turner's "frontier" thesis—that the ability to expand into new geographical frontiers sustained American democracy—meant that opinion leaders were acutely susceptible to Bush's argument.

15. VB began privately advancing plans for permanent postwar support of civilian research in late 1943. While he labored in secret to line up support, he published the essence of his thinking in the Dec. 31 issue of *Science* magazine:

> I think that federal subsidy of the independent science of universities and other institutions in this country may be beneficial, and it may in fact be necessary in the years to come if we are to preserve that preeminence in science as a nation which is essential to our progress. But, unless we have the wisdom to extend support wisely, it may do much more harm than good. If federal support is to be given, strenuous efforts will be necessary to ensure that it furthers the work of the best and most brilliant scientific minds, and does not merely increase the bulk of mediocre work. And the support should be divorced from governmental control of the scientists and laboratories themselves, or it will stifle rather than expedite their true accomplishments.

Both the appeal and the weakness of VB's core philosophy lay in its skillful avoidance of binary choices. Many contemporaries thought in terms of civilian versus military research, and government-funded versus privately funded research. VB thought that the key question had to do with *control*, not the source of the funding. Government funding was fine provided it didn't limit the researcher's independence; similarly, civilians could attack military problems as long as the *ends* of their investigations were not dictated by the military. To VB, research policy should be fundamentally determined by the *process* of discovery and invention, not by the imperatives of the financial or organizational sponsors of research.

By avoiding the central dichotomies bedeviling the role of research in the national enterprise, VB presented a seemingly pain-free program. This program concealed political tensions; it did not resolve them.

16. Sherry, *Shadow*, 74–75.

17. Overy, 1–3, 331–32; Ellis, John, *Brute Force: Allied Strategy and Tactics in the Second World War* (1990), xviii–xix, 485–95.

18. VB to Charles A. Thomas, March 11, 1946 (LOC, VBP, 111); VB testimony, Oct. 15, 1945, hearings on science legislation, U.S. Senate subcommittee of the Committee on Military Affairs.

At the end of World War II, VB was so sure of U.S. industry's dominance that he told a Senate hearing: "No people has ever approached our own in the ability to take new knowledge and apply it to immediate tangible and practical ends. Whether the ends be the commerce of peace or the weapons of war, we have to date no equals in our industrial or productive genius."

19. Millis, *Arms*, 302; Sherry, *Shadow*, 70–71, 97.

20. VB, Nov. 4, 1946, foreword to one of the official OSRD histories, *Organizing Scientific Research for War*; Cohen, I. B., "Science and the Civil War," *Technology Review*, Jan. 1946; Roland, "Science and War," 264–65.

The two most vexing historical questions arising from World War II are the following: first, did the new relations between government, industry and academia constitute a revolution in

American life or simply a continuation of a trend that began at least during World War I or perhaps even in the 19th century?

Second, was the Allied victory due more to its prodigious production capability or to its overwhelming lead in radar and atomic technologies?

Regarding the first question: while the form of the military-industrial-academic complex was apparent during World War I, the scale of this complex in World War II and thereafter made for a qualitatively different effect. The arguments against this assertion are numerous and even if decisive are irrelevant since, regardless of the antecedents, a postwar military-industrial-academic complex was undeniable.

Regarding the second question, the answer is less settled. The Japanese surrender after two atomic attacks prompted many to credit the Manhattan Project with winning the war. Before Hiroshima and Nagasaki, some scientists were ready to crown radar as the winning weapon. Almost as soon as the war ended, however, military analysts and historians questioned whether *any* technological achievement had been decisive. Just six months after Hiroshima and Nagasaki, I. B. Cohen, a leading historian of technology, observed in an article about the Civil War that "it can hardly be said that scientific or technological innovation was the sole cause of winning or losing any war—even that which was recently concluded." In a 1985 essay on science and war, historian Alex Roland concluded that World War II, like World War I, was decided by "industrial production. The other great innovations of the war, such as operations research, radar, the proximity fuze and jet aircraft, were remarkable achievements . . . but they did not determine the war's outcome." Roland allowed that the A-bomb represents the "one possible exception to this generalization," but after reviewing the evidence found that "just because the atomic bombs ended the war does not prove they caused its end."

21. VB to E. B. Wilson, March 23, 1945 (NARA, RG 227, Director OSRD, Reports to President, 4); VB to W. R. Brode, March 13, 1945 (NARA, RG 227, Director OSRD, GR, 54).

22. VB undated memo to *Atlantic Monthly,* c. May 1945 (LOC, VBP, 118).

23. Sherry, *Air Power,* 196–97, 230–31.

24. Sherry, *Air Power,* 186–87; Gorn, 97; von Karman, Theodore, *The Wind and Beyond* (1967), 271; Coffey, 207–8; Franklin, 156–57.

To be sure, Arnold long had listened more closely to scientists than other senior officers, undoubtedly because air war tended to be more technology-intensive than land or sea war. After Arnold had invited General Marshall to join him for lunch with VB and other scientists in late 1939 or early 1940, Marshall asked, "What on earth are you doing with people like that?" Arnold replied: "Using them. Using their brains to help us."

Then Arnold had practical aid in mind. He wanted the "brains" to help solve existing problems and improve existing weapons. By the end of the war, however, Arnold's inclinations ran decidedly to the futuristic. Within weeks of the atomic bombing of Hiroshima and Nagasaki, Arnold gave a series of press interviews in which he described a future "pushbutton" war using missiles of tremendous range and destructive power. He even predicted the advent of defensive missiles that would automatically seek out and destroy enemy rockets. While Arnold wasn't alone in pushing such predictions, he was certainly among those most captivated by fantasy weapons. He prompted *Life* magazine to sketch out for its Nov. 19, 1945, issue a horrifying scenario for a "36-Hour War" that began with an atomic bombardment of key U.S. cities. The U.S. won Arnold's imaginary missile war, but at a cost of 40 million lives.

25. Emphasis added: von Karman, Theodore, *The Wind and Beyond* (1967), 271–72, 294.

26. E. Bowles memo, March 31, 1945 (NARA, RG 107, 81/89).

27. Sherry, *Air Power,* 187; Coffey, 353.

Even after the end of World War II, the War Department still coveted Bowles's services. After Stimson's departure in September 1945, Robert Patterson became the new war secretary. One of his early decisions was to ask Bowles to "continue as Special Consultant" because he "regard[ed] it as a matter of the highest importance that the War Department do its utmost on scientific [research]" (R. Patterson to E. Bowles, Dec. 29, 1945, NARA, 107).

28. "The Evolution of the Office of Naval Research," *Physics Today,* Aug. 1961; Furer diary, April 9, 1943 (LOC, JFP, 1); Sapolsky, 19–22; interview, Bruce S. Old.

29. Interview, Oscar Ruebhausen.

30. Stewart, 299–305; interview, Bruce Old; VB to Bundy, Aug. 8, 1944 (NARA, RG 227, Historian Office, SF, 3); VB report to FDR, Aug. 28, 1944 (LOC, VBP, 7).

VB repeatedly stated that OSRD was a temporary agency. Critics and allies alike credited him for not seeking to turn his emergency outfit into a permanent one. Yet his insistence on rapidly closing down OSRD bordered on the irrational. He officially gave FDR his unilateral plan for closing his agency a year before the end of the war. Such foresight only made sense if VB was trying to get the president and his cabinet to give more thought to postwar control of the bomb and the organization of the military. Another explanation for his drive to shut OSRD is that he sought to give the successor agency, which he hoped would be the organization he would describe in his report to FDR, a chance to take root and grow.

In his report to FDR, VB stated that as "a war agency," OSRD "goes out of existence at the end of the war." This bald statement obscured the reality that it was ultimately the president's decision whether to close or keep open OSRD. FDR had no obligation, legal or otherwise, to shut down OSRD. This was the bluff VB always knew FDR could call.

31. "Conference to Consider Needs for Post-War Research and Development for the Army and Navy," April 26, 1944 (NARA, RG 227, Director OSRD, RG 55).

Some of the officers in attendance showed an awareness that money would only be a down payment on the hearts and minds of civilian scientists. They also would have to cosset these temperamental experts. But the proper appeals ought to work because at bottom scientists desired the status and respect conferred by advising the powerful. In his comments, L. H. Campbell, chief of Army ordnance, hinted at the possibility that scientists merely played hard to get and that their self-righteous purity would be abandoned in favor of worldly rewards. Suggesting that scientists were a kind of new priesthood that had to be treated worshipfully, or at least deferentially, Campbell noted, "When I need spiritual guidance, I don't have any trouble getting it if I go to church. When I need it, that's where I go. I think in this research and development, in my experience, there has been no time . . . that when I have gone to scientific bodies, have gone to Dr. Bush and to Dr. Jewett among others, that I have not gotten the very best advice and guidance."

General S. G. Henry, chief of the War Department's New Developments Division, made it plain that the armed services had no choice but to accommodate the scientists, because "regardless of our individual views in this matter . . . the people and our Congress is going to insist on a better coordination and cooperating after this emergency is over." Henry warned that if the services dropped the ball, "somebody is going to take it [research control] away from us." But the implication was clear: if somehow the Army and Navy grasped the right approach to research, they would keep control.

The growth in military influence would invariably mean further duplication, expense and counterproductive research—the very reasons VB had pushed a civilian monitor in the first place. But as Henry indicated, the Balkanization of the services meant the fragmentation of research was inevitable. "I am in complete agreement with the thoughts of the individual that the ideal way would be to carry on, and in *our separate compartments,*" he said (emphasis added).

32. "The Evolution of the Office of Naval Research," *Physics Today,* Aug. 1961; interview with Bruce Old; Sapolsky, 19–22, 26–27; Forrestal quotation in "Bushwhacked! Vannevar Bush and the Dilemmas of Science Policy," unpublished paper delivered by Michael Dennis at American Physical Society Meetings, April 19, 1994.

Legislation for the ONR cleared Congress in August 1946.

Even as the Navy moved to forge links with university researchers, it also pondered ways to improve the morale and quality of its researchers in uniform. Like VB, the bird dogs assumed no self-respecting and talented scientist would willingly work on the sorts of development projects that typified in-house naval research. Some in the service, however, rejected this assumption. Testifying before a House committee on postwar policy on Nov. 24, 1944, Rear Admiral A. H. Van Keuren, director of the Naval Research Laboratory, wanted "to sound a note of warning" not to lose sight of the value of the government's staff scientists. Contrary to the impression left by advocates of OSRD-style contracting, the Navy and other government agencies had "real scientists" in their labs, Van Keuren declared. "Should it turn out that [the government] decided to give the cream of fundamental research along military lines to civilian rather than government laboratories, it would be nearly fatal to the morale of our Naval scientists. The Army and Navy representatives on the Board should zealously guard against such a contingency. Our men have willingly and unsparingly done any kind of work they have been called on to do to win the war, but many of them will not be contented until they get back to more creative work than they are doing now—back to the kind of work for which they were trained and which they can do as well as anyone in civil life. We do not claim to have all the geniuses at the Naval Research Laboratory, but we have our share of them." To aid these government scientists, Van Keuren called for them to receive higher salaries and different treatment from other military personnel. In calling for these perks, the admiral inadvertently proved VB's point: no matter how it was sliced, the deal for scientists must be special (NARA, RG 227, Director OSRD, GR, 54).

33. Kevles, *The Physicists,* 343; Maddox, 128–29; NYT quotation in Greenberg, 102–3.

The formal name of Truman's committee was the Senate Special Committee to Investigate the National Defense Program.

34. Maddox, Robert F., "The Politics of World War II Science," *West Virginia History,* Fall 1979, 26.

35. Ibid., 30; F. Jewett to VB, Nov. 16, 1942; VB to F. Jewett, Nov. 18, 1942 (LOC, VBP, 56); Maddox, 72–73.

VB kept Kilgore at a distance in 1942–43, reacting coolly to advice from Frank Jewett, the National Academy president, that he meet with the senator. In Nov. 1942, Jewett met with Kilgore and found him to be "a man of intelligence and extremely reasonable and easy to talk to. He is clearly trying to do something constructive in a sector where he thinks help is indicated." VB replied that he "would be glad to talk with [Kilgore], of course, if he would like to have me do so." Rather than meet with Kilgore, however, VB bluntly detailed his objections to Kilgore's initiatives in a letter to the senator dated Dec. 7, 1942, and published June 17, 1943, along with testimony on the senator's technological mobilization bill. VB complimented Kilgore for paying attention to technological matters, which were "a fascinating and exceedingly important aspect of the entire war effort." He then charged that Kilgore's bill was based on "inaccurate premises," accusing the senator of harboring "some feeling that we are far behind the enemy [in technology], that there are serious bottlenecks in our technical effort, and that in general we have a serious situation upon our hands which warrants radical changes." VB disagreed, of course, with these premises and offered a detailed description of OSRD and the administration's whole approach to mobilizing research. He concluded that "no such reorganization [of research efforts] is at present called for."

A year later, VB launched another broadside against Kilgore and his legislative aims, this time publishing an open letter to the senator in the Dec. 31, 1943, issue of *Science* magazine. VB observed that the current organization for war research was adequate and that "it would be ill-advised and dangerous to throw a monkey wrench into such finely meshed machinery at this late date." VB singled out for special criticism Kilgore's proposals on patents, arguing that patents needed by the government for war purposes could be requisitioned. He admitted to sharing with Kilgore "the same end; namely, the furtherance of science for the benefit of the country." But he still insisted that the "underlying philosophy" of Kilgore's bill "was wrong . . . and not needed."

36. Maddox, 30–31.

37. Ibid.; Kilgore statement, "Full Employment—The First Objective of Domestic Policy," Aug. 13, 1945 (Kilgore Papers, FDRL); VB, "The Kilgore Bill," *Science,* Dec. 31, 1943.

38. Kevles, Daniel, "Principles and Politics in Federal R&D Policy, 1945 to 1990: An Appreciation of the Bush Report," from the National Science Foundation's 1990 reissue of *The Endless Frontier.*

39. VB to I. Bowman, Feb. 6, 1945 (LOC, VBP, 13).

40. Interview, P. Abelson.

It is worth remembering that many Americans found new hope and expectations in World War II, following the pessimism and despair of the 1930s. "The war was a good thing," Andrew Hacker reminded in *The End of the American Era* (1970, 10). "If its manifest purpose was to defeat America's enemies, its more enduring by-product was a population with a sense of individual dignity and pride."

41. "The Great Science Debate," *Fortune,* June 1946, 236.

42. VB to L. Groves, April 2, 1945 (B-C, 18); VB memo, Sept. 22, 1944 (B-C, 37).

VB also played a bit part in an Anglo-American attempt to monopolize the world's supply of uranium and thorium, essential for making an A-bomb. Spearheaded on the U.S. side by Groves, the effort was considered at best short-term protection. Along with Conant, VB advised Stimson in September 1944 that "control of supply" of A-bomb raw materials could not be counted on over a period of a decade to stop other countries from building A-bombs. For details on this effort and VB's views, see Helmreich, 248.

43. Sherwin, 136.

44. VB to J. Conant, Feb. 13, 1945 (B-C, 37); Hewlett and Anderson, 333.

45. VB to FDR, Feb. 22, 1945 (NARA, RG 227, Office of Chairman NDRC & Director OSRD, GR, 54).

46. Sherry, *Air Power,* 170; "U.S. Was Prepared to Combat Axis in Poison-Germ Warfare," Jan. 4, 1946, NYT; Bernstein, Barton J., "America's Biological Warfare Program in the Second World War," *Journal of Strategic Studies,* Sept. 1988, 292–317; Overy, 10.

During World War II, the definition of "biological warfare" covered the use of chemicals against plants. The U.S. geared up in April 1942 when Stimson proposed a substantial program, explaining to FDR, "Biological warfare is a dirty business. We must be prepared."

47. JNW minutes, Feb. 19, 1944 (NARA, RG 218, JNW); Bernstein, Barton J., "America's Biological Warfare Program in the Second World War," *Journal of Strategic Studies,* Sept. 1988, 292–317.

48. Sherry, *Air Power,* 171; VB–J. Conant to H. Stimson, Oct. 27, 1944 (B-C, 36).

49. Stimson and Bundy, 615–16.

Chapter Eleven: "After peace returns"

Epigraph: VB's "summary" introduction to *Science—The Endless Frontier,* July 1945, 2.

1. Catton, 304–5.

2. VB interview transcript, Jan. 5, 1965, NBC News (MIT); VB to R. G. Sproul, Sept. 21, 1948 (LOC, VBP, 107).

3. VB POA, 292; VB to Elliot Roosevelt, May 23, 1950 (LOC, VBP, 99).

VB briefly reviewed his relationship with FDR in the 1950 letter to his son.

4. The day after FDR's death, VB wrote Truman, pledging to continue his program "unaltered" unless the new president wished otherwise. Fifteen days later, Truman replied, "My dear Dr. Bush. Many thanks for . . . your pledge of loyal support. This assurance means much to me in these trying days. I want to commend you and your fellow scientists on your excellent work. I shall be very glad to have all of you carry on" (HSTL, OF 53).

5. VB POA, 293; VB to R. G. Sproul, Sept. 21, 1948 (LOC, VBP, 107); Truman, *Year of Decisions,* 11.

The date of VB's first meeting with Truman is not clear. VB recalled that it occurred after FDR's death and before an April 25 meeting about the bomb between Truman, Stimson and Groves.

Implicit in VB's sympathy for Truman was a poor opinion of the new president's talents. VB confided to Sproul that he had "developed a personal liking for [Truman's] courage and his willingness to carry on an exceedingly tough assignment to the best of his ability."

6. Stimson, Henry L., "The Decision to Use the Atomic Bomb," *Harpers,* Feb. 1947; Hewlett and Anderson, 342–43.

7. VB to H. Bundy, April 25, 1945 (B-C, 19); Hewlett and Anderson, 344.

8. VB to J. Conant, May 14, 1945 (B-C, 19).

9. VB to H. Stimson, May 4, 1945; VB to Conant, May 14, 1945 (B-C, 19).

10. Quotation by Gorden Arneson, secretary of the Interim Committee in Giovannitti, L., and Freed, F., *The Decision to Drop the Bomb* (1965), 60.

11. Hewlett and Anderson, 356–61; "Notes of the Interim Committee Meeting," May 31, 1945 (HSTL).

At the meeting, VB did not directly address the question of whether to use the bomb and under what circumstances. The notes indicate that Stimson "expressed the conclusion, on which there was general agreement, that we could not give the Japanese any warning; that we could not concentrate on a civilian area; but that we should seek to make a profound psychological impression on as many of the inhabitants as possible."

12. Giovannitti, L., and Freed, F., *The Decision to Drop the Bomb* (1965), 99.

Of the three estimates, VB's was the closest. The Soviet Union detonated its first atomic bomb in August 1949, four years and three months after the Interim Committee meeting.

13. VB Phi Beta Kappa address at Tufts, Nov. 1, 1935 (CIW); VB to Eric Hodgins, April 10, 1941 (LOC, VBP, 50).

14. O'Neill, *Democracy at War,* 410.

15. Arnold, Henry, *Global Mission* (1949), 243; VB to Arnold, Oct. 13, 1944, cited in Sherry, *Air Power,* 230.

VB did question the practice of firebombing *before* the first raid. In October 1944, as the raids were planned, VB told Arnold that the tactic raised "humanitarian aspects" for which a decision "will have to be made at a high level if it has not been done already." Nothing came of VB's advice, Sherry notes, because Arnold already believed he had FDR's sanction.

16. Emphasis added. VB POA, 62–63; VB, *Modern Arms and Free Men,* 39.

VB told Oscar Ruebhausen, a wartime associate, on Oct. 24, 1949: "I like to make the comparison between the atomic bomb and the fire bombs over Japan. The latter did a lot more damage and were really much more terrible weapons in the ways they were used. Yet their use occasioned no public interest and certainly no protests of any moment. It was fear of the future that brought all of the reactions in connection with the bomb" (LOC, VBP, 100/2309).

17. VB POA, 62; VB to J. R. Chapman, March 26, 1945 (LOC, VBP, 23/510); *Boston Globe,* Oct. 1, 1945; VB to R. D. Mershon, May 14, 1945 (LOC, VBP, 76/1757).

The battle for Okinawa and its effect on the decision to use the bomb is examined in detail by George Feifer in *Tennozan* (1992), esp. 566–84.

Cirumstances made it impossible for VB to keep close tabs on either of his sons. Of John, he wrote in March: "I have not heard much from him recently, but I judge this is because he has been exceedingly busy."

18. VB OH (MIT), 802–3.

No historical question is debated more fiercely than the "inevitability" of the use of atomic weapons against Hiroshima and Nagasaki. The latest scholarship, prompted in part by the

50th anniversary of the bombings in August 1995, reveals new vistas on a debate that shows no signs of flagging. In *The Decision to Use the Atomic Bomb* (1995), Gar Alperovitz demolished the notion that an invasion of Japan's home island would have cost upward of 250,000 lives. Casualty estimates were actually far lower, making the price of an invasion—and thus the avoidance of atomic war—much lower. In the same study, Alperovitz persuasively made the case that the Soviet entry into the war against Japan, coupled with unrelenting conventional bombing by the U.S., could well have triggered Japan's surrender.

The big question remains just how long Japan would have held out in the face of combined U.S. and Soviet attacks. Absent an A-bomb strike, it seems reasonable to conclude that Japanese resistance could have persisted for weeks or even months, given the evidence from Japanese sources presented by Herbert P. Bix. In his anniversary essay ("Japan's Delayed Surrender: A Reinterpretation," *Diplomatic History,* Spring 1995, 197–225), Bix made two compelling points that support VB's belief in the inevitability of A-bomb use. First, "Even immediately after the dropping of the atomic bombs and the Soviet declaration of war, [Japanese] people generally clung to the hope of a final victory and thus the belief that their 'divine land' was indestructible." Second, "When political leaders in Washington said that the Japanese were likely to fight to the death rather than surrender unconditionally, they were not exaggerating what the Japanese government itself was saying." Had the U.S. dropped its demand for an unconditional surrender, it is possible—but not certain—that Japan would have given up without either an invasion or an A-bomb strike. Bix concluded: "It was not so much the Allied policy of unconditional surrender that prolonged the Pacific war, as it was the unrealistic and incompetent actions of Japan's highest leaders."

19. VB OH (MIT), 802–3; VB POA, 63.

VB was not alone in doubting the necessity of an invasion of Japan. William Leahy, the president's military aide, voiced skepticism on the need for any invasion. Even after approving initial preparations, Truman asked that he be alerted before plans reached the no-return stage, thus allowing him a chance to change his mind. In his memoirs, Leahy recalled, "The invasion itself was never authorized" (Bird, 245–46).

20. JNW minutes, Feb. 19, 1944 (NARA, RG 218, JNW).

21. Hewlett and Anderson, 357.

Not only wasn't there a clear leader in biochemical weapons, but no nation had found sure ways of delivering these weapons against an enemy without dissipating their effects or causing self-inflicted wounds. VB referred to such technical limitations in the Oct. 27, 1944, memo on biological weapons he penned with Conant. Surveying the current state of biological weapons, they concluded, "Since this weapon has not been introduced in this war, it seems evident that our enemies have not yet solved the extremely difficult problem of producing a munition which will disseminate biological agents in the form of a dust or cloud which will . . . behave in general like a cloud of poison gas" (B-C, 36).

22. VB to J. Conant, April 5, 1945 (B-C, 37); H. Smith to FDR, March 23, 1945 (FDRL, HSP, 3); H. Smith to J. Conant, March 23, 1945 (FDRL, HSP, 4); H. Smith to FDR, March 30, 1945 (FDRL, HSP, 4); Smith, Harold, "The Budget as an Instrument of Legislative Control and Executive Management," *Public Administration Review* (Summer 1944).

FDR created the WPA by executive order in 1935. As late as 1938, 3.3 million jobless Americans received WPA assistance. FDR ordered the program phased out in Dec. 1941.

23. W. H. Shapley to Baker, March 24, 1945 (NARA, RG 51, 39.19, 62).

24. Stewart, 306–8.

25. VB memo, April 5, 1945, on meeting with DuBridge (B-C, 37); L. DuBridge memo, Feb. 22, 1945 (NARA, RG 227, Office of Historian, 5).

DuBridge said that the projects "on which there is urgent pressure for us to begin work" included improved and new types of airborne air interception equipment for Pacific night fighters, equipment to detect sources of mortar gun and rocket fire, ship radar to cope better with Japanese suicide attacks, and radar specifically designed for ground control of air operations.

On the horizon, DuBridge said, were the problems of "radar control of pilotless aircraft or missiles, the large problem of aircraft navigation . . . to say nothing of great improvements in all of the existing equipments and techniques which lie ahead of the present in radar."

26. VB interview, Dec. 31, 1973 (courtesy Mrs. Esther Edgerton).

The Rad Lab's independence was a natural consequence of the growing sophistication of radar and its wide-ranging applications. That DuBridge was a forceful, nimble leader made OSRD control problematic and insured that "the atmosphere" between VB and DuBridge "became explosive," recalled Edward Bowles, who noted that at tense moments VB "had a way of making clear in no uncertain terms who was boss" (E. Bowles to H. Johnson, Oct. 10, 1974, LOC, Bowles Papers, 5).

27. Interview, Oscar Ruebhausen.

28. VB to H. Smith, April 25, 1945 (NARA, RG 51, 47.3, E5-1 thru E6-3, 3).

29. H. Smith to VB, April 27, 1945 (NARA, RG 51, 47.3, E5-1 thru E6-3, 3).

30. Willis H. Shapley to Baker, May 3, 1945 (NARA, RG 51, 39.19, 62).

31. Officials who made decisions on contracts early in the war adopted a rule that "no member would participate in the discussion of, or vote upon, any proposal affecting the institution from which he received his salary." As a practical matter, however, it was impossible for Bush, Compton, Harvard president James Conant and others not to steer contracts to their respective employers.

32. England, 19, 21; VB to S. I. Rosenman, June 1, 1945 (HSTL, OF 53).

33. HST to secretary of war, June 8, 1945 (NARA, RG 51, 39.27, "Sci R&D," 82); Kevles, "Control of Postwar Defense Research," 20–47; Kevles, *The Physicists,* 353.

Smith had written FDR on March 31, 1945: "A matter as crucial to the national interest as the direction of research on weapons of war should be carried on by an agency responsible to the Commander-in-Chief."

On Aug. 22, 1945, Truman definitively informed VB of his desire that OSRD remain active "until the Congress actually establishes a permanent Federal research agency," and he instructed VB "to maintain in their present status those major projects which are now under OSRD control" (HST to VB, Aug. 22, 1945 [HSTL, 53]).

Truman's letter came in response to an Aug. 16, 1945, letter from VB, outlining a program for termination of OSRD (Stewart, 310–11).

34. VB to HST, June 12, 1945 (NAS, RBNS, 1945).

Scientists weren't the only experts who made Truman uncomfortable. He wasn't crazy about economists either. His crack about wanting a one-armed economist—who would not say, "On the one hand and on the other hand"—became a legendary putdown of this profession. As part of the Employment Act of 1946, Congress created the Council of Economic Advisers, but of its three members Truman relied most on a lawyer, Leon Keyserling, who had no degree in economics (Stein, Herbert, "A Successful Accident: Recollections and Speculations About the CEA," *Journal of Economic Perspectives,* Summer 1996, 4–5).

35. Wallace diary, May 3, 1945 (Wallace OH, Columbia, 3736–37).

Wallace's best guess was that Smith suspected VB of mismanaging funds earmarked for secret defense projects. The thought of so much money spent with so little oversight enraged Smith, who once told Wallace that VB "was not intellectually honest and that he had . . . become somewhat irrational."

36. H. Smith to J. Furer, June 14, 1945 (LOC, Furer, 7).

37. R. Millikan to VB, April 2, 1945 (Caltech, RAMP, 9.6); VB to R. Millikan, April 5, 1945 (NARA, RG 227, OSRD Reports to President, 3).

38. F. Jewett to VB, June 5, 1945 (NARA, RG 227, OSRD Reports to President, 3).

39. F. Jewett to VB, June 8, 1945 (NARA, RG 227, OSRD Reports to President, 3).

In a June 7, 1945, reply to the first Jewett letter, VB let pass Jewett's principal objection that VB ignored the possibility of private research funding. VB probably considered it impossible to persuade Jewett that an allegiance to private funding was anachronistic. If he had wanted to make his differences with Jewett explicit, VB might have borrowed an observation later supplied by historian Carroll Pursell, who wrote: "The basic point missed by Jewett, although grasped by Bush, was that American life was already being revolutionized—the only question was whether the blueprint for the future should be drawn up by politicians answerable to the people or by private enterprises answerable only to themselves." While VB was uneasy about his role in this putative revolution—so much so that he would not have parroted Pursell's formulation in 1945—he certainly believed that experts ought to chart the nation's future (VB to F. Jewett, June 7, 1945, NARA, RG 227, OSRD Reports to President, 3; Pursell, "Science Agencies," 374).

40. England, 25–27.

41. Interview with Oscar Ruebhausen; Maddox, 166.

42. Interview with Oscar Ruebhausen. As OSRD general counsel, OR helped to write the Mills-Magnuson bill; England, 25.

43. Ibid.

44. VB to B. Baruch, Oct. 24, 1945, cited in Kevles, *The Physicists,* 347.

45. VB to O. Buckley, July 25, 1945 (NARA, RG 227, Director OSRD, Reports to the President, 3).

46. The National Science Foundation's budget for fiscal year 1954 totalled $8 million and doubled two years later (England, 211).

47. England, 25.

48. *Science—The Endless Frontier,* 13, 17.

VB opposed the formal integration of government-funded research into overall economic planning. He insisted that the fruits of research were best harvested by private industry and had no desire for the government to address the sometimes yawning gap between academic research and commercial technology. "How will we find ways to make better products at lower costs?" VB asked in *Science—The Endless Frontier.* "The answer is clear. There must be a stream of scientific knowledge to turn the wheels of private and public enterprise." But critics could rightly point out that the wheels did not always turn on their own because sometimes individual companies or even entire industries lacked the incentives to embrace new science-based technologies or the expertise to make sense of them. Even VB noted that "basic research is essentially noncommercial in nature" and that "we cannot expect industry adequately to fill the gap" in research funding. But as far as the gap between the lab and the marketplace, VB had no desire for the government to step in. "Industry will fully rise to the challenge of applying new knowledge to new products," he wrote. "The commercial incentive can be relied upon for that."

49. VB to FG Fassett, Feb. 9, 1945 (LOC, VBP 37/886).

50. J. Furer to Charles West, June 30, 1945 (LOC, JFP, 7).

51. England, 21; VB to Fassett, June 19, 1945 (LOC, VBP 37/886); O. Ruebhausen memo, June 16, 1945; "Endless Frontier" advance memo, July 17, 1945 (NARA, RG 227, OSRD Reports to President, 3–4).

Truman signed off on *Science—The Endless Frontier* in mid-June, and on June 19 VB predicted it would be publicly released ten days later.

Those who received advance copies of the report included U.S. senators, Supreme Court Justices, military officers, columnist Walter Lippmann and *New York Times* reporters James Reston and Hanson Baldwin.

52. Lyman Chalkley to VB, July 20, 1945 (NARA, RG 227, OSRD, Chairman & Dir. Reports to the President, 4); Chalkley to C. Wilson, July 20, 1945, quoted in Kevles, *The Physicists,* 347; *Business Week,* July 21, 1945; NYT, July 21, 1945; Jones, K. M., "The Endless Frontier," *Prologue,* Spring 1976, 43–44.

A transcript of Swing's July 19, 1945, broadcast is in Henry Pringle's papers. Swing also applauded VB for his "strong democratic" support of college scholarships for science students in financial need. "If ability, and not the circumstances of family fortune, is made to determine who shall receive higher education in science, then we shall be assured of constantly improving quality at every level of scientific ability," he quoted VB as saying.

In closing his broadcast, Swing made the TR comparison: "It took a Theodore Roosevelt to stir this country to the conservation of its limited natural resources, and it needed a long fight for his policy to be adopted. Now Dr. Bush is appealing for the conservation of our limited and essential human resources. One hopes it will not need as long a fight to have his belated and modest program adopted" (LOC, Pringle Papers, 27).

53. *Science—The Endless Frontier,* 12–13.

While the relationship between civilian research and national security was the cornerstone of his report, VB accepted that some people would view "Endless Frontier" as simply a call for financial aid to the research universities. "It is my belief that Bush did not contradict [such] faulty interpretations of his report because he thought basic science would have a difficult time in obtaining federal funds," recalled Bruce Old, a naval officer active in science politics at the end of the war. "Therefore if his report could assist the universities, he decided to leave well enough alone" (B. Old to author, Jan. 15, 1996).

VB wasn't alone in insisting that the military could improve existing weapons but could not spawn new classes of them. Frank Jewett observed in May 1945 that military officers "tend to see the future as largely the problem of an improved present—better and more powerful guns, etc., rather than as a future of new and hitherto unthought-of things and methods such as might come out of free imaginative research. While this tendency is, I think, basic to all professional military systems, I suspect that it is greater in a country like ours where war is looked on as a defensive thing to be avoided if possible rather than as an offensive thing to be fostered as a tool of state policy" (F. Jewett to Waldemar Kaempffert, May 14, 1945, NAS, RBNS, 50.8).

54. *Science—The Endless Frontier,* 22–23; VB to H. W. Prentis, June 9, 1945 (LOC, VBP, 93).

55. NYT, July 21, 1945.

For some time, the editorial writers on the *Times* had paid close attention to the evolving relations between science and the military. In August 1944, the newspaper had recommended high levels of spending on military research following the war, asserting: "Military research is one form of insurance against war, and it is time that Congress should so regard it and take some step to exploit it in in the interest of world peace" (NYT, Aug. 29, 1944).

56. Agronsky broadcast, July 20, 1945 (NARA, RG 227, Director OSRD, Reports to the President, 2).

For industry, the military and science to keep in "intelligent touch," this triad had to remain in balance, which implied that scientists must have ultimate authority over advanced weapons. Read in this context, *Science—The Endless Frontier* was part of a subterranean struggle between different wings of the national-security elite spawned during World War II. Both wings agreed that in the postwar world, technological supremacy was of paramount importance, yet they could not agree on how best to achieve this. In this struggle, Bush's proposed foundation was "part of [his] strategy for containing the social and political power of the military in the postwar world" (Dennis, Michael, "Bushwhacked! Vannevar Bush and the Dilemmas of Science Policy," presented to the American Physical Society, April 19, 1994).

57. VB, *Science—The Endless Frontier,* 6.

58. Ibid., 7.

59. H. Smith to C. L. Wilson, July 18, 1945, cited in Reingold, 329.

Chapter Twelve: "As we may think"

Epigraph: "Dr. Bush Sees a Boundless Future for Science," NYT, Sept. 2, 1945.

1. Nyce, James, and Kahn, Paul, "Innovation, Pragmaticism and Technological Continuity: Vannevar Bush's Memex," *Journal of the American Society For Information Science,* May 1989, 216.

VB found the near-simultaneous publication of "Endless Frontier" and "As We May

Think" to be "a strange coincidence which has its advantages." The timing was strange indeed, since the two articles proved to be VB's most durable writings and continued to be cited and studied into the 1990s (VB to F. Fassett, June 19, 1945, LOC, VBP, 37/886).

2. Yates, Frances A., *The Art of Memory* (1966), 4–6, 129–59, 370–73.

3. "As We May Think," *Atlantic Monthly,* July 1945, 108.

4. VB OH, 11 (Columbia, 1967).

5. Rheingold, 175–76.

6. "As We May Think: A Top U.S. Scientist Foresees a Possible Future World in Which Man-Made Machines Will Start to Think," *Life,* Sept. 10, 1945, 112–24.

Self-conscious about his writing, VB's own view of "As We May Think" was somewhat subdued. He described the article as "reasonably good," but he wished he had had "more time to really work on it" (VB to F. Fassett, June 19, 1945, LOC, VBP, 37/886).

7. Wells, H. G., *World Brain* (1938), 86–87; Buckland, Michael, "Emanuel Goldberg, Electronic Document Retrieval, and Vannevar Bush's Memex," *Journal of the American Society for Information Science,* 43–4 (1992), 284–94; Rayward, Boyd, "Visions of Xanadu: Paul Otlet (1868–1944) and Hypertext," *Journal of the American Society for Information Science* (1994).

8. "A Machine That Thinks," *Time,* July 23, 1945.

9. N. Wiener to VB, Sept. 20, 1940; N. Wiener to VB, Sept. 23, 1940; VB to N. Wiener, Sept. 24, 1940; VB to N. Wiener, Oct. 19, 1940; L. F. Safford to N. Wiener, Dec. 19, 1940; VB to N. Wiener, Dec. 31, 1940 (MIT, NWP, 1).

10. Wiener, 232.

11. Edwards, 48–51; Flamm, 48; Pursell, *The Machine,* 279; Ceruzzi, Paul, "An Unforeseen Revolution: Computers and Expectations, 1935–1985," in Corn, Joseph J., ed., *Imagining Tomorrow: History, Technology, and the American Future* (1986), 191, 195; Ceruzzi, Paul E., *Reckoners: The Prehistory of the Digital Computer, 1935–1945* (1983), 123–26.

From 1945 to 1950, the ENIAC was the only working electronic computer in the U.S. It could do the equivalent of about 30 million operations a day, or roughly the work of 75,000 people. Eventually, the ENIAC's operators kept it running for more than 20 hours a day, seven days a week.

Of the fears about the ENIAC's unreliability, Flamm cites the opinion of Samuel Caldwell, VB's collaborator on MIT's differential analyzers, who wrote of the ENIAC proposal, "The reliability of electronic equipment required great improvement before it could be used with confidence for computation purposes."

Pursell asserts that the OSRD also withheld "important data" from the ENIAC development team. After the war, the first task given the new computer was the creation of a mathematical model of a hydrogen bomb.

12. Hartree, D. R., *Calculating Machines* (1947), 9; Wiener, 235.

For a view of the shift from analog to digital that argues it "neither clear-cut nor simple," see: Owens, Larry, "Where Are We Going, Phil Morse? Changing Agendas and the Rhetoric of Obviousness in the Transformation of Computing at MIT, 1939–1957," *IEEE Annals of the History of Computing,* vol. 18, no. 4, 1996.

13. Aspray, 186–94; Heims, Steve J., *The Cybernetics Group* (1991) 31–51; "The Brain Is a Machine," *Newsweek,* Nov. 15, 1948.

These ideas first surfaced after the war in von Neumann's model of self-replicating automata, but they were implicit in the mathematician's wartime work on digital computers. Even William Aspray, author of the definitive study on von Neumann's contributions to digital computing, can't date for certain when he began to consider the similarities between the brain and the computer. In June 1946, von Neumann gave informal lectures at Princeton University on the topics that composed his influential Hixon Lecture in September 1948, "The Logic of Analogue Nets and Automata." Von Neumann also participated in the first meeting, in March 1946, of the Macy Conferences on Feedback Mechanisms and Circular Causal Systems in Biology and the Social Sciences. Warren S. McCulloch, a psychiatrist the nervous system, chaired the meetings and became the foremost proponent brain thesis. In "Of Digital Computers Called Brains," he cowrote a popular exp ideas with science writer John Pfeiffer (*Scientific Monthly,* Dec. 1949).

14. Daniels, George H., "Big Questions in the History of American Technolog *and Culture,* 11 (1970), 1–21.

The corollary to this maxim by technology historian George Daniels is that what people do "for other reasons determines the nature of their future technology." In the battle against technological determinism, VB bet that social factors would decisively influence the course of computing over time.

15. D. Engelbart to VB, May 24, 1962 (MIT, VBP); Rheingold, Howard, *Tools for Thought* (1985), 175–78.

16. "The Inscrutable Past," *Technology Review,* Jan. 1933. The article was reprinted in a collection of VB's essays, *Endless Horizons* (1946).

17. VB to F. P. Keppel, March 5, 1939, cited in Nyce and Kahn, 215.

18. VB to E. Hodgins, Dec. 7, 1939, cited in Nyce and Kahn, 215; VB to C. L. Wilson, Jan. 8, 1940 (LOC, VBP, 96/2208); VB to E. Hodgins, April 10, 1941 (LOC, VBP, 50); VB to F. Fassett, Sept. 1944 (LOC, VBP, 37/886).

19. Brinkley, D., *Washington,* 109.

20. VB to F. Fassett, Aug. 19, 1944 (LOC, VBP, 37/886).

21. F. Fassett to VB, Oct. 10, 1944; VB to F. Fassett, Oct. 12, 1944 (LOC, VBP, 37/886); VB to T. Barbour, Oct. 24, 1944 (LOC, VBP, 9/194).

Barbour, of the publishing house Houghton Mifflin, had asked VB to consider penning an entire book on the memex instead of a mere magazine article. But VB was also considering a book on military affairs, which "would be warranted only if the whole matter of reorganization of our military affairs were bogged down by reason of public apathy." VB ultimately wrote such a book, *Modern Arms and Free Men,* published in 1949.

22. VB and S. H. Caldwell, "A New Type of Differential Analyzer," *Journal of the Franklin Institute,* vol. 240 (1945), 255–326; Goldstine, 97–99; Shurkin, 78; Ceruzzi, Paul E., *Reckoners: The Prehistory of the Digital Computer, 1935–1945* (1983), 107.

23. Goldstine, 135; Shurkin, 80; Edwards, 51; Ceruzzi, *Reckoners,* 106–7.

24. "M.A.C. Outlines," April 1947 (NSA FOIA).

All records of the Navy's OP-20-G branch, including any relating to the Comparator, now reside with the National Security Agency, the top U.S. codebreaking agency based in Fort Meade, Md. The branch was established in 1935 to manage the security of U.S. naval communications and to attack foreign codes.

25. Burke, 173–74, 208, 212; H. Hazen to A. F. Sulzer, Nov. 22, 1940 (LOC, VBP, 48).

By agreement with the Navy, the old Comparator was shipped to MIT, where a new one was to be designed under the leadership of an engineer named John Howard. Bush also arranged for three graduate students who worked on the rapid selector to receive security clearances in order to work on the Comparator.

26. Burke, 215; Kahn, David, *The Codebreakers* (1973), 394.

27. Burke, 220, 226, 231–32, 257; VB OH, 113; "Brief Descriptions of RAM Equipment," U.S. Naval Communications, October 1947 (NSA, FOIA); "M.A.C. Outlines," April 1947 (NSA, FOIA).

OP-20-G also designed a related line of codebreaking equipment, inspired by Bush's earlier ideas, called the Rapid Machine. As of 1947, another version of the Comparator using 70-millimeter punch tape had been in active use since August 1944. This Comparator could compare two messages or two portions of text, each 1 to 2,000 characters.

28. VB to H. Hazen, Jan. 21, 1944 (LOC, VBP, 48/1184).

29. VB to Archibald MacLeish, May 8, 1940 (MIT, AC4, 42, VBP, 1940).

30. Burke, 154.

31. Burke, 119; VB to Hazen, Jan. 21, 1944 (LOC, VBP, 48/1184).

32. VB to Hazen, Jan. 21, 1944 (LOC, VBP, 48/1184); Burke, 190–91.

VB came to realize his naivete about coding, though too late to do anything about it. After reviewing in 1964 a study of various types of rapid selectors, VB confessed to an MIT professor his surprise "not at the progress made [in rapid selectors] but the lack of it. One would have thought that, with all the new gadgets . . . there would now be available something rather striking. But I do not find it."

"Strangely enough, in spite of all the work that has gone on for 25 years, I think the old machine I built at MIT had the most useful ideas [on rapid selectors] that have appeared," VB wrote a former aide at the Carnegie Institution the same month.

VB to Carl Overhage, Oct. 6, 1964 (MIT, RS); VB to G. P. Bauer, Oct. 5, 1964 (MIT, RS).

33. VB to Carl Overhage, Oct. 6, 1964 (MIT, RS).

34. J. E. Hoover to VB, July 30, 1940; VB to E. P. Coffey, July 22, 1940 (LOC, VBP, 37/891).

35. "Electronic Filing Machine Sorts Millions of Government Records: Agriculture Department Unveils Its 'Rapid Selector,' Cutting Storage Space as Many as 4,000 Times," NYT, June 23, 1949.

The Agriculture Department's rapid selector cost $75,000 to build and contained many changes from earlier versions. One key improvement came in the coding scheme. As *The New York Times* described it: "When an item is photographed [for storage on microfilm], whether a page from a scientific work or stray memorandum of a few lines mimeographed on a page of white paper, the photographing machine also is 'fed' a code series. These consist of dots in long multiple rows. The 10,000,000 possible arrangements of the dots provide every conceivable combination of cross-index code for any subject ranging . . . from hybrid corn to penicillin and its multiple experiments and applications. . . . An operator seeking material on a subject simply puts into a slot in the front of the machine a blank piece of paper perforated with the code arrangement of the records to be copied."

As late as 1961, the story of the rapid selector was reviewed in a technical study by the National Bureau of Standards, entitled "Information Selection Systems Retrieving Replica Copies: A State-Of-The-Art Report," by Thomas C. Bagg and Mary E. Stevens (Dec. 31, 1961).

The postwar history of the rapid selector remains to be written. For efforts by private industry and government, see "Status Report on the Rapid Selector," by Thomas C. Bagg and P. F. Ordung (June 1957); "The Rapid Selector and Other NBS Document Retrieval Studies," in *Proceedings of 11th Annual Meeting of National Microfilm Assoc.* (1962). For VB's frustration with the lack of progress in the device, see VB to Carl Overhage, Oct. 6, 1964; VB to Max Tishler, Oct. 5, 1964; VB to H. D. Brown, March 26, 1965 (MIT, RS).

36. The author is grateful to Michael Buckland for this explanation.

37. VB to W. Weaver, Oct. 3, 1950 (LOC, VBP, 117); VB to von Neumann, Oct. 31, 1949 (LOC, VBP, 47); Ceruzzi, Paul E., "An Unforeseen Revolution: Computers and Expectations, 1935–1985," in *Imagining Tomorrow* (1986), J. Corn, ed., 189–90.

VB was not alone in worrying about a glut of electronic computers. In the late 1940s Howard Aiken, Harvard University's computer guru, predicted that less than five computers could satisfy the computing needs of the entire nation. Other respected estimates ran only slightly higher.

38. Owens, Larry, "Where Are We Going, Phil Morse? Changing Agendas and the Rhetoric of Obviousness in the Transformation of Computing at MIT, 1939–1957," *IEEE Annals of the History of Computing*, vol. 18, no. 4, 1996, 2; VB to F. Cooper, Nov. 17, 1952 (LOC, VBP, 28).

39. Owens, "Where Are We Going, Phil Morse?"

40. VB to F. Cooper, April 1, 1955 (LOC, VBP, 28); "minimizing" quotation in Smith, Linda, "Memex as an Image of Potentiality Revisited," in Nyce and Kahn, 266.

41. VB to F. Cooper, Nov. 17, 1952 (LOC, VBP, 28).

42. VB OH, 113–14.

A year after the end of the war, the Navy prepared a patent application in VB's name for his design of the Comparator. When asked to review the application, VB raised concerns about the potential breach of secrecy and assured the Navy that "I am entirely willing to assign this invention to the government, irrespective of whether the arrangement I had with the Navy at that period called for such action or not."

Though he often teased security-minded officers about their zealousness, VB favored some restrictions on access to the military's technical information. He considered cryptographic matters "exceedingly confidential" and believed his work on the Comparator rightly fell under "a very close cloak of secrecy" (VB to Robert Conrad, Oct. 9, 1946, LOC, VBP, 88/1924).

Chapter Thirteen: "A carry-over from the war"

Epigraph: VB OH, 799.

1. Kevles, *The Physicists,* 332–33; Conant, James, "Notes on the 'Trinity' Test," in Hershberg, 758–61; Hershberg, 231–34; VB OH (CIW), 421–22; VB OH MIT (Q 26); *Life,* "Atomic Power," Aug. 12, 1946; Groves, 288–304; NBC interview, VB, Jan. 1965 (MIT, VBP).

The first atomic test, code-named Trinity, came at about 5:30 on a Monday morning. VB had arrived at the test site near Alamogordo the night before, at about 8:00, with Conant and Leslie Groves. After dinner and conversations with Oppenheimer and a few other physicists, VB went to bed in a tent shared with Conant. VB had slept little the preceding two or three nights and would sleep little that night. At about 3:15, after a rain shower, a physicist woke VB and Conant, telling them the test had been pushed back an hour from the original schedule of 4:00 A.M. and might be postponed. VB got dressed, wandered around with Conant, then saw an empty tent with a cot in it. He rolled in to get some sleep, but the wind and rain came up and blew the tent down, leaving VB soaked.

2. Hershberg, 231–34; VB OH, 422–23; VB to J. R. Page, Oct. 6, 1945 (CIW, VBP).

VB's comments about Oppenheimer came in a letter to Page of Caltech, which was seeking a successor to Robert Millikan as president. VB recommended that Caltech consider Lee DuBridge, the head of the Radiation Lab and Caltech's ultimate choice in 1946. Among the other candidates endorsed by VB was Oppenheimer, although VB doubted that the physicist wanted to continue as an administrator. He also noted that Oppenheimer leaned "strongly to the left in his political philosophy" and "is Jewish I believe." VB allowed that "I do not know whether this makes any difference or not under your circumstances; it certainly did not interfere with his effectiveness during the war, and personally to me it does not make the slightest difference in any way."

In late July 1945 the U.S. had two finished bombs, with just one more nearly ready (Weinberg, 889).

3. Groves, Leslie, *Now It Can Be Told* (1962), 303; VB testimony on April 23, 1954, "In the Matter of J. Robert Oppenheimer," 561.

In a speech on May 23, 1945, VB spoke of the "furious" fighting on Okinawa and waxed proud over improved techniques of treating wounded Marines (NARA, RG 227, Office of Historian, 8.)

With the aid of hindsight, VB divined other benefits of the bomb. In answering a question about the bomb's significance at the Atomic Energy Commission's Oppenheimer security hearing, VB added that the bomb "was also delivered on time so that there was no necessity for any concessions to Russia at the end of the war. It was on time in the sense that after the war we had the principal deterrent that prevented Russia from sweeping over Europe after we demobilized. It is one of the most magnificent performances of history in any development to have that thing on time."

4. VB OH, 423; interviews, Catherine Bush, Edna Haskins; "Townspeople of Dennis Say Dr. Bush Is 'Regular Feller,'" n.d., 1945, *Boston Globe* (Boston Globe archives); VB interview, NBC, June 5, 1965 (MIT, VBP).

Phoebe could not always quiet her curiosity. Edna Haskins recalled one occasion near the end of war when she asked VB about the existence of secret weapons. "Don't ask me that again," VB growled.

In a 1956 television interview, Phoebe Bush, with VB sitting next to her, told Ed Murrow that she knew nothing of the bomb before Hiroshima. "He didn't tell me anything."

VB's hasty return to Washington was dictated by a scheduled meeting of the Interim Committee, which at 10:00 A.M., on July 19, would discuss proposed atomic energy legislation. He also expected legislation for his proposed National Research Foundation to be introduced in the Senate.

5. Sherwin, 223–24, 227, 231; Weinberg, 888–89.

6. Murrow interview (MIT, VBP); Rhodes, 734.

Radiation sickness and other injuries from the bombing of Hiroshima increased the number killed over time. The deaths totaled an estimated 140,000 by the end of 1945; over five years, the toll reached an estimated 200,000.

A few months after Hiroshima, Phoebe told the *Boston Globe:* "Personally I believe it was wonderful we were able to be ahead of the enemy and use the bomb first. Mothers who have sons in the Pacific will understand that" (Oct. 1, 1945, *Boston Globe*).

7. McCullough, 454–55.

8. VB memo, Aug. 6, 1945 (RG 298, GC, 1941–45, 80).

VB displayed no guilt or regrets to his staff, observing that the bombing of Hiroshima "will form one more basis on which we can take pride in the accomplishments of American Science, and in the success of our part of the war effort."

9. Camus, 110.

10. VB to J. Tate, Aug. 13, 1945 (CIW, VBP, AE), cited in Hershberg, James, "Reconsidering the Nuclear Arms Race: The Past as Prelude?" in Martel, Gordon, ed., *American Foreign Relations Reconsidered, 1890–1993* (1994), 194.

11. VB to R. D. Mershon, Aug. 25, 1945 (LOC, VBP, 76/1757); Hershberg, 244.

A few days later, VB sent Conant a note on a possible atomic statement in which he echoed the sentiments in the Mershon letter. "Devastating though the atom bomb is, it does not compare in horror with other weapons which we declined to use in the war," VB wrote. This "hideous catalog," he added, ranged from "plant hormones to be sprayed from airplanes to kill all the enemy's crops and impose man-made famine on an entire population" to "poison gases of such penetrative power that no protection of masks or special clothing can withstand them."

12. VB to J. Conant, July 18, 1945 (B-C, 19); NYT, "Bush Holds War Is Barred by Bomb," Dec. 4, 1945.

Though VB had nothing tangible since Sept. 1944 to show for his advocacy of international cooperation, he continued to work with Conant to promote the idea that an accord with the Soviets on atomic weapons was both conceivable and beneficial. Three weeks before the first atomic bombing, VB and Conant advised the Interim Committee that the U.S. should urge the new United Nations to form a scientific office to the Security Council that could act as a medium of exchange of basic atomic information. The UN would have the power to inspect atomic facilities anywhere in the world, with the U.S. off-limits for the first five years. The deal was a rough one for the Soviet Union, but reflected the harder line in favor with the Truman administration. As VB and Conant wrote:

> The one hope of preventing major devastation to industrialized nations in another war lies in the United Nations. It is a bad start for the organization if a division between the Big Five powers occurs at the outset based on the fact that two [U.S. and U.K.] are bound together by a treaty involving a new and perhaps decisive weapon. To avoid a secret armament race and strengthen [UN] . . . must be the prime objective of every sane man. It is too Utopian

to hope for complete interchange with the Russians now. We must test out their good faith. The scheme above does so.

If the proposed arrangement should go through, the [US] would be giving to Russia and the rest of the world technical information which might cut down the time of their armament by perhaps two years assuming a 10-year total time. In return we would be gaining a good chance of following the course of their development and in addition we should be building the basis for a peaceful future. If this scheme fails after a year or two of operation because of the bad faith of one country, the [US] would still be in the lead in [atomic weapons] . . . and the bad faith of the other country would be clear and apparent. The third world war would be just over the horizon. But the situation would be clear.

Of the two, Conant was far more pessimistic than VB, who consistently said that another world war was avoidable. Less than four months after this memo, for instance, he repeated publicly that "the coming of the atomic bomb will stop great wars."

13. Catton, 305–6; Diggins, 53; Kaltenborn quotation in Boyer, 2–8; Conant quotation in Hershberg, 244.

Catton's observation came in a book published in 1948 and echoed sentiments expressed in the days and weeks after Hiroshima. In its first postwar issue, *Time* noted, "The knowledge of victory was as charged with sorrow and doubt as with joy and gratitude." In the most memorable postwar editorial, printed in the *Saturday Review* four days after Japan's surrender, Norman Cousins speculated that an ever-present irrational fear of death "has become intensified, magnified. It has burst out of the subconscious and into the conscious, filling the mind with primordial apprehensions."

The A-bomb was an important source of this fear, but not the only one. As historian Michael C. C. Adams has written, "World War II could not create a world free from fear, because there was never a realistic possibility that the democracies and the Soviets could form a lasting partnership or that the Communists would move after the war to a more representative form of government" (Adams, 139).

While VB never anticipated the advent of the Cold War, in 1945 he saw the Soviet Union as the likeliest challenger to U.S. hegemony over military technology.

14. NYT, Sept. 2, 1945; interview, Richard Bush; Adams, 154; Franklin, 145–46.

VB and his family were spared any violent retaliation for his role in the creation and use of atomic weapons. In a letter dated March 16, 1951, an anonymous writer threatened to spit in VB's face, if he ever met him, and likened VB's "puny brain" to "a hungry rat imprisoned in a steel-lined room—only death can open its eyes." VB turned the letter over to the FBI, which never found the writer, who closed his missive by placing "a curse on you & all your marks" (FBI, VB file).

The extent of VB's involvement in atomic matters was publicly known, and to some extent, the press actually exaggerated his role. In its profile of VB, the NYT noted, somewhat inaccurately, that the bomb was planned in his Carnegie office. "There around a long table a dozen scientists and military men discussed the splitting of the atom and a new form of warfare" the *Times* wrote. "At the head of the table sat [VB] . . . who laid the plans and assigned the investigations which led to the epochal bombing."

H. Bruce Franklin has argued that the bomb destroyed the core of democracy by placing huge power in the hands of the few who would decide whether to use nuclear weapons. He cites the prediction following Hiroshima by physicist Harold Urey that in "five years or perhaps less" it might be necessary for the U.S. to adopt a dictatorial form of government in order to act quickly against an atomic threat. Urey, who worked on the bomb, said, "I do not see any way to keep our democratic form of government if everybody has atomic bombs."

Other historians, notably Gabriel Kolko, place the bomb within the context of an overall decline of genuine democracy within modern America and the shrinking power of individuals to control the circumstances of their lives.

15. Boyer, 183–84.

Public opinion polls after Hiroshima and Nagasaki found upward of 80 percent of Americans favored the bombings. In one poll, just 5 percent of the respondents said no bombs should have been dropped; four times the number of respondents expressed disappointment that *more* bombs had not been dropped on Japanese cities.

16. NYT, Sept. 2, 1945; *New York Mirror,* Aug. 14, 1945; *Business Week,* Aug. 25, 1945; *Life,* Aug. 12, 1946; Reingold, "MGM Meets the Atom Bomb," in *Science: American Style,* 336; VB to J. Conant, Nov. 4, 1946 (LOC, VBP, 27); VB to J. Conant, Dec. 6, 1946 (LOC, VBP, 172); *Time,* Feb. 24, 1947; Hershberg, 287–91; Lifton, R. J., and Mitchell, G., *Hiroshima in America: A Half Century of Denial* (1995), 361–68.

17. F. Jewett, Oct. 19, 1945, testimony at Joint Hearings of Subcommittees of the Senate's Committees of Commerce and Military Affairs; J. Furer to F. Jewett, June 29, 1945 (LOC, Furer Papers, 7); Compton, 64; W. Weaver interview, April 6, 1966 (courtesy Daniel Greenberg); interviews, Caryl Haskins, Richard Bush; Edward Bowles, "Office Diary," Feb. 9, 1947, 15 (Bowles Papers, courtesy Martin Collins).

In his diary, Bowles recounted in early 1947 a statement by Julius Stratton, an influential professor at MIT, that VB could "have" the presidency if he wanted it.

18. Hughes, 246–48; *The Nation,* Dec. 22, 1945.

19. Quotation by physicist Samuel Allison: Greenberg, 95.

20. VB to Charles A. Thomas, Feb. 26, 1946 (LOC, VBP, 111).

21. VB speech, Oct. 24, 1962, emphasis added (CIW); VB to Alexander Magoun, Dec. 15, 1945 (LOC, VBP, 68); Hughes, 440.

The historian Thomas Hughes has noted the way in which the atomic bomb seemed to dwarf other historic accomplishments:

> In the memories of the participants, in official histories, scholarly publications, and newspapers and television reports, the Manhattan Project became history's most impressive technical, scientific, industrial and organizational achievement. The Manhattan Project—symbolized by the Hanford Piles and the mushroom cloud over Trinity—cast its shadow over the great construction and production achievements of the past, such as the pyramids, medieval cathedrals, Renaissance Venice, Baroque palaces, Industrial Revolution canals and railroads, regional electric-power systems, Ford's River Rouge plant, the Soviet Magnitogorsk, and the TVA.

22. Hewlett and Anderson, 415.

23. May 26, 1942, meeting of the Joint Committee on New Weapons and Equipment (NARA, RG 218, 15); Weaver interview (Greenberg).

24. *Fortune,* "The Great Science Debate," June 1946; VB POA, 40; VB OH, 142–48; McCullough, 291.

25. Hewlett and Anderson, 409.

26. Ibid, 412.

27. Ibid, 413–15; "Notes of Interim Committee Meeting," July 19, 1945 (HSTL).

28. Herken, 27; Hewlett and Anderson, 422–23; A. Smith, 89–97.

29. VB to J. Conant, Oct. 1, 1945 (LOC, VBP, 27/614).

30. VB to H. Stimson, April 21, 1948 (LOC, VBP, 109); Herken, 29–33, 37; Hewlett and Anderson, 420–21; Forrestal diary, Sept. 21, 1945 (Princeton, Forrestal Papers).

Vinson and Clark officially were Treasury secretary and attorney general, respectively.

31. VB memo, dated Sept. 25, 1945, quoted in Truman, *Year of Decisions,* 525–27, for full text see NARA, AEC, HDN, 217; Herken, 61; Holloway, 127–33.

Like many commentators on the bomb, VB lapsed into simplistic dichotomies when speaking of the atomic future: "Down one path lies a secret arms race . . . down the other international collaboration and possibly ultimate control. Both paths are thorny, but we live in a new world and have to choose." VB sometimes conceded that the future might be less binary, but in his presidential memo he played down such thoughts, aware of Truman's penchant for black-and-white choices.

Stark analyses of the bomb contained the seeds of paralysis. Like many giving advice on arms control at the time, VB made vague suggestions but did not really chart a practical path between idealistic and unacceptable alternatives, or "world state or world doom," as journalist Max Lerner put it (Sherry, *Shadow*, 134).

32. Weigley, *American Way*, 363–65; Nuclear Weapons Databook, 1983.

The U.S. stockpile grew to 13 in 1947 and 50 in 1948.

33. Hershberg, 249; Forrestal Diary, Sept. 18, 1945 (Princeton, Forrestal Papers).

Forrestal mistrusted the Russians. When Truman told him before the Sept. 21 cabinet meeting that "his present disposition was to disclose the principles of atomic energy to the Russians and others of the United Nations but not the method of making the bomb," Forrestal retorted that he was "violently opposed to any disclosure" and thought the country would be too.

34. VB to J. Conant, Sept. 24, 1945 (LOC, VBP, 27/614).

35. VB to J. Conant, Sept. 24, 1945 (LOC, VBP, 27/614); Hewlett and Anderson, 420–21; Herken, 27–31.

In his reference to "no powder in the gun," VB may also have been alluding to the tiny U.S. atomic arsenal.

36. VB testimony, Oct. 9, 1945 before the House Committee on Military Affairs; Kevles, *The Physicists*, 350–51; VB to H. Hoover, Sept. 25, 1945 (LOC, VBP, 51/1261); Truman's message to Congress, Oct. 3, 1945, found in Truman, *Year of Decision*, 530–33; Lanouette, William, "Atomic Energy, 1945–1985," *Wilson Quarterly*, Winter 1985; Zachary, G. Pascal, "Where Did Big Science Take Us," *Upside*, April 1992.

VB had thought for some time about the economic benefits of atomic power. During the war he worried about Britain's commercial designs on atomic power and spearheaded U.S. steps to maintain favorable control over patents in this area. The postwar debate over domestic control of atomic power brought out the optimist in many. In 1946, Enrico Fermi, who had engineered the first chain reaction four years earlier, predicted that atomic power would be in widespread use within 20 to 30 years. In 1947, *Business Week* concluded that atomic-power reactors might be five years away. But VB's view had its supporters, too. In a 1947 book, *The Atomic Story*, science writer John W. Campbell concluded that the much-hyped economic benefits of atomic energy had been "badly oversold."

37. Hewlett and Anderson, 430–31; VB testimony, Oct. 9, 1945, to the House Committee on Military Affairs.

38. Hershberg, 260, 268; J. Snyder to HST, Nov. 14, 1945 (LOC, Byron S. Miller Papers, 1).

Complaints against May-Johnson, from scientists and other professionals, were "still pour[ing] in at an unabated rate" six weeks after the bill's debut.

39. VB to L. DuBridge, Oct. 26, 1945 (American Philosophical Society, H. D. Smyth Papers, VB folder); VB to F. Jewett, Nov. 6, 1945 (LOC, VBP, 56); VB to J. Conant, Nov. 7, 1945) (LOC, VBP, 27/614).

To make sure the views he expressed to DuBridge trickled out, Bush sent copies of his letter to ten others scientists, including Arthur Compton, a leader of the recalcitrant Chicago physicists.

40. VB to Representative James W. Wadworth, April 3, 1946 (LOC, VBP, 116); VB to Alexander Magoun, Dec. 15, 1945 (LOC, VBP, 68).

As if speaking about children, VB insisted that younger scientists had no grasp of day-to-day politics, writing to Magoun: "Of course the thing they do not seem to understand is that no such bill is ever written by one individual, nor is it ever possible to set up simple ideals and expect to pursue them without alteration through the entire maze of government."

41. VB POA, 294–95; A. Smith, 82–83; Herken, 35–36; Truman, *Year of Decisions*, 532–33.

The atomic report, written by physicist Henry DeWolf Smyth, gave an account of the origin and history of the Manhattan Project, described the scientific principles behind the bomb and the separation of fissionable materials and outlined the various methods used to achieve atomic explosives. VB argued that the report gave no benefit to the Russians, but showed the U.S. commitment to liberal exchange of scientific information. It also might reduce unfounded speculation about the bomb, he thought. Many scientists welcomed the move, but were perplexed by what historian Smith described as "the sudden and unexplained departure from the rigid wartime restrictions." After all, the information in the Smyth report wasn't trivial; many scientists on the Manhattan Project had such compartmentalized knowledge that they learned "a great deal of technical interest."

42. J. Snyder to HST, Nov. 14, 1945 (LOC, Byron S. Miller Papers, 1); VB to HST, Oct. 13, 1945 (HSTL, OF 53); Hewlett and Anderson, 436–39; Donovan, *Conflict,* 15, 116–17; Kevles, *The Physicists,* 351; Newman, James, "America's Most Radical Law: The Atomic Revolution," *Harpers,* May 1947.

While Snyder refrained from personally criticizing Bush, he clearly blamed him, Conant and the War Department for the mess May-Johnson had become. "Whatever the intention of the draftsmen," Snyder wrote Truman, "the bill tends to restrict and stunt development of nuclear physics rather than to simplify and encourage it."

In forming his opinions, Snyder relied heavily on James Newman, his assistant in the Office of War Mobilization and Reconversion. A lawyer by training and an authority on mathematics, Newman once wrote, "This new force offers enormous possibilities for improving public welfare, for revamping our industrial methods and for increasing the standard of living."

VB was dubious of such claims, partly on technical grounds and partly because he thought Newman and other liberals were merely seeking a fresh pretext to intervene in the economy. VB believed that if atomic power had economic value, private corporations ultimately would discover that. Newman mistrusted the marketplace and argued that the commission overseeing national military issues ought also to have the responsibility for making applications of atomic discoveries available to the whole population. That this meant some new form of public power was clear, since Newman himself in a 1947 article said that the nation's atomic-energy act "sets up an island of socialism in the midst of a free enterprise economy."

It was Newman, along with colleague Byron Miller, who drafted McMahon's bill and advised him during the debate over it.

43. VB to HST, Oct. 13, 1945 (HSTL, OF 53).

44. Truman News Conference, Oct. 25, 1945 (HSTL); VB to Jewett, Nov. 6, 1945 (LOC, VBP, 56).

45. VB to F. Jewett, Nov. 6, 1945 (LOC, VBP, 56).

46. VB to J. Conant, Oct. 20, 1945 (LOC, VBP, 27/614); "Natural Aristocracy Asked by Dr. Bush," Feb. 20, 1955, *Washington Post;* on his political leanings kept from presidents, see VB OH, 1967, Columbia, 52.

47. Lewin, Ronald, *Hitler's Mistakes* (1984), 99; VB, "Roosevelt" speech, Oct. 27, 1945 (LOC, VBP, 99); VB to F. Keppel, March 3, 1939 (VB, 61/1462).

48. VB to L. DuBridge, Oct. 26, 1945 (American Philosophical Society, H. D. Smyth Papers, VB folder); "Bush Holds War Is Barred by Bomb," NYT, Dec. 4, 1945.

49. Herken, 59–66; VB to J. Byrnes, Nov. 5, 1945 (NARA, AEC, HDN 218); Gaddis, *Origins of the Cold War,* 270–71.

Byrnes feared the Russians might ask for a share of atomic technology. VB presumed the U.S. could simply say no.

50. Interview, Oscar Ruebhausen; draft, VB to HST, Oct. 13, 1945 (NARA, RG227, Director OSRD, Reports to the President, 1); Gaddis, *Origins of the Cold War,* 272.

VB often called for new, younger leaders for the research community and expressed a willingness to step aside. But actually doing so was another matter. In a draft of his letter to Truman dated Oct. 13, 1945, VB wrote: "For my own part, I should perhaps make it clear that when the work of the OSRD is closed out, it is my strong belief that this type of work shoud thereafter be the responsibility of a younger man and a fresh mind. It would be a mistake for the direction of the affairs of science to continue without the infusion of the vigor that can come from new leadership. Regardless of the capacity in which you choose to have me serve you for the remaining period of the OSRD's life, therefore, I must insist that I be relieved of responsibility for the further direction of scientific research when the OSRD's job is fully done." In the final text of his letter, Bush cut the references to the need for new scientific leadership.

51. Herken, 61; VB to J. Conant, Nov. 7, 1945 (LOC, VBP, 27/614); VB to William Emerson, Nov. 6, 1945 (LOC, VBP 36/850).

52. Herken, 63–65, 351; Hewlett and Anderson, 461–69; VB memo, Nov. 15, 1945 (NARA, AEC, HDN, 218); VB to J. Conant, Nov. 10, 1945 (LOC, VBP, 27/614); VB memo, Nov. 13, 1945 (B-C, 12); VB memo, Nov. 14, 1945 (B-C, 19); VB to H. Stimson, Nov. 13, 1945 (LOC, VBP, 109).

53. Kaempffert, Waldermar, "For a Hierarchy of Scientists," March 17, 1946, NYT; Hewlett and Anderson, 534–58; Kennan quotation in Hodgson, 28; VB to Conant, Oct. 21, 1945, LOC, VBP, 27; Baruch quotation in Herken, 160–61; Donovan, *Conflict,* 204–6.

The State Department plan—the so-called Acheson-Lilienthal Report—called for controls on atomic energy to prevent clandestine production of bombs and left nations to respond on their own to a violation. Baruch's plan included methods for sanctions against violators. Byrnes's deputy, Dean Acheson, feared that the Russians, who were building their own bombs, would perceive Baruch's harder line as an attempt to turn the UN against Russia.

VB initially dismissed Baruch's ideas, refusing even to brief some of his aides on scientific points. VB refrained from making alarmist statements about Russian intentions that spouted from Forrestal and Kennan, but he matter-of-factly sided with hardliners as time went on. In October 1946, he privately defended Baruch's tougher line, telling David Lilienthal, Truman's nominee to chair the Atomic Energy Commission, that "I hope we sit tight as long as necessary" on atomic bargaining with the Russians. He called Baruch's proposal "perfectly reasonable" and noted, "I would be decidedly disturbed if" the U.S. showed "an inclination to soften up" in dealing with Russia (VB to C. L. Wilson, Oct. 15, 1946, LOC, VBP, 120). A week later VB wrote that "the only place where I differ" with Baruch "is on the question whether we ought to make concessions in the negotiations. . . . As I see it, the Russians, when we first sprung this plan, looked behind it for sinister motives and felt it must conceal something. I believe that they have recovered from that now." VB allowed that the U.S. must walk a tough line without letting "the negotiations break down completely, but I feel sure the Russians have no idea of letting them break in any case and I look forward to a period of several years of wrangling of one sort or another" (VB to Conant, Oct. 21, 1945, LOC, VBP, 27).

54. VB speech, Feb. 22, 1946 (LOC, VBP, 139); VB to J. Conant, Jan. 2, 1946 (LOC, VBP, 27/614).

In early January, VB admitted to Conant, "I judge that I am now out of the atomic energy matter except as I may be called on for comment on plans or the like at various times."

55. VB to J. Snyder, Nov. 23, 1945; J. Snyder to VB, Nov. 30, 1945 (LOC, Byron Miller Papers, 3); VB to J. Conant, Nov. 7, 1945 (LOC, VBP, 27/614); Kevles, *The Physicists,* 351–52; Hewlett and Anderson, 513.

Harold Smith, VB's old nemesis, also contributed to Truman's abandonment of May-Johnson. Smith, the budget chief who had tangled with Bush for years about the scope of authority to be granted independent experts, convinced Truman that the bill posed a threat to presidential power. On Oct. 18, Truman publicly backed away from May-Johnson. As an administration initiative, the bill presumably had the president's backing, but Truman nonetheless told a reporter this wasn't so. Claiming to not have studied the bill carefully, he said he had not made up his mind whether to sign it if sent to him by Congress (Donovan, *Conflict,* 133–34).

56. Donovan, *Conflict,* 269, 207; J. Snyder memo, Feb. 20, 1946 (HSTL, OF 53); VB to J. Conant, Nov. 7, 1945 (LOC, VBP, 27/614).

57. VB to Matthew J. Connelly, Jan. 19, 1946 (OF 53 [46–53], HSTL).

Connelly made no reply to VB, or at least left no record.

58. Interview, Warren Weaver, April 6, 1966 (courtesy Daniel Greenberg).

59. *Fortune,* "The Great Science Debate," June 1946.

Chapter Fourteen: "So doggone weary"

Epigraph: VB address to Joint Air Conference, March 30, 1946 (LOC, VBP, 128/3070).

1. VB to J. Conant, Jan. 2, 1946 (LOC, VB, 27/614); interview, Richard Bush; VB to Nelson Rockefeller, Sept. 4, 1953 (Rockefeller Family Archives, RG 4, Projects, 19/198); VB POA, 152.

VB first toyed with solar power in the 1930s, while at MIT, but the war held up his investigations. The goal of his solar device—he called it "a tin can affair"—was to take "heat from a collector and use it to pump water for irrigation."

Vacations often turned into hobby-fests for VB. "He was seldom idle, even on a vacation," his son Richard said.

Nostalgia for the Old West grew in the late 1940s, and dude ranches—western spas for the well-off—did brisk business (Graebner, 60).

2. *Newsweek,* "Bush at the Crossroads," March 11, 1946; VB to B. O'Brien, March 16, 1946 (LOC, VBP, 89/1968).

On reviving dormant hobbies, VB wrote O'Brien: "The war being over and things beginning to settle into position a bit, I have returned to some of my old interests, and in particular I am getting my photographic equipment into shape and am putting myself together a small shop."

3. VB to J. Killian, March 5, 1946 (AC4, MIT); VB to J. Killian, March 25, 1946 (VB, 62/1471); VB to Georges F. Doriot, Dec. 15, 1947 (LOC, VBP, 34/770); VB solar memo, Aug. 13, 1946 (MIT, AC4, Office of the President, 1930–58, 43).

None of these devices appear ever to have been commercialized.

To Doriot, an early venture capitalist to whom he often supplied technical tips, VB wrote, "If you are really itching for ideas around your place [you can] . . . try this gag out."

Of his solar pump, VB wrote that he sought an inexpensive collector, citing the relatively high cost of this piece of the machine as the reason "many previous attempts to use solar energy economically have failed." It isn't clear whether VB made significant strides toward a cheaper solar collector, since he probably did not even build a prototype of his design.

4. Interview, Caryl Haskins; VB memo, Sept. 30, 1948 (LOC, VBP, 159); VB to Sawyer, Sept. 12, 1947 (LOC, VBP, 123/2351); VB to Bradley Dewey, Oct. 7, 1947 (LOC, VBP, 125); VB to S. Lovell, July 7, 1947; VB to S. Lovell, Oct. 17, 1947 (LOC, VBP, 66).

The pond greatly concerned VB, who managed its affairs from faraway Washington. "I think next week I may send you by express a few fish in a can," he wrote the New Hampshire neighbor who looked after his pond. "If I do will you please dump them into the pond? Whether they will stand the winter or not I am not sure, but this seems to be a good time to find out." VB went on to describe the principal fish he planned to send, Golden Orfs. "They live on algae and seem to be able to thrive on it, but they also take insects when they can get them. They multiply very rapidly." If the Orfs thrived, "We could then introduce brook trout in the upper pond and brown trout into the lower pond and have quite a show. We would want to do a little fertilizing of the pond to insure a vigorous growth of algae, but not too much. If, on the other hand, they do not prosper, we can turn to Dace as the subordinate species to eat the vegetation and supply food for the trout. . . . Incidentally, I had an idea on how to clear that upper pond. If you leave the water level low until the pond freezes it should then not be difficult to clear away all the debris of one sort or another."

5. VB, "Shop" memo, Dec. 7, 1945 (LOC, VBP, 125); *Popular Science,* "Top Scientist Has

Own Shop," June 1948; VB to B. O'Brien, March 16, 1946 (LOC, VBP, 89/1968); VB to B. Dewey, Oct. 7, 1947 (LOC, VBP, 125).

VB completed his workshop in December 1946, after a year of toil. He was aided by 16 friends from around the country who collectively were called VB's "Shop Group." As a token of their affection, the men—some wartime colleagues, others prewar friends—sent him a scroll with a motto in Latin, which translated meant, "From this delightful workshop nothing shall go out except what is wholly pleasing or agreeable to me." Touched by the gift, VB wrote "Shop Group" members that the scroll would "serve as inspiration in the moments when one looks up from the lathe and possibly to serve as admonition for those (I hope) very infrequent moments when a slip ruins a job 99 percent near perfection" (VB to E. G. Farrand, July 22, 1946, LOC, VBP, 125).

6. VB to William G. Thompson, April 4, 1947 (LOC, VBP, 111).

VB found fishing a good excuse for idling about outdoors. He told Thompson, "The thought of getting out into the open is grand, and while I hope we catch some fish this is really secondary to having a reasonable excuse for spending a day in the open."

7. VB to J. Conant, Oct. 20, 1945 (LOC, VBP, 27/614).

8. Yergin, 199–200, 217; Hodgson, 28.

9. Sherry, *Shadow,* 83; Huntington, *Soldier,* 317–18; R. Patterson to VB, April 23, 1946 (LOC, VBP, 93/2120).

Many years later, VB described his wartime relations with the JCS in bizarre terms. "Occasionally, I sat with them, but very rarely indeed," he recalled in the mid-1960s. "The atmosphere was quite different. I was not sitting in while they were discussing a major matter of policy or strategy where I could interpose technical points. I was there as a sort of witness. When they made real decisions they made them with the absence of everybody except themselves. As far as I know they didn't even have subordinate officers at the times when they made real decisions. Of course, in addition, we have to say that those decisions were limited to things where they were unanimous. They did not resolve in the Joint Chiefs points where there was difference of opinion. Those had to go to the President for decision if they . . . got decided at all. I think the President refrained from making such decisions" (VB OH [MIT], 818).

10. Transcript of JNW, Jan. 21, 1946 (NARA, RG 218, JCS, JNW, 15–17).

11. VB's "dictum" in McDougall, 89; VB to B. McMahon, Jan. 22, 1946 (CIW: AE file).

12. Rearden, 12.

Military forces shrank during roughly the same period. In the 22 months after Japan's surrender, the Army went from 97 divisions to 12, the Navy went from 1,166 combat ships to 343, the Air Forces went from 213 combat groups to 11 fully active ones.

13. E. Bowles to D. Eisenhower, April 12, 1946; J. Devers to E. Bowles, April 18, 1946; J. Hull to E. Bowles, April 20, 1946; D. Eisenhower, "Scientific and Technological Resources as Military Assets" (HSTL, Bowles Papers).

Bowles wrote his initial memo to Eisenhower on April 12, 1946, prompted by the War Department's approval of a directorate for research and development. In a cover letter to the memo, Bowles descibed his goal for the new office as "a means of coupling our professional military resources with those outside of the Army."

Eisenhower's staff liked his idea. A slightly revised version of Bowles's memo was returned to him on April 20. Seven days later, the memo was issued under Eisenhower's signature. It retained Bowles's original title and his main points.

For the comparison of the technological strengths of the U.S. and Russia, Bowles turned to W. B. Shockley in late 1945. Shockley, who in 1948 would be celebrated as one of three coinventors of the transistor, found the U.S. ahead in the categories of aircraft, ordnance and electronics, with Russia leading in tanks by one to three years. In missiles, Shockley could not make a judgment because of lack of data.

Project Rand ultimately became The Rand Corporation (Bowles Papers, Courtesy Martin Collins).

14. The Bowles-Eisenhower statement naturally influenced thinking within the Army but it also caught the attention of industry, which stood to benefit from the military's desire for closer peacetime relations with corporations, especially those peddling advanced technology. In July 1946, the director of General Electric's research lab congratulated Bowles on how he had "been wonderfully successful in selling science where it was formerly not very welcome." Bowles's "present activity," he added, "indicates to me that your ideas have become thoroughly established in high places" (C. G. Suits to E. Bowles, July 10, 1946, NARA, 107, Bowles corresp.).

15. "OSRD in War," VB to Truman, October 12, 1945 (LOC, VBP, 139); VB to J. Killian, March 2, 1955; J. Killian to VB, March 22, 1955 (MIT, AC4, Office of the President, 1930–58, 43); Bowles office diary, Jan. 27, 1947 (LOC, Bowles Papers).

A decade after the release of the Bowles-Eisenhower statement, VB ran across a copy of it, prompting him to write: "This is quite a document . . . it might do quite a lot of good if attention was again drawn to this statement. What do you think I ought to do with it?" Killian replied: "I agree that the Eisenhower statement of 1946 is an exceptionally fine and prescient document."

16. VB quotation in Hall, R. Cargill, "Early U.S. Satellite Proposals," *Technology and Culture,* Fall 1963; McDougall, 98.

17. VB to R. Patterson, Jan. 3, 1947 (LOC, VBP, 91/2150).

VB reiterated these complaints in an interview with *The New York Times,* published Jan. 9, 1947 as "Push-Button Ideas of War Hit by Bush." VB told the *Times:* "This talk [of push-button war] has done a lot of harm. The trouble is that the American people get to thinking in terms of our pushing the buttons, and lose sight of the fact that, if there were a war tomorrow, it would be the same tough slugging match that the last one was. . . . Some of this 'Buck Rogers' thinking is a lot of hooey."

Arnold did not take kindly to VB's outbursts, seeing them as veiled attacks on the Air Forces' independence. When Edward Bowles reported on signs of Bush's antagonism toward the service, Arnold replied, "I am not surprised at the action of Van Bush. You remember I have preached the gospel of the [Air Forces] standing on its own feet, scientifically, just as soon as it is able. . . . There is no reason why the AAF cannot take care of itself, after a few years of special preparation." (H. Arnold to E. Bowles, March 3, 1947, LOC, Bowles Papers).

18. VB to R. Patterson, Jan. 3, 1947 (LOC, VBP, 91/2150); McDougall, 87–88; Huie, W. B., "The Backwardness of the Navy Brass," *The American Mercury,* June 1946; E. L. Eubank to H. Kerr, May 3, 1946 (LOC, Carl Spaatz Papers, 263, R&D 2).

General Eubank, writing to the secretary-general of the Air Board, found that because "there is no single agency charged with the complete responsibility for the development of guided missiles," the field "is characterized by a great deal of disjointed activity and an even greater amount of confusion. Perhaps some healthy competition between agencies exists, but for the most part it appears that the controversies over responsibility for development and over responsibility for the ultimate employment of guided missiles have been detrimental to progress." Eubank noted that coordinating agencies, such as VB's Joint New Weapons Committee, had failed to exert sufficient influence to overcome the obstacles imposed by interservice struggles for control.

19. VB to I. Bowman, cited in Dennis, "Bushwacked!: Vannevar Bush and the Dilemmas of Science Policy" (unpublished paper); VB to J.F., June 3, 1946 (LOC, VBP, 63).

Perhaps trying to re-create his war experience, when he had a direct line to Roosevelt, VB asked Truman on June 3, 1946, for "guidance" regarding the board post. In no hurry to bless the assignment, Truman did not reply to VB's plea. After 18 days without hearing from the president, VB pushed ahead with plans for the new board but still hoped to receive "an indication" that Truman "wishes me to undertake this responsibility." By the first meeting of the JRDB, in early July, VB had still not received a word from Truman on the matter (VB to HST, June 3, 1946; VB to W. Kenney, June 21, 1946, LOC, VBP, 63).

20. JRDB memo, Oct. 7, 1946 (NARA, RG 330, R&D Bd, 1946–1953, 2); K. Compton memo, Oct. 9, 1946 (MIT, AC4, Office of the President, 1930–58, 42).

Compton hoped only to "tide over the situation" until VB found a permanent leader.

21. "Head" quotation: Bowles office diary, Jan. 31, 1947 (LOC, Bowles Papers); Mc-Dougall, 100.

Bowles attributed the quotation to aerospace entrepreneur Donald Douglas.

22. McDougall, 101–3; Needell, Allan A., "Lloyd Berkner, Merle Tuve, and the Federal Role in Radio Astronomy," *Osiris* (1987), 3, 263–64; Hall, R. Cargill, "Early U.S. Satellite Proposals," *Technology and Culture,* Fall 1963; "Satellite Vehicle Program," Committee on Guided Missiles, Research and Development Board, March 29, 1948 (NARA, RDB Papers, courtesy Allan Needell); VB, *Modern Arms,* 85; Burrows, 58–59, 79–80.

By 1949, neither the Navy nor the Army had significant satellite efforts. The Air Force was the lone service to actively support satellite studies, mainly through its Rand think-tank.

With the Russian launch of *Sputnik* in 1957, the propaganda and military value of satellites was proven to be as monumental as Rand had predicted. *Sputnik* raised questions about the competence of the U.S. technological elite, but the alarm had been sounded years before. In 1954, Rand engineers urged the Air Force to build and use efficient spy satellites "at the earliest possible date . . . as a matter of vital strategic interest to the United States." The Air Force responded in 1955 by calling for proposals from industry, but working satellites were still years away.

23. W. Lalor, JCS memo, April 14, 1947 (NARA, RG 218, CDF 1948–1950, 334 R&D Bd., 73); W. Carey memo, Aug. 14, 1947 (NARA RG 51, 39.19, 105); VB to J. Conant, Oct. 11, 1946 (LOC, VBP, 27/614); H. Aurand, "Meeting with Secretary of War on 22 May 1947," May 23, 1947 (LOC, VBP); C. Nimitz to JCS, July 12, 1947 (NARA, RG 218, CDF 1948–1950, 334 R&D Bd., 73); *Washington Post,* "Research Chief," Oct. 8, 1948.

VB's reputation for conservatism caused resentment among scientists as well as soldiers, the *Post* reported.

24. Bowles office diary, July 8, 1947 (LOC, Bowles Papers); C. LeMay to C. Spaatz, May 22, 1946 (LOC, Spaatz Papers, 263, R&D, 2).

25. VB, "Military Organization for the United States," May 1, 1946 (LOC, VBP, 93/2120).

VB's nuanced argument for planned competition has generally been ignored by historians who lump him together with advocates of simple-minded centralization. Bush knew that duplication, while costly, yielded superior weaponry in World War II and that his own OSRD tolerated overlaps and redundancies among its divisions. Harvey Sapolsky, among others, has convincingly argued that over decades U.S.-style pluralism in weapons research produced better results than Soviet centralization. But it carried a price.

26. VB speech, Armed Forces Staff College, April 15, 1947 (CIW, VB speeches).

In officially describing the first-year activities of the JRDB, VB put on a brave face, writing on July 30, 1947, that "genuine progress has been made . . . the Armed Services are getting together on their programs; and these programs are receiving the benefits of impartial examination and review." Major problems remained, however, in guided missiles, and the board itself, which relied on 400 part-time consultants, had "perhaps not as yet accomplished much of great moment," VB admitted, but added that the board's "principal accomplishment, though intangible, is real—the existing movement toward a sound, integrated, national program of military" research. He concluded that the JRDB "experiment . . . is well worth continuing for the present" (JRDB draft annual report, June 30, 1947, NARA, RG 330, R&D Bd., 1946–1953, 2).

27. VB, "Military Organization for the United States," May 1, 1946; VB to W. Leahy, April 29, 1946 (LOC, VBP, 93/2120).

VB received extensive comments from the secretaries of war and the Navy and General

Spaatz, commanding general of the Army Air Forces (R. Patterson to VB, April 23, 1946; J. Forrestal to VB, April 26, 1946; C. Spaatz to VB, April 30, 1946; LOC, VBP, 93/2120).

28. Sherry, *Shadow*, 137; Hoopes and Brinkley, 349.

As part of the National Security Act, the services retained their own secretaries; each served on the National Security Council, but no longer sat in the president's cabinet.

29. S. Lovell to HST, April 2, 1947 (HSTL, OF 53, 46–53); NYT, July 16, 1947; Hoopes and Brinkley, 350; E. Bowles, "Office Diary," June 23, 1947 (Bowles Papers, courtesy Martin Collins); E. Bowles to H. Johnson, Oct. 10, 1974 (LOC, Bowles Papers, 5).

30. The journalist was James E. Cassidy, who wrote about VB in his *Passing Parade* column of Sept. 11, 1947 (LOC, VBP, 22). In the syndicated *Washington Merry-Go-Round* column published earlier in September, Robert Allen also reported Truman's indecision about appointing VB, attributing this to a host of factors having to do with VB's wartime performance and postwar politics. Allen reported that Truman privately told his "defense chiefs" that "if I should decide to give Dr. Bush this job, I can tell you that we're going to keep a very close eye on him." Of the two, Cassidy's simpler explanation seemed to fit the facts better. After reading Allen's column, VB complained to Conant that the journalist "tried to throw a monkey wrench into the works," but that he thought "there is really no hard feeling" between him and the president, though he noted the joke Truman made "in a very light vein" to Forrestal. VB added: "Forrestal wants me to go on as [RDB] chairman, and I have a certain obligation to do so" (VB to J. Conant. Sept. 5, 1947, LOC, VBP, 27/614).

31. Transcript of Jan. 21, 1946, meeting of Joint Committee on New Weapons and Equipment (NARA, RG 218); VB to HST, Dec. 19, 1946 (LOC, VBP, 112/2675); Donovan, *Tumultuous Years*, 192–93; Brint, 130.

Truman ignored VB's advice, but the engineer had spied the drift of history in calling for an expanded presidential staff. From the end of Roosevelt's first term to the administration of George Bush, the total White House staff grew from 37 to 900. The same trend played out in Congress, where legislative staffs grew sixfold from 1947 to 1980.

32. VB testimony, March 7, 1947, House Committee on Interstate and Foreign Commerce (LOC, VBP, 87/1912).

33. Sherry, *Shadow*, 131–32.

34. McDougall, 79–80; interview, Philip Morrison.

35. Emphasis added. VB to R. Patterson, Sept. 5, 1947 (LOC, VBP, 91); VB to H. Stimson, April 21, 1948 (LOC, VBP, 109).

VB told Stimson he especially remembered "the evenings we occasionally spent quietly, Phoebe and I with you and Mrs. Stimson, when we took a deep breath, and were a little more ready for the turmoil because of living reasonably for a moment."

VB was treated to similar expressions of nostalgia by war colleagues, especially junior members of his OSRD staff. Two years after the Japanese surrender, one wrote: "I miss the 'Bush says—' days" (C. Norcross to VB, Oct. 3, 1947, LOC, VBP, 88/1964).

36. Accounts of OSRD farewell party: VB PoA, 136; interview, John Connor, a former OSRD attorney who attended the party; Hershberg, 310; VB correspondence (LOC, VBP, 125); VB to E. Hodgins, Dec. 16, 1946 (LOC, VBP, 50/1236).

37. Wilson quoted in Lapp, *Priesthood*, 1; Balogh, 13–14; Schlesinger quotation in Dickson, 265–66.

For an exposition on the role of self-governance by expert communities in VB's politics, see Hollinger, David, "Free Enterprise and Free Inquiry: The Emergence of Laissez-Faire Communitarianism in the Ideology of Science in the United States," *New Literary History* (1990), 897–919.

Enthusiasm for the new politics of expertise peaked in the early 1960s. Social critic Daniel Bell predicted that political decisions would increasingly be made by a "New Class" of professionals whose aim was to replace ideology with expertise. And Kennedy became the first president to frankly embrace experts. As Arthur Schlesinger, Jr., wrote not long after Kennedy's assasination, "the governing attitude of the White House was that public policy is no longer a matter of ideology but of technocratic management."

The Vietnam War, the counterculture's critique of the Establishment and the rising awareness of environmental problems associated with such elite technologies as nuclear power and pesticides combined to spawn a backlash against experts in the 1970s.

38. VB repeated Forrestal's comment in a letter to B. Elliot, Dec. 15, 1947 (LOC, VBP, 36/843).

Though he exaggerated his aloofness from practical politics, VB took perverse pride in not staying current with political trends because he liked to present political activity as too unsavory for the professional man. He told Elliot: "As a scientist and engineer of course I refrain from becoming involved in the political maelstrom, and I am not even remotely connected with their affairs in the minds of the political group."

39. England, 41–43, 45–49, 78.

The National Science Foundation's official historian, Merton England, judged that the new bill, S. 1850, was "fundamentally the same" as Kilgore's 1945 bill. It called for Truman to appoint the foundation's chief, who would serve no fixed term but at the president's pleasure, and the nine-member board. The bill also retained Kilgore's geographical provision and patent language allowing for public use of inventions spawned by federal funds. Bush's only clear victory was negative: he succeeded in essentially gutting support for the social sciences.

Jewett continued to attack from the right. In the summer of 1947, he made another concerted effort to stall a foundation, prompting syndicated columnist Marquis Childs to call him a "one-man crusade" against the bill.

40. Sapolsky, 41, 44–45; "Science Dons a Uniform," *Business Week,* Sept. 14, 1946; interview, Emanuel R. Piore.

"It is doubtful scientists had ever encountered a more accommodating patron," historian Harvey Sapolsky has concluded.

41. *Newsweek,* Jan. 5, 1948, 50; *Business Week,* "Science Dons a Uniform," Sept. 14, 1946; Kleinman, 32; Sapolsky, 45.

42. *Newsweek,* Jan. 6, 1947, 44; Admiral Paul Lee to Detlev Bronk, June 5, 1948 (NARA, RG 330/341, R&D Bd., 93).

43. Wiener, N., "A Scientist Rebels," *Atlantic Monthly,* Jan. 1947; *Fortune,* "The Great Science Debate," June 1946.

Predictably, revulsion by scientists against war research rose in the aftermath of Hiroshima. Scores of scientists quit military projects, with some—like Wiener—refusing to ever again do such work. But the ardor for rebellion cooled with time, making Wiener's position—which he stuck to for the rest of his life—a lonely one.

44. Ridenour, Louis, "The Scientist Fights for Peace," *Atlantic Monthly,* May 1947.

Wiener and Ridenour squared off in a forum entitled "Should the Scientists Resist Military Intrusion?" published in the Spring 1947 issue of *The American Scholar.* In the forum, VB sided with Ridenour. Without directly rebutting the moral issue raised by Wiener, VB noted that as a practical matter even when research funding came from the military "a very substantial part of the basic research thus financed is as free of restrictions as ever research has been."

45. VB comment, *The American Scholar,* Spring 1947, 219–20; VB to J. Connor, Sept. 5, 1947 (LOC, VBP, 27/620); VB to Homer Smith, Jan. 27, 1948 (LOC, VBP, 104).

VB wrote Smith that the "amount of money flowing into universities from the Navy at the present time is not negligible, and is having a very great result indeed in my opinion."

He told Connor, a friend and former OSRD counsel, that even without the NSF "we now have as much research going on in this country as can be carried by available personnel."

46. Donovan, *Conflict,* 269–70.

Steelman began handling science matters in June 1946, at first gathering memos on atomic energy legislation from Byron Miller and James Newman, administration staffers who greatly influenced the language of the atomic legislation (see J. R. Newman to Steelman, June 26, 1946; B. S. Miller to J. Steelman, July 5, 1946, LOC, Miller Papers, 1). In early July, Steelman even began compiling a list of possible commissioners of the new Atomic Energy Commission (B. S. Miller to J. Steelman, July 8, 1946, LOC, Miller Papers, 1). Steelman tended to rely heavily on others, "delegating most of this [science work] to his staff" (interview, Elmer Staats).

47. J. Steelman to HST, Oct. 7, 1946 (Steelman personal files).

48. R. Neustadt to E. Staats, Nov. 8, 1946 (NARA, RG 51, 39.19, 105); England, 62–63.

VB complained that he had not been consulted about the decision to create the board; neither had the War or Navy departments.

49. W. D. Carley, memo on President's Science Research Board, Feb. 17, 1947 (RG 51, 39.19, 105).

50. VB OH, 465–66.

Like Smith, Steelman looked askance at "the general attitude among scientists at the time . . . they didn't want to be subjected to the usual, normal government controls," Elmer Staats, a senior budget official, later recalled. "There was an attitude that they'd be contaminated if the President had usual controls" (interview, Elmer Staats).

Steelman's refusal to view scientists as special certainly lowered his standing in VB's eyes. But even some nonscientists thought Steelman unfairly gave scientists short shrift. William Golden, a Wall Street investor and an AEC consultant who in 1950 conducted a survey of the science community for the budget bureau, said of Steelman: "He himself didn't know much, certainly not much about science" (interview, William Golden).

Staats said Steelman had the good sense "to assemble a credible staff," but he conceded that in mid-1947 "it was obvious that there was a strained relationship between the scientific community and the White House." Staats believed this state of affairs reflected, not Steelman's own agenda, but "his understanding of what he thought the President wanted" (interview, Elmer Staats).

Steelman's exchanges with VB were perfunctory. When Steelman asked for advice in preparing a new science report, VB supplied a condescending eight-page letter in which he prefaced his comments by noting that Steelman's questions were "of such magnitude and importance that a complete reply would require the writing of a treatise rather than a letter."

51. VB to Homer Smith, Jan. 27, 1948 (LOC, VBP, 104); VB to Alan Valentine, Sept. 11, 1947 (LOC, VBP, 114). VB to J. Forrestal, Sept. 10, 1947 (LOC, VBP, 59); *Science and Public Policy*, Report by the President's Scientific Research Board, Aug. 27, 1947; B. Smith, 85.

VB's abuse of *Science and Public Policy* was unwarranted, especially since the report parroted many of VB's views about the centrality of research in modern America. Like VB, Steelman extolled the virtues of basic science and highlighted its relation to national security, calling the scientist "the indispensable warrior." Also like Bush, Steelman paid little attention to industrial innovation, accepting the premise that market forces would insure that technical advances translated into economic growth. These were big areas of shared perspective, but VB parted ways with Steelman on the devilish details. Steelman proposed a National Science Foundation that was responsive to the president, not the science community. He also endorsed government spending on basic research at levels well above what VB thought prudent. *Science and Public Policy* called on the U.S. to spend $250 million annually on basic research by 1957 and suggested that overall research spending by all private and public sources in the U.S. should total at least 1 percent of the national income by that year. Steelman also called for sharper increases in spending on medical research than VB wanted. As it turned out, government spending on research and development, from military and nonmilitary sources, reached even higher levels than those advocated by Steelman.

52. Technically, Truman killed the bill through a pocket veto, since Congress had adjourned less than ten days after the bill reached his desk.

53. "Memorandum of Disapproval," Aug. 6, 1947 (HSTL, George Elsey Papers, 88); VB to J. Connor, Sept. 5, 1947 (LOC, VBP, 27/620); interview, Don Price; England, 80–82.

VB had received a warning from John Connor "that Steelman and his boys probably will advise the President to veto the Science Foundation Bill" a full month before Truman's action (J. Connor to VB, July 9, 1947, LOC, VBP, 27). VB suspected that this was what happened, but Merton England, the NSF's official historian, concluded that "whatever influence Steelman . . . may have had on Truman's decision, the advice of the Budget Bureau was probably decisive." James Webb, the new budget director, carried the torch for former director Harold Smith, a VB critic, by arguing that vesting part-time officials with "full administrative and political responsibility" and "the virtual nullification of the President's appointment power" contradicted "established notions of responsible government in a democracy." Still it bothered William Carley, a senior budget official, that Truman had not clearly indicated in advance that certain provisions in the bill would prompt a veto. He also warned that Truman's action might prompt such scientific leaders as Karl Compton and James Conant to resign from federal advisory boards, since it implied that the administration "cannot trust them to serve the public interest first." This fear proved unfounded.

In his memoir, *Year of Decisions,* Truman insisted that he had told Senator Alexander Smith, the chief backer of the bill in the Senate, that he would veto the science legislation if it infringed on presidential powers. "I think I know a little more about the Constitution than you do, Senator, and as long as I am here I am going to support it as I have sworn to do," Truman recalled saying (Truman quotation in England, 378).

Elmer Staats, a budget official, backed up Truman's recollections, writing that "the veto should have been no surprise to Bush and other supporters of the Foundation since there had been adequate warning from the White House" (letter to author, July 28, 1993).

Despite these "warnings," Truman never laid out his crucial objections to the bill during the lengthy congressional debate or publicly thereafter. His coy approach further confused an already difficult matter. Immediately after Truman's pocket veto, Senator Smith, a sponsor of the bill, advanced a more cynical explanation, charging the president with making "a political football" out of the legislation.

54. VB to Alan Valentine, Sept. 11, 1947 (LOC, VBP, 114).

Bruce Smith, a scholar at the Brookings Institution, has accorded signal importance to Truman's veto, suggesting that had the president approved the bill the resulting National Science Foundation "would have had one of the wartime scientific leaders as its director, and there would have been a board of equal distinction. The funding patterns of the mission agencies and the division of responsibilities between NSF and the rest of the government could have evolved differently. The NSF might have expanded its function more rapidly and approximated the coordinating and leadership role envisioned by Bush and Kilgore" (Smith, 50).

While such a scenario was plausible, the odds were far greater that the NSF would have played a secondary role to the services no matter when it had been formed. Perhaps rapid passage of a foundation bill in 1945 would have served to preempt the Navy's move into academic research. By 1947, the Navy was too entrenched in academia to withdraw, and all three services had gone a long way toward reassuring all but the most diehard opponents of military funding that it would not in itself corrupt research. In favoring civilian control of research, "Bush was probably right as a matter of principle," Philip Morrison, a physicist and liberal critic of VB, later said. "But it turned out that military dominance didn't have a highly negative effect on science" (author interview). Many science administrators, meanwhile, worried that a new NSF would have so little clout in Congress that it would be starved for funds; they suspected that only a military rationale for research would convince lawmakers to consistently approve large appropriations for basic science. As early as Dec. 1946, Warren Weaver, chair of the Navy's science advisory group, worried about precisely this possibility. Suggesting to a fellow naval adviser that sudden passage of a foundation bill could lead to "drastic cuts" in the ONR's budget, he painted a disastrous scenario. This "would give the whole national science foundation idea a black eye and two lame legs. It would take years to build up, again, the situation of increasing confidence which now exists between scientists and universities, on the one hand, and government on the other." Weaver's letter came after he spent two days with the ONR's chief, Admiral Lee (W. Weaver to Naval Research Advisory Committee, Dec. 20, 1946, NAS, ONR file).

55. VB to J. Connor, Sept. 5, 1947 (LOC, VBP, 27/620); Sherry, *Shadow,* 137.

VB's ability to juggle his support for both civilian and military management of research illustrated the thinness of this distinction, especially immediately after the war. The services relied increasingly on civilian managers, and the putatively independent academics increasingly fed from the military's trough. As the lines blurred, it became hard for VB to maintain the civil-military distinction that loomed so large in the early reaction to *Science—the Endless Frontier.* This dilemma was not lost on VB's critics, who frequently accused him of promoting military aims under the guise of civilian forms. But as VB himself often complained, these critics missed the point. The fights over atomic energy and science legislation were in retrospect more about style than substance. "The struggle over civilian control," historian Michael Sherry has written, "obscured how civilian elites matched the zeal of military officers in pursuing national security." VB rarely acknowledged this core reality of postwar research policy, though he acted as if the civilian-military distinction was an arbitrary one. Such a view undergirded his belief that research was too important to leave to either soldiers or scientists, but rather an amalgamation of the two classes was required.

56. Sapolsky, 34.

57. VB to K. Compton, July 7, 1947 (LOC, VBP, 26).

58. On plans for holiday: VB to F. Jewett, July 2, 1947 (LOC, VBP, 56); VB to Seeley G. Mudd, Sept. 5, 1947 (LOC, VBP, 79); Karl Compton to Lee DuBridge, Oct. 27, 1947 (Caltech, DuBridge Papers, 109.2); Cleveland Norcross to VB, Oct. 3, 1947 (LOC, VBP, 88/1964).

59. VB memo, Sept. 24, 1947 (LOC, VBP, 39/954); VB to J. Forrestal, Oct. 9, 1947 (LOC, VBP, 26).

60. D. Lilienthal, Sept. 26, 1947 (Princeton, Lilienthal Papers, 121); J. Furer to VB, Sept. 27, 1947 (LOC, VBP, 40/988); F. Hovde to VB, Sept. 29, 1947 (LOC, VBP, 52/1291).

VB had been at odds with Lilienthal in the mid-1930s when the latter ran the Tennessee Valley Authority and MIT tried unsuccessfully to win a research contract from the agency. The two moved on friendlier terms after working together on the Acheson-Lilienthal report on atomic control in early 1946. Lilienthal typified the earnest do-gooder who often irritated VB, but he seems to have found VB charming, writing in his journal on March 9, 1946, of "our obvious liking for each other" (Lilienthal, vol. 2, 28).

On the possibility of VB refusing the RDB job: Cleveland Norcross, a former OSRD staffer and now assistant director of the American Institute of Physics, wrote after VB's appointment: "I'm just one of many who felt you were the only qualified person for the job and who had no doubt that the President would ask you to serve. But many of us were concerned for fear you might decline" (C. Norcross to VB, Oct. 3, 1947, LOC, VBP, 88/1964).

61. VB to M. H. Dodge (LOC, VBP, 33/762).

62. VB to J. L. Devers, April 8, 1948 (NARA, RG 218, CDF 1948–1950, 334 RDB, 102); Price, *Government and Science,* 150–51; Rearden, 99–101; W. Weaver to R. Waterman, Sept. 12, 1948 (RAC, Weaver diaries); Baldwin, Hanson, "Defense 'Front' Develops," NYT, Jan. 23, 1950; Baldwin, Hanson, "Defense Action Urgent," NYT, Feb. 8, 1950 (Office of Secretary of Defense, History Office); Golden, William, "Reorganization of the Research and Development Board," draft, April 24, 1951 (NARA, RG 51, 47.3, E5-1 to E6.3); B. Smith, 49–50.

The RDB proved incapable of reforming, despite a series of attempts in the four years after VB's departure. It wasn't until early 1950 that it made what Hanson Baldwin called "fairly extensive changes in the board's informal method of doing business." The effort followed defense legislation that gave more authority to the RDB's chairman. The reorganization, however, proved too limited, and a year later an internal study of the RDB found that the chairman needed more power still; notably, the analyst advised stripping board members of their votes. This helped briefly, but problems resurfaced. The board was finally abolished in 1953 and replaced by an assistant secretary of defense for research. This structure basically persists to this day.

63. Rearden, 98–99; "Directive: RDB," Dec. 18, 1947 (Office of Secretary of Defense, History Office).

64. Hoopes and Brinkley, 362; Rearden 392–93; VB to J. Ohly, April 30, 1948 (HSTL, Ohly Papers); VB to J. Forrestal, Oct. 13, 1948 (NAS, courtesy Allan Needell); VB testimony in McDougall, 97–99.

65. Donovan, *Tumultuous Years,* 53; Leffler, 202–6; Hoopes and Brinkley, 370–72; Kofsky, 103–19; Steel, 450–52.

The war scare was real enough that Forrestal and Air Secretary Symington suggested the U.S. might have to launch a preventive attack against Russia, striking its key cities with atomic bombs.

66. Quesada, Elwood, "How to Make Peace at the Pentagon," *Colliers,* Aug. 3, 1956; VB to E.R., Sept. 29, 1952 (LOC, VBP, 99).

67. Carter, 230–32.

The story, "Project Hush" by William Tenn, was published in 1954.

68. Atomic weapons offered a prime example of this. The services were still wrestling with the implications of the bomb. Civilians, meanwhile, found the bomb so disquieting that they refused to give physical custody of the country's stockpile to the military. Forrestal endorsed the military's call for possession of the weapons in the event "instant use" was required. But the misuse of atomic bombs was precisely what worried civilians. In July, Truman ruled out the transfer of bombs to the military. The president privately told Forrestal that "political considerations" made any transfer impossible until after the November election (Hoopes and Brinkley, 378–80).

69. "Report on Weapons Systems Evaluation," May 11, 1948 (NARA, RG 330/341, RDB, 29/150).

70. "Report of the Scientific Advisors," Dec. 20, 1947 (NARA, RG 218, CDF 1958–50, 334 RDB, 102).

71. Rearden, 401–2; J. Forrestal to JCS, Feb. 4, 1948 (NARA, RG 218, JCS, 335.14, 6-6-42, 104).

In advocating VB's idea, Forrestal informed the Joint Chiefs: "There remains a need for a centrally located, impartial and highly qualified group, which from a technical standpoint can objectively analyze each component program, and examine the programs of each department in their relationships to the programs of the others."

72. VB distilled his motives for backing the weapons evaluation group in an Aug. 5, 1948, memo to Forrestal (NARA, RG 330/341, RDB, 93).

73. Fairchild, Muir S., "JCS Meeting on Establishment of Weapons Systems Evaluation Group," July 16, 1948 (LOC, Fairchild Papers, 3).

The mastermind behind this clever plot was Muir S. Fairchild, Air Force vice-chief of staff. Fairchild hatched the plan when, during a meeting with top officers, Forrestal stepped out of the room for a few moments. Fairchild took the opportunity to tell the generals present, "Regardless of what action is taken, a necessary part of it must be the setting up of an Evaluation Group with the JCS. If the original directive, or any modification thereof, should go to Dr. Bush, the Joint Chiefs will still need a scientific Evaluation Group to check and advise them directly."

74. VB to J. Forrestal, Aug. 5, 1948 (NARA, RG 330/341, RDB, 93).

Forrestal officially ordered the establishment of the Weapons Systems Evaluation Group on Dec. 11, 1948. He considered his order "among the most important taken since the passage of the National Security Act."

Anticipating the order, VB wrote Shockley, one of his consultants, in late September, saying that the weapons group should "become generally accepted as an important adjunct to planning" and ultimately "be taken under the wing of the [JCS] where it very properly belongs as a permanent effort" (VB to W. Shockley, Sept. 28, 1948, NARA, RG 330/341, RDB, 93). In his July explosion to Forrestal, however, VB suggested it could be a long time indeed before the JCS could handle weapons evaluation, declaring, "I am as convinced as I ever was that JCS cannot now adequately handle the whole affair. Ultimately I hope they can and will. But I would like to see them demonstrate determination and capability to use such an analytical group effectively before trusting its entire future in their hands."

75. VB to J. Forrestal, July 23, 1948 (NARA, RG 330/341, RDB, 93); VB to F. Jewett, Aug. 3, 1948 (CIW, VB correspondence); VB OH, 805.

VB was so unhappy with the RDB that practically his last act before setting off on a holiday in mid-August was to ask Irvin Stewart, who had been the administrative backbone of OSRD, to chair a committee to study the desirability of reviving the wartime agency. This was either an act of nostalgia or poor judgment on Bush's part; the clock could not be turned back. None of the services wanted to lose control of their ability to innovate advanced weapons. Curiously, VB claimed that his action was prompted by a request from a naval officer who wondered what role the RDB would play in the event of a shooting war. Stewart's committee recommended that the secretary of defense have a science adviser who would also have contact with the president (VB to I. Stewart, Aug. 13, 1948, NARA, RG 330/341, RDB, 78).

76. Interview, Richard Bush; R. Bush to VB, Feb. 2, 1948 (LOC, VBP, 123/414); interview, Richard Bush; Killian, *Education*, 90; VB to P. Scherer, July 13, 1948 (LOC, VBP, 21); VB to K. Compton, Oct. 28, 1948 (LOC, VBP, 26); VB OH, 182.

The only silver lining in VB's condition was that his eyesight improved with the purchase of new glasses.

77. M. Tuve OH, 39 (American Institute of Physics, May 1967).

78. Stiles, C. S., and Leffler, M. P., "James Forrestal: The Tragic End of a Successful Entrepreneur," *Leadership and Innovation*, Doig, J., and Hargrove, E. C., eds. (1987), 369; *The Nation*, "Ulcers and History," June 18, 1949.

No one else claimed to suffer from a brain tumor, as did VB, but so many government officials had ulcers that George Marshall, the former secretary of state, bemoaned the situation, saying at an Overseas Press Club dinner in 1949, "Ulcers have had a strange effect upon the history of our times. In Washington, I had to contend with, among other things, the ulcers of Bedell Smith in Moscow and the ulcers of Bob Lovett and Dean Acheson in Washington." It was Forrestal's suicide, however, that caused the most concern. *The Nation* article ran within a month of Forrestal's "tragic death," saying it had "brought under scrutiny, with some dramatic urgency, the problem of ailing statesmen in office." VB's illness was not reported in the media at the time.

79. On VB's dislike for hospital and doctors: interviews, Richard and Catherine Bush. On his fears of mental incapacitation: interview, Caryl Haskins.

80. Interview, Caryl Haskins.

Haskins recalled being invited to Montana on short notice by Bush, but Bush had thought about making a trip to Montana in the last two weeks of August at least since late May (VB to A. Roe, May 18, 1948, LOC, VBP, 124).

Chapter Fifteen: "The grim world"

Epigraph: VB, "Organization for Strength," Oct. 11, 1952 (CIW, VBP).

1. VB to E. Hodgins, Sept. 7, 1948 (LOC, VBP, 141); VB to J. Forrestal, Sept. 8, 1948 (NARA, RG 218, CDF 1948–1950, 334 RDB, 102).

In his letter to Forrestal, VB complained that the armed services had held a series of meetings about their research budgets without informing the RDB staff. Since the RDB had been authorized to review those budgets, VB considered this an underhanded practice—all the more so because it took place while he was out of Washington. Afraid that the RDB's reputation had sunk within the military establishment, he warned Forrestal that the services were encouraging a "conviction among military personnel generally that the affairs of the [RDB] are controlled by the" Joint Chiefs. Such an understanding is "clearly contrary to the intent of Congress, and your own directives," he said. "It is essential that the Joint Chiefs lean over backwards in order to avoid such an impression; that they emphasize at every opportunity that JCS and RDB are coequal bodies."

Though VB's impending resignation lent a pathetic quality to his objections, the military responded anyway, losing no occasion to put VB in his place.

2. *Newsweek,* Sept. 13, 1948.

3. VB to R. Loeb, Sept. 19, 1948; VB to R. Loeb, Sept. 21, 1948 (LOC, VBP, 66); VB to A. N. Richards, Sept. 29, 1948 (LOC, VBP, 97).

4. NYT, "Dr. Bush Resigns, Karl Compton to Head U.S. Military Research," Oct. 6, 1948; *Washington Post,* "Dr. Bush Quits as Military Research Head," Oct. 6, 1948; VB to R. Loeb, Oct. 6, 1948 (LOC, VBP, 66/1590).

The *Post* noted that VB "has not been in good health for some time," but did not cite this as the reason for his departure. Less than two weeks after Compton took the RDB post, the NYT published a lengthy profile of the new "top man in American science" (Oct. 17, 1948).

5. "Chief of U.S. Brain Power," *American Magazine,* June 1948.

6. Newspaper accounts of VB's resignation (LOC, VBP, 96/2213); "Chief of U.S. Brain Power," *American Magazine,* June 1948.

7. Among the press, only a trade publication, *Chemical and Engineering News,* even reported the tip of the iceberg. In its issue of Oct. 18, 1948, CEN reported that Bush departed because of "opposition to his programs from other government agencies" (LOC, VBP, 96/2213).

Weaver, who held out the highest hopes for VB in the field of public service, was possibly the most surprised by his fall. A few days after VB's return from Montana, Weaver learned of his intention to leave the government. Seeing VB as a victim of the vicious bureaucratic politics that would increasingly dominate the Washington scene, Weaver spent a few hours discussing his plight with the research chief of the Office of Naval Research. In his diary, Weaver deftly boiled down the story line to a few words: "Bush very unpopular with the services. Headaches and resignation."

8. Interservice rivalries were so intense that they warped civil-military relations well into the 1950s. For one contemporary report, see "How to Make Peace at the Pentagon" (*Colliers,* Aug. 3, 1956). The author, retired Air Force general Elwood "Pete" Quesada, noted that enmity between the services made it too risky to adopt the most effective approach to defense: a single service and single command structure. "For their own special reasons, the Army, Navy and Air Force would resist it to the death, and the Marines perhaps most bitterly of all," he concluded. "As a practical matter, the turmoil which such a revolutionary change would create would weaken our defenses, perhaps for years."

9. Reingold, 285.

10. Interviews, Richard Bush, John Connor; VB to K. Compton, Oct. 28, 1948 (LOC, VBP, 26); VB to Laurence Marshall, Oct. 26, 1948, (LOC, VBP, 69/1701); VB to D. Lilienthal, Oct. 25, 1948 (Princeton, Lilienthal Papers, 129); VB to J. Victory, Oct. 22, 1948, found in J. Victory to J. Hunsaker, Oct. 25, 1948 (NASA Archives, VB files); VB OH, 181–83.

VB told Compton that he was "sure" his persistent insomnia would "disappear as soon as the present turmoil gets out of my head."

On how VB's original physician could mistake his condition for a brain tumor, Loeb surmised that "little scars . . . [of] no real significance" in VB's brain—caused by a childhood fever or other youthful illnesses—showed up on his X-rays and were "hard to distinguish from a real brain tumor" (VB OH, 182).

11. VB to J. Conant, Oct. 20, 1945 (CIW, VB correspondence); VB, *Modern Arms,* 2.

VB dedicated the book to Henry Stimson, the former war secretary who was his mentor during the war.

12. VB, "Scientific Weapons and a Future War," *Life,* Nov. 14, 1949; "Our Moral Armor: A Famous Man of Science Explodes the Marxist Myth of 'Scientific Materialism,'" *Life,* Dec. 5, 1949; *Scientific American,* Jan. 1950.

13. *Washington Evening Star,* Dec. 24, 1949; *Yale Review,* Spring 1950 (LOC, VBP, 29/654); NYT, Nov. 20, 1949; *Washington Post,* Nov. 25, 1949; *Bulletin of Atomic Scientists,* 1950 (typescript copy, LOC, W. S. Parson Papers, 1).

The film documentary of VB's book was released under the ominous title "If Moscow Strikes," in May 1952.

Perhaps as a reflection of VB's image in the late 1940s as chiefly a military mind, the *Times* assigned Hanson Baldwin, its defense correspondent, to review *Modern Arms.* He called VB's atomic chapter "dated" and found it impossible to justify VB's statement that "we have more breathing time than we once thought."

14. Moley quotation in Boyer, 336; Hewlett and Duncan, 362–69; Burrows, 61–62.

15. VB to D. Lilienthal, Sept. 26, 1949 (Princeton, Lilienthal Papers, 142).

In *Modern Arms,* VB bows to his critics, conceding: "I am not much of a prophet; there is a great deal of guesswork inevitably involved when we attempt to predict just what applied science may still do to our lives" (2).

16. Forrestal Diary, Sept. 21, 1945 (Princeton, Forrestal Papers); Holloway, 115, 220–22; Rhodes, 770; O'Neill, *American High,* 228.

German-born Fuchs, a naturalized British citizen, gave secret information to the Russians from 1942 to 1949. Among other things, he gave Stalin an advance warning of the first U.S. test in 1945 and the news that the bomb would be used against Japan if it worked. Fuchs was arrested in January 1950, imprisoned for nine years in Britain and then released to East Germany, where he died in 1988.

17. VB, *Modern Arms,* 2–3.

18. On VB's desire to "take important matters 'out of politics,'" see Appleby, Paul, "Civilian Control of a Department of National Defense," in Kerwin, Jerome G., ed., *Civil-Military Relationships in American Life* (1948), 66; "Bush Cites Danger in U.S. 'Handouts,'" NYT, Dec. 6, 1949; Diggins, 101–9; Matusow, Allen J., *Farm Policies & Politics in the Truman Years* (1967), 170, 185; Zieger, Robert, *American Workers, American Unions* (1994), 104.

In turbulent 1946, strikes involved about 4.5 million workers.

19. VB to L. Cutler, Oct. 4, 1949 (LOC, VBP, 30/677).

20. Diggins, 100–101; "Bush Cites Danger in U.S. 'Handouts,'" NYT, Dec. 6, 1949.

Russia's success in building an atomic bomb shook VB's confidence in the natural superiority of democracy in technological affairs. Ideologically, he equated freedom with the spirit of scientific inquiry. The experience of World War II bolstered his belief. After the war, he crowed often and loudly that the U.S. victory showed the advantage of freeing scientists and engineers from political interference. He blamed Germany's failure to create an atomic bomb, for instance, on the Reich's unwillingness to give its scientists and engineers the latitude enjoyed by U.S. scientists. "All other things being equal a democracy can outclass any despotism in bringing to bear on the struggle the combined efforts of science, industry, and military might," Bush said in a speech delivered on October 27, 1945. "Here, then, lies the profound significance of the renewed faith in our governmental philosophy which our wartime experience has given us. We have seen written in sharp clarity on the record of history the fact that free men, free groups, free institutions can merge their wills and their energies and by so doing can remain free and strong as they beat down brutality and aggression. That same spirit of mutual support will sustain our hands in the great task ahead." Significantly, Bush extolled the willingness of "free men" and "free institutions" to meet great challenges, playing down the role of the federal government in cajoling, directing and even coercing the citizenry into effective action (LOC, VBP, 99).

21. VB, *Modern Arms,* 224, 259.

22. Lasch, Christopher, *The Revolt of the Elites* (1995), 74–77, 84, 168–69; VB, *Modern Arms,* 222, 224; Sherry, *Shadow,* 167.

23. "Natural Aristocracy Asked by Dr. Bush," Feb. 20, 1955, *Washington Post.*

24. Carter, 152–53; Balogh, 18; VB, *Modern Arms,* 223.

25. Redmond, Louis, "The Face of Greatness," *Coronet,* April 1952.

26. Diggins, 119; Donovan, *Tumultuous Years,* 243–47.

27. VB, "A Few Quick," Nov. 5, 1951 (LOC, VBP, 139); VB to O. Bradley, Nov. 5, 1951; O. Bradley to VB, Nov. 6, 1951 (LOC, VBP, 14); Zeitlin, Jonathan, "Flexibility and Mass Production at War: Aircraft Manufacture in Britain, the United States and Germany, 1939–1945," *Technology and Culture,* Jan. l995, 49.

In advance of circulating the "few quick" essay, VB alerted General Bradley that his aim was not to publicize his views but to get "results." He wondered if "the article as an internal document could really produce more genuine effect than if it appeared publicly," adding "perhaps instead of putting this article before the public it ought to be put before the Defense Establishment." Bradley sent VB an encouraging note—"You have certainly given us much good food for thought in your article"—and VB kept the paper within the military family.

28. "Discussion of Dr. Vannevar Bush's Paper, 'A Few Quick,'" F. W. Barnes to Air Force Secretary, Dec. 19, 1951 (NARA, RG 307, 17, SAC, 1951); Duffield, Eugene S., "Organizing for Defense," *Harvard Business Review,* Sept.–Oct. 1953.

Even three years after VB first circulated his essay, it was still being read. "I, among many, have always endorsed the principles of 'a few quick,'" an Army official wrote VB in May 1954. "Whereas the resistance of existing Defense-wide organizations tends to defeat its application, perhaps these [few quick] principles can be put in practice [by the Army] to support your experimental needs" (VB to J. Davis, May 27, 1954, LOC, VBP, 31/701).

29. Lewis, Eugene, "Admiral Hyman Rickover: Technological Entrepreneurship in the U.S. Navy," in *Leadership and Innovation* (1987), Doig, J., and Hargrove, E., eds., 96–123; Hughes, 426–35; Hewlett and Duncan, 189–93, 219–20.

Rickover's achievement seemed most notable for how he succeeded in spite of administrative conflict and confusion. But this success came because his technical judgments were sound. No manner of public entrepreneur would long succeed if his technological judgments were flawed. This was the ultimate source of VB's authority during World War II. It was Rickover's during the 1950s. His subs performed sensationally. By contrast, a concurrent effort to build a nuclear-powered airplane foundered for many technical reasons, notably the difficulty of shielding the plane's crew from lethal radiation. VB doubted the value of the "Nuclear Energy Propulsion for Aircraft" project from its inception in 1946. In this case his doubts proved correct. Despite lobbying from the Air Force, he never changed his view (Hewlett and Duncan, 71–74, 106–7, 120–21, 189, 419–20; Whittemore, Gilbert, "A Crystal Ball in the Shadows of Nuremberg and Hiroshima: The Ethical Debate Over Human Experimentation to Develop a Nuclear-Powered Bomber," *Science, Technology and the Military* [1988], E. Mendelsohn, M. R. Smith, P. Weingart, eds.).

30. VB, "Planning," Sept. 26, 1952 (CIW, VB speeches).

31. Lippmann, "Military Setup," Sept 29, 1952; *Washington Post,* Sept. 30, 1952; *Christian Science Monitor,* Sept. 27, 1952; *Minneapolis Star,* Sept. 27, 1952 (CIW, VBP).

32. VB to L. DuBridge, Sept. 29, 1952 (LOC, VBP); VB to J. Killian, Oct. 8, 1952 (MIT, AC4, Office of the President, 1930–58, 42).

Though publicly VB disavowed any desire to personally embarrass military chiefs, privately he repeated the phrase "my private war" when telling friends of his activities. The phrase left the impression that VB sought a measure of revenge and vindication for past transgressions. He wrote to Edwin Cox on Oct. 9, 1952, "It will probably not surprise you in the slightest degree that I have been renewing my private war" with the JCS (LOC, VBP, 29/653).

33. VB, Tufts speech, "Organization for Strength," Oct. 11, 1952 (CIW, VB speeches); *New York Herald-Tribune,* Nov. 5, 1952 (CIW, VB newsclippings).

34. Divine, *Blowing,* 14–15; Holloway, 294–303; Donovan, *Tumultuous Years,* 148–57; O'Neill, *American High,* 226–28; Acheson, 345–49.

The first Russian A-bomb was a copy of the first U.S. bomb (courtesy of Fuchs), but its first H-bomb embodied an original design. David Holloway, the preeminent scholar on Soviet nuclear weapons, has written that the Russian decision "occasioned none of the soul-searching that took place in the United States."

35. Interviews, McGeorge Bundy, Richard Bush; VB testimony, April 23, 1954, "In the Matter of J. Robert Oppenheimer," Transcript of Hearing Before Personnel Security Board of the Atomic Energy Commission, 562–63; Truman, *Years of Trial,* 312–14; Divine, *Blowing,* 16, York, Herbert F., *The Advisors: Oppenheimer, Teller, and the Super Bomb* (1989), 82–83; Hewlett and Duncan, 590–93; Donovan, *Tumultuous Years,* 157; Minutes, Panel of Consultants, May 16–18, 1952 (LOC, R. Oppenheimer Papers, 191); Panel of Consultants, "The Timing Of The Thermonuclear Test," Department of State, Foreign Relations of the United States, 1952–54, vol. 2, 994–1008; Bernstein, Barton J., "A Missed Opportunity to Stop the H-Bomb?" *International Security,* Fall 1989, 132–60; Hirsch, Daniel, and Mathews, William G., "The H-Bomb: Who Really Gave Away the Secret," *Bulletin of Atomic Scientists,* Jan—Feb. 1990.

On the extent to which fallout from the Mike test aided H-bomb programs in other countries, scholars differ. David Holloway, in a conversation with the author, said VB's "fear was not realized," because Russian scientists simply did not need to analyze "specific information and pointers" for their H-bomb and did not do so. But VB was correct in assuming that the data existed for the Russians and others to analyze.

On VB's relations to Truman in the 1950s, see William Golden's Oct. 24, 1950, interview with VB: "Though President Truman is very cordial to him [VB] he does not call upon him for advice, though Dr. Bush has pointed this out to him [Truman] on several occasions. He [VB] feels this is not because of any personal dislike but rather because President Truman just doesn't operate in this way—the contrast between President Truman and President Roosevelt is very strong in this respect. It is evident that Dr. Bush, who had a very close working relationship with President Roosevelt, does not approve of the present state of affairs" (HSTL, Golden Papers).

36. VB to M. Bundy, Nov. 3, 1952 (LOC, VBP, 108); *Washington Post,* Nov. 24, 1952 (CIW, VBP, newsclips).

37. VB to J. Conant, March 29, 1954 (LOC, VBP, 27).

38. VB testimony, April 23, 1954, "In the Matter of J. Robert Oppenheimer," Transcript of Hearing Before Personnel Security Board of the Atomic Energy Commission, 562.

Once they understood the terrible power of the H-bomb, many people wished that the U.S. and Soviets would halt development of this weapon. What made VB's complaints about the H-bomb more striking was that he did not favor general nuclear disarmament and even imagined various scenarios in which the U.S. would be forced into using nuclear weapons. "I would be exceedingly reluctant to use A-bombs in a minor war," but he would not rule out their use, he wrote in mid-1954 (VB to John C. Slessor, July 28, 1954, LOC, VBP, 104).

39. VB to T. K. Glennan, Nov. 14, 1952 (LOC, VBP, 43); VB to K. Compton, Nov. 21, 1952 (LOC, VBP, 26).

40. R. Lovett to HST, Nov. 18, 1952 (OSD HO) Condit, Dorothy, *History of the Office of the Secretary of Defense: The Test of War, 1950–1953* (1988), 525–31; "Wilson Appoints 7 For 'Streamlining,'" NYT, Feb. 20, 1953; R. Johnson to N. Rockefeller (Rockefeller Family Archives, RG4, Reorganization Advisory Committee, 49/438).

On VB's invitation to serve: VB to P. Scherer, Feb. 18, 1953 (LOC, VBP, 22). On overlap between VB's and Lovett's defense views: "The Task Before Wilson," Alsop, J., and Alsop, S., *Washington Post,* Jan. 18, 1953. The Alsops concluded that Lovett, Bush and others "who have studied this problem seriously have all reached approximately the same conclusion."

41. Report of the Rockefeller Committee on Department of Defense (OSD HO) Reorganization Advisory Committee, April 11, 1954; Eisenhower message to Congress, April 30, 1953 (OSD HO).

Joining VB on the committee were Lovett, General Omar Bradley, the president's brother Milton Eisenhower, Arthur Flemming and David Sarnoff.

42. VB to Don K. Price, April 6, 1953 (Rockefeller Family Archives, RG 4, Reorganization Advisory Committee, 49/438); Duffield, Eugene S., "Organizing for Defense," *Harvard Business Review*, Sept.–Oct. 1953.

43. VB to C. Dollard, Nov. 12, 1952 (LOC, VBP, 34/767).

44. VB to J. Conant, June 17, 1954 (LOC, VBP, 27); VB to N. Rockefeller, May 26, 1954 (LOC, VBP).

In his complaint to Rockefeller, VB also noted: "The President does not yet understand that military planning at the present time is lacking in certain essential respects, and it is hardly possible he will find that out through his normal contacts. In fact the lack of appreciation by the President of the subtlety of some of the things that are now going on disturbs me more than I could tell you. . . . I wish I could discover that I am wrong in this opinion, but I fear that he does not read the essential things."

45. Watson, Robert J., *The Joint Chiefs of Staff and National Policy, 1953–1954* (1986), 188–89.

46. Holloway, 306–7; Ambrose, *Eisenhower: The President,* 131–34.

The Russian H-bomb test, coming within a year of the U.S. test, disabused VB of any lingering notions he had about Russian scientific capabilities. In an article headlined "Dr. Bush Feels This Way Now," VB said, "We have now had some pretty convincing demonstrations of the success of Russian applied science. . . . Despite all the handicaps [of a Communist system] they've built atom and hydrogen bombs, and some pretty doggone good jet planes. After all, if the full effort of Russian technology is devoted to military goals, you can expect results" (*Newsweek,* March 1, 1954, 46).

47. Ambrose, *Eisenhower: The President,* 135; "At 80, Scientist Bush Looks Back at Eventful Years," *Boston Globe,* undated, probably March 1970 (Boston Globe library); Eisenhower message to Congress, April 30, 1953 (OSD HO).

48. Sherry, *Shadow,* 203–5; Diggins, 139–41.

49. Gaddis, J. L., *Strategies of Containment* (1982), 149.

50. VB to A. Dulles, Feb. 16, 1954 (LOC, VBP, 35/789).

51. VB to J. Connor, Feb. 24, 1954 (LOC, VBP, 27); "Dr. Bush Feels This Way Now," March 1, 1954, *Newsweek,* 46; Lee Anna Embrey attachment to VB's notes on NSF talk, Nov. 23, 1953 (CIW, VBP); "This Is Prejudice: Dr. Bush Calls Gradualism Best Way of Erasing It," *Baltimore Sun,* Oct. 22, 1954.

VB accepted segregated schools, provided equal resources were given to black and white institutions. "I think we have every reason to suppose that segregation with equal facilities according to race is—for the present—the surest way to promote the continued harmonious progress toward the sound education of both races," he told a Baltimore newspaper in Oct. 1954. He added that he thought the South was making "excellent progress" in improving race relations and applauded southerners for "doing it the slow way."

VB's contradictory ideas about scientific manpower befuddled an aide, who wrote: "Dr. Bush also expressed the belief that in spite of all the talk about shortages of scientists and engineers, there is some danger that we may overproduce engineers and scientists. He pointed out that it had taken a long time for industry to appreciate the potentialities of research and development, but now that industry understands these processes can help it make more money, it tends to overemphasize [R&D]; there is a tendency to rush into research because it pays" (CIW, VBP).

By 1961, VB realized he had mistakenly predicted a glut of engineers and scientists. In pursuit of new talent, he told educators, "We cannot afford to leave a stone unturned. We need to seek out, wherever they may be located, youngsters of appropriate talents and see to it that they receive all the education they can usefully absorb." Yet this effort apparently still did not apply to African-American students (VB speech at MIT, "Education, Wisdom and Happiness," April 10, 1961, MIT, VBP, 19).

For the record at least, VB never revised his views on segregated schools or his presumption that blacks had inferior intelligence. In 1963, he told a journalist: "Many [Negroes] are children in their attitudes, and I believe that burning intellectual ambition is comparatively rare among them, although there are notable exceptions" (*Baltimore Sun,* Sept. 8, 1963).

52. England, 109; Truman statement, May 10, 1950 (HSTL, Public Papers of the President); Hershberg, 560 ("another contestant"), 577 (Conant quotation).

53. Interview, Emanuel R. Piore; Waterman diary, Oct. 21, 1952 (LOC, Waterman Papers, 1); VB to A. Waterman, March 24, 1954 (LOC, VBP, 117); England, 211–12, 347–51; VB speech on Waterman retirement, June 21, 1963 (CIW, VBP).

Three full years into its life, the NSF had a budget of just $8 million. In the mid-1950s, NSF funding rose sharply. For the 1957 fiscal year, the agency's budget totaled $40 million, just a tiny fraction of the budget for defense research.

VB's words to Waterman closed as follows: "We respect you for the devotion, integrity, and wisdom with which you have carried out a great undertaking over the years. You have rendered the name of scientist in government halls a name of honor and worthiness. You have accomplished a thing which is rare: you have molded the course of science in this country in a salutary manner for many years, and at the same time made yourself a host of devoted friends. We salute you and wish you well. As long as this country has men of your caliber in its service, we need fear no rocks."

It wasn't until the late 1960s that the NSF began realizing some of its original potential. With the backlash against military funding of academic research at that time, the NSF emerged as a credible alternative for government support. But as a sponsor of visionary science and engineering projects, the Pentagon's Advanced Research Projects Agency far surpassed it. By the 1990s, the NSF was in the forefront of a broad range of technologies, though its biggest contributions tended to come in the practical fields of engineering and electronics rather than in "pure" research.

54. VB notes for NSF talk, Nov. 23, 1953 (CIW, VBP).

55. VB to R. Patterson, March 24, 1947 (LOC, VBP, 91/2050).

VB's belief in the integrity of capitalist critics withstood the ravages of the Red Scare and even the rise of the New Left in the 1960s. "I think a man can have Communist sympathies and be an utterly honest man," he said in 1967 (VB OH, Columbia).

56. Wang, Jessica, "Science, Security, and the Cold War: The Case of E. U. Condon," *Isis,* 83 (1992), 238–69.

57. Hershberg, 420.

Though VB never flatly said he disliked Condon, he gave that impression, at least to Condon, who felt VB was irritated by his appointment (by Commerce Secretary Henry Wallace) to the top post at the National Bureau of Standards. "Bush never welcomed me to Washington, or acted as if he knew I existed," Condon later said (AIP, Condon OH, 185, 1973).

58. A. N. Richards, "Conversation With Dr. Bush," March 11, 1948; A. N. Richards to VB, March 16, 1948; VB to A. N. Richards, March 20, 1948; VB to A. N. Richards, March 25, 1948 (NAS, Congress: 1948, Committee: Un-American Activities, Condon Case).

59. Gellhorn, 56–57; Nichols to Tolson, Dec. 8, 1949 (FBI, VB file); VB to J. Conant, May 31, 1949 (LOC, VBP, 145).

The bureau apparently never asked VB to give a graduation speech.

60. SAC to director, June 5, 1948; memo on VB, Dec. 20, 1949 (FBI, VB file).

The FBI could not even confirm that Bush had ever received an invitation to the forum, no less that he wished to attend.

61. David, Caute, *The Great Fear* (1978), 472.

62. Hershberg, 316–19.

63. Hershberg, 677; L. Garrison to J. Conant, March 12, 1954 (Harvard, Pusey Library, Conant Papers, Special Subject File, 8).

Hershberg, Conant's biographer, has written that VB told Conant about Oppenheimer's situation "during an automobile ride—perhaps to evade surveillance."

64. VB to L. Strauss, April 19, 1954 (LOC, VBP, 109).

65. Ambrose, *Eisenhower: The President,* 166–67.

66. Major, John, *The Oppenheimer Hearing* (1971), 9, 12.

67. VB testimony, April 23, 1954, "In the Matter of J. Robert Oppenheimer," Transcript of Hearing Before Personnel Security Board of the Atomic Energy Commission, 560–67.

68. VB to L. Strauss, April 28, 1954 (LOC, VBP, 109); VB to K. Darrow, April 30, 1954 (LOC, VBP, 30).

69. Nichols to Jones, June 28, 1954 (FBI, VB file); VB notes to St. Botolph Club talk, undated June 1954 (CIW, VBP).

In response to the FBI incident, the bureau decided that VB "is not to be contacted on any occasion in the future unless prior Bureau approval has been obtained."

70 VB, "If We Alienate Our Scientists," June 13, 1954, NYT, cited in Major, John, *The Oppenheimer Hearing* (1971), 190–91; VB to D. Lilienthal, June 14, 1954 (Princeton, Lilienthal Papers, 377).

71. Hodgson, 72; VB to J. Conant, June 17, 1954 (LOC, VBP, 27).

A year later, VB was still railing against excessive security precautions, which struck him as absurd given the open flow of information in U.S. society. "We have an obsession in this country on secrecy," he wrote T. K. Glennan on April 7, 1955. "We go to great extremes and do a lot of damage trying to prevent leaks. At the same time we publish about everything the Russians want to know anyway" (LOC, VBP, 43).

72. "The Oppenheimer Case: 'Grave Danger,'" and "The Atom: The Door Slams Shut," *Newsweek,* July 12, 1954, 24.

73. *Washington Post,* July 25, 1954.

Chapter Sixteen: "Crying in the wilderness"

Epigraph: VB, "Science Pauses," *Fortune,* May 1965.

1. VB to J. Gardner, Feb. 14, 1955 (LOC, VBP, 41).

2. "Vannevar Bush Accuses," Jan. 10, 1955, *Los Angeles Times.*

3. L. DuBridge, Nov. 7, 1955 (Caltech, DuBridge Papers, 108.9).

4. Nelson Rockefeller, Dec. 10, 1955 (Rockefeller Family Archives, RG 4, Projects, 19/198); VB to E. Root, Jr., Sept. 28, 1955 (LOC, VBP, 99).

5. For an account of VB's assets, see his correspondence with The New England Trust Co., especially M. Standish to VB, Aug. 8, 1950, and VB to M. Standish, Sept. 28, 1955 (LOC, VBP, 124).

6. VB to M. Tuve, April 3, 1956 (MIT, VBP); on Millipore, interview, John Bush; VB to T. Brown, Feb. 24, 1954 (LOC, VBP, 124); VB to T. Brown, May 4, 1954 (LOC, VBP, 124); VB to J. Bush, June 17, 1954 (LOC, VBP, 123); Millipore company history (Millipore Corp.).

VB wrote to Tuve: "I miss Washington too. And it is not that I miss . . . the old office and all of its conveniences. It is rather that I miss a few of my good friends very much."

7. On the engine: VB to J. Bell, Sept. 21, 1954; VB to J. Bell, Dec. 2, 1954; VB to J. Bell, Dec. 13, 1954 (LOC, VBP, 11); "Bush Redesigns Free-Piston System: This Could Revolutionize Auto," *Boston Herald,* Jan. 31, 1965 (Boston Globe library); POA, 211–12. On hydrofoils: VB to G. Cabot, March 14, 1949 (LOC, VBP, 18); "Panel on Hydrofoil Ships," April 21, 1950, Research and Development Board (NARA, RG 330/341, RDB, 41/270); VB to W.

Holmes, Jr., March 3, 1950 (LOC, VBP, 51/1252); VB to Charles Lindbergh, Aug. 25, 1953 (LOC, VBP, 65/1572); VB to Nelson Rockefeller, Sept. 4, 1953 (Rockefeller Family Archives, RG 4, Projects, 19/198); VB to J. McCone, Jan. 17, 1957 (MIT, VBP); C. Glover III to VB, "The Hydrofoil Corporation," Feb. 13, 1958; VB to C. Glover III, Feb. 27, 1958 (MIT, VBP); VB POA, 226–28.

VB pressed the Navy to support research on hydrofoils starting in the late 1940s. In Dec. 1949, VB formed the Hydrofoil Corporation with a CIW aide, Paul Scherer. Oscar Cox, the Washington attorney who aided VB throughout World War II, served as the first president and chief executive of the company. VB contributed his designs, and 11 investors invested $44,000. While mainly friends, investors included Juan Trippe, the aviation industry pioneer.

In 1950 the Research and Development Board wrote a favorable survey of hydrofoils and found that the Navy was spending about $150,000 a year researching the field. In 1953, VB's optimism peaked; he excitedly wrote Charles Lindbergh that his hydrofoil project "is a very interesting affair" and offered to show him his prototype boat in action. By 1957 VB had grown discouraged by his lack of progress and the company was essentially moribund. But he still hoped "even if the whole subject does not prosper . . . the patents will at least be worth something significant."

A decade later, a 78-year-old VB remained obsessed with the hydrofoil and displayed the classic symptoms of the unstoppable inventor, ever willing to learn from experience and begin anew even as he faced the failing of his physical powers. "I am wondering just what to do about the craft at the Cape," he wrote to Scherer in 1968. "Last summer was a disappointment since I got no tests. The reason is clear enough. I had hired a couple of MIT boys, and they turned out to be far from what I expected. I wonder, even if I got good tests would they mean much? You know the reason I built the craft, for I had some ideas which seemed worthwhile when I started. But they do not look the same now, for the whole art has changed. I suppose common sense would indicate that I just plain forget the matter. It is not a matter of money. I spent quite a lot on the craft, but I am perfectly willing to charge that up to experience. This winter I have rebuilt some parts, which incidentally the boys wrecked. I suppose it is pride or stubbornness that has kept me at it. But I cannot, today, work at it myself, or make the tests myself, so it is a different game with much of the fun gone" (VB to P. Scherer, March 1, 1968, MIT, VBP, 6).

Yet VB stayed a believer in hydrofoils, writing to L. Hafstad August 22, 1972: "What the Navy really needs are craft that can maintain, in rough water, speeds well above those of submarines. It can be done with submerged hydrofoils. I have maintained this for years, at some personal expense, and lots of disappointment" (MIT, VBP, 6).

8. Interviews, Richard Bush, John Bush: "A Scientist Makes Surgeon's Valves," *Boston Herald,* June 17, 1956 (MIT, VBP); VB to R. Bush, Dec. 7, 1955; R. Bush to VB, Dec. 3, 1955 (LOC, VBP 123).

VB's foray into valves resulted in the joint publication of one academic paper, "The Surgical Correction of Calcific Aortic Stenosis in Adults." His coauthors were D. Harken, H. Black, W. Taylor, W. Thrower and H. Soroff, *American Journal of Cardiology,* 4 (1959), 135–46.

9. VB, "Can Men Live Without War?" *Atlantic Monthly,* Feb. 1956; VB, "For Man to Know," *Atlantic Monthly,* Aug. 1955; "Bush: Shaper of American Science Policy," *New Scientist,* June 2, 1960; VB, "Science Pauses," *Fortune,* May 1965; VB, "The Art of Management," in *Science Is Not Enough* (1967); CIW annual report, president's message, Dec. 1955; "culture" quotation, *Saturday Review,* Sept. 19, 1970, 40; "stalemate" quotation, May 8, 1955, CBS transcript, "VE Plus 10" (CIW, VBP).

Many of VB's more rarefied essays were collected in the 1967 volume *Science Is Not Enough.* His views on conservation would jibe with those of the leading thinkers on natural resources in the 1990s, such as Lester Brown of Worldwatch Institute.

Remaining optimistic about solar power in the face of commercial resistance, VB anticipated the widespread acceptance of this technology beginning in the 1970s. As early as 1953, he concluded: "One can now make a solar operated irrigation plant that is economically sound" (VB to Nelson Rockefeller, Sept. 4, 1953, Rockefeller Family Archives, RG 4, Projects, 19/198).

10. VB interview, Edward R. Murrow, *Person to Person,* broadcast, June 1956; May 8, 1955, CBS transcript, "VE Plus 10" (CIW, VBP).

VB deduced that since wars were now too costly to be fought, they would not be fought. Hence, his statement to CBS, "No one can now any more win a great war." He worried about "minor wars" escalating to the point where nuclear weapons would be used, but said, "I hope we have the good sense not to use A-bombs in minor wars." He cited as evidence the Korean War, which passed with threats of a nuclear attack by the U.S. but no actual bombings.

Given some of his views, VB decided it was best not to speak publicly too much. "Once in a while I speak out when I feel that I have something important to say," he told a television journalist, turning down an interview request. "But I have found that those who talk frequently become ignored, and that if I speak only at long intervals, I get plenty of attention" (VB to Lawrence E. Spivak, Nov. 5, 1957, MIT, VBP).

11. Wittner, Lawrence S., *Rebels Against War: The American Peace Movement, 1933,* 213–75; Carter, 285.

12. "The Limitations of Science," *Time,* May 7, 1965; Carter, 96–97; Hacker, 38–39.

13. Hodgson, 76–81; VB, "Notes for Talk to Freshman," March 13, 1963 (MIT, VBP); Carter, 97.

14. VB to G. Merck, Jan. 23, 1950 (LOC, VBP, 72/1751); *Values & Visions: A Merck Century* (1991), 87.

15. G. Merck to VB, Jan. 31, 1950 (LOC, VBP, 72/1751); VB to E. Hodgins, Feb. 19, 1965 (MIT, VBP).

16. VB to E. Hodgins, Oct. 5, 1964 (MIT, VBP); Patterson, 342–43.

17. Divine, *Sputnik Challenge,* introduction; Killian, *Sputnik, Scientists and Eisenhower,* introduction.

18. "Bush Talks," *Newsweek,* Oct. 21, 1957.

Far from indicating Russian success at long-range missiles, *Sputnik* may have been "a deception to cover up" Russian difficulties in that area (Diggins, 316).

19. Emphasis added. Hearings Before the U.S. Senate Committee on Armed Services, *Inquiry Into Satellites and Missiles,* Nov. 25, 1957, 57–89; "creative" quotation in Beard, 74.

Even vulnerability to Russian missiles did not erode security, since the U.S. had the capacity, through the Strategic Air Command, to strike the Russian heartland with H-bombs. "Our safety lies primarily in having such a strong offense that no one will ever dare to challenge us."

20. VB's most extended defense of his postwar technology decisions came ten years after *Sputnik,* in a new foreword to a paperback edition of *Modern Arms and Free Men:*

> When I wrote this book in 1948–49, I held that intercontinental ballistic missiles were then possible but impractical. I did so because officers of the Air Force were proclaiming loudly that such missiles were just around the corner, that we would have them in a year or two, and that they must be controlled by the Air Force which would proceed to build them. This pronouncement was part of a rather disgraceful competition between the services for appropriation and control, from which we have now rather well recovered. I said that if the ICBM appeared at all, it would be after many years, and I was dead right. We did the right thing, put our efforts into bombs and means of destroying them, and did not divert our efforts prematurely. Later the ICBM did become practicable, and we built them promptly and well. I recite all this because I have often been accused of opposing the whole program, and I dislike to be taken to task when I was, for once, right (MIT, VBP, 21).

21. Divine, *Sputnik Challenge,* 100–101; Diggins, 313–16; Bilstein, 211; Barber, Richard J., "The Advanced Research Projects Agency, 1958–1974," Dec. 1975 (DOD: 903-74-C-0096), II-III.

22. "Astronaut Plan Termed a 'Stunt': Bush Says Project Has Little Value," NYT, April 7, 1960; VB to T. K. Glennan, April 18, 1960 (NASA, HO); VB, "Four Aspects of Space Research," *Technology Review,* June 1960.

23. Murray, Charles, and Cox, C.B., *Apollo: The Race to the Moon* (1989), 60–61, 76–83; McDougall, 315–24.

24. NYT, May 24, 1961.

25. VB to T. K. Glennan, April 18, 1960 (NASA, HO); VB to T. Voorhees, Jan. 11, 1965 (MIT, VBP).

26. VB to J. Webb, April 11, 1963 (NASA, HO); J. Webb to VB, May 9, 1963 (HSTL, Webb Papers); VB to E. Hodgins, Nov. 4, 1963 (MIT, VBP).

27. NYT, Nov. 17, 1963.

28. VB to C. Wilson, June 14, 1963 (MIT, VBP).

29. VB to P. Stocker, Aug. 24, 1964 (MIT, VBP); McDougall, 444.

30. VB OH, 101; interview, I. B. Cohen; Daniel J. Kevles presentation, May 30, 1991, MIT, Vannevar Bush Centennial Symposium; VB to Gerald Holton, Jan. 13, 1969 (MIT, VBP, 6).

31. VB to J. M. Barker, Dec. 16, 1946 (LOC, VBP, 125).

32. NYT, Sept. 27, 1970; *Los Angeles Times,* Dec. 25, 1970; Brooks, Harvey, "To Rise Above the Commonplace: Pieces of the Action," *Technology Review,* April 1971; VB interview, Dec. 31, 1971 (courtesy Mrs. Harold Edgerton); *Saturday Review,* Sept. 19, 1970.

A full account of the making of *Pieces of the Action* can be found in Owens, Larry, "Vannevar Bush: An Engineer Builds a Book," *Science As Culture* 5 (3) (1996), 373–99.

33. Compton, 64.

34. Frank Press, President Carter's science adviser, coined the "Bush era" term in an April 1992 speech to the National Academy of Sciences, "Science and Technology Policy for a New Era."

35. Greenberg, 292; Dickson, 217–31; Zachary, G. Pascal, "Where Did Big Science Take Us," *Upside,* April 1992.

The partisan cast of scientific research grew even more pronounced in the 1980s and 1990s. For an extended study, see Cohen, Linda, and Noll, Roger, *The Technology Pork Barrel* (1991).

36. Berkner, Lloyd, *The Scientific Age: The Impact of Science on Society* (1964), 53–55.

37. Terman quotation in Owens, "Counterproductive Management," 553; Zeitlin, Jonathan, "Flexibility and Mass Production at War: Aircraft Manufacture in Britain, the United States, and Germany, 1939–1945," *Technology and Culture,* Jan. 1995, 48; Flamm, 1–79, 251–55; Kleinman, 188–95.

38. D. F. Carpenter, chief of Munitions Board, to K. Compton, chairman of Research and Development Board, Feb. 29, 1949 (NARA, RG 330/341, RDB, 37/209).

39. Koistinen, 5–20, 97–103.

40. Brooks quotation (emphasis added); B. Smith, 85.

A vast literature exists on the enervating effects of military research and the divergence between civilian and military technology. For the basic argument, see Kaldor, Mary, *The Baroque Arsenal* (1982); Dickson, David, *The New Politics of Science* (1988); Markusen, Ann, and Yudkin, Joel, *Dismantling the Cold War Economy* (1992); and Pursell, Carroll, *The Machine in America: A Social History of Technology* (1995), 299–319.

One sign of the frustration with VB's model of science funding was the 1978 decision by the National Science Foundation to fund for the first time cooperative ventures between university and industrial researchers (Dickson, 69).

41. Interview, Frank Press.

42. M. Fischer to VB, Dec. 5, 1966 (MIT, VBP, 1).

43. The author thanks Thomas Hughes for suggesting this interpretation.

44. Interview, Douglas Engelbart; on Engelbart's influence on computer networking, see

Hafner, K., and Lyon, M., *Where Wizards Stay Up Late: The Origins of the Internet* (1996), 72–78; Licklider cited in Nyce and Kahn, 136–37.

MIT Press, Licklider's publisher, sent VB a book only after he asked for a copy.

45. Nelson, Theodor, "As We Will Think," paper delivered Sept. 1972, reprinted in Nyce and Kahn, 245–60.

For an exhaustive account of the many people who have adapted VB's memex to their own purposes, see Smith, Linda C., "Memex as an Image of Potentiality Revisited," in Nyce and Kahn, 261–86.

46. Interview, John Connor; descriptions of "As We May Think" in Smith, Linda C., "Memex as an Image of Potentiality Revisited," in Nyce and Kahn, 264–65.

47. VB quotation in Greenberger, Martin, ed., *Computers and the World of the Future* (1962), 307.

48. Emphasis added. VB, "Communications: Where Do We Go From Here?" Feb. 16, 1955 (MIT, VBP, 19).

49. VB, "Science Pauses," *Fortune,* May 1965; VB, "Memex II," Aug. 27, 1959 (MIT, VBP).

50. "Memex Revisited" appeared in VB's *Science Is Not Enough* (1967).

51. VB to T. Voorhees, June 27, 1967 (MIT, VBP).

52. Leslie, 233–41.

53. Interviews, Catherine Bush, Richard Bush; L. Embrey to P. Putnam, Aug. 1, 1969; V. A. Manganelli, note on VB's health, March 18, 1966; VB to H. Wilson, March 11, 1968; VB to L. Hafstad, Dec. 30, 1970 (MIT, VBP); interview, I. B. Cohen.

The loss of bodily control drove VB's inventive mind in new directions. From February to June 1969, he wrote a series of letters to a friend about a design for a movable seat for bathtubs. "I see two uses," he wrote. "First, nurses in hospitals would welcome it. They are told to get their patients sitting up in chairs; then sometimes they have difficulty getting them on their feet again. Second, patients at home, sometimes alone, have real difficulty." VB first got "tripped up" over fitting the seat to all sizes of tubs and then found pitfalls with the seat's inner tube. By June, however, he considered his design in good shape and reported that "Sears is interested" in marketing the seat. This may have been wishful thinking since no further records of the seat exist (VB to John Hastings, Feb. 13, 1969, April 23, 1969, May 22, 1969, June 5, 1969, MIT, VBP, 6).

54. VB, Oct. 9, 1970 (MIT, VBP).

55. Lapp, *Priesthood,* 3.

56. VB OH, 1967, Columbia, 52.

57. Balogh, 20; VB to B. Hallowell, Sept. 27, 1967 (MIT, VBP).

On the paradox of expert hegemony, historian Brian Balogh may well have had Bush in mind when he noted: "The inexorable expansion of expert debate in America's porous political environment was perhaps the most crucial prerequisite to public participation. No matter how insulated the origins of policy, nor how self-conscious the effort to cement expert consensus, access to the crucial ingredients in prominist politics—professionals and administrative bases—was simply too easy to procure by the late 1960s to prevent the trend toward public participation. How ironic that the unrivaled (and unexamined) political success of America's first generation of prominstrators, who for the most part were committed to restricting public debate, virtually ensured this dramatic change in the nature of expert debate and public participation."

Postscript: "Earlier than we think"

Epigraph: VB POA, 311.

1. Mary Corrigan to J. H. Jackson, Feb. 4, 1973 (MIT, VBP, 7); VB interview, Dec. 31, 1971, conducted by Harold Edgerton and Karl Wildes (courtesy Mrs. Harold Edgerton).

2. On VB's animus toward being taped, see VB to E. Bowles, Sept. 17, 1970: "Don't bring a recorder. I have talked so much before recorders that when I see one I involuntarily freeze, that is I become careful of what I say. And that is no way to recall old times" (LOC, Bowles Papers, 5/16).

In the early 1960s, VB had taped dozens of hours of recollections to use as the basis for his memoirs. The tapes were transcribed, then sealed and later stored in MIT's archives.

3. VB interview, Dec. 31, 1971 (courtesy Mrs. Harold Edgerton).

4. NYT, *Washington Post, Associated Press, Boston Globe:* June 30, 1974.

5. Not everyone remembered VB's triumphs. Days after the memorial service at MIT, Edward Bowles, an old rival, wrote the president of MIT, Howard Johnson, that the "tragedy" of VB was that "many of Van's methods . . . provoked a gut feeling of distrust." Bowles went on to catalog VB's failures in dealing with military officers, concluding that "he simply did not have their confidence. It was this that kept him from his greatest ambition," serving as Pentagon chief (E. Bowles to H. Johnson, Oct. 10, 1974, LOC, Bowles Papers, 5).

Despite his accomplishments, Bowles had for decades been obsessed with VB, often imputing dark motives to his mundane actions. In diary entries from 1947 to 1948, Bowles repeatedly criticized VB, whom he described as power-mad, jealous, selfish and overrated. When dealing with VB, Bowles stayed alert for "what power it is he is trying to seize" and once expressed shock over his "guile and . . . subtle brazenness." Not surprisingly, Bowles seemed ungrateful for the aid VB had lent him at crucial points during World War II. He also seemed unaware of the degree to which he distorted VB's image. That others leveled this claim against him did nothing to make him conscious of his relentless criticism of VB. After one colleague suggested that he ease off VB, saying, "Some of us paint Bush blacker than he really is," Bowles swept aside the complaint, noting bitterly: "I might say that black is hardly qualifiable." Bowles's secretary after the war, Heddy Redheffer, recalled that the mere mention of VB's name uncommonly agitated her boss (LOC, Bowles Papers, Office Diary, Jan. 27, 1947, Feb. 4, 1947, May 8, 1948; interview, H. Redheffer).

6. VB used this phrase as the title for one of the essays in *Science Is Not Enough.* On his affection for the phrase, see VB to F. Fassett, April 2, 1967 (MIT, Fassett Papers).

Principal Sources

PERSONAL PAPERS CITED

H. H. Arnold (LOC)
Henry Barton (AIP)
Edward Bowles (LOC)
Vannevar Bush (CIW)
Vannevar Bush (LOC)
Vannevar Bush (MIT)
Vannevar Bush (Tufts)
Karl T. Compton (MIT)
James B. Conant (Harvard University)
Oscar Cox (FDRL)
Frederic Delano (FDRL)
Lee DuBridge (Caltech)
Muir Fairchild (LOC)
F. Fassett (MIT)
James Forrestal (Princeton)
Julius Furer (LOC)
William Golden (HSTL)
Samuel Goudsmit (AIP)
Stanford C. Hooper (LOC)
D. C. Jackson (MIT)
Frank Jewett (NAS)
Harley Kilgore (FDRL)
James Killian (MIT)
Ernest Lawrence (Bancroft Library, University of California)
David Lilienthal (Princeton)
Byron Miller (LOC)
Robert A. Millikan Papers (Caltech)
John Ohly (HSTL)
J. Robert Oppenheimer (LOC)
Henry F. Pringle (LOC)

Nelson Rockefeller (Rockefeller Family Archives)
Franklin Delano Roosevelt (FDRL)
Glenn T. Seaborg (Lawrence Berkeley Laboratory)
Harold Smith (FDRL)
Harold Smith (HSTL)
H. D. Smyth (American Philosophical Society)
Carl Spaatz (LOC)
Harry S. Truman (HSTL)
Theodore von Karman (Caltech)
John von Neumann (Princeton)
Alan Waterman (LOC)
James Webb (HSTL)
Norbert Wiener (MIT)
C. L. Wilson (MIT)

BOOKS AND ARTICLES

Acheson, Dean, *Present at the Creation* (1969).

Adams, Michael C. C., *The Best War Ever: America and World War II* (1994).

Alperovitz, Gar, *Atomic Diplomacy: Hiroshima & Potsdam* (1985).

Ambrose, Stephen E.
   —*Eisenhower: Soldier, General of the Army, President-Elect* (1983).
   —*Eisenhower: The President* (1984).

Aspray, William, *John Von Neumann and the Origins of Modern Computing* (1990).

Baldwin, Ralph, *The Deadly Fuze: The Secret Weapon of World War II* (1980).

Balogh, Brian, *Chain Reaction: Expert Debate & Public Participation in American Commercial Nuclear Power, 1945–1975* (1991).

Barnow, Erik, *Tube of Plenty* (1975).

Baxter, James Phinney, *Scientists Against Time* (1946).

Beard, Edmund, *Developing The ICBM* (1976).

Bilstein, Roger, *Flight in America: From the Wrights to the Astronauts* (1984).

Bird, Kai, *The Chairman: John J. McCloy, the Making of the American Establishment* (1992).

Blum, John Morton, *V Was for Victory: Politics and American Culture During World War II* (1976).

Bowen, Harold G., *Ships, Machinery and Mossbacks* (1954).

Boyer, Paul, *By the Bomb's Early Light: American Thought and Culture at the Dawn of the Atomic Age* (1985).

Brinkley, Alan
   —"Prosperity, Depression, and War, 1920–1945," in *The New American History*, Benson, S. P., Brier, S., and Rosenzweig, R., eds. (1990).
   —*The End of Reform: New Deal Liberalism in Recession and War* (1995).

Brinkley, David, *Washington Goes to War* (1988).

Brint, Steven, *In an Age of Experts: The Changing Role of Professionals in Politics and Public Life* (1994).

Buderi, Robert, *The Invention That Changed The World* (1996).

Bundy, McGeorge, *Danger and Survival: Choices About the Bomb in the First Fifty Years* (1988).

Burke, Colin, *Information and Secrecy: Vannevar Bush, Ultra, and the Other Memex* (1994).

Burns, James MacGregor, *Roosevelt: The Soldier of Freedom, 1940–1945* (1970).

Burrows, William E., *Deep Black: Space Espionage and National Security* (1986).

Bush, Vannevar
   —*Endless Horizons* (1946).

—*Modern Arms and Free Men: A Discussion of the Role of Science in Preserving Democracy* (1949).

—*Pieces of the Action* (1970).

Camus, Albert, *Between Hell and Reason: Essays from the Resistance Newspaper Combat, 1944–1947* (1991).

Carter, Paul A., *Another Part of the Fifties* (1983).

Catton, Bruce, *The War Lords of Washington* (1948).

Cochrane, Rexmond C., *The National Academy of Sciences* (1978).

Coffey, Thomas M., *Hap: The Story of the U.S. Air Force and the Man Who Built It* (1982).

Commager, Henry Steele, *The American Mind* (1950).

Compton, Arthur, *Atomic Quest* (1956).

Conant, James, *My Several Lives* (1970).

Condit, Doris M., *The Test of War: History of the Office of the Secretary of Defense, 1950–1953* (1984).

Conkin, Paul K., *The New Deal* (1975).

Cooling, Benjamin F., ed., *War, Business, and American Society: Historical Perspectives on the Military Industrial Complex* (1979).

Dickson, David, *The New Politics of Science* (1988).

Diggins, John Patrick, *The Proud Decades: America in War and Peace, 1941–1960* (1988).

Divine, Robert A.

—*Blowing on the Wind: The Nuclear Test Ban Debate, 1954–1960* (1978).

—*The Sputnik Challenge: Eisenhower's Response to the Soviet Satellite* (1993).

Donovan, Robert

—*Conflict and Crisis: The Presidency of Harry S. Truman, 1945–1948* (1977).

—*Tumultuous Years: The Presidency of Harry S. Truman, 1949–1953* (1982).

Douglas, Susan J., *Inventing American Broadcasting, 1899–1922* (1987).

Dower, John W., *War Without Mercy: Race & Power in the Pacific War* (1986).

Dupree, A. Hunter, *Science in the Federal Government: A History of Politics and Activities to 1940* (1957).

Edwards, Paul, *The Closed World: Computers and the Politics of Discourse in Cold War America* (1996).

England, J. Merton, *A Patron for Pure Science: The National Science Foundation's Formative Years, 1945–57* (1982).

Farkas-Conn, Irene S., *From Documentation to Information Science* (1990).

Flamm, Kenneth, *Creating the Computer: Government, Industry, and High Technology* (1988).

Forman, Paul, "Behind Quantum Electronics: National Security as a Basis for Physical Research in the United States, 1940–1960," *Historical Studies in the Physical and Biological Sciences,* 18:1 (1987).

Franklin, H. Bruce, *War Stars: The Superweapon and the American Imagination* (1988).

Gaddis, John L.

—*The United States and the Origins of the Cold War, 1941–1947* (1972).

—*The Long Peace: Inquiries into the History of the Cold War* (1987).

Gannon, Robert, *Hellions of the Deep: The Development of American Torpedoes in World War II* (1996).

Geiger, Roger L.

—*To Advance Knowledge: The Growth of American Research Universities, 1900–1940* (1986).

—*Research and Relevant Knowledge: American Research Universities Since World War II* (1993).

Gellhorn, Walter, *Security, Loyalty & Science* (1950).

Goldberg, Stanley, "Inventing a Climate of Opinion: Vannevar Bush and the Decision to Build the Bomb," *Isis,* 83 (1992).

Goldstine, Herman H., *The Computer: From Pascal to Von Neumann* (1972).

Gorn, Michael, *The Universal Man: Theodore von Karman's Life in Aeronautics* (1992).

Graebner, William S., *The Age of Doubt: American Thought and Culture in the 1940s* (1991).

Greenberg, Daniel S., *The Politics of Pure Science* (1971).

Hacker, Andrew, *The End of the American Era* (1970).

Heilbron, J. L., and Seidel, R. W., *Lawrence and His Laboratory* (1989).

Helmreich, Jonathan E., *Gathering Rare Ores: The Diplomacy of Uranium Acquisition, 1943–1954* (1986).

Herken, Gregg, *The Winning Weapon: The Atomic Bomb in the Cold War, 1945–1950* (1981).

Hersey, John, *Hiroshima* (1946).

Hershberg, James, *James B. Conant: Harvard to Hiroshima and the Making of the Nuclear Age* (1993).

Hewlett, Richard G., and Anderson, Oscar E., Jr., *The New World: A History of the United States Atomic Energy Commission* (1962).

Hewlett, Richard G., and Duncan, Francis, *Atomic Shield: A History of the United States Atomic Energy Commission* (1962).

Hodgson, Godfrey, *America in Our Time* (1976).

Hollinger, David, "Free Enterprise and Free Inquiry: The Emergence of Laissez-Faire Communitarianism in the Ideology of Science in the United States," *New Literary History,* 21 (1990).

Holloway, David, *Stalin and the Bomb* (1994).

Hooks, Gregory, *Forging the Military-Industrial Complex: World War II's Battle of the Potomac* (1991).

Hoopes, Townsend, and Brinkley, Douglas, *Driven Patriot: The Life and Times of James Forrestal* (1992).

Howarth, Stephen, *To Shining Sea: A History of the United States Navy* (1991).

Hughes, Thomas P., *American Genesis: A Century of Technological Enthusiasm* (1989).

Huntington, Samuel P.
— *The Soldier and the State: The Theory and Politics of Civil-Military Relations* (1957).
— *The Common Defense: Strategic Programs in National Politics* (1961).

Isaacson, Walter, and Thomas, Evan, *The Wise Men: Six Friends and the World They Made* (1986).

Karl, Barry D., *The Uneasy State: The United States from 1915 to 1945* (1983).

Keegan, John
— *A History of Warfare* (1993).
— *The Second World War* (1990).

Kevles, Daniel J.
— "Scientists, the Military, and Control of Postwar Defense Research: The Case of the Research Board for National Security, 1944–46," *Technology and Culture,* Jan. 1975.
— *The Physicists: The History of a Scientific Community in Modern America* (1977).
— "Cold War and Hot Physics: Science, Security, and the American State, 1945–56," *Historical Studies in the Physical Sciences,* 20:2 (1990).

Killian, James R., Jr.
— *Sputnik, Scientists and Eisenhower* (1977).
— *The Education of a College President: A Memoir* (1985).

Kistiakowsky, George B., *A Scientist at the White House* (1976).

Kleinman, Daniel, L., *Politics on the Endless Frontier: Postwar Research Policy in the United States* (1995).

Klingaman, William K., *1941: Our Lives in a World on the Edge* (1988).

Kofsky, Frank, *Harry S. Truman and the War Scare of 1948* (1993).

Koistinen, Paul A. C., *The Military-Industrial Complex: A Historical Perspective* (1980).

Landis, James M., *The Administrative Process* (1938).

Lapp, Ralph
— *The New Priesthood: The Scientific Elite and the Uses of Power* (1965).
— *The Weapons Culture* (1968).

Larrabee, Eric, *Commander In Chief: Franklin Delano Roosevelt, His Lieutenants, and Their War* (1987).

Leffler, Melvyn P., *A Preponderance of Power: National Security, the Truman Administration, and the Cold War* (1992).

Leslie, Stuart W., *The Cold War and American Science* (1993).

Leuchtenberg, William, *The Perils of Prosperity, 1914–32* (1958).

Lilienthal, David, *The Journals*, vols. 1–2 (1964).

Lovell, Stanley P., *Of Spies and Stratagems* (1963).

Mackenzie, Donald, *Inventing Accuracy: A Historical Sociology of Nuclear Missile Guidance* (1990).

Maddox, Robert F., *The Senatorial Career of Harley Martin Kilgore* (1981).

McCullough, David, *Truman* (1992).

McDougall, Walter A., *The Heavens and the Earth: A Political History of the Space Age* (1985).

McMahon, A. Michal, *The Making of a Profession: A Century of Electrical Engineering in America* (1984).

Meigs, Montgomery, *Slide Rules and Submarines: American Scientists and Subsurface Warfare in World War II* (1990).

Millett, Allan R., and Maslowski, Peter, *For the Common Defense: A Military History of the United States of America* (1994).

Millis, Walter
— *Arms and Men: A Study in American Military History* (1956).
— *American Military Thought* (1966).
— (ed.) *The Forrestal Diaries* (1951).

Morison, Elting E.
— *Turmoil and Tradition: A Study in the Life and Times of Henry L. Stimson* (1960).
— *From Know-How to Nowhere: The Development of American Technology* (1974).

Morse, Philip M., *In at the Beginnings: A Physicist's Life* (1977).

Neufeld, Jacob, *Ballistic Missiles in the United States Air Force, 1945–1960* (1989).

Neufeld, Michael J., *The Rocket and the Reich: Peenemunde and the Coming of the Ballistic Missile Era* (1995).

Noble, David F.
— *America By Design* (1977).
— *Forces of Production: A Social History of Industrial Automation* (1984).

Norton, Hugh S., *The Quest for Economic Stability: Roosevelt to Bush* (1991).

Nyce, James M., and Kahn, Paul, *From Memex to Hypertext: Vannevar Bush and the Mind's Machine* (1991).

O'Neill, William L.
— *American High: The Years of Confidence, 1945–1960* (1986).
— *A Democracy at War: America's Fight at Home & Abroad in World War II* (1993).

Overy, Richard, *Why the Allies Won* (1995).

Owens, Larry
— "Vannevar Bush and the Differential Analyzer: The Text and Context of an Early Computer," *Technology and Culture*, 27:1 (1986).
— "MIT and the Federal 'Angel': Academic R&D and Federal-Private Cooperation Before World War II," *Isis*, 81 (1990).

—"Patents, The 'Frontiers' of American Invention and the Monopoly Committee of 1939: Anatomy of a Discourse," *Technology and Culture*, 32 (1991).

—"The Counterproductive Management of Science in the Second World War," *Business History Review*, 68 (1994).

—"Vannevar Bush: An Engineer Builds a Book," *Science As Culture*, 5:3 (1996).

—"Where Are We Going, Phil Morse? Changing Agendas and the Rhetoric of Obviousness in the Transformation of Computing at MIT, 1939–1957," *IEEE Annals of the History of Computing*, vol. 18, no. 4 (1996).

Parker, R. A. C., *Struggle for Survival: The History of the Second World War* (1989).

Patterson, James T., *Grand Expectations: The United States, 1945–1974* (1996).

Perret, Geoffrey, *A Country Made by War: From the Revolution to Vietnam—The Story of America's Rise to Power* (1989).

Polenberg, Richard, *War and Society: The United States, 1941–1945* (1972).

Powers, Thomas, *Heisenberg's War: The Secret History of the German Bomb* (1993).

Price, Don K.
—*Government and Science* (1954).
—*The Scientific Estate* (1965).

Pursell, Carroll W., Jr.
—"Science Agencies in World War II: The OSRD and Its Challengers," in *The Sciences in the American Context* (1979).
—*The Machine in America: A Social History of Technology* (1995).
—(ed.), *Technology in America* (1990).

Rearden, Steven L., *The Formative Years: History of the Office of the Secretary of Defense, 1947–1950* (1984).

Reingold, Nathan, *Science: American Style* (1991).

Reynolds, Terry S., ed., *The Engineer in America* (1991).

Rheingold, Howard, *Tools for Thought* (1985).

Rhodes, Richard, *The Making of the Atomic Bomb* (1986).

Rigden, John S., *Rabi: Scientist and Citizen* (1987).

Rogow, Arnold A., *James Forrestal: A Study of Personality, Politics and Policy* (1963).

Roland, Alex
—*Model Research: The National Advisory Committee for Aeronautics, 1915–1958* (1985).
—"Science and War," in *Historical Writing on American Science*, Sally G. Kohlstedt and Margaret Rossiter, eds. (1985).

Rovere, Richard H.
—*The Eisenhower Years* (1956).
—*The American Establishment* (1962).

Sapolsky, Harvey, *Science and the Navy: The History of the Office of Naval Research* (1990).

Saxenian, Annalee, *Regional Advantage: Culture and Competition in Silicon Valley and Route 128* (1994).

Schaffel, Kenneth, *The Emerging Shield: The Air Force and the Evolution of the Continental Air Defense, 1945–1960* (1991).

Scott, Otto J., *The Creative Ordeal: The Story of Raytheon* (1974).

Sherry, Michael
—*Preparing for the Next War: American Plans for Postwar Defense, 1941–45* (1977).
—*The Rise of American Air Power: The Creation of Armageddon* (1987).
—*In the Shadow of War: The United States Since the 1930s* (1995).

Sherwin, Martin, *A World Destroyed: The Atomic Bomb and the Grand Alliance* (1975).

Sherwood, Robert E., *Roosevelt and Hopkins: An Intimate History* (1950).

Shurkin, Joel, *Engines of the Mind: From Abacus to Apple—The Men and Women Who Created the Computer* (1985).

Smith, Alice K., *A Peril and a Hope: The Scientists' Movement in America 1945–47* (1970).

Smith, Bruce L. R., *American Science Policy Since World War II* (1990).

Steel, Ronald, *Walter Lippmann and the American Century* (1980).

Stewart, Irvin, *Organizing Scientific Research for War: The Administrative History of the Office of Scientific Research and Development* (1948).

Stimson, Henry L., and Bundy, McGeorge, *On Active Service in Peace and War* (1948).

Stone, I. F., *The Haunted Fifties* (1963).

Sulzberger, C. L., *A Long Row of Candles: Memoirs & Diaries, 1934–1954* (1969).

Thiesmeyer, Lincoln R., and Burchard, John E., *Combat Scientists* (1947).

Truman, Harry S.
— *Year of Decisions* (1955).
— *Years of Trial and Hope* (1956).

Vander Meulen, Jacob, *The Politics of Aircraft: Building an American Military Industry* (1991).

Vatter, Harold G., *The U.S. Economy in World War II* (1985).

Warner, Sam Bass, *The Province of Reason* (1984).

Weigley, Russell F.
— *History of the United States Army* (1967).
— *The American Way of War: A History of United States Military Strategy and Policy* (1973).

Weinberg, Gerhard L., *A World at Arms: A Global History of World War II* (1994).

Wiener, Norbert, *I Am a Mathematician* (1956).

Wildes, Karl L., and Lindgren, Nilo A., *A Century of Electrical Engineering and Computer Science at MIT, 1882–1982* (1985).

Wright, Gordon, *The Ordeal of Total War, 1939–1945* (1968).

Yergin, Daniel, *Shattered Peace: The Origins of the Cold War and the National Security State* (1977).

UNPUBLISHED DISSERTATIONS

Cornell, Thomas D., "Merle Tuve and His Program of Nuclear Studies at the Department of Terrestrial Magnetism" (1986, Johns Hopkins University).

Dennis, Michael, "A Change of State: the Political Cultures of Technical Practice at the MIT Instrumentation Laboratory and the Johns Hopkins University" (1990, Johns Hopkins University).

Hart, David M., "Competing Conceptions of the Liberal State and the Governance of Technological Innovation in the U.S., 1933–1953" (1995, Massachusetts Institute of Technology).

Meigs, Montgomery C., "Managing Uncertainty: Vannevar Bush, James B. Conant and the Development of the Atomic Bomb, 1940–1945" (1982, University of Wisconsin, Madison).

Owens, Larry W., "Straight-Thinking Vannevar Bush and the Culture of American Engineering" (1987, Princeton University).

INTERVIEWS

Philip H. Abelson
Tom Bagg
David Beckler
Hans Bethe
Lee Anna (Embrey) Blick

Edmund A. Bowles
George Brown
Gordon Brown
McGeorge Bundy
Catherine Bush
John Bush
Richard Bush
Susan Cantrill
I. Bernard Cohen
Russell C. Coile
Elmer Colcord
John Connor
Mary Connor
Lloyd Cutler
James Ebert
Esther E. Edgerton
Douglas Engelbart
Stacey French
John Gardner
David Ginsberg
William Golden
Truman S. Gray
Daniel Greenberg
Caryl Haskins
Edna Haskins
Katherine Hazen
Richard Hewitt
Antonie Knoppers
Norman Krim
David Langmuir
Ralph Lapp
Lorna Marshall
Dorothy McDonald
Elting Morison
Philip Morrison
Bruce S. Old
Arthur Porter
Emanuel R. Piore
Frank Press
Don Price
Heddy Redheffer
Oscar Ruebhausen
Claude E. Shannon
Glenn T. Seaborg
Elmer Staats
Charles Stauffaucher
H. Guy Stever
Jerome Wiesner
Walter Wriston
Herbert York

Acknowledgments

In any lengthy project, an author incurs great debts. I hold my share. My research into Vannevar Bush's life and times began in 1988 while on a fellowship at the University of Michigan. Charles Eisendrath invited me to Michigan for eight splendid months and gave me the time to conceive this biography. In 1991, Rich Karlgaard, then-editor of *Upside* magazine, published my first two articles on Bush, which emboldened me to try a longer work. Joyce Seltzer, my original editor at The Free Press, saw the need for a biography of Bush and gave me the chance to write the first one. Paul Steiger and G. Christian Hill, my editors at *The Wall Street Journal,* granted me three separate leaves of absence to complete this book.

I am also grateful for the help given by professional historians whom I met during my research. Some deserve special thanks. Martin Sherwin gave me the confidence to tackle a Bush biography. Allan Needell introduced me to the various archives in Washington, D.C., and twice invited me to discuss my research at the Smithsonian Air & Space Museum's history seminar. Early in this project, Alex Pang, Michael Buckland and James Hershberg gave me wise counsel. Larry Owens, the author of many scholarly articles on Bush and a professor at the University of Massachusetts, generously allowed me to read early versions of his papers. The late Stanley Goldberg questioned my original notions about Bush, challenging me to build my views from the ground up. The late Elting Morison and I. Bernard Cohen graciously shared their memories of encounters with Bush. Martin Collins shared the postwar diary of Edward Bowles, his interview with Bowles and other papers. David Guston invited me to address an annual meeting of the American Association for the Advancement of Science (the occasion, in 1995, was all the more satisfying because it brought together people who had known Bush with scholars who studied him). Daniel Kevles invited me to speak on Bush at Caltech, and his crisp advice was a beacon for me throughout this project. Richard Nelson and Columbia University made possible my participation in two seminars marking the 50th anniversary of Bush's *Science—The Endless Frontier* report.

My account of Bush's life was pieced together from many sources: archives, interviews, memoirs, government documents and historical studies. Archivists at MIT, the Library of Congress, the National Archives, the National Academy of Sciences, the Truman Library and the Rockefeller Archives were especially helpful in obtaining documents. Many other people supplied crucial materials. Katherine Hazen shared with me photographs and writings of her husband, Harold, a close associate of Bush at MIT. Maxine Singer, president of the Carnegie Institution, allowed me to roam the building where Bush once worked and encouraged me to continue this project even when I seemed unlikely to finish it. Bush's son Richard and Richard's wife, Catherine, patiently answered my many queries, shared personal materials and allowed me to tour their home in Belmont, Massachusetts (originally built by Vannevar and Phoebe). Norman Krim located many essential documents from Raytheon's archives and generously shared them. John Wilke reviewed the Bush files at *The Boston Globe*. Jim Theriault found many copies of speeches by R. Perry Bush, Vannevar's father, and vital records on other Bush ancestors. Elizabeth Bianco tracked down scores of books and articles from libraries in Berkeley, California. Michael Yeates, of the MIT Museum, generously provided photographs.

Various people read portions of the manuscript, correcting errors and suggesting ways to sharpen my narrative and deepen my interpretations. They deserve credit for what is worthy in this book but should be spared any blame for its flaws: Brian Balogh, Barton Bernstein, Michael Buckland, Bob Buderi, Paul Ceruzzi, Alfred Chandler, Paul Edwards, Daniel Greenberg, Barton Hacker, Dennis Hayes, David Hollinger, Thomas Hughes, Zachary Karabell, Mark Kavanagh, Derek Leebaert, Wendy Lustbader, Michael Neufeld, Alex Pang, Steven Rearden, Alex Roland and Michael Sherry. My editor at The Free Press, Bruce Nichols, extended my deadline three times and, when a manuscript finally arrived, edited it with sympathy. Sean Devlin and Bill Brazell copyedited a long manuscript with verve and determination.

Finally, the support of my family lightened the often lonely experience of researching and writing this book. Honorah Curran, my wife, helped in ways too numerous to list. I owe her a great debt. My son, Liam, and daughter, Oona, were born during this lengthy project, adding to its many pleasures and complications. While their interruptions sometimes slowed my work, my children reminded me that history, in the end, is written for the living.

Index